The three main themes of this book — probability theory, differential geometry and the theory of integrable systems — reflect the broad range of mathematical interests of Henry McKean, to whom it is dedicated.

Written by experts in probability, geometry, integrable systems, turbulence and percolation, the seventeen papers included here demonstrate a wide variety of techniques that have been developed to solve various mathematical problems in these areas. The topics are often combined in an unusual and interesting fashion to give solutions outside of the standard methods. The papers contain some exciting results and offer a guide to the contemporary literature on these subjects.

T0214911

Mathematical Sciences Research Institute
Publications

55

Probability, Geometry and Integrable Systems

Mathematical Sciences Research Institute Publications

Probability, Geometry and Integrable Systems

For Henry McKean's Seventy-Fifth Birthday

Edited by

Mark Pinsky
Björn Birnir

CAMBRIDGE
UNIVERSITY PRESS

CAMBRIDGE UNIVERSITY PRESS

Cambridge, New York, Melbourne, Madrid, Cape Town, Singapore,
São Paulo, Delhi, Dubai, Tokyo, Mexico City

Cambridge University Press
32 Avenue of the Americas, New York, NY 10013-2473, USA

www.cambridge.org
Information on this title: www.cambridge.org/9780521175401

© Mathematical Sciences Research Institute 2008

This publication is in copyright. Subject to statutory exception
and to the provisions of relevant collective licensing agreements,
no reproduction of any part may take place without the written
permission of Cambridge University Press.

First published 2008
First paperback edition 2010

A catalog record for this publication is available from the British Library

Library of Congress Cataloging in Publication data

Probability, geometry, and integrable systems / edited by Mark Pinsky, Björn Birnir.
   p.   cm. – (Mathematical Sciences Research Institute publications ; 55)
"For Henry McKean's Seventy-Fifth Birthday." Includes bibliographical references
and index.
   ISBN 978-0-521-89527-9 (hardback)
   1. Probabilities. 2. Geometry, Differential. 3. Hamiltonian systems. I. Pinsky, Mark
A., 1940- II. Birnir, Björn. III. McKean, Henry P.
QA273.P79536   2008
519.2–dc22   2007052192

ISBN 978-0-521-89527-9 Hardback
ISBN 978-0-521-17540-1 Paperback

Cambridge University Press has no responsibility for the persistence or
accuracy of URLs for external or third-party internet websites referred to in
this publication, and does not guarantee that any content on such websites is,
or will remain, accurate or appropriate.

Probability, Geometry and Integrable Systems
MSRI Publications
Volume **55**, 2008

Contents

Probability, Geometry and Integrable Systems
MSRI Publications
Volume 55, 2008

Preface

This volume is dedicated to Henry McKean, on the occasion of his seventy-fifth birthday. His wide spectrum of interests within mathematics is reflected in the variety of theory and applications in these papers, discussed in the Tribute on page xv. Here we comment briefly on the papers that make up this volume, grouping them by topic. (The papers appear in the book alphabetically by first author.)

Since the early 1970s, the subject of completely integrable systems has grown beyond all expectations. The discovery that the Kortweg – de Vries equation, which governs shallow-water waves, has a complete system of integrals of motion has given rise to a search for other such evolution equations. Two of the papers in this volume, one by **Boutet de Monvel and Shepelsky** and the other by **Loubet**, deal with the completely integrable system discovered by Camassa and Holm. This equation provides a model describing the shallow-water approximation in inviscid hydrodynamics. The unknown function $u(x, t)$ refers to the horizontal fluid velocity along the x-direction at time t. The first authors show that the solution of the CH equation in the case of no breaking waves can be expressed in parametric form in terms of the solution of an associated Riemann–Hilbert problem. This analysis allows one to conclude that each solution within this class develops asymptotically into a train of solitons.

Loubet provides a technical *tour de force*, extending previous results of McKean on the Camassa–Holm equation. More specifically, he gives an explicit formula for the velocity profile in terms of its initial value, when the dynamics are defined by a Hamiltonian that is the sum of the squares of the reciprocals of a pair of eigenvalues of an associated acoustic equation. The proof depends on the analysis of a simpler system, whose Hamiltonian is defined by the reciprocal of a *single* eigenvalue of the acoustic equation. This tool can also solve more complex dynamics, associated to several eigenvalues, which eventually leads to a new proof of McKean's formula for the Fredholm determinant. The paper concludes with an asymptotic analysis (in both past and future directions), which allows partial confirmation of statements about soliton genesis and interaction that were raised in an earlier CPAM paper.

Meanwhile we have a contribution from **Gibbons, Holm and Tronci** of a more geometric nature. This deals with the Vlaslov equation, which describes the evolution of the single-particle probability density in the evolution of N point particles. Specifically, they study the evolution of the p-th moments when the dynamics is governed by a quadratic Hamiltonian. The resulting motion takes place on the manifold of symplectomorphisms, which are smooth invertible maps acting on the phase space. The singular solutions turn out to be closely related to integrable systems governing shallow-water wave theory. In fact, when these equations are "closed" at level p, one retrieves the peaked solitons of the integrable Camassa–Holm equation for shallow-water waves!

Segur's paper provides an excellent overview of the development of our understanding of integrable partial differential equations, from the Boussinesq equation (1871) to the Camassa–Holm equation (1993) and their relations to shallow-water wave theory. In physical terms, an integrable system is equivalent to the existence of *action-angle variables*, where the *action variables* are the integrals of motion and the angle variables evolve according to simple ordinary differential equations. Since each of these PDEs also describes waves in shallow water, it is natural to ask the question: *Does the extra mathematical structure of complete integrability provide useful information about the behavior of actual physical waves in shallow water?* The body of the paper takes up this question in detail with many illustrations of real cases, including the tsunami of December 26, 2004. A video link is provided, for further documentation.

Previato's paper contains a lucid account of the use of theta functions to characterize lines in abelian varieties. Strictly speaking, "line" is short for linear flow, since an abelian variety cannot properly contain a (projective) line. More than twenty years ago Barsotti had proved that, on any abelian variety, there exists a direction such that the derivatives of sufficiently high order of the logarithim of the theta function generate the function field of the abelian variety. The purpose of the present paper is to use the resulting differential equations to characterize theta functions, a generalization of the KP equations, introduced by Kadomtsev and Petviashvili in 1971, as well as to study spectra of commutative rings of partial differential operators.

Arov and **Dym** summarize their recent work on inverse problems for matrix-valued ordinary differential equations. This is related to the notion of reproducing kernel Hilbert spaces and the theory of J-inner matrix functions. A time-independent Schrödinger equation is written as a system of first-order equations, which permits application of the basic results.

Cruzeiro and **Malliavin** study a first-order Burgers equation in the context of flows on the group of diffeomorphisms of the circle, an infinite-dimensional Riemannian manifold. The L^2 norm on the circle defines the Riemannian metric, so that the Burgers equation defines a geodesic flow by means of an ordinary differential equation which flows along the Burgers trajectories. The authors compute the connection coefficients for both the Riemannian connection defined by the parallel transport and the algebraic connection defined by the right-invariant parallelism. These computations are then used to solve a number of problems involving stochastic parallel transport, symmetries of the noise, and control of ultraviolet divergence with the help of an associated Markov jump process.

The paper of **Cambronero**, **Ramírez** and **Rider** describes various links between the spectrum of random Schrödinger operators with particular emphasis on Hill's equation and random matrix theory. The unifying theme is the utility of the Riccati map in converting problems about second-order differential operators and their matrix analogues into questions about one-dimensional diffusions. The paper relies on functional integration to derive many interesting results on the spectral properties of random operators. The paper provides an excellent overview of results on Hill's equation, including the exploitation of low-lying eigenvalues. The penultimate section of the paper describes the recent and exciting developments involving Tracy–Widom distributions and their far-reaching generalizations to all positive values of the inverse temperature β.

Birnir's paper provides an excellent overview of his recent results on unidirectional flows. This is a special chapter in the theory of turbulence, and is not commonly presented. This type of modeling is important in the study of fluvial sedimentation that gives rise to sedimentary rock in petroleum reservoirs. The flow properties of the rock depend strongly on the topological structure of the meandering river channels. The methods developed can also be applied to problems of atmospheric turbulence. Contrary to popular belief, the turbulent temperature variations in the atmosphere may be highly anisotropic, nearly stratified. Thus, the scaling first developed in the case of a river or channel may have a close analogue in the turbulent atmosphere.

Halfway between probability theory and classical physics is the subject of statistical mechanics. In the paper of **Costeniuc**, **Ellis**, **Touchette** and **Turkington**, the Gaussian ensemble is introduced, to complement the micro-canonical and macro-canonical ensembles that have been known since the time of Gibbs. It is demonstrated that many minimization problems in statistical physics are most effectively expressed in terms of the Gaussian ensemble.

Grinevich and **Novikov** present a lucid overview of their work on finding formulas for the topological charge and other quantities associated with the sine-Gordon equation, which describes immersions of negatively curved surfaces into $\mathbb{R}^3$. Non-singular real periodic finite-gap solutions of the sine-Gordon equation are characterized by a genus g hyperelliptic curve whose branch points are either real positive or form complex conjugate pairs. The authors describe the admissible branch points as zeros of a meromorphic differential of the third kind, which in turn is defined by a real polynomial $P(\lambda)$ of degree $g - 1$. This leads to a formula for the topological charge of these solutions, which was first given in a 1982 paper of Dubrovin and Novikov. The proof relates the topological charge to a set of certain integer characteristics of the polynomial $P(\lambda)$. The methods developed here can be applied, with suitable modifications, to the KdV equation, the defocusing nonlinear Schrödinger equation and the Kadomtsev–Petvishvilii equation.

The paper of **Ercolani** and **McLaughlin** investigates a system of equations which originate in the physics of two-dimensional quantum gravity, the so-called loop equations of random matrix theory. The analysis depends on an asymptotic formula, of the large deviation type, for the partition function in the Unitary Ensemble of random matrix theory. The loop equations are satisfied by the coefficients in a Laurent expansion expressing certain Cauchy-like transforms in terms of a quadratic expression in the derivatives. The final paragraph of the paper suggests some open problems, such as the following: to use the loop equations to find closed form expressions for the expansion coefficients of the logarithm of the partition function when the dimension $N \to \infty$; it is also anticipated that the loop equations could be used to determine qualitative behavior of these coefficients.

Manna and **Moll** offer a beautiful set of generalizations of the classical Landen transform, which states that a certain elliptic integral of the first kind containing two parameters, when expressed in trigonometric form, is invariant under the transformation defined by replacing the parameters by their arithmetic (resp. geometric) means. Later Gauss used this to prove that the limit of the iterates of this transformation exist and converge to the reciprocal of this elliptic integral, suitably modified. This limit is, by definition, the *arithmetic-geometric mean* of the initial conditions. This idea is generalized in several directions, the first of which is to a five-parameter set of rational integrals, where the numerator is of order four and the denominator of order six. The requisite Landen transformation has a simple geometric interpretation in terms of doubling the angle of the cotangent function in the trigonometric form of the integrand. The remainder of the paper describes generalizations to higher-order rational integrands, where the doubling of the cotangent function is replaced by a magnification of order $m \geq 2$. It is proved that the limit of the Landen transforms exists and can be represented as a suitable integral. All of the models studied in this paper can be considered as discrete-time (partially) integrable dynamical systems where the conserved quantity is the definite integral that is invariant under the change of parameters. The simplest of these is the elliptic integral studied by Landen, where the dynamics is defined by the arithmetic-geometric mean substitution.

Varadhan has contributed a lucid overview of his recent joint work on homogenization. In general, this theory leads to approximations of solutions of a differential equation with rapidly varying coefficients in terms of solutions of a closely related equation with constant coefficients. A model problem is the second-order parabolic equation in one space dimension, where the constant coefficient is the harmonic mean of the given variable coefficient equation. This analytical result can be expressed as a limit theorem in probability, specifically an ergodic theorem for the variable coefficient, when composed with the diffusion process defined by the parabolic equation; the normalized invariant measure is expressed in terms of the harmonic mean. Having moved into the probabilistic realm, one may just as well consider a second-order parabolic equation with random coefficients, where the results are a small variation of those obtained in the non-random case. With this intuitive background, it is natural to expect similar results when one begins with a d-dimensional equation of Hamilton–Jacobi type, defined by a convex function and with a small noise term. In joint work with Kosygina and Rezhakanlou, it is proved that the noise disappears and the HJ solution converges to the solution of a well-determined first-order equation, where the homogenized convex function is determined by a convex duality relation. The reader is invited to pursue the details, which are somewhat parallel to the model case described above.

Camia and **Newman**'s paper relies on the stochastic Schramm–Loewner equation (SLE), which provides a new and powerful tool to study scaling limits of critical lattice models. These ideas have stimulated further progress in understanding the conformally invariant nature of the scaling limits of several such models. The paper reviews some of the recent progress on the scaling limit of two-dimensional critical percolation, in particular the convergence of the exploration path to chordal SLE and the "full" scaling limit of cluster interface loops. The results on the full scaling limit and its conformal invariance are presented here for the first time. For site percolation on the triangular lattice, the results are fully rigorous and the main ideas are explained.

Grünbaum's paper proposes a new spectral theory for a class of discrete-parameter Markov chains, beginning with the case of the birth-death process, studied by Karlin/McGregor in the 1950s. More generally, for each Markov chain there is a system of orthogonal polynomials which define the spectral decomposition. Explicit computation of the relevant orthogonal polynomials is available for other Markov chains, such as random walk on the N-th roots of unity and the processes associated with the names of Ehrenfest and Tchebychev.

A principal emphasis here, due originally to M. G. Krein, is the formulation of *matrix-valued* orthogonal polynomials. Several solved examples are presented, while several natural open problems are suggested for the adventurous reader.

Van Moerbeke provides a masterful account of the close connection between nonintersecting Brownian motions and integrable systems, where the connection is made in terms of the theory of orthogonal polynomials, using previous results of Adler and Van Moerbeke. If N Brownian particles are started at p definite points and required to terminate at q other points in unit time, the object is to study the distribution as $N \to \infty$, especially in the short time limit $t \downarrow 0$ and the unit time limit $t \uparrow 1$. When the supports of these measures merge together, we have a *Markov cloud*, defined as an infinite-dimensional diffusion process depending only on the nature of the various possible singularities of the equilibrium measure. The connection between non-intersecting Brownian motion and orthogonal polynomials begins with a formula of Karlin and McGregor which expresses the non-intersection probability at a fixed time as the N-dimensional integral of the product of two determinants. Numerous special cases are provided to illustrate the general theory.

Acknowledgments

We thank Hugo Rossi for both conceiving the idea of an MSRI workshop and for the encouragement to prepare this edited volume. The editorial work was assisted by Björn Birnir, who also served on the Organizing Committee of the MSRI Workshop, together with Darryl Holm, Kryll Vaninsky, Lai-Sang Young and Charles Newman.

We also thank Professor Anne Boutet de Monvel for making available to us information documenting McKean's honorary doctorate from the Université Paris VII in May 2002.

Mark A. Pinsky
Department of Mathematics
Northwestern University
Evanston, IL 60208-2370, USA

Probability, Geometry and Integrable Systems
MSRI Publications
Volume **55**, 2008

Henry McKean

A TRIBUTE BY THE EDITORS

The significance of McKean's work

Henry McKean has championed a unique viewpoint in mathematics, with good taste and constant care toward a balance between the abstract and the concrete. His great influence has been felt through both his publications and his teaching. His books, all of which have become influential, can be recognized by their concise, elegant and efficient style.

His interests include probability theory, stochastic processes, Brownian motion, stochastic integrals, geometry and analysis of partial differential equations, with emphasis on integrable systems and algebraic curves, theta functions, Hill's equation and nonlinear equations of the KdV type. He also was a pioneer in the area of financial mathematics before it became a household word.

The importance of stochastic models in modern applied mathematics and science cannot be overestimated. This is well documented by the large number of diverse papers in this volume that are formulated in a stochastic context. Henry McKean was one of the early workers in the theory of diffusion processes, as documented in his classic work with K. Itô, *Diffusion Processes and Their Sample Paths* (Springer, 1965). This was followed by his *Stochastic Integrals* (Academic Press, 1969). This book led the way to understanding the close connections between probability and partial differential equations, especially in a geometric setting (Lie groups, Riemannian manifolds).

On January 6, 2007, McKean was awarded the Leroy P. Steele Prize for Lifetime Achievement, presented annually by the American Mathematical Society. The prize citation honors McKean for his "rich and magnificent mathematical career" and for his work in analysis, which has a strong orientation towards probability theory; it states further that "McKean has had a profound influence on his own and succeeding generations of mathematicians. In addition to the important publications resulting from his collaboration with Itô, McKean has written several books that are simultaneously erudite and gems of mathematical exposition. As his long list of students attest, he has also had enormous impact on the careers of people who have been fortunate enough to study under his direction."

McKean's published work includes five books and more than 120 articles, in such journals as the *Annals of Mathematics*, *Acta Mathematica* and *Inventiones Mathematicae*, to name a few. To illustrate the richness of the mathematics he has been involved with, we take a brief look at published reviews of his books and then discuss the main threads of his research articles by subject.

1. Diffusion Processes and Their Sample Paths (with Kiyosi Itô). *Grundlehren der mathematischen Wissenschaften* **125**, Springer, New York, 1965; second (corrected) printing, 1974; reprinted in *Classics in Mathematics*, 1996.

We adapt the comments by T. Watanabe in *Mathematical Reviews*: Feller's work on linear diffusion was primarily of an analytic character. This spurred some outstanding probabilists (including the authors) to reestablish Feller's results by probabilistic methods, solving some conjectures of Feller and studying profoundly the sample paths of one-dimensional diffusion. Their purpose is to extend the theory of linear diffusion to the same level of understanding which Paul Lévy established for Brownian motion. This is completely realized in this book by combining special tools such as Brownian local time with the general theory of Markov processes. This book is the culmination of a ten-year project to obtain the general linear diffusion from standard Brownian motion by time change and killing involving local times.

On the occasion of the book's republishing in Springer's *Classics in Mathematics* series, the cover blurb could boast without the least exaggeration: "Since its first publication ... this book has had a profound and enduring influence on research into the stochastic processes associated with diffusion phenomena. Generations of mathematicians have appreciated the clarity of the descriptions given of one or more dimensional diffusion processes and the mathematical insight provided into Brownian motion."

2. *Stochastic Integrals.* Academic Press, New York, 1969.

From comments by E. B. Dynkin in *Mathematical Reviews:* "This little book is a brilliant introduction to an important interface between the theory of probability and that of differential equations. The same subject was treated in the recent book of I. I. Gihman and A. V. Skorokhod. The author's book is smaller, contains more examples and applications and is therefore much better suited for beginners. Chapter 1 is devoted to Brownian motion. Chapter 2 deals with stochastic integrals and differentials. Chapter 3 deals with one-dimensional stochastic integral equations. In Chapter 4, stochastic integral equations on smooth manifolds are investigated. Winding properties of planar Brownian motion (about one or two punctures) are deduced from the study of Brownian motion on Riemann surfaces. The last three sections are devoted to constructing Brownian motion on a Lie group, starting from Brownian motion on the Lie algebra. by means of the so-called product integral and the Maljutov–Dynkin results about Brownian motion with oblique reflection. In treating the applications of stochastic integrals, the author frequently explains the main ideas by means of typical examples, thus avoiding exhausting generalities. This remarkable book will be interesting and useful to physicists and engineers (especially in the field of optimal control) and to experts in stochastic processes."

3. *Fourier Series and Integrals* (with Harry Dym). Academic Press, New York, 1972.

Elliot Lieb of Princeton University wrote: "In my opinion, the book of Dym and McKean is unique. It is a book on analysis at an intermediate level with a focus on Fourier series and integrals. The reason the book is unique is that books on Fourier analysis tend to be quite abstract, or else they are applied mathematics books which give very little consideration to theory. This book is a solid mathematics book but written with great fluency and many examples. In addition to the above considerations, there is also the fact that the literary style of the authors is excellent, so that the book has a readability that is rarely found in mathematics books, especially in modern texts."

4. *Stationary Gaussian Processes* (with Harry Dym). Academic Press, New York, 1976.

From comments by S. Kotani in *Mathematical Reviews:* "This is a monograph on stationary Gaussian processes in one dimension. Given a mean zero Gaussian process $x(t), t \in \mathbb{R}$, one may ask the following questions: (i) to predict the future given the past $-\infty < t < 0$; (ii) to predict the future given the finite segment of the past $-2T < t < 0$; (iii) to predict $x(t)$ for $|t| < T$ given $x(t)$ for $|t| > T$; (iv) the degree of dependence of the future on the past; (v) the degree of mixing of the process $x(t)$. This book contains a clear and concise introduction to the subject, often original presentations of known results in addition to several new results. The solution to problem (i) goes back to Kolmogorov, while the solution of (iii) is due to M. G. Krein in 1954. The authors re-work the solution of Krein and from that point they completely solve problem (ii). They also make important contributions to the understanding of problems (iv) and (v)."

5. *Elliptic Curves: Function Theory, Geometry and Arithmetic* (with Victor Moll). Cambridge University Press, Cambridge, 1997.

From W. Kleinert's review in *Zentralblatt für Mathematik*: "Altogether, this highly non-standard textbook provides the reader with ... a deep insight into historically known mathematical interrelations and references to modern developments in the analysis, geometry and arithmetic of elliptic curves. The book reflects the authors' profound knowledge and deep devotion to the historical development of the theory of elliptic curves. In these days it is certainly very profitable for the mathematical community to have such a book among the increasing number of others on the subject ... this book is not only a perfect primer for beginners in the field, but also an excellent source for researchers in various areas of mathematics and physics."

Peter Sarnak agrees: "Very unusual in covering the important aspects of elliptic curves (analytic, geometric and arithmetic) and their applications — in a single reasonably sized volume. This account of the subject, in the style of the original discoverers is, in my opinion, the best way to present the material in an introductory book."

Works by subject

Given his wide spectrum of interest, it is difficult to summarize McKean's contributions in a few paragraphs. We mention some of the principal themes:

Questions in the theory of probability and stochastic processes, beginning with Brownian motion and leading to the completion of W. Feller's program, to found the theory of one-dimensional diffusion on a probabilistic basis rather than the analytical foundation (Hille–Yosida theorem) that was standard before the 1950s. One probabilistic approach is to use the natural scale and speed measure to obtain the diffusion process directly from Brownian motion. This approach is well documented in Itô–McKean. The other approach is to develop Itô's theory of stochastic differential equations (SDE) to obtain the diffusion as the image of a Brownian motion process via a nonlinear mapping that defines the solution operator of the SDE. This approach, which works equally well in higher dimensions, is well documented in his *Stochastic Integrals*.

Geometry and analysis of partial differential equations, especially integrable systems and algebraic curves, theta functions and Hill's equation. Hill's equation is the one-dimensional Schrödinger equation

$$-y'' + q(x)y = \lambda y$$

where the potential function $q(x)$ is periodic. The spectrum of the operator $L = -d^2/dx^2 + q$ associated to such an equation is generally made up of an infinite number of intervals. But it can happen that, in limiting cases, the spectrum is formed of a finite number of n intervals and an infinite interval $\lambda > \lambda_{2n}$. For which potentials does this occur? Several authors have shown that this phenomenon requires that the potential be a solution of some auxiliary differential equations. Then Peter Lax showed in 1972/1974 the relation with certain solutions of the KdV equation. McKean and van Moerbeke solved the problem by establishing a close relation with the classic theory of hyperelliptic functions. Then McKean and Trubowitz in 1976/1978 extended the results to the case $n = \infty$, showing that the periodic spectrum of the Hill operator is infinite. This work proved to be the starting point of a series of new developments associated with the names of McKean and Trubowitz, Feldman, Knörrer, Krichever and Merkl.

Geometry of KdV equations. The Korteweg–de Vries equation

$$\frac{\partial q}{\partial t} = 3q\,\frac{\partial q}{\partial x} - \frac{1}{2}\frac{\partial^3 q}{\partial x^3}$$

describes the propagation of a wave $q(x,t)$ in a shallow canal. It has certain common features with other nonlinear partial differential equations: KdV has the structure of the equations of Hamiltonian dynamics; it has certain "solitary solutions" rather than wave train solutions; finally, it has a rich set of constants of motion related to the spectra of certain associated equations. These equations appear as "isospectral deformations" of some natural operators. This series of articles, devoted to the geometry of KdV, is based on tools from algebraic geometry (in dimension ∞), especially hyperelliptic curves, their Jacobians, theta functions and their connections with the spectral theory of Hill's equation.

Nonlinear equations. The nonlinear equations that interested McKean are the classical commutation relations

$$\frac{\partial L}{\partial t} = [L, K],$$

where L is a differential operator of first order whose coefficients are 2×2 matrices of class C^∞ and where K is an antisymmetric operator of the same type. The KdV equation

$$\frac{\partial q}{\partial t} = 3q\,\frac{\partial q}{\partial x} - \frac{1}{2}\frac{\partial^3 q}{\partial x^3}$$

is of this type where L is the Hill operator

$$L := -\Delta + q,$$
$$K := 2D^3 - \tfrac{3}{2}(qD + Dq),$$

where $D = \partial/\partial x$ and $\Delta = D^2$. In addition to these equations, McKean studied the sine-Gordon equation

$$\frac{\partial^2 u}{\partial t^2} = \frac{\partial^2}{\partial x^2} - \sin q$$

and the equations of Boussinesq and Camassa–Holm, written, respectively, as

$$\frac{\partial^2 q}{\partial t^2} = \frac{\partial^2}{\partial x^2}\left(\frac{4}{3}q^2 + \frac{1}{3}\frac{\partial^2 q}{\partial x^2}\right),$$

$$\frac{\partial u}{\partial t} + 3u\frac{\partial u}{\partial x} = \alpha^2\left(\frac{\partial^3 u}{\partial x^2 \partial t} + 2\frac{\partial u}{\partial x}\frac{\partial^2 u}{\partial x^2} + u\frac{\partial^4 u}{\partial x^4}\right).$$

Financial mathematics. In 1965, following a suggestion of the economist Paul Samuelson, McKean wrote a short article on the problem of "American options". This was first published as an appendix to a treatise by Samuelson on financial economics. It anticipated by eight years the famous Black–Scholes–Merton formula. In this paper, McKean shows that the price of an American option can be computed by solving a free boundary problem for a parabolic partial differential equation. This was the first application of nonlinear partial differential equations in financial mathematics.

Some personal tributes

Paul Malliavin: In 1972 I was working in the theory of functions of several complex variables, more specifically on the characterization of the set of zeros of a function of the Nevanlinna class in a strictly pseudo-convex domain of $\mathbb{C}^n$. In the special case of the ball, by using semisimple harmonic analysis, I computed exactly the Green function and obtained the desired characterization. For the general case I was completely stuck; I looked at the canonical heat equation associated to an exhaustion function; a constructive tool to evaluate the Green function near the boundary was urgently needed. Then I started to study Itô's papers on SDEs; their constructive essence seemed to me quite appropriate for this estimation problem.

Then McKean's *Stochastic Integrals* appeared: I was fully rescued in my efforts to grasp from scratch the theory of SDEs. McKean's book is short, with carefully written concrete estimates; it presents with sparkling clarity a conceptual vision of the theory. In a note in the *Proceedings of the National Academy* I found the needed estimate for the Green function. From there I started to study the case of weakly pseudoconvex domains, where hypoelliptic operators of Hörmander type appeared; from that time onward (for the last thirty years) I became fully involved in probability theory. So Henry has had a key influence on this turning point of my scientific interests.

From the middle seventies Henry has kindly followed the different steps of my career, supporting me at every occasion. For instance, when I was in the process of being fired from Université Paris VI in 1995, he did not hesitate to cross the Atlantic in order to sit on a special committee, specially constituted by the president of Université Paris VI, in order to judge if my current work was then so obsolete that all the grants for my research associates had to be immediately eliminated.

With gratitude I dedicate our paper to Henry; also with admiration, discovering every day more and more the breadth and the depth of his scientific impact.

Daniel Stroock: I have known Henry since 1964, the year he visited Rockefeller Institute and decided to leave MIT and New England for the city of New York.

I do not know any details of Henry's childhood, but I have a few impressions which I believe bear some reasonable resemblance to the truth. Henry was the youngest child of an old New England family and its only son. He grew up in the North Shore town of Beverly, Massachusetts, where he developed a lasting love for the land, its inhabitants and the way they pronounced the English language. I gather that Henry was younger than his siblings, so he learned to fend for himself from an early age, except that he was taught to ski by his eldest sister, who was a superb athlete. He was sent to St. Paul's for high school where, so far as I know, his greatest distinction was getting himself kicked out for smoking. Nonetheless, he was accepted to Dartmouth College, where, under the influence of Bruce Knight, he began to develop a taste for mathematics. His interests in mathematics were further developed by a course which he took one summer from Mark Kac, who remained a lasting influence on Henry.

In fact, it was Kac who invited Henry to visit and then, in 1966, move to Rockefeller Institute. Since I finished my PhD the spring before Henry joined the faculty, I do not know many details of life in the Rockefeller mathematics-physics group during the late 1960s. However I do know one story from that period which, even if it's not totally true, nicely portrays Henry's affection for Kac, the founding father of the Rockefeller Mathematics Department. As such, it was his job to make it grow. For various reasons, not the least of which was his own frequent absence, Kac was having limited success in this enterprise; at one faculty meeting Kac solicited suggestions from his younger colleagues. Henry's suggestion was that they double Kac's salary in order to have him there on a full-time basis.

Henry's mathematical achievements may be familiar to anyone who is likely to be reading this book, with one proviso: most mathematicians do not delve into the variety of topics to which Henry has contributed. Aside from the constant evidence of his formidable skills, the property shared by all of Henry's mathematics is a strong sense of *taste*. Whether it is his early collaboration with Itô, his excursion into Gaussian prediction theory, or his interest in completely integrable systems and spectral invariance, Henry has chosen problems because they interest him and please his sense of aesthetics. As a result, his mathematics possesses originality all of its own, and a beauty that the rest of us can appreciate.

Srinivasa Varadhan, former director of the Courant Insitute: Henry has been my colleague for nearly thirty-five years. I have always been impressed by the number of students Henry has produced. I checked the Genealogy project. It lists him with nearly fifty students. It is even more impressive that he has nearly three hundred descendants, which means he has taught his students how to teach.

Henry is known for meticulous attention to detail. When Dan Stroock and I wrote our first paper on the martingale problem we gave the manuscript to Henry for comments. The paper was typewritten, before the days of word processors and xerox machines. Henry gave it back to us within days and the comments filled out all the empty space between the lines on every page. The typist was not amused.

Henry's own work drifted in the seventies out of probability theory into integrable systems and then back again at some point into probability. Talking to Henry about any aspect of mathematics is always fun. He has many interesting but unsolvable problems and will happily share them with you. If I come across a cute proof of something Henry is the first one I will think of telling.

Charles Newman, member and former Director of the Courant Institute: Henry's earliest papers (*Ann. Math.* 1955) concerned sample functions of stable processes, together with several papers on Brownian local time (*J. Math. Kyoto* 1962, *Adv. Math.* 1975). These topics and the general subject of one-dimensional diffusions were of course analyzed in great detail in Henry's celebrated 1965 book with Itô. I mention these all because they play a major role in my own much later work (jointly with Fontes and Isopi, *Ann. Prob.* 2002) on scaling limits of random walks with random rates to singular diffusions (with random speed measures).

There are a number of other connections between papers of Henry's and my own — typically with a multidecade gap. For example, there are close connections between Henry's paper "Geometry of differential space" (*Ann. Prob.* 1973) and my 2003 paper with D'Aristotile and Diaconis, "Brownian motion and the classical groups".

The paper by Camia and myself in this volume is less directly related to Henry's work. However, the general subject of Schramm–Loewner evolutions, to which our paper belongs, combines many of the same themes that have permeated Henry's work: Brownian motions and related processes, complex variable theory, and statistical mechanics.

Probability, Geometry and Integrable Systems
MSRI Publications
Volume **55**, 2008

Bitangential direct and inverse problems for systems of differential equations

DAMIR Z. AROV AND HARRY DYM

ABSTRACT. A number of results obtained by the authors on direct and inverse problems for canonical systems of differential equations, and their implications for certain classes of systems of Schrödinger equations and systems with potential are surveyed. Connections with the theory of J-inner matrix valued and reproducing kernel Hilbert spaces, which play a basic role in the original developments, are discussed.

1. Introduction

In this paper we shall present a brief survey of a number of results on direct and inverse problems for canonical integral and differential systems that have been obtained by the authors over the past several years. We shall not attempt to survey the literature, which is vast, or to compare the methods surveyed here with other approaches. The references in [Arov and Dym 2004; 2005b; 2005c] (the last of which is a survey article) may serve at least as a starting point for those who wish to explore the literature.

The differential systems under consideration are of the form

$$y'(t,\lambda) = i\lambda y(t,\lambda)H(t)J, \quad 0 \leqslant t < d, \tag{1.1}$$

where $H(t)$ is an $m \times m$ locally summable mvf (matrix valued function) that is positive semidefinite a.e. on the interval $[0, d)$, J is an $m \times m$ signature matrix, i.e., $J = J^*$ and $J^*J = I_m$, and $y(t,\lambda)$ is a $k \times m$ mvf.

The matrizant or fundamental solution, $Y_t(\lambda) = Y(t,\lambda)$, of (1.1) is the unique locally absolutely continuous $m \times m$ solution of (1.1) that meets the initial condition $Y_0(\lambda) = I_m$, i.e.,

$$Y_t(\lambda) = I_m + i\lambda \int_0^t Y_s(\lambda)H(s)\,ds\,J \quad \text{for } 0 \leq t < d. \tag{1.2}$$

Standard estimates yield the following properties:

(1) $Y_t(\lambda)$ is an entire mvf that is of exponential type in the variable λ for each fixed $t \in [0, d)$.

(2) The identity

$$\frac{J - Y_t(\lambda)JY_t(\omega)^*}{-2\pi i(\lambda - \overline{\omega})} = \frac{1}{2\pi} \int_0^t Y_s(\lambda)H(s)Y_s(\omega)^* ds \qquad (1.3)$$

holds for each $t \in [0, d)$ and for every pair of points $\lambda, \omega \in \mathbb{C}$.

(3) $Y_t(\lambda)$ is J-inner (in the variable λ) in the open upper half plane $\mathbb{C}_+ = \{\lambda \in \mathbb{C} : \lambda + \overline{\lambda} > 0\}$ for each fixed $t \in [0, d)$. This means that

$$J - Y_t(\omega)JY_t(\omega)^* \geq 0 \quad \text{for } \omega \in \mathbb{C}_+ ,$$

with equality if $\omega \in \mathbb{R}$.

(4) $J - Y_t(\omega)JY_t(\overline{\omega})^* = 0 \quad \text{for } \omega \in \mathbb{C}$.

(5) The kernel

$$K_\omega^t(\lambda) = \frac{J - Y_t(\lambda)JY_t(\omega)^*}{-2\pi i(\lambda - \overline{\omega})}$$

is positive in the sense that

$$\sum_{i,j=1}^n u_i^* K_{\omega_j}^t(\omega_i)u_j \geq 0$$

for every choice of the points $\omega_1, \ldots, \omega_n$, vectors $u_1, \ldots, u_n$ and every positive integer n.

(6) $Y_{t_1}^{-1}Y_{t_2}$ is also an entire J-inner mvf for $0 \leq t_1 \leq t_2 < d$.

Every $m \times m$ signature matrix $J \neq \pm I_m$ is unitarily equivalent to the matrix

$$j_{pq} = \begin{bmatrix} I_p & 0 \\ 0 & -I_q \end{bmatrix}, \quad p + q = m,$$

with

$$p = \text{rank}\,(I_m + J) \geq 1 \quad \text{and} \quad q = \text{rank}\,(I_m - J) \geq 1.$$

The main examples of signature matrices, apart from $\pm j_{pq}$, are $\pm j_p$, $\pm J_p$ and $\pm \mathcal{I}_p$, where

$$J_p = \begin{bmatrix} 0 & -I_p \\ -I_p & 0 \end{bmatrix}, \quad j_p = \begin{bmatrix} I_p & 0 \\ 0 & -I_p \end{bmatrix}, \quad \mathcal{I}_p = \begin{bmatrix} 0 & -iI_p \\ iI_p & 0 \end{bmatrix}.$$

In the last three examples $q = p$ (so that $2p = m$). The signature matrices J_p and j_p are connected by the signature matrix

$$\mathfrak{V} = \frac{1}{\sqrt{2}} \begin{bmatrix} -I_p & I_p \\ I_p & I_p \end{bmatrix}, \quad \text{i.e.,} \quad \mathfrak{V} J_p \mathfrak{V} = j_p \quad \text{and} \quad \mathfrak{V} j_p \mathfrak{V} = J_p.$$

There is a natural link between each of the three principal signature matrices and each of the following three classes of mvf's that are holomorphic in $\mathbb{C}_+$, the open upper half plane:

(1) The Schur class $\mathcal{S}^{p \times q}$ of $p \times q$ mvf's $\varepsilon(\lambda)$ that are holomorphic in $\mathbb{C}_+$ and satisfy the constraint $I_q - \varepsilon(\lambda)^* \varepsilon(\lambda) \geq 0$, since

$$I_q - \varepsilon(\lambda)^* \varepsilon(\lambda) \geq 0 \Longleftrightarrow [\varepsilon(\lambda)^* \quad I_q] j_{pq} \begin{bmatrix} \varepsilon(\lambda) \\ I_q \end{bmatrix} \leq 0. \tag{1.4}$$

(2) The Carathéodory class $\mathcal{C}^{p \times p}$ of $p \times p$ mvf's $\tau(\lambda)$ that are holomorphic in $\mathbb{C}_+$ and satisfy the constraint $\tau(\lambda) + \tau(\lambda)^* \geq 0$, since

$$\tau(\lambda) + \tau(\lambda)^* \geq 0 \Longleftrightarrow [\tau(\lambda)^* \quad I_p] J_p \begin{bmatrix} \tau(\lambda) \\ I_p \end{bmatrix} \leq 0. \tag{1.5}$$

(3) The Nevanlinna class $\mathcal{R}^{p \times p}$ of $p \times p$ mvf's $\tau(\lambda)$ that are holomorphic in $\mathbb{C}_+$ and satisfy the constraint $(\tau(\lambda) - \tau(\lambda)^*)/i \geq 0$, since

$$(\tau(\lambda) - \tau(\lambda)^*)/i \geq 0 \Longleftrightarrow [\tau(\lambda)^* \quad I_p] \mathcal{J}_p \begin{bmatrix} \tau(\lambda) \\ I_p \end{bmatrix} \leq 0, \tag{1.6}$$

A general $m \times m$ mvf $U(\lambda)$ is said to be J-inner with respect to the open upper half plane $\mathbb{C}_+$ if it is meromorphic in $\mathbb{C}_+$ and if

(1) $J - U(\lambda)^* J U(\lambda) \geq 0$ for every point $\lambda \in \mathfrak{h}_U^+$ and
(2) $J - U(\mu)^* J U(\mu) = 0$ a.e. on $\mathbb{R}$,

in which $\mathfrak{h}_U^+$ denotes the set of points in $\mathbb{C}_+$ at which U is holomorphic. This definition is meaningful because every mvf $U(\lambda)$ that is meromorphic in $\mathbb{C}_+$ and satisfies the first constraint automatically has nontangential boundary values. The second condition guarantees that $\det U(\lambda) \not\equiv 0$ in $\mathfrak{h}_U^+$ and hence permits a pseudo-continuation of $U(\lambda)$ to the open lower half plane $\mathbb{C}_-$ by the symmetry principle

$$U(\lambda) = J\{U^\#(\lambda)\}^{-1} J \quad \text{for } \lambda \in \mathbb{C}_-,$$

where $f^\#(\lambda) = f(\bar{\lambda})^*$.

The symbol $\mathfrak{U}(J)$ will denote the class of J-inner mvf's considered on the set $\mathfrak{h}_U$ of points of holomorphy of $U(\lambda)$ in the full complex plane $\mathbb{C}$ and $\mathcal{E} \cap \mathfrak{U}(J)$ will denote the class of entire J-inner mvf's.

If $U \in \mathfrak{U}(I_m)$, then $U \in \mathcal{S}^{m \times m}$ and $U(\lambda)$ is said to be an $m \times m$ inner mvf. The set of $m \times m$ inner mvf's will be denoted $\mathcal{S}_{in}^{m \times m}$ and the set of outer $m \times m$

mvf's in $\mathcal{S}^{m \times m}$ will be denoted $\mathcal{S}_{out}^{m \times m}$. (It is perhaps useful to recall that if $s \in \mathcal{S}^{m \times m}$, then $s \in \mathcal{S}_{out}^{m \times m}$ if and only if $\det s \in \mathcal{S}_{out}^{1 \times 1}$, and that if $s \in \mathcal{S}^{1 \times 1}$, then

$$s \in \mathcal{S}_{out}^{1 \times 1} \iff \ln|s(i)| = \frac{1}{\pi} \int_{-\infty}^{\infty} \frac{\ln|s(\mu)|}{1 + \mu^2} d\mu .)$$

If $A \in \mathcal{U}(J_p)$, there exists a pair of $p \times p$ inner mvf's $b_3(\lambda)$ and $b_4(\lambda)$ that are uniquely characterized in terms of the blocks of $B(\lambda) = A(\lambda)\mathfrak{V}$ and the set

$$\mathcal{N}_{out}^{p \times p} = \left\{ \frac{g}{h} : g \in \mathcal{S}_{out}^{p \times p} \quad \text{and} \quad h \in \mathcal{S}_{out}^{1 \times 1} \right\}$$

by the constraints

$$b_{21}^{\#} b_3 \in \mathcal{N}_{out}^{p \times p} \quad \text{and} \quad b_4 b_{22} \in \mathcal{N}_{out}^{p \times p},$$

up to a constant $p \times p$ unitary multiplier on the left of $b_3(\lambda)$ and a constant $p \times p$ unitary multiplier on the right of $b_4(\lambda)$. Such a pair will be referred to as an *associated pair of the second kind* for $A(\lambda)$ and denoted

$$\{b_3, b_4\} \in ap_{II}(A).$$

(There is also a set of associated pairs $\{b_1, b_2\}$ of the first kind that is more convenient to use in some other classes of problems that will not be discussed here.) The pairs $\{b_3^t, b_4^t\} \in ap_{II}(A_t)$ that are associated with the matrizant $A_t(\lambda), 0 \le t < d$, of a canonical system of the form (1.1) with $J = J_p$ are entire $p \times p$ inner mvf's that are *monotonic* in the sense that

$$(b_3^{t_1})^{-1} b_3^{t_2} \quad \text{and} \quad b_4^{t_2} (b_4^{t_1})^{-1}$$

are $p \times p$ entire inner mvf's for $0 \le t_1 \le t_2 < d$. Moreover, they are uniquely specified by imposing the *normalization* conditions $b_3^t(0) = b_4^t(0) = I_p$ for $0 \le t < d$.

2. Reproducing kernel Hilbert spaces

If $U \in \mathcal{U}(J)$ and
$$\rho_\omega(\lambda) = -2\pi i (\lambda - \overline{\omega}),$$

then the kernel

$$K_\omega^U(\lambda = \frac{J - U(\lambda) J U(\omega)^*}{\rho_\omega(\lambda)}$$

is positive on $\mathfrak{h}_U \times \mathfrak{h}_U$ in the sense that $\sum_{i,j=1}^{n} u_i^* K_{\omega_j}^U(\omega_i) u_j \ge 0$ for every set of vectors $u_1, \ldots, u_n \in \mathbb{C}^m$ and points $\omega_1, \ldots, \omega_n \in \mathfrak{h}_U$; see [Dym 1989], for example. Therefore, by the matrix version of a theorem of Aronszajn [1950], there is an associated RKHS (reproducing kernel Hilbert space) $\mathcal{H}(U)$ of $m \times 1$ mvf's defined and holomorphic in $\mathfrak{h}_U$ with RK (reproducing kernel) $K_\omega^U(\lambda)$. This means that for every choice of $\omega \in \mathfrak{h}_U$, $u \in \mathbb{C}^m$ and $f \in \mathcal{H}(U)$,

(1) $K_\omega u \in \mathcal{H}(U)$ and

(2) $\langle f, K_\omega u \rangle_{\mathcal{H}(U)} = u^* f(\omega)$.

Thus, item (5) in the list of properties of the matrizant implies that there is a RKHS $\mathcal{H}(Y_t)$ of entire $m \times 1$ mvf's with RK $K_\omega^t(\lambda)$ for each $t \in [0, d)$, i.e., for each choice of $\omega \in \mathbb{C}$, $u \in \mathbb{C}^m$ and $f \in \mathcal{H}(Y_t)$,

(a) $K_\omega^t u \in \mathcal{H}(Y_t)$ and

(b) $\langle f, K_\omega^t u \rangle_{\mathcal{H}(Y_t)} = u^* f(\omega)$.

Moreover, property (6) of the matrizant Y_t implies that if $0 \le t_1 \le t_2 < d$, then $\mathcal{H}(Y_{t_1}) \subseteq \mathcal{H}(Y_{t_2})$ as sets and

$$\| f \|_{\mathcal{H}(Y_{t_2})} \le \| f \|_{\mathcal{H}(Y_{t_1})}$$

for every $f \in \mathcal{H}(U_{t_1})$.

In this short review we shall restrict attention to canonical systems with signature matrices $J = J_p$ and shall denote the matrizant of such a system by $A_t(\lambda)$ and the corresponding RK by $K_\omega^{A_t}(\lambda)$. Thus,

$$K_\omega^{A_t}(\lambda) = \frac{J_p - A_t(\lambda) J_p A_t(\omega)^*}{\rho_\omega(\lambda)}.$$

Let

$$N_2^* = \sqrt{2}\,[\,0 \quad I_p\,], \quad B_t(\lambda) = A_t(\lambda)\mathfrak{V}$$

and

$$\mathcal{E}_t(\lambda) = N_2^* B_t(\lambda) = [\,E_-(t, \lambda) \quad E_+(t, \lambda)\,]$$

with $p \times p$ components $E_\pm(t, \lambda)$. Then, since

$$N_2^* K_\omega^{A_t}(\lambda) N_2 = -\frac{\mathcal{E}_t(\lambda) j_p \mathcal{E}_t(\omega)^*}{\rho_\omega(\lambda)},$$

the kernel

$$K_\omega^{\mathcal{E}_t}(\lambda) = -\frac{\mathcal{E}_t(\lambda) j_p \mathcal{E}_t(\omega)^*}{\rho_\omega(\lambda)} = \frac{E_+(t, \lambda) E_+(t, \omega)^* - E_-(t, \lambda) E_-(t, \omega)^*}{\rho_\omega(\lambda)}$$

is also positive and defines a RKHS of entire $p \times 1$ entire mvf's that we shall denote $\mathcal{B}(\mathcal{E}_t)$. These spaces will be called *de Branges spaces*, since they were introduced and extensively studied by L. de Branges; see e.g., [de Branges 1968b; 1968a] and, for additional applications and expository material, [Dym and McKean 1976; Dym 1970; Dym and Iacob 1984]. They can be characterized in terms of the blocks $E_\pm(t, \lambda)$ by the following criteria:

$$f \in \mathcal{B}(\mathcal{E}_t) \iff (E_+^t)^{-1} f \in H_2^p \quad \text{and} \quad (E_-^t)^{-1} f \in K_2^p,$$

where H_2^p denotes the vector Hardy space of order 2 and K_2^p denotes its orthogonal complement with respect to the standard inner product

$$\langle g, h \rangle_{st} = \int_{-\infty}^{\infty} h(\mu)^* g(\mu) \, d\mu \tag{2.1}$$

in $L_2^p(\mathbb{R})$. Moreover, if $f \in \mathcal{B}(\mathscr{E}_t)$, then

$$\|f\|_{\mathcal{B}(\mathscr{E}_t)}^2 = \langle (E_+^t)^{-1} f, (E_+^t)^{-1} f \rangle_{st}.$$

It turns out that with each matrizant $A_t(\lambda)$, there is a unique associated pair $b_3^t(\lambda)$ and $b_4^t(\lambda)$ of $p \times p$ entire inner mvf's that meet the normalization conditions $b_3^t(0) = I_p$ and $b_4^t(0) = I_p$ and corresponding sets of RKHS's $\mathscr{H}(b_3^t)$ and $\mathscr{H}_*(b_4^t)$ with RK's

$$k_\omega^{b_3^t}(\lambda) = \frac{I_p - b_3^t(\lambda)b_3^t(\omega)^*}{\rho_\omega(\lambda)} \quad \text{and} \quad \ell_\omega^{b_4^t}(\lambda) = \frac{b_4^t(\lambda)b_4^t(\omega)^* - I_p}{\rho_\omega(\lambda)},$$

respectively.

3. Linear fractional transformations

The linear fractional transformation T_U based on the four block decomposition

$$U(\lambda) = \begin{bmatrix} u_{11}(\lambda) & u_{12}(\lambda) \\ u_{12}(\lambda) & u_{22}(\lambda) \end{bmatrix},$$

of an $m \times m$ mvf $U(\lambda)$ that is meromorphic in $\mathbb{C}_+$ with diagonal blocks $u_{11}(\lambda)$ of size $p \times p$ and $u_{22}(\lambda)$ of size $q \times q$ is defined on the set

$$\mathscr{D}(T_U) = \{p \times q \text{ meromorphic mvf's } \varepsilon(\lambda) \text{ in } \mathbb{C}_+$$
$$\text{such that } \det\{u_{21}(\lambda)\varepsilon(\lambda) + u_{22}(\lambda)\} \not\equiv 0 \text{ in } \mathbb{C}_+\}$$

by the formula

$$T_U[\varepsilon] = (u_{11}\varepsilon + u_{12})(u_{21}\varepsilon + u_{22})^{-1}.$$

If $U_1, U_2 \in \mathcal{U}(J)$ and if $\varepsilon \in \mathscr{D}(T_{U_2})$ and $T_{U_2}[\varepsilon] \in \mathscr{D}(T_{U_1})$ then

$$T_{U_1 U_2}[\varepsilon] = T_{U_1}[T_{U_2}[\varepsilon]].$$

The notation

$$T_U[E] = \{T_U[\varepsilon] : \varepsilon \in E\} \quad \text{for } E \subseteq \mathscr{D}(T_U)$$

will be useful.

The principal facts are that

(1) $U \in \mathcal{U}(j_{pq}) \implies \mathcal{S}^{p \times q} \subseteq \mathscr{D}_{T_U}$ and $T_U[\mathcal{S}^{p \times q}] \subseteq \mathcal{S}^{p \times q}$.
(2) $U \in \mathcal{U}(J_p) \implies T_U[\mathscr{C}^{p \times p} \cap \mathscr{D}_{T_U}] \subseteq \mathscr{C}^{p \times p}$.

Moreover, if

$$B(\lambda) = A(\lambda)\mathfrak{V},$$

then

$$T_A[\mathscr{C}^{p \times p} \cap \mathscr{D}(T_A)] \subseteq T_B[\mathscr{S}^{p \times p} \cap \mathscr{D}(T_B)] \subseteq \mathscr{C}^{p \times p},$$

where the first inclusion may be proper. The set

$$\mathscr{C}(A) = T_B[\mathscr{S}^{p \times p} \cap \mathscr{D}(T_B)].$$

is more useful than the set $T_A[\mathscr{C}^{p \times p} \cap \mathscr{D}(T_A)]$.

We remark that

$$\mathscr{S}^{p \times p} \subseteq \mathscr{D}(T_B) \iff b_{22}(\omega)b_{22}(\omega)^* > b_{21}(\omega)b_{21}(\omega)^* \qquad (3.1)$$

for some (and hence every) point $\omega \in \mathfrak{h}_A^+$; see Theorem 2.7 in [Arov and Dym 2003a].

4. Restrictions and consequences

In addition to fixing the signature matrix $J = J_p$ in the canonical system (1.1), we shall assume that $\mathcal{H}(A_t) \subset L_2^m$ for every $t \in [0, d)$, i.e., (in our current terminology) A_t belongs to the class $\mathcal{U}_{rSR}(J_p)$ of *right strongly regular J-inner mvf's*. (In our earlier papers the set $\mathcal{U}_{rSR}(J_p)$ was designated $\mathcal{U}_{SR}(J_p)$.) One of the important consequences of this assumption rests on the fact that

$$\mathcal{H}(A_t) \subset L_2^m \iff \mathscr{C}(A_t) \cap \overset{\circ}{\mathscr{C}}{}^{p \times p} \neq \varnothing, \qquad (4.1)$$

where

$$\overset{\circ}{\mathscr{C}}{}^{p \times p} = \{c \in \mathscr{C}^{p \times p} : c \in H_\infty^{p \times p} \quad \text{and} \quad \Re c(\mu) \geq \delta_c I_p \quad \text{a.e. on } \mathbb{R}\} \quad (4.2)$$

and $\delta_c > 0$. Other characterizations of the class $\mathcal{U}_{rSR}(J)$ in terms of the Treil–Volberg matrix version of the Muckenhoupt $(A)_2$ condition presented in [Treil and Volberg 1997] are furnished in [Arov and Dym 2001] and [Arov and Dym 2003b].

If $A_t \in \mathcal{U}_{rSR}(J_p)$ for every $t \in [0, d)$, the following conclusions are in force:

(1) The unique normalized monotonic chain of $p \times p$ entire inner mvf's

$$\{b_3^t, b_4^t\} \in ap_{II}(A_t)$$

consists of continuous functions of t on the interval $0 \leq t < d$ for each fixed point $\lambda \in \mathbb{C}$.

(2) The RKHS $\mathcal{H}(A_{t_1})$ is isometrically included in the RKHS $\mathcal{H}(A_{t_2})$ for $0 \leq t_1 \leq t_2 < d$.

(3) The de Branges spaces $\mathcal{B}(\mathcal{E}_t)$ based on the de Branges matrix

$$\mathcal{E}_t(\lambda) = \sqrt{2}\,[\,0 \quad I_p\,]\,A_t(\lambda)\mathfrak{V}, \quad \text{for } 0 \leq t < d$$

are nested by isometric inclusion, i.e., $\mathcal{B}(\mathcal{E}_{t_1})$ is isometrically included in $\mathcal{B}(\mathcal{E}_{t_2})$ for $0 \leq t_1 \leq t_2 < d$.

(4) $\mathcal{B}(\mathcal{E}_t) = \mathcal{H}(b_3^t) \oplus \mathcal{H}_*(b_4^t)$ as Hilbert spaces with equivalent norms.

(5) The mapping $N_2^* : f \in \mathcal{H}(A_t) \longrightarrow N_2^* f \in \mathcal{B}(\mathcal{E}_t)$ is unitary for every $t \in [0, d)$.

(6) If $\mathcal{S}^{p \times p} \subseteq \mathcal{D}(T_{B_{t_0}})$ for some $t_0 \in [0, d)$, then:

(a) $\mathcal{S}^{p \times p} \subseteq \mathcal{D}(T_{B_t})$ for every $t \in [t_0, d)$.

(b) $\mathcal{C}(A_{t_2}) \subseteq \mathcal{C}(A_{t_1})$ for $t_0 \leq t_1 \leq t_2 < d$.

(c) $\bigcap_{t_0 \leq t < d} \mathcal{C}(A_t) \neq \varnothing$.

(d) $\{c(\omega) : \omega \in \mathbb{C}_+ \quad \text{and} \quad c \in \bigcap_{t_0 \leq t < d} \mathcal{C}(A_t)\}$ is a (Weyl–Titchmarsh) matrix ball with left and right semiradii $R_\ell(\omega)$ and $R_r(\omega)$ with

$$\operatorname{rank} R_\ell(\omega) = \operatorname{rank}\{\lim_{t \uparrow \infty} b_3^t(\omega)b_3^t(\omega)^*\} \tag{4.3}$$

and

$$\operatorname{rank} R_r(\omega) = \operatorname{rank}\{\lim_{t \uparrow \infty} b_4^t(\omega)^* b_4^t(\omega)\}. \tag{4.4}$$

Moreover, these ranks are independent of the choice of the point $\omega \in \mathbb{C}_+$.

An mvf $c(\lambda)$ that belongs to the set

$$\mathcal{C}_{\mathrm{imp}}(H) = \bigcap_{t_0 \leq t < d} \mathcal{C}(A_t)$$

is called an input impedance (or Weyl function) of the system (1.1). If $H \in L_1^{m \times m}$, then, without loss of generality, it may be assumed that $d < \infty$. In this case, $A_d(\lambda)$ is the monodromy matrix of the system (1.1), $\mathcal{C}_{\mathrm{imp}}(H) = \mathcal{C}(A_d)$ and the semiradii $R_\ell(\omega)$ and $R_r(\omega)$ are both positive definite.

5. Inverse problems for canonical systems

Inverse problems for the canonical system (1.1) aim to recover $H(t)$, given some information about the solution of the system. In this direction it is convenient to first consider inverse problems for the canonical integral system

$$y(t, \lambda) = y(0, \lambda) + i\lambda \int_0^t y(s, \lambda)\,dM(s)\,J_p \quad \text{for } 0 \leq t < d, \tag{5.1}$$

in which the mass function $M(t)$, $0 \le t < d$ is a continuous nondecreasing $m \times m$ mvf on the interval $[0, d)$ with $M(0) = 0$. Then the matrizant $A_t(\lambda) = A(t, \lambda)$ of this system is a continuous solution of the equation

$$A_t(\lambda) = I_m + i\lambda \int_0^t A_s(\lambda) \, dM(s) \, J_p \quad \text{for } 0 \le t < d \qquad (5.2)$$

and consequently,

$$\left. \frac{\partial A_t}{\partial \lambda}(\lambda) \right|_{\lambda=0} = \lim_{\lambda \to 0} \frac{A_t(\lambda) - I_m}{\lambda} = i M(t) J_p. \qquad (5.3)$$

Thus, $M(t)$, $0 \le t < d$, can be recovered from the matrizant $A_t(\lambda)$, $0 \le t < d$. The main tool is the following result, which, for a given triple $b_3 \in \mathcal{S}_{in}^{p \times p}$, $b_4 \in \mathcal{S}_{in}^{p \times p}$ and $c \in \mathcal{C}^{p \times p}$, is formulated in terms of the sets

$$\mathcal{C}(b_3, b_4; c) = \{ \widetilde{c} \in \mathcal{C}^{p \times p} : b_3^{-1}(\widetilde{c} - c)b_4^{-1} \in \mathcal{N}_+^{p \times p} \}$$

and

$$\mathcal{N}_+^{p \times p} = \left\{ \frac{g}{h} : g \in \mathcal{S}^{p \times p} \quad \text{and} \quad h \in \mathcal{S}_{out}^{1 \times 1} \right\}.$$

The set $\mathcal{C}(b_3, b_4; c)$ was identified as the set of solutions of a generalized Carathéodory interpolation problem that is formulated in terms of the three given mvf's b_3, b_4 and c in [Arov 1993] and connections with the class $\mathcal{U}(J_p)$ were studied there. These results were developed further in [Arov and Dym 1998] in the case that $b_3(\lambda)$ and $b_4(\lambda)$ are also entire mvf's. In that special case, the interpolation problem is equivalent to a bitangential Krein extension problem in a class of helical mvf's. Krein understood the deep connections between such extension problems and inverse problems for canonical systems. Theorem 5.2, below, illustrates the Krein strategy of identifying the solution of an inverse problem with an appropriately defined chain of extension problems; see [Arov and Dym 2005b] for more details.

THEOREM 5.1. *Let $b_3(\lambda)$, $b_4(\lambda)$ be a pair of entire $p \times p$ inner mvf's and let $c \in \mathcal{C}^{p \times p}$. Then there exists at most one mvf $A \in \mathcal{E} \cap \mathcal{U}(J_p)$ such that*

(1) $\mathcal{C}(A) = \mathcal{C}(b_3, b_4; c)$.
(2) $\{b_3, b_4\} \in ap_{II}(A)$.
(3) $A(0) = I_m$.

Moreover, if

$$\mathcal{C}(b_3, b_4; c) \cap \overset{\circ}{\mathcal{C}}{}^{p \times p} \ne \varnothing,$$

there exists exactly one mvf $A \in \mathcal{E} \cap \mathcal{U}(J_p)$ for which these three conditions are met and it is automatically right strongly regular.

Correspondingly, in our formulation of the inverse impedance problem (inverse spectral problem) for the canonical integral system (5.1) we shall specify a continuous monotonic normalized chain of entire inner $p \times p$ entire inner mvf's $\{b_3^t(\lambda), b_4^t(\lambda)\}$, in addition to an mvf $c \in \mathscr{C}^{p \times p}$ (or a spectral function $\sigma(\mu)$). Spectral functions and the inverse spectral problem are introduced in Section 9.

THEOREM 5.2. *Let* $\{b_3^t(\lambda), b_4^t(\lambda)\}, 0 \leq t < d$, *be a normalized monotonic continuous chain of pairs of entire inner* $p \times p$ *mvf's and let* $c \in \mathscr{C}^{p \times p}$. *Then there exists at most one Hamiltonian* $H(t), 0 \leq t < d$, *such that the matrizant* $A_t(\lambda)$ *of the corresponding canonical system meets the following conditions for every* $t \in [0, d)$:

(1) $\mathscr{C}(A_t) = \mathscr{C}(b_3^t, b_4^t; c)$.
(2) $\{b_3^t, b_4^t\} \in ap_{II}(A_t)$.
(3) $A_t(0) = I_m$.

There exists exactly one continuous nondecreasing mvf $M(t)$ *on the interval* $[0, d)$ *with* $M(0) = 0$ *such that the matrizant* A_t *of the integral system* (5.1) *meets these conditions if*

$$\mathscr{C}(b_3^t, b_4^t; c) \cap \mathring{\mathscr{C}}^{p \times p} \neq \varnothing \text{ for every } t \in [0, d) . \tag{5.4}$$

PROOF. See Theorem 7.9 in [Arov and Dym 2003a]. $\qquad\qquad\square$

6. Description of the RKHS's $\mathscr{H}(A)$ and $\mathscr{B}(\mathfrak{e})$ for $A \in \mathcal{U}_{rsR}(J_p)$

THEOREM 6.1. *If* $A \in \mathcal{U}_{rsR}(J_p)$, $\{b_3(\lambda), b_4(\lambda)\} \in ap_{II}(A)$ *and* $c \in \mathscr{C}(A) \cap H_\infty^{p \times p}$, *then*

$$\mathscr{H}(A) = \left\{ \begin{bmatrix} -\Pi_+ c^* g + \Pi_- ch \\ g + h \end{bmatrix} : g \in \mathscr{H}(b_3) \text{ and } h \in \mathscr{H}_*(b_4) \right\},$$

where Π_+ *denotes the orthogonal projection of* L_2^p *onto the Hardy space* H_2^p, $\Pi_- = I - \Pi_+$ *denotes the orthogonal projection of* L_2^p *onto* $K_2^p = L_2^p \ominus H_2^p$,

$$\mathscr{H}(b_3) = H_2^p \ominus b_3 H_2^p \quad and \quad \mathscr{H}_*(b_4) = K_2^p \ominus b_4^* K_2^p .$$

Moreover,

$$f = \begin{bmatrix} -\Pi_+ c^* g + \Pi_- ch \\ g + h \end{bmatrix} \Longrightarrow \langle f, f \rangle_{\mathscr{H}(A)} = \langle (c + c^*)(g + h), g + h \rangle_{st} ,$$

where $g \in \mathscr{H}(b_3), h \in \mathscr{H}_*(b_4)$ *and* $\langle \cdot, \cdot \rangle_{st}$ *denotes the standard inner product* (2.1) *in* L_2^p.

PROOF. See Theorem 3.8 in [Arov and Dym 2005a]. $\qquad\qquad\square$

THEOREM 6.2. *If $A \in \mathcal{U}_{rsR}(J_p)$, $\{b_3, b_4\} \in ap_{II}(A)$ and $\mathfrak{E}(\lambda) = N_2^* B(\lambda)$, then*

$$\langle f, f \rangle_{\mathcal{H}(A)} = 2\|[0 \quad I_p] f\|_{\mathcal{B}(\mathfrak{E})}^2$$

for every $f \in \mathcal{H}(A)$ and

$$\mathcal{B}(\mathfrak{E}) = \mathcal{H}(b_3) \oplus \mathcal{H}_*(b_4) \quad \text{as Hilbert spaces with equivalent norms}.$$

PROOF. See Theorem 3.8 in [Arov and Dym 2005a].   □

REMARK 6.3. If $\mathcal{B}(\mathfrak{E}) = \mathcal{H}(b_3) \oplus \mathcal{H}_*(b_4)$ as linear spaces, the two norms in these spaces are equivalent, i.e., there exist a pair of positive constants γ_1, γ_2 such that

$$\gamma_1 \|f\|_{st} \leq \|f\|_{\mathcal{B}(\mathfrak{E})} \leq \gamma_2 \|f\|_{st}$$

for every $f \in \mathcal{B}(\mathfrak{E})$. This follows from the closed graph theorem and the fact that $\mathcal{B}(\mathfrak{E})$ and $\mathcal{H}(b_3) \oplus \mathcal{H}_*(b_4)$ are both RKHS's.

7. A basic conclusion

In order to apply Theorem 5.2, we need to know when condition (5.4) is in force. In particular, the condition (5.4) is satisfied if $c \in \overset{\circ}{\mathcal{C}}^{p \times p}$. However, if the given matrix $c \in \mathcal{C}^{p \times p} \cap \mathcal{W}_+^{p \times p}(\gamma)$, i.e., if $c(\lambda)$ admits a representation of the form

$$c(\lambda) = \gamma_c + \int_0^\infty e^{i\lambda t} h_c(t)\, dt \quad \text{with } \gamma_c \in \mathbb{C}^{p \times p} \quad \text{and} \quad h_c \in L_1^{p \times p}([0, \infty)),$$
(7.1)

then condition (5.4) will be in force if $\gamma_c + \gamma_c^* > 0$, even if $\det \Re c(\mu) = 0$ at some points $\mu \in \mathbb{R}$; see Theorem 5.2 in [Arov and Dym 2005a]. Moreover, if either

$$\lim_{\nu \uparrow \infty} b_3^{t_0}(i\nu) = 0 \quad \text{or} \quad \lim_{\nu \uparrow \infty} b_4^{t_0}(i\nu) = 0$$

for some point $t_0 \in [0, d)$, the condition $\gamma + \gamma^* > 0$ is necessary for (5.4) to be in force and hence for the existence of a canonical system (1.1) with a matrizant $A_t(\lambda)$, $0 \leq t < d$, that meets the conditions (1) (2) and (3) in Theorem 5.2; see Theorem 5.4 in [Arov and Dym 2005a].

REMARK 7.1. The method of solution depends upon the interplay between the RKHS's that play a role in the parametrization formulas presented in Theorem 6.1 and their corresponding RK's. This method also yields the formulas for $M(t)$ and the corresponding matrizant $A_t(\lambda)$ that are discussed in the next section. It differs from the known methods of Gelfand–Levitan, Marchenko and Krein, which are not directly applicable to the bitangential problems under consideration.

8. An algorithm for solving the inverse impedance problem

In this section we shall assume that an mvf $c \in \mathscr{C}^{p \times p}$ and a normalized monotonic continuous chain of pairs $\{b_3^t(\lambda), b_4^t(\lambda)\}$, $0 \le t < d$, of entire inner $p \times p$ mvf's that meet the condition (5.4) have been specified. Then there exists an mvf

$$c^t \in \mathscr{C}(b_3^t, b_4^t; c) \cap H_\infty^{p \times p} \tag{8.1}$$

for every $t \in [0, d)$ and hence, the operators

$$\Phi_{11}^t = \Pi_{\mathscr{H}(b_3^t)} M_{c^t}\Big|_{H_2^p}, \quad \Phi_{22}^t = \Pi_- M_{c^t}\Big|_{\mathscr{H}_*(b_4^t)}, \quad \Phi_{12}^t = \Pi_{\mathscr{H}(b_3^t)} M_{c^t}\Big|_{\mathscr{H}_*(b_4^t)}, \tag{8.2}$$

$$Y_1^t = \Pi_{\mathscr{H}(b_3^t)}\{M_{c^t} + (M_{c^t})^*\}\Big|_{\mathscr{H}(b_3^t)} = 2\Re\big(\Phi_{11}^t\big|_{\mathscr{H}(b_3^t)}\big) \tag{8.3}$$

and

$$Y_2^t = \Pi_{\mathscr{H}_*(b_4^t)}\{M_{c^t} + (M_{c^t})^*\}\Big|_{\mathscr{H}_*(b_4^t)} = 2\Re\big(\Pi_{\mathscr{H}_*(b_4^t)}\Phi_{22}^t\big) \tag{8.4}$$

are well defined. Moreover, they do not depend upon the specific choice of the mvf c^t in the set indicated in formula (8.1). In order to keep the notation relatively simple, an operator T that acts in the space of $p \times 1$ vvf's will be applied to $p \times p$ mvf's with columns $f_1, \ldots, f_p$ column by column: $T[f_1 \cdots f_p] = [Tf_1 \cdots Tf_p]$.

We define three sets of $p \times p$ mvf's $\hat{y}_{ij}^t(\lambda)$, $\hat{u}_{ij}^t(\lambda)$ and $\hat{x}_{ij}^t(\lambda)$ by the following system of equations, in which $\tau_3(t)$ and $\tau_4(t)$ denote the exponential types of $b_3^t(\lambda)$ and $b_4^t(\lambda)$, respectively,

$$e_t(\lambda) = e^{i\lambda t} \quad \text{and} \quad (R_0 f)(\lambda) = \{f(\lambda) - f(0)\}/\lambda :$$

$$\hat{y}_{11}^t(\lambda) = i\big(\Phi_{11}^t(R_0 e_{\tau_3(t)} I_p)\big)(\lambda), \qquad \hat{y}_{12}^t(\lambda) = -i(R_0 b_3^t)(\lambda),$$
$$\hat{y}_{21}^t(\lambda) = i\big((\Phi_{22}^t)^*(R_0 e_{-\tau_4(t)} I_p)\big)(\lambda), \quad \hat{y}_{22}^t(\lambda) = i(R_0 (b_4^t)^{-1})(\lambda), \tag{8.5}$$

THEOREM 8.1. *In the setting of this section, there exists exactly one mvf $A_t \in \mathscr{E} \cap \mathscr{U}(J_p)$ for each $t \in [0, d)$ such that:*

(1) $\mathscr{C}(A_t) = \mathscr{C}(b_3^t, b_4^t; c)$.
(2) $\{b_3^t, b_4^t\} \in ap_{II}(A_t)$.
(3) $A_t(0) = I_m$.

The RK $K_\omega^t(\lambda)$ of the RKHS $\mathscr{H}(A_t)$ evaluated at $\omega = 0$ is given by the formula

$$K_0^t(\lambda) = \frac{1}{2\pi}\left[\begin{array}{cc} \hat{x}_{11}^t(\lambda) + \hat{x}_{21}^t(\lambda) & \hat{x}_{12}^t(\lambda) + \hat{x}_{22}^t(\lambda) \\ \hat{u}_{11}^t(\lambda) + \hat{u}_{21}^t(\lambda) & \hat{u}_{12}^t(\lambda) + \hat{u}_{22}^t(\lambda) \end{array}\right], \tag{8.6}$$

where:

(1) *The $\widehat{u}_{ij}^t(\lambda)$ are $p \times p$ mvf's such that the columns of $\widehat{u}_{1j}^t(\lambda)$ belong to $\mathcal{H}(b_3^t)$ and the columns of $\widehat{u}_{2j}^t(\lambda)$ belong to $\mathcal{H}_*(b_4^t)$. The $\widehat{u}_{ij}^t(\lambda)$ may be defined as the solutions of the systems of equations:*

$$Y_1^t \widehat{u}_{1j}^t + \Phi_{12}^t \widehat{u}_{2j}^t = \widehat{y}_{1j}^t(\lambda)$$
$$(\Phi_{12}^t)^* \widehat{u}_{1j}^t + Y_2^t \widehat{u}_{2j}^t = \widehat{y}_{2j}^t(\lambda) , \quad j = 1, 2 . \tag{8.7}$$

(2) *The mvf's $\widehat{x}_{ij}^t(\lambda)$ are defined by the formulas*

$$\widehat{x}_{1j}^t(\lambda) = -(\Phi_{11}^t)^* \widehat{u}_{1j}^t ,$$
$$\widehat{x}_{2j}^t(\lambda) = \Phi_{22}^t \widehat{u}_{2j}^t , \quad j = 1, 2 . \tag{8.8}$$

PROOF. This theorem is Theorem 4.2 of [Arov and Dym 2005a]. $\square$

REMARK 8.2. In the one-sided cases when either $b_4^t(\lambda) = I_p$ or $b_3^t(\lambda) = I_p$, the formulas for recovering $M(t)$ are simpler:

If, for example, $b_4^t(\lambda) = I_p$, then $\tau_4(t) = 0$ and $\mathcal{H}_*(b_4^t) = \{0\}$ and hence equations (8.7) and (8.8) simplify to

$$Y_1^t \widehat{u}_{1j}^t = \widehat{y}_{1j}^t(\lambda) \quad \text{and} \quad \widehat{x}_{1j}^t = -(\Phi_{11}^t)^* \widehat{u}_{1j}^t \quad \text{for } j = 1, 2 , \tag{8.9}$$

and

$$\widehat{u}_{2j}^t = 0 \quad \text{and} \quad \widehat{x}_{2j}^t = 0 \quad \text{for } j = 1, 2 .$$

THEOREM 8.3. *Let $\{c(\lambda); b_3^t(\lambda), b_4^t(\lambda), 0 \leq t < d\}$ be given where $c \in \mathbb{C}^{p \times p}$, $\{b_3^t(\lambda), b_4^t(\lambda)\}, 0 \leq t < d$, is a normalized monotonic continuous chain of pairs of entire inner $p \times p$ mvf's and let assumption (5.4) be in force. Then the unique solution $M(t)$ of the inverse input impedance problem considered in Theorem 5.2 is given by the formula*

$$M(t) = 2\pi K_0^t(0)$$
$$= \int_0^{\tau_3(t)} \begin{bmatrix} x_{11}^t(a) & x_{12}^t(a) \\ u_{11}^t(a) & u_{12}^t(a) \end{bmatrix} da + \int_{-\tau_4(t)}^0 \begin{bmatrix} x_{21}^t(a) & x_{22}^t(a) \\ u_{21}^t(a) & u_{22}^t(a) \end{bmatrix} da \tag{8.10}$$

and the corresponding matrizant may be defined by the formula

$$A_t(\lambda) = I_m + 2\pi i \lambda K_0^t(\lambda) J_p , \tag{8.11}$$

where $K_0^t(\lambda)$ is specified by formula (8.6) and $x_{ij}^t(a)$ and $u_{ij}^t(a)$ designate the inverse Fourier transforms of $\widehat{x}_{ij}^t(\lambda)$ and $\widehat{u}_{ij}^t(\lambda)$, respectively.

PROOF. Formula (8.11) follows from the definition of the RK $K_0^t(\lambda) = K_0^{A_t}(\lambda)$ and the fact that $A_t(0) = I_m$. Formula (8.10) follows from (5.3), (8.6) and (8.11). $\square$

9. Spectral functions

The term *spectral function* is defined in two different ways: The first definition is in terms of the generalized Fourier transform

$$(\mathcal{F}_2 f)(\lambda) = [0 \quad I_p] \frac{1}{\sqrt{\pi}} \int_0^d A(s, \lambda) \, dM(s) \, f(s) \tag{9.1}$$

based on the matrizant of the canonical system (5.1) applied initially to the set of $f \in L_2^m(dM s; [0, d))$ with compact support inside the interval $[0, d)$.

A nondecreasing $p \times p$ mvf $\sigma(\mu)$ on $\mathbb{R}$ is said to be a *spectral function* for the system (5.1) if the Parseval equality

$$\int_{-\infty}^{\infty} (\mathcal{F}_2 f)(\mu)^* d\sigma(\mu)(\mathcal{F}_2 f)(\mu) = \int_0^d f(t)^* dM(t) f(t) \tag{9.2}$$

holds for every $f \in L_2^m(dM s; [0, d))$ with compact support. The notation $\Sigma_{sf}^d(M)$ will be used to denote the set of spectral functions of a canonical system of the form (5.1).

REMARK 9.1. The generalized Fourier transform introduced in formula (9.1) is a special case of the transform

$$(\mathcal{F}^L f)(\lambda) = L^* \frac{1}{\sqrt{\pi}} \int_0^d A(s, \lambda) \, dM(s) \, f(s) \tag{9.3}$$

that is based on a fixed $m \times p$ matrix L that meets the conditions $L^* J_p L = 0$ and $L^* L = I_p$. The mvf $y(t, \lambda) = L^* A_t(\lambda)$ is the unique solution of the system (5.1) that satisfies the initial condition $y(0, \lambda) = L^*$. Spectral functions may be defined relative to the transform $\mathcal{F}^L$ in just the same way that they were defined for the transform $\mathcal{F}_2$. Direct and inverse spectral problems for these spectral functions are easily reduced to the corresponding problems based on $\mathcal{F}_2$.; see Sections 4 and 5 of [Arov and Dym 2004] and Section 16 of [Arov and Dym 2005c] for additional discussion.

The second definition of spectral function is based on the Riesz–Herglotz representation

$$c(\lambda) = i\alpha - i\beta\lambda + \frac{1}{\pi i} \int_{-\infty}^{\infty} \left\{ \frac{1}{\mu - \lambda} - \frac{\mu}{1 + \mu^2} \right\} d\sigma(\mu), \quad \lambda \in \mathbb{C}_+, \tag{9.4}$$

which defines a correspondence between $p \times p$ mvf's $c \in \mathcal{C}^{p \times p}$ and a set $\{\alpha, \beta, \sigma\}$, in which $\sigma(\mu)$ is a nondecreasing $p \times p$ mvf on $\mathbb{R}$ that is normalized to be left continuous with $\sigma(0) = 0$ and is subject to the constraint

$$\int_{-\infty}^{\infty} \frac{d \operatorname{trace} \sigma(\mu)}{1 + \mu^2} < \infty, \tag{9.5}$$

and α and β are constant $p \times p$ matrices such that $\alpha = \alpha^*$ and $\beta \geq 0$.

The mvf $\sigma(\mu)$ in the representation (9.4) will be referred to as the *spectral function* of $c(\lambda)$. Correspondingly, if $\mathcal{F} \subseteq \mathcal{C}^{p \times p}$, then

$$(\mathcal{F})_{sf} = \{\sigma : \sigma \text{ is the spectral function of some } c \in \mathcal{F}\}.$$

If $A \in \mathcal{U}_{rsR}(J_p)$ and $c(\lambda) = T_A[I_p]$, then $\beta = 0$ and $\sigma(\mu)$ is absolutely continuous with $\sigma'(\mu) = \Re c(\mu)$ a.e. on $\mathbb{R}$; see Lemma 2.2 and the discussion following Lemma 2.3 in [Arov and Dym 2005a]. Moreover, if $A \in \mathcal{U}_{rsR}(J_p)$ and $\mathcal{S}^{p \times p} \subseteq \mathcal{D}(T_{A\mathfrak{V}})$, then for each $\sigma \in (\mathcal{C}(A))_{sf}$, there exists at least one $p \times p$ Hermitian matrix α such that

$$c^{(\alpha)}(\lambda) = i\alpha + \frac{1}{\pi i} \int_{-\infty}^{\infty} \left\{ \frac{1}{\mu - \lambda} - \frac{\mu}{1 + \mu^2} \right\} d\sigma(\mu) \tag{9.6}$$

belongs to $\mathcal{C}(A)$; see Theorem 2.14 in [Arov and Dym 2004].

We shall also make use of the following condition on the growth of the mvf $\chi_1^t(\lambda) = b_4^t(\lambda) b_3^t(\lambda)$:

$$\|\chi_1^a(re^{i\theta} + \omega)\| \leq \gamma < 1 \quad \text{on the indicated ray } r \geq 0 \text{ in } \mathbb{C}_+, \tag{9.7}$$

i.e., the inequality holds for some fixed choice of $\theta \in [0, \pi]$, $\omega \in \mathbb{C}_+$, $a \in (0, d)$ and all $r \geq 0$. It is readily checked that if this inequality is in force for some point $a \in (0, d)$, then it holds for all $t \in [a, d)$.

REMARK 9.2. The condition (9.7) will be in force if

$$e_{-a}\chi_1^{t_0}(\lambda) \in \mathcal{S}_{in}^{p \times p}$$

for some choice of $a > 0$ and $t_0 \in (0, d)$.

These observations leads to the following conclusion:

LEMMA 9.3. *If the matrizant $A_t(\lambda)$ of the canonical differential system* (1.1) *with $J = J_p$ satisfies the condition*

$$\mathcal{C}(A_t) \cap \overset{\circ}{\mathcal{C}}{}^{p \times p} \neq \varnothing \quad \text{for every } t \in [0, d) \tag{9.8}$$

and if condition (9.7) *is in force for some $a \in [0, d)$ when $\{b_3^t, b_4^t\} \in ap_{II}(A_t)$ for $t \in [0, d)$ and if $c \in \mathcal{C}(A_a)$, then $\beta = 0$ in the representation* (9.4).

In view of the fact that

$$A \in \mathcal{U}_{rsR}(J_p) \iff \mathcal{C}(A) \cap \overset{\circ}{\mathcal{C}}{}^{p \times p} \neq \varnothing, \tag{9.9}$$

the conditions (5.4) and (9.8) are equivalent if $\mathcal{C}(A_t) = \mathcal{C}(b_3^t, b_4^t; c)$ for every $0 \leq t < d$. In particular, these conditions are satisfied if $\mathcal{C}_{imp}^d(M) \cap \overset{\circ}{\mathcal{C}}{}^{p \times p} \neq \varnothing$. They are also satisfied if there exists an mvf $c \in \mathcal{C}_{imp}^d(M)$ of the form (7.1) with $\gamma + \gamma^* > 0$, by Theorem 5.2 in [Arov and Dym 2005a].

Direct problems. Two direct problems for a given canonical system with mass function $M(t)$ on the interval $[0, d)$ are to describe the set of input impedances $\mathscr{C}^d_{\text{imp}}(M)$ and the set $\Sigma^d_{sf}(M)$ of spectral functions of the system.

THEOREM 9.4. *Let $A_t(\lambda)$ denote the matrizant of a canonical integral system* (5.1) *and suppose that the two conditions* (5.4) *and* (9.7) *are met. Then*

(1) $\Sigma^d_{sf}(M) = (\mathscr{C}^d_{\text{imp}}(M))_{sf}$.

(2) *To each $\sigma \in \Sigma^d_{sf}(M)$ there exists exactly one mvf $c(\lambda) \in \mathscr{C}^d_{\text{imp}}(M)$ with spectral function $\sigma(\mu)$. Moreover, this mvf $c(\lambda)$ is equal to one of the mvf's $c^{(\alpha)}(\lambda)$ defined by formula* (9.6) *for some Hermitian matrix α.*

(3) *If $d < \infty$ and* trace $M(t) < \delta < \infty$ *for every $t \in [0, d)$, equation* (5.1) *and the matrizant $A_t(\lambda)$ may be considered on the closed interval $[0, d]$ and $\mathscr{C}^d_{\text{imp}}(M) = \mathscr{C}(A_d)$.*

PROOF. See Theorem 2.21 in [Arov and Dym 2004]. □

A spectral function $\sigma \in \Sigma^d_{sf}(M)$ of the canonical integral system (5.1) with $J = J_p$ is said to be *orthogonal* if the isometric operator that extends the generalized Fourier transform $\mathscr{F}_2$ defined by formula (9.1) maps $L^m_2(dM; [0, d))$ onto $L^p_2(d\sigma; \mathbb{R})$.

THEOREM 9.5. *Let the canonical integral system* (5.1) *with mass function $M(t)$ and matrizant $A_t(\lambda)$ be considered on a finite closed interval $[0, d]$ (so that* trace $M(d) < \infty$) *and let $A(\lambda) = A_d(\lambda)$, $B(\lambda) = A(\lambda)\mathfrak{V}$ and $\mathfrak{E}(\lambda) = \sqrt{2}[0 \quad I_p]B(\lambda)$. Suppose further that*

(a) $(\mathscr{C}(A))_{sf} = \Sigma^d_{sf}(M)$ *and*

(b) $K^{\mathfrak{E}}_\omega(\omega) > 0$ *for at least one (and hence every) point $\omega \in \mathbb{C}_+$.*

Then:

(1) $\mathscr{P}^{p \times p} \subseteq \mathscr{D}(T_B)$.

(2) *The spectral function $\sigma(\mu)$ of the mvf $c(\lambda) = T_B[\varepsilon]$ is an orthogonal spectral function of the given canonical system if ε is a constant $p \times p$ unitary matrix.*

PROOF. The first assertion is equivalent to condition (b); see (3.1). The proof of assertion (2) will be given elsewhere. □

10. The bitangential inverse spectral problem

In our formulation of the *bitangential inverse spectral problem* the given data $\{\sigma; b^t_3, b^t_4, \ 0 \le t < d\}$ is a $p \times p$ nondecreasing mvf $\sigma(\mu)$ on $\mathbb{R}$ that meets the constraint (9.5) and a normalized monotonic continuous chain $\{b^t_3, b^t_4\}, 0 \le t < d$, of pairs of entire inner $p \times p$ mvf's. An $m \times m$ mvf $M(t)$ on the interval

$[0, d)$ is said to be a solution of the bitangential inverse spectral problem with data $\{\sigma(\mu); b_3^t(\lambda), b_4^t(\lambda),\ 0 \leq t < d\}$ if $M(t)$ is a continuous nondecreasing $m \times m$ mvf on the interval $[0, d)$ with $M(0) = 0$ such that the matrizant $A_t(\lambda)$ of the corresponding canonical integral system (5.1) meets the following three conditions:

(i) $\sigma(\mu)$ is a spectral function for this system, i.e., $\sigma \in \Sigma_{sf}^d(M)$.

(ii) $\{b_3^t, b_4^t\} \in ap_{II}(A_t)$ for every $t \in [0, d)$.

(iii) $A_t \in \mathcal{U}_{rsR}(J_p)$ for every $t \in [0, d)$.

The constraint (ii) defines the class of canonical integral systems in which we look for a solution of the inverse problem for the given spectral function $\sigma(\mu)$. Subsequently, the condition (iii) guarantees that in this class there is at most one solution.

The solution of this problem rests on the preceding analysis of the bitangential inverse input impedance problem with data $\{c^{(\alpha)}; b_3^t, b_4^t,\ 0 \leq t < d\}$, where $c^{(\alpha)}(\lambda)$ is given by formula (9.6).

THEOREM 10.1. *If the data* $\{\sigma; b_3^t, b_4^t,\ 0 \leq t < d\}$ *for a bitangential inverse spectral problem meets the conditions* (9.5) *and* (9.7) *and the mvf* $c(\lambda) = c^{(0)}(\lambda)$ *satisfies the constraint* (5.4), *the following conclusions hold:*

(1) *For each Hermitian matrix* $\alpha \in \mathbb{C}^{p \times p}$, *there exists exactly one solution* $M^{(\alpha)}(t)$ *of the bitangential inverse input spectral problem such that* $c^{(\alpha)}(\lambda)$ *is an input impedance for the corresponding canonical integral system* (5.1) *with* $J = J_p$ *based on the mass function* $M^{(\alpha)}(t)$.

(2) *The solutions* $M^{(\alpha)}(t)$ *are related to* $M^{(0)}(t)$ *by the formula*

$$M^{(\alpha)}(t) = \begin{bmatrix} I_p & i\alpha \\ 0 & I_p \end{bmatrix} M^{(0)}(t) \begin{bmatrix} I_p & 0 \\ -i\alpha & I_p \end{bmatrix}. \qquad (10.1)$$

The corresponding matrizants are related by the formula

$$A_t^{(\alpha)}(\lambda) = \begin{bmatrix} I_p & i\alpha \\ 0 & I_p \end{bmatrix} A_t^{(0)}(\lambda) \begin{bmatrix} I_p & -i\alpha \\ 0 & I_p \end{bmatrix}. \qquad (10.2)$$

(3) *If* $M(t)$ *is a solution of the bitangential inverse spectral problem, then* $M(t) = M^{(\alpha)}(t)$ *for exactly one Hermitian matrix* $\alpha \in \mathbb{C}^{p \times p}$.

(4) *The solution* $M^{(0)}(t)$ *and matrizant* $A_t^{(0)}(\lambda)$ *may be obtained from the formulas for the solution of the bitangential inverse input impedance problem with data* $\{c^{(0)}; b_3^t, b_4^t, 0 \leq t < d\}$ *that are given in Theorem 8.3.*

PROOF. See Theorem 2.20 in [Arov and Dym 2004]. □

The condition (5.4) is clearly satisfied if $c^{(0)} \in \mathring{\mathscr{C}}^{p \times p}$. However, this condition is far from necessary. If, for example, $c^{(0)}$ is of the form (7.1) with $\gamma + \gamma^* > 0$, then, as noted earlier, condition (5.4) holds if $\gamma + \gamma^* > 0$, even if $\det\{\Re c(\mu)\} = 0$ on some set of points $\mu \in \mathbb{R}$.

EXAMPLE. If $c^{\circ}(\lambda) = I_p$, i.e., if $\sigma(\mu) = \mu$, then the unique solution of the inverse input impedance problem based given data $\{b_3^t, b_4^t; c^{\circ}\}$ is

$$M^{(0)}(t) = \mathfrak{V} \begin{bmatrix} m_3(t) & 0 \\ 0 & m_4(t) \end{bmatrix} \mathfrak{V},$$

where

$$m_3(t) = -i \frac{\partial b_3^t}{\partial \lambda}\Big|_{\lambda=0} \quad \text{and} \quad m_4(t) = -i \frac{\partial b_4^t}{\partial \lambda}\Big|_{\lambda=0}. \qquad (10.3)$$

Moreover, in this case

$$A_t^{(0)}(\lambda) = \mathfrak{V} \begin{bmatrix} b_3^t(\lambda) & 0 \\ 0 & (b_4^t)^{\#}(\lambda) \end{bmatrix} \mathfrak{V}, \quad B_t^{(0)}(\lambda) = \frac{1}{\sqrt{2}} \begin{bmatrix} -b_3^t(\lambda) & (b_4^t)^{\#}(\lambda) \\ b_3^t(\lambda) & (b_4^t)^{\#}(\lambda) \end{bmatrix},$$

$\mathfrak{E}^t(\lambda) = [b_3^t(\lambda) \ (b_4^t)^{\#}(\lambda)]$ and $\mathfrak{B}(\mathfrak{E}^t) = \mathscr{H}(b_3^t) \oplus \mathscr{H}_*(b_4^t)$ as equivalent RKHS's. If $\|b_4^s(\omega)b_3^s(\omega)\| < 1$ for some $\omega \in \mathbb{C}_+$ and $s \in (0, d)$, then

$$\mathfrak{C}(A_t^{(0)}) = \{T_{\mathfrak{V}}[b_3^t \varepsilon b_4^t] : \varepsilon \in \mathscr{S}^{p \times p}\} \quad \text{for } t \geq s.$$

Although the choice $c^{\circ}(\lambda) = I_p$ in the preceding example is very special, the exhibited one-to-one correspondence between monotonic normalized continuous chains of $p \times p$ entire inner mvf's $\{b_3^t(\lambda), b_4^t(\lambda)\}$ and pairs $\{m_3(t), m_4(t)\}$ of continuous nondecreasing $p \times p$ mvf's on the interval $[0, d)$ with $m_3(0) = m_4(0) = 0$ exhibited in (10.3) is completely general. Moreover, the mvf's $b_3^t(\lambda)$ and $b_4^t(\lambda)$ are the unique continuous solutions of the integral equations

$$b_3^t(\lambda) = I_p + i\lambda \int_0^t b_3^s(\lambda) \, dm_3(s) \quad \text{and} \quad b_4^t(\lambda) = I_p + i\lambda \int_0^t b_4^s(\lambda) \, dm_4(s),$$

respectively, for $0 \leq t < d$.

11. Differential systems with potential

The results referred to above have implications for differential systems of the form

$$u'(t, \lambda) = i\lambda u(t, \lambda) N J + u(t, \lambda) \mathscr{V}(t), \quad 0 \leq t < d, \qquad (11.1)$$

with an $m \times m$ signature matrix J, a constant $m \times m$ matrix N such that

$$N \geq 0 \qquad (11.2)$$

and an $m \times m$ matrix valued potential $\mathcal{V}(t)$ such that

$$\mathcal{V} \in L_{1,loc}^{m \times m}([0,d)) \quad \text{and} \quad \mathcal{V}(t)J + J\mathcal{V}(t)^* = 0 \quad \text{a.e. on } [0,d). \quad (11.3)$$

It is readily checked that the matrizant $U_t(\lambda) = U(t,\lambda)$, $0 \le t < d$, of this system satisfies the identity

$$\{U_t(\lambda)JU_t(\omega)^*\}' = i(\lambda - \overline{\omega})U_t(\lambda)NU_t(\omega)^* \quad \text{for } 0 \le t < d, \quad (11.4)$$

and hence that

$$\frac{J - U_t(\lambda)JU_t(\omega)^*}{\rho_\omega(\lambda)} = \frac{1}{2\pi} \int_0^t U_s(\lambda)NU_s(\omega)^* ds \quad \text{for } 0 \le t < d. \quad (11.5)$$

This in turn leads easily to the conclusion that

$$U_t \in \mathcal{U}(J) \quad \text{for every } t \in [0,d). \quad (11.6)$$

In particular, $U_t(0)$ is J-unitary and so invertible. Moreover, the mvf

$$Y_t(\lambda) = U_t(\lambda)U_t(0)^{-1} \quad \text{for } 0 \le t < d,$$

is the matrizant of the canonical differential system

$$y'(t,\lambda) = i\lambda y(t,\lambda)H(t)J, \quad 0 \le t < d, \quad (11.7)$$

with Hamiltonian

$$H(t) = U_t(0)NU_t(0)^*, \quad 0 \le t < d. \quad (11.8)$$

THEOREM 11.1. *If*

$$NJ = JN, \quad (11.9)$$

then the matrizants $U_t(\lambda)$ and $Y_t(\lambda)$ of the systems (11.1) *and* (11.7) *are both right strongly regular*:

$$U_t \in \mathcal{U}_{rsR}(J) \quad \text{and} \quad Y_t \in \mathcal{U}_{rsR}(J) \quad \text{for every } t \in [0,d), \quad (11.10)$$

and

$$\bigcap_{0 \le t < d} \mathcal{C}(Y_t) = \bigcap_{0 \le t < d} \mathcal{C}(U_t).$$

PROOF. See Section 3 in [Arov and Dym 2005b]. $\qquad \square$

In particular, the condition $NJ = JN$ is met if N is a convex combination of the orthogonal projections

$$P_J = \frac{I_m + J}{2} \quad \text{and} \quad Q_J = \frac{I_m - J}{2},$$

or, even more generally, if $N = N_{\gamma,\delta}$, where

$$N_{\gamma,\delta} = \gamma P_J + \delta Q_J \quad \text{with} \quad \gamma \ge 0, \ \delta \ge 0 \quad \text{and} \quad \kappa = \gamma + \delta > 0. \quad (11.11)$$

In [Arov and Dym 2004; 2005c; 2005b], the direct and inverse impedance spectral problems are considered under the assumption that

$$\text{rank } P_J = \text{rank } Q_J = p,$$

i.e., that J is unitarily equivalent to J_p. If $\gamma = \delta$, then the system (11.1) is called a Dirac system: if $\gamma = 0$ or $\delta = 0$, it is called a Krein system. In the sequel we shall take $J = J_p$ in order to simplify the exposition.

The *generalized Fourier transform* for the system (11.1) with $J = J_p$ is defined by the formula

$$g^\triangle(\lambda) = [0 \quad I_p] \frac{1}{\sqrt{\pi}} \int_0^d U(s, \lambda) N g(s)\, ds \tag{11.12}$$

for every $g \in L_2^m(Nds; [0, d))$ with compact support in $[0, d)$. Correspondingly, a nondecreasing $p \times p$ mvf $\sigma(\mu)$ on $\mathbb{R}$ for which the Parseval equality

$$\int_\infty^\infty g^\triangle(\mu)^* d\sigma(\mu) g^\triangle(\mu) = \int_0^d g(s)^* N g(s)\, ds \tag{11.13}$$

holds for every $g \in L_2^m(Nds; [0, d))$ with compact support in $[0, d)$ is called a *spectral function* for the system (11.1), and the symbol $\Sigma_{sf}^d(\mathcal{V})$ denotes the set of spectral functions for this system. The generalized Fourier transform (9.1) for the canonical system with $H(t) = U(t) N U(t)^*$ for $0 \le t < d$, is related to the transform (11.12):

$$(\mathcal{F}_2 f)(\lambda) = [0 \quad I_p] \frac{1}{\sqrt{\pi}} \int_0^d U(s, \lambda) U(s, 0)^{-1} H(s) f(s)\, ds \tag{11.14}$$

$$= [0 \quad I_p] \frac{1}{\sqrt{\pi}} \int_0^d U(s, \lambda) N U(s, 0)^* f(s)\, ds \tag{11.15}$$

for $f \in L_2^m(H(s)\, ds; [0, d))$ with compact support in $[0, d)$.

The direct problem. The following results on the direct problem are established in [Arov and Dym 2005b]:

THEOREM 11.2. *Let $A_t(\lambda) = A(t, \lambda), 0 \le t < d$, be the matrizant of the system* (11.1) *with $J = J_p$ and $N = N_{\gamma, \delta}$ and assume that the potential $\mathcal{V}(t)$ meets the conditions in* (11.3) *and that $\mathcal{V}(t) = \mathcal{V}(t)^*$ a.e. on the interval $[0, d)$. Then:*

(1) *$A_t \in \mathcal{U}_{rsR}(J_p)$ for every $t \in [0, d)$.*

(2) *$\{e_{\gamma t} I_p, e_{\delta t} I_p\} \in ap_{II}(A_t)$ for every $t \in [0, d)$.*

(3) *The de Branges spaces $\mathcal{B}(\mathfrak{E}_t)$ based on $\mathfrak{E}_t(\lambda) = \sqrt{2} A_t(\lambda) \mathfrak{V}$ are independent of the potential $\mathcal{V}(t)$ as linear topological spaces, i.e.,*

$$\mathcal{B}(\mathfrak{E}_t) = \left\{ \int_{-\delta t}^{\gamma t} e^{i\lambda s} h(s)\, ds : h \in L_2^p([-\delta t, \gamma t]) \right\}$$

as linear spaces and for each $t \in [0, d)$, *there exist a pair of positive constants* $k_1 = k_1(t)$ *and* $k_2 = k_2(t)$ *such that*

$$k_1 \|f\|_{st} \leq \|f\|_{\mathcal{B}(\mathfrak{E}_t)} \leq k_2 \|f\|_{st}$$

for every $f \in \mathcal{B}(\mathfrak{E}_t)$.

(4) $\mathcal{S}^{p \times p} \subseteq \mathcal{D}(T_{A_t \mathfrak{V}})$ *for every* $t \in (0, d)$.

(5) $\mathcal{C}(A_t) = T_{A_t \mathfrak{V}}[\mathcal{S}^{p \times p}]$ *for every* $t \in (0, d)$ *and* $\beta = 0$ *in the integral representation* (9.4) *of every mvf* $c \in \mathcal{C}(A_t)$, $0 < t < d$.

(6) *The set of input impedances* $\mathcal{C}^d_{\text{imp}}(\mathcal{V}) = \bigcap_{0 \leq t < d} \mathcal{C}(A_t)$ *is not empty.*

(7) $\Sigma^d_{sf}(\mathcal{V}) = (\mathcal{C}^d_{\text{imp}}(\mathcal{V}))_{sf}$ *and the integral representation* (9.4) *defines a one-to-one correspondence between these two sets.*

(8) *If* $d < \infty$ *and* $\mathcal{V} \in L_1([0, d])$, *then* $\mathcal{C}^d_{\text{imp}}(\mathcal{V}) = \mathcal{C}(A_d)$.

(9) *If* $d = \infty$, *the set* $\mathcal{C}^d_{\text{imp}}(V)$ *contains exactly one mvf* $c(\lambda)$. *(This is the Weyl limit point case.) If also* $V \in L_1^{m \times m}([0, \infty))$, *then this mvf* $c(\lambda)$ *admits a representation of the form* (7.1) *with* $\gamma_c = I_p$.

PROOF. A proof is supplied in [Arov and Dym 2005b]. □

REMARK 11.3. In the preceding theorem, the sets $\mathcal{C}(A_t)$ depend only upon the potential $\mathcal{V}(t)$ and the positive number $\kappa = \gamma + \delta$ and not on the particular choices $\gamma \geq 0$ and $\delta \geq 0$. This follows from the fact that the mvf $e^{i\delta_0 \lambda t} U_t(\lambda)$ is the matrizant of the system (11.1) with potential $\mathcal{V}(t)$ that is independent of δ_0 and with $N = (\gamma + \delta_0) P_J + (\delta - \delta_0) Q_J$ for every number δ_0 in the interval $-\gamma \leq \delta_0 \leq \delta$. Consequently, the sets $\mathcal{C}^d_{\text{imp}}(\mathcal{V})$ and $\Sigma^d_{sf}(\mathcal{V})$ depend only upon the potential $\mathcal{V}(t)$ and the number κ. Thus, any system of the form (11.1) with $N = N_{\gamma,\delta}$ may be reduced to a Dirac system as well as to a Krein system.

The inverse input impedance problem. The data for the inverse input impedance problem for differential systems of the form (11.1) on an interval $[0, d)$ is an mvf $c \in \mathcal{C}^{p \times p}$ and the right hand endpoint d, $0 < d \leq \infty$, of the interval and the problem is to find a locally summable potential $\mathcal{V}(t)$ of the prescribed form on $[0, d)$ such that $c \in \mathcal{C}^d_{\text{imp}}(\mathcal{V})$, the class of input impedances of the system. In the setting of Theorem 11.2, it is not necessary to specify a chain $\{b_3^t, b_4^t\}$, $0 \leq t < d$, to solve this inverse problem, because, as noted in (2) of that theorem, it is automatically prescribed by the choice $N = N_{\gamma,\delta}$.

THEOREM 11.4. *Let an mvf* $c \in \mathcal{C}^{p \times p}$, *a number* d, $0 < d < \infty$ *and an* $m \times m$ *matrix* N *of the form* (11.11) *be given. Then:*

(1) *There exists at most one differential system of the form* (11.1) *with the given* N, $J = J_p$, *and potential* $\mathcal{V}(t) = \mathcal{V}(t)^*$ *a.e. on* $[0, d]$ *that meets the condition* (11.3) *such that* $c \in \mathcal{C}^d_{\text{imp}}(\mathcal{V})$.

(2) *If $c \in \mathcal{C}^{p \times p}$ admits a representation of the form*

$$c(\lambda) = \gamma_c + \int_0^\infty e^{i\lambda t} h_c(t) \, dt \, , \qquad (11.16)$$

in which $h_c \in L_1^{p \times p}([0, \infty))$ and $h_c(t)$ is continuous on the interval $[0, \kappa d]$ and if $\Re \gamma_c$ is positive definite and $\kappa > 0$, there exists exactly one locally summable potential $\mathcal{V}(t)$, $0 \leq t \leq d$, such that

$$\mathcal{V}(t) = \mathcal{V}(t)^* \quad and \quad \mathcal{V}(t) J_p + J_p \mathcal{V}(t)^* = 0$$

$$a.e. \ on \ the \ interval \ [0, d] \quad (11.17)$$

and $c \in \mathcal{C}_{\mathrm{imp}}^d(\mathcal{V})$. Moreover, this potential $\mathcal{V}(t)$ is continuous on the interval $[0, d]$ and is of the form

$$\mathcal{V}(t) = \mathfrak{V} \begin{bmatrix} 0 & a(t) \\ a(t)^* & 0 \end{bmatrix} \mathfrak{V} \quad for \ 0 \leq t \leq d \, . \qquad (11.18)$$

(3) *If $\kappa = 1$ and $c(\lambda)$ is given by formula (11.16) with $\gamma_c = I_p$, then $a(t) = \gamma^t(t, 0)$, where $\gamma^t(a, b)$ is the unique solution of the integral equation*

$$\gamma^t(a, b) - \int_0^t h_c(a - c) \gamma^t(c, b) \, dc = h_c(a - b) \quad for \ 0 \leq a, b \leq t \leq d \, . \quad (11.19)$$

If $d = \infty$, analogous conclusions hold on the open interval $[0, \infty)$, if $\gamma_c = I_p$ in formula (11.16).

(The modifications needed for general γ_c are discussed in Theorem 5.5 of [Arov and Dym 2005a].)

PROOF. Assertion (1) follows from (1) and (2) of Theorem 11.2, Theorem 7.9 of [Arov and Dym 2003a] and the fact that the set $\mathcal{C}_{\mathrm{imp}}^d(\mathcal{V})$ for the system (11.1) coincides with the set $\mathcal{C}_{\mathrm{imp}}^d(H)$ for the corresponding cannonical differential system (11.7) with Hamiltonian (11.8). Assertion (2) follows from Theorem 5.13 of [Arov and Dym 2005a], Remark 11.3 and the connection between the systems (11.1) and (11.7). □

If $c \in H_\infty^{p \times p} \cap \mathcal{C}_{\mathrm{imp}}^d(\mathcal{V})$ and $d = \infty$, then $\mathcal{C}_{\mathrm{imp}}^d(\mathcal{V}) = \{c\}$ for every N of the form (11.11) with $\gamma + \delta > 0$, i.e., the limit point case prevails for all such $\kappa = \gamma + \delta$. This follows from the formulas for the ranks of the left and right semiradii of the Weyl balls that are given in formulas (4.3) and (4.4). In this case, $\mathcal{V} \notin L_1^{m \times m}([0, \infty))$ if c does not admit a representation of the form (7.1). Moreover, if $\gamma > 0$, the values of the input impedance $c(\lambda)$ may be characterized by the Weyl–Titchmarsh property:

$$[\xi^* \quad \eta^*] V U_t(\bar{\lambda}) V^* \in L_2^{m \times m} \iff \eta = c(\lambda) \xi$$

for every point $\lambda \in \mathbb{C}_+$.

The inverse spectral problem. The data for the inverse spectral problem is a $p \times p$ nondecreasing mvf $\sigma(\mu)$ on $\mathbb{R}$ that meets the condition (9.5). The special form of N in (11.11) automatically insures that the matrizant will be strongly regular and prescribes the associated pair of the matrizant in accordance with (1) and (2) of Theorem 11.2. Moreover, for a fixed pair of nonnegative numbers $\gamma \geq 0, \delta \geq 0$ with $\kappa = \gamma + \delta > 0$, there is at most one mvf $c \in \mathscr{C}^d_{\mathrm{imp}}(\mathscr{V})$ with the spectral function in its Riesz–Herglotz representation (9.4) equal to the given spectral function $\sigma(\mu)$. This is established in Theorem 23.4 of [Arov and Dym 2005c] for the case $\gamma = \delta$. The case $\gamma \neq \delta$ may be reduced to the case $\gamma = \delta$ by invoking Remark 11.3. Thus, Theorem 11.4 yields exactly one solution for the inverse spectral problem for a system of the form (11.1) with $N = N_{\gamma, \delta}$ as in (11.11).

12. Spectral problems for the Schrödinger equation

The matrizant (or fundamental matrix) $U_t(\lambda) = U(t, \lambda)$, $0 \leq t < d$, of the Schrödinger equation

$$-u''(t, \lambda) + u(t, \lambda)q(t) = \lambda u(t, \lambda), \quad 0 \leq t < d, \quad (12.1)$$

with a $p \times p$ matrix valued potential $q(t)$ of the form

$$q(t) = v'(t) + v(t)^2 \quad \text{for every } t \in [0, d), \quad (12.2)$$

where

$$v(t) = v(t)^* \quad \text{is locally absolutely continuous on the interval } [0, d), \quad (12.3)$$

enjoys the following properties:

(1) $U_t \in \mathscr{U}(-\mathscr{J}_p)$ for every $t \in [0, d)$.

(2) $\displaystyle \limsup_{r \uparrow \infty} \frac{\ln \max\{\|U_t(\lambda)\| : |z| \leq r\}}{r^{1/2}} = \limsup_{\mu \downarrow -\infty} \frac{\ln \|U_t(\mu)\|}{|\mu|^{1/2}} = t$.

In particular, $U_t \notin \mathscr{U}_{rsR}(-\mathscr{J}_p)$ and therefore, the results discussed in the preceding sections are not directly applicable. Nevertheless, it turns out that for Schrödinger equations with potential of the given form, the mvf $A_t(\lambda)$ that is defined by the formulas

$$A_t(\lambda) = L_\lambda Y_t(\lambda^2) L_\lambda^{-1} \quad \text{for } 0 \leq t < d, \quad \text{where} \quad L_\lambda = \begin{bmatrix} I_p & 0 \\ 0 & \lambda I_p \end{bmatrix} \quad (12.4)$$

and

$$Y(t, \lambda) = \begin{bmatrix} I_p & v(0) \\ 0 & -i I_p \end{bmatrix} U(t, \lambda) \begin{bmatrix} I_p & -i v(t) \\ 0 & i I_p \end{bmatrix} \quad \text{for } 0 \leq t < d, \quad (12.5)$$

is a solution of the Cauchy problem

$$A'_t(\lambda) = i\lambda A_t(\lambda)J_p + A_t(\lambda)\mathcal{V}(t), \quad 0 \le t < d,$$

and

$$A_0(\lambda) = I_m,$$

with potential

$$\mathcal{V}(t) = \begin{bmatrix} v(t) & 0 \\ 0 & -v(t) \end{bmatrix}, \quad 0 \le t < d.$$

Thus, $A_t(\lambda)$ is the matrizant of a differential system of the form (11.1), with $N = I_m$, $J = J_p$ and a potential

$$\mathcal{V}(t) = \mathcal{V}(t)^*$$

that meets the constraints (11.3). Therefore, by Theorem 11.2, $A_t \in \mathcal{U}_{rsR}(J_p)$ for every $t \in [0, d)$ and hence, Theorem 11.4 is applicable to A_t.

Let $\psi(t, \lambda)$ and $\varphi(t, \lambda)$ be the unique solutions of equation (12.1) that meet the initial conditions

$$\psi(0, \lambda) = I_p, \quad \psi'(0, \lambda) = 0, \quad \varphi(0, \lambda) = 0 \quad \text{and} \quad \varphi'(0, \lambda) = I_p,$$

respectively, and let

$$U(t, \lambda) = \begin{bmatrix} \psi(t, \lambda) & \psi'(t, \lambda) \\ \varphi(t, \lambda) & \varphi'(t, \lambda) \end{bmatrix} \tag{12.6}$$

be the matrizant (fundamental matrix) of equation (12.1).

A nondecreasing $p \times p$ mvf $\sigma(\mu)$ on $\mathbb{R}$ is said to be a spectral function of (12.1) with respect to the transform

$$g^\triangle(\lambda) = \frac{1}{\sqrt{\pi}} \int_0^d \varphi(s, \lambda)g(s) \, ds \tag{12.7}$$

(of vvf's $g \in L_2^p([0, d))$) with compact support in $[0, d)$), if the Parseval equality

$$\int_{-\infty}^\infty g^\triangle(\mu)^* d\sigma(\mu) g^\triangle(\mu) = \int_0^d g(s)^* g(s) \, ds \tag{12.8}$$

holds for every $g \in L_2^p([0, d))$ with compact support in $[0, d)$. The symbol $\Sigma_{sf}^d(q)$ will be used to denote the set of all spectral functions of (12.1) with respect to this transform.

Formulas (12.4) and (12.5) imply that

$$-iu_{21}(s, \lambda) = \frac{a_{21}(s, \sqrt{\lambda})}{\sqrt{\lambda}} = -i\varphi(s, \lambda)$$

for $s \in [0, d)$. This connection permits one to reduce the spectral problem for the Schrödinger equation (12.1) to a spectral problem for Dirac systems. This strategy was initiated by M. G. Krein in [Kreĭn 1955].

de Branges spaces. Let

$$\mathfrak{E}_t^A(\lambda) = \sqrt{2} \, [0 \quad I_p] \, A_t(\lambda) \mathfrak{V}$$
$$= [a_{22}(t, \lambda) - a_{21}(t, \lambda) \quad a_{22}(t, \lambda) + a_{21}(t, \lambda)]$$

denote the de Branges matrix based on the matrizant $A_t(\lambda)$ of equation that was defined by formulas (12.4) and (12.5). Then the corresponding de Branges space

$$\mathcal{B}(\mathfrak{E}_t^A) =$$
$$\left\{ \frac{1}{\sqrt{\pi}} \int_0^t [a_{21}(s, \lambda) \quad a_{22}(s, \lambda)] f(s) \, ds : f \in L_2^m([0, t]) \text{ for every } t \in [0, d) \right\}$$

with norm

$$\langle f^\Delta, f^\Delta \rangle_{\mathcal{B}(\mathfrak{E}_t^A)} = \int_{-\infty}^{\infty} f^\Delta(\mu)^* \Delta_t^A(\mu) f^\Delta(\mu) \, d\mu \, ,$$

where, upon writing $f = \text{col}[g \, h]$, with components $g, h \in L_2^p([0, t])$,

$$f^\Delta(\mu) = \frac{1}{\sqrt{\pi}} \int_0^t [a_{21}(s, \lambda) \quad a_{22}(s, \lambda) f(s) \, ds$$
$$= \frac{1}{\sqrt{\pi}} \int_0^t a_{21}(s, \lambda)g(s) + a_{22}(s, \lambda)h(s) \, ds$$

and

$$\Delta_t^A(\mu)^{-1} = (a_{22}(t, \mu) + a_{21}(t, \mu))(a_{22}(t, \mu) + a_{21}(t, \mu))^*$$
$$= a_{22}(t, \mu)a_{22}(t, \mu)^* + a_{21}(t, \mu)a_{21}(t, \mu)^* \, ,$$

since $A_t \in \mathfrak{U}(J_p)$. Moreover, formula (12.5) implies that $a_{21}(t, \lambda)$ is an odd function of λ, whereas $a_{22}(t, \lambda)$ is an even function of λ. Thus,

$$\mathcal{B}(\mathfrak{E}_t^A) = \mathcal{B}(\mathfrak{E}_t^A)_{\text{odd}} \oplus \mathcal{B}(\mathfrak{E}_t^A)_{\text{ev}} \, ,$$

where

$$\mathcal{B}(\mathfrak{E}_t^A)_{\text{odd}} = \left\{ \int_0^t a_{21}(s, \lambda)g(s) \, ds : g \in L_2^p([0, t]) \right\} \, ,$$
$$\mathcal{B}(\mathfrak{E}_t^A)_{\text{ev}} = \left\{ \int_0^t a_{22}(s, \lambda)g(s) \, ds : g \in L_2^p([0, t]) \right\} \, .$$

At the same time, Theorem 23.1 of [Arov and Dym 2005c] implies that

$$\mathcal{B}(\mathfrak{C}_t^A) = \left\{ \int_{-t}^{t} e^{i\lambda s} g(s)\, ds : g \in L_2^p([-t, t]) \right\}$$

and hence that

$$\mathcal{B}(\mathfrak{C}_t^A)_{\text{odd}} = \left\{ \int_0^t \sin(s\lambda) g(s)\, ds : g \in L_2^p([0, t]) \right\},$$

$$\mathcal{B}(\mathfrak{C}_t^A)_{\text{ev}} = \left\{ \int_0^t \cos(s\lambda) g(s)\, ds : g \in L_2^p([0, t]) \right\}.$$

Thus, as

$$y_{21}(s, \lambda) = \frac{a_{21}(s, \sqrt{\lambda})}{\sqrt{\lambda}} = -i\varphi(s, \lambda),$$

we obtain the following conclusion:

THEOREM 12.1. *If the potential* $q(t) = v'(t) + v^2(t)$ *of the Schrödinger equation* (12.1) *is subject to the constraints* (12.3), *the de Branges space* $\mathcal{B}(\mathfrak{C}_t^A)$ *equals*

$$\left\{ \frac{1}{\sqrt{\pi}} \int_0^t \frac{\sin\sqrt{\lambda}s}{\sqrt{\lambda}} g(s)\, ds : g \in L_2^p([0, t]) \quad \text{for every } t \in [0, d) \right\}, \quad (12.9)$$

as linear spaces and hence these spaces do not depend upon the potential.

In view of the indicated connection between Dirac systems and Schrödinger equations, Theorems 23.2 and 23.4 of [Arov and Dym 2005c] may be applied to yield existence and uniqueness theorems for the inverse input impedance problem and the inverse spectral problem for the latter when the potential is of the form (12.2), as well as recipes for the solution. A detailed analysis will be presented elsewhere.

REMARK 12.2. The identification (12.9) for the scalar case $p = 1$ was obtained in [Remling 2002] under less restrictive assumptions on the potential $q(t)$ of the Schrödinger equation than are imposed here.

References

[Aronszajn 1950] N. Aronszajn, "Theory of reproducing kernels", *Trans. Amer. Math. Soc.* **68** (1950), 337–404.

[Arov 1993] D. Z. Arov, "The generalized bitangent Carathéodory-Nevanlinna-Pick problem and (j, J_0)-inner matrix functions", *Izv. Ross. Akad. Nauk Ser. Mat.* **57**:1 (1993), 3–32. In Russian; translated in *Russian Acad. Sci. Izvest* **42** (1994), 1–26.

[Arov and Dym 1998] D. Z. Arov and H. Dym, "On three Krein extension problems and some generalizations", *Integral Equations Operator Theory* **31**:1 (1998), 1–91.

[Arov and Dym 2001] D. Z. Arov and H. Dym, "Matricial Nehari problems, J-inner matrix functions and the Muckenhoupt condition", *J. Funct. Anal.* **181**:2 (2001), 227–299.

[Arov and Dym 2003a] D. Z. Arov and H. Dym, "The bitangential inverse input impedance problem for canonical systems. I. Weyl-Titchmarsh classification, existence and uniqueness", *Integral Equations Operator Theory* **47**:1 (2003), 3–49.

[Arov and Dym 2003b] D. Z. Arov and H. Dym, "Criteria for the strong regularity of J-inner functions and γ-generating matrices", *J. Math. Anal. Appl.* **280**:2 (2003), 387–399.

[Arov and Dym 2004] D. Z. Arov and H. Dym, "The bitangential inverse spectral problem for canonical systems", *J. Funct. Anal.* **214**:2 (2004), 312–385.

[Arov and Dym 2005a] D. Z. Arov and H. Dym, "The bitangential inverse input impedance problem for canonical systems. II. Formulas and examples", *Integral Equations Operator Theory* **51**:2 (2005), 155–213.

[Arov and Dym 2005b] D. Z. Arov and H. Dym, "Direct and inverse problems for differential systems connected with Dirac systems and related factorization problems", *Indiana Univ. Math. J.* **54**:6 (2005), 1769–1815.

[Arov and Dym 2005c] D. Z. Arov and H. Dym, "Strongly regular J-inner matrix-valued functions and inverse problems for canonical systems", pp. 101–160 in *Recent advances in operator theory and its applications*, Oper. Theory Adv. Appl. **160**, Birkhäuser, Basel, 2005.

[de Branges 1968a] L. de Branges, "The expansion theorem for Hilbert spaces of entire functions", pp. 79–148 in *Entire functions and related parts of analysis* (La Jolla, CA, 1966), Amer. Math. Soc., Providence, R.I., 1968.

[de Branges 1968b] L. de Branges, *Hilbert spaces of entire functions*, Prentice-Hall, Englewood Cliffs, NJ, 1968.

[Dym 1970] H. Dym, "An introduction to de Branges spaces of entire functions with applications to differential equations of the Sturm-Liouville type", *Advances in Math.* **5** (1970), 395–471.

[Dym 1989] H. Dym, *J contractive matrix functions, reproducing kernel Hilbert spaces and interpolation*, CBMS Regional Conference Series in Mathematics **71**, Published for the Conference Board of the Mathematical Sciences, Washington, DC, 1989.

[Dym and Iacob 1984] H. Dym and A. Iacob, "Positive definite extensions, canonical equations and inverse problems", pp. 141–240 in *Topics in operator theory systems and networks* (Rehovot, Israel, 1983), edited by H. Dym and I. Gohberg, Oper. Theory Adv. Appl. **12**, Birkhäuser, Basel, 1984.

[Dym and McKean 1976] H. Dym and H. P. McKean, *Gaussian processes, function theory, and the inverse spectral problem*, Probability and Mathematical Statistics **31**, Academic Press, New York, 1976.

[Kreĭn 1955] M. G. Kreĭn, "Continuous analogues of propositions on polynomials orthogonal on the unit circle", *Dokl. Akad. Nauk SSSR (N.S.)* **105** (1955), 637–640.

[Remling 2002] C. Remling, "Schrödinger operators and de Branges spaces", *J. Funct. Anal.* **196**:2 (2002), 323–394.

[Treil and Volberg 1997] S. Treil and A. Volberg, "Wavelets and the angle between past and future", *J. Funct. Anal.* **143**:2 (1997), 269–308.

DAMIR Z. AROV
DEPARTMENT OF MATHEMATICAL ANALYSIS
SOUTH UKRANIAN PEDAGOGICAL UNIVERSITY
65020 ODESSA
UKRAINE

HARRY DYM
DEPARTMENT OF MATHEMATICS
THE WEIZMANN INSTITUTE OF SCIENCE
REHOVOT 76100
ISRAEL
dym@wisdom.weizmann.ac.il

Probability, Geometry and Integrable Systems
MSRI Publications
Volume **55**, 2008

Turbulence of a unidirectional flow

BJÖRN BIRNIR

Dedicated to Henry P. McKean, a mentor and a friend

ABSTRACT. We discuss recent advances in the theory of turbulent solutions of the Navier–Stokes equations and the existence of their associated invariant measures. The statistical theory given by the invariant measures is described and associated with historically-known scaling laws. These are Hack's law in one dimension, the Batchelor–Kraichnan law in two dimensions and the Kolmogorov's scaling law in three dimensions. Applications to problems in turbulence are discussed and applications to Reynolds Averaged Navier Stokes (RANS) and Large Eddy Simulation (LES) models in computational turbulence.

1. Introduction

Everyone is familiar with turbulence in one form or another. Airplane passengers encounter it in wintertime as the plane begins to shake and is jerked in various directions. Thermal currents and gravity waves in the atmosphere create turbulence encountered by low-flying aircraft. Turbulent drag also prevents the design of more fuel-efficient cars and aircrafts. Turbulence plays a role in the heat transfer in nuclear reactors, causes drag in oil pipelines and influence the circulation in the oceans as well as the weather.

In daily life we encounter countless other examples of turbulence. Surfers use it to propel them and their boards to greater velocities as the wave breaks and becomes turbulent behind them and they glide at great speeds down the unbroken face of the wave. This same wave turbulence shapes our beaches and carries enormous amount of sand from the beach in a single storm, sometime to dump it all into the nearest harbor. Turbulence is harnessed in combustion engines in cars and jet engines for effective combustion and reduced emission of pollutants. The flow around automobiles and downtown buildings is controlled by turbulence and so is the flow in a diseased artery. Atmospheric turbulence is important in remote sensing, wireless communication and laser beam propagation through the atmosphere; see [Sølna 2002; [2003]]. The applications of turbulence await us in technology, biology and the environment. It is one of the major problems holding back advances of our technology.

Turbulence has puzzled and intrigued people for centuries. Five centuries ago a fluid engineer by the name of Leonardo da Vinci tackled it. He did not have modern mathematics or physics at his disposal but he had a very powerful investigative tool in his possession. He explored natural phenomena by drawing them. Some of his most famous drawings are of turbulence.

Leonardo called the phenomenon that he was observing "la turbolenza" in 1507 and gave the following description of it:

Observe the motion of the surface of the water, which resembles that of hair, which has two motions, of which one is caused by the weight of the hair, the other by the directions of the curls; thus the water has eddying motions, one part of which is due to the principal current, the other to the random and reverse motion.

This insightful description pointed out the separation of the flow into the average flow and the fluctuations that plays an important role in modern turbulence theory. But his drawings also led Leonardo to make other astute observations that accompany his drawings, in mirror script, such as where the turbulence of water is generated, where it maintains for long, and where it comes to rest. These three observations are well-known features of turbulence and they are all illustrated in Leonardo's drawings.

One reason why turbulence has not been solved yet is that the mathematics or the calculus of turbulence has not been developed until now. This situation is analogous to the physical sciences before Newton and Leibnitz. Before the physical sciences could bloom into modern technology the mathematics being the language that they are expressed in had to be developed. This was accomplished by Newton and Leibnitz and developed much further by Euler. Three centuries later we are at a similar threshold regarding turbulence. The mathematics of turbulence is being born and the technology of turbulence is bound to follow.

The mathematics of turbulence is rooted in stochastic partial differential equations. It is the mathematical theory that expresses the statistical theory of turbulence as envisioned by the Russian mathematician Kolmogorov, one of the fathers of modern probability theory, in 1940. The basic observation is that turbulent flow is unstable and the white noise that is always present in any physical system is magnified in turbulent flow. In distinction, in laminar flow the white noise in the environment is suppressed. The new mathematical theory of turbulence expresses how the noise is magnified and colored by the turbulent fluid. This then leads to a computation or an approximation of the associated invariant measure for the stochastic partial differential equation. The whole statistical theory of Kolmogorov can be expressed mathematically with this invariant measure in hand.

The problems that mathematicians have with proving the existence of solutions of the Navier–Stokes equations in three dimensions has lead to the mistaken impression that turbulence is only a three dimensional phenomenon. Nothing is further from the truth. Turbulence thrives in one and two dimensions as well as in three dimensions. We will illustrate this by describing one dimensional turbulence in rivers.

Although we will coach it in terms of river flow in his paper, this type of modeling and theory have many other applications. One such application is to the modeling of fluvial sedimentation that gives rise to sedimentary rock in petroleum reserves. The properties of the flow through the porous rock turn out to depend strongly on the structure of the meandering river channels; see [Holden et al. 1998]. Another application is to turbulent atmospheric flow. Contrary to popular belief, in the presence of turbulence, the temperature variations in the atmosphere my be highly anisotropic or stratified. Thus the scaling of the fluid model corresponding to a river or a channel may have a close analog in the turbulent atmosphere; see [Sidi and Dalaudier 1990].

Two dimensionless numbers the Reynolds number and the Froude number are used to characterize turbulent flow in rivers and streams. If we model the river as an open channel with x parameterizing the downstream direction, y the horizontal depth and U is the mean velocity in the downstream direction, then the Reynolds number

$$R = \frac{f_{\text{turbulent}}}{f_{\text{viscous}}} = \frac{Uy}{\nu}$$

is the ratio of the turbulent and viscous forces whereas the Froude number

$$F = \frac{f_{\text{turbulent}}}{f_{\text{gravitational}}} = \frac{U}{(gy)^{1/2}}$$

is the ratio of the turbulent and gravitational forces. ν is the viscosity and g is the gravitational acceleration. Other forces such as surface tension, the centrifugal force and the Coriolis force are insignificant in streams and rivers.

The Reynolds number indicates whether the flow is laminar or turbulent with the transition to turbulence starting at $R = 500$ and the flow usually being fully turbulent at $R = 2000$. The Froude number measures whether gravity waves, with speed $c = (gy)^{1/2}$ in shallow water, caused by some disturbance in the flow, can overcome the flow velocity and travel upstream. Such flow are called tranquil flows, $c > U$, in distinction to rapid or shooting flows, $c < U$, where this cannot happen; they correspond to the Froude numbers

(i) $F < 1$, subcritical, $c > U$,
(ii) $F = 1$, critical, $c = U$,
(iii) $F > 1$, supercritical, $c < U$.

Now for streams and rivers the Reynolds number is typically large $(10^5 - 10^6)$, whereas the Froude numbers is small typically $(10^{-1} - 10^{-2})$; see [Dingman 1984]. Thus the flows are highly turbulent and ought to be tranquil. But this is not the whole story, as we will now explain.

In practice streams and rivers have varied boundaries which are topologically equivalent to a half-pipe. These boundaries are rough and resist the flow and this had lead to formulas involving channel resistance. The most popular of these are Chézy's law, where the average velocity V is

$$V = u_c C r^{1/2} s_o^{1/2}, \quad u_c = 0.552 m/s$$

and Manning's law, with

$$V = u_m \frac{1}{n} r^{2/3} s_0^{1/2}, \quad u_m = 1.0 m/s$$

where s_o is the slope of the channel and r is the hydraulic radius. C is called Chézy's constant and measures inverse channel resistance. The number n is Manning's roughness coefficient; see [Dingman 1984]. We get new effective Reynolds and Froude numbers with these new averaged velocities V,

$$R^* = \frac{g}{3u_c^2 C^2} R, \quad F^* = \left(\frac{g}{u_c^2 C^2 s_o} \right)^{1/2} F$$

from Chézy's law.

It turns out that in real rivers the effective Froude number is approximately one and the effective Reynolds number is also one, when $R = 500$ for typical channel roughness $C = 73.3$. Thus the transition to turbulence typically occurs in rivers when the effective turbulent forces are equal to the viscous forces.

The reason for the transition to turbulence is that at this value of R^* the amplification of the noise that grows into fully developed turbulence is no longer damped by the viscosity of the flow. The damping by the effective viscosity is overcome by the turbulent forces.

Now let us ignore the boundaries of the river. The point is that in a straight segment of a reasonably deep and wide river the boundaries do not influence the details of the river current in the center, except as a source of flow disturbances. We will simply assume that these disturbances exist, in the flow at the center of the river and not be concerned with how they got there. For theoretical purposes we will conduct a thought experiment where we start with an unstable uniform flow and then put the disturbances in as small white noise. Then the mathematical problem is to determine the statistical theory of the resulting turbulent flow. The important point is that this is now a theory of the water velocity $u(x)$ as a function of the one-dimensional distance x down the river. Thus if u is turbulent it describes one-dimensional turbulence in the downstream direction of the river.

The flow of water in streams and rivers is a fascinating problem with many application that has intrigued scientists and laymen for many centuries; see [Levi 1995]. Surprisingly it is still not completely understood even in one or two-dimensional approximation of the full three-dimensional flow. Erosion by water seems to determine the features of the surface of the earth, up to very large scales where the influence of earthquakes and tectonics is felt, see [Smith et al. 1997a; 1997b; 2000; Birnir et al. 2001; 2007a; Welsh et al. 2006]. Thus water flow and the subsequent erosion gives rise to the various scaling laws know for river networks and river basins; see [Dodds and Rothman 1999; 2000a; 2000b; 2000c; 2000d].

One of the best known scaling laws of river basins is Hack's law [1957] that states that the area of the basin scales with the length of the main river to an exponent that is called Hack's exponent. Careful studies of Hack's exponent (see [Dodds and Rothman 2000d]) show that it actually has three ranges, depending on the age and size of the basin, apart from very small and very large scales where it is close to one. The first range corresponds to a spatial roughness coefficient of one half for small channelizing (very young) landsurfaces. This has been explained, see [Birnir et al. 2007a] and [Edwards and Wilkinson 1982], as Brownian motion of water and sediment over the channelizing surface. The second range with a roughness coefficient of $\frac{2}{3}$ corresponds to the evolution of a young surface forming a convex (geomorphically concave) surface, with young rivers, that evolves by shock formation in the water flow. These shocks are called bores (in front) and hydraulic jumps (in rear); see [Welsh et al. 2006]. Between them sediment is deposited. Finally there is a third range with a roughness coefficient $\frac{3}{4}$. This range that is the largest by far and is associated with what is called the mature landscape, or simply the landscape because it persists for a long time, is what this paper is about. This is the range that is associated with turbulent flow in rivers and we will develop the statistical theory of turbulent flow in rivers that leads to Hack's exponent.

Starting with the three basic assumption on river networks: that the their structure is self-similar, that the individual streams are self-affine and the drainage density is uniform (see [Dodds and Rothman 2000a]), river networks possess several scalings laws that are well documented; see [Rodriguez-Iturbe and Rinaldo 1997]. These are self-affinity of single channels, which we will call the meandering law, Hack's law, Horton's laws [1945] and their refinement Tokunaga's law, the law for the scaling of the probability of exceedance for basin areas and stream lengths and Langbein's law. The first two laws are expressed in terms of the meandering exponent m, or fractal dimension of a river, and the Hack's exponent h. Horton's laws are expressed in terms of Horton's ratio's of link numbers and link lengths in a Strahler ordered river network, Tokunaga's law is expressed in term of the Tokunaga's ratios, the probability of exceedance is expressed by decay exponents and Langbein's law is given by the Langbein's exponents [Dodds and Rothman 2000a].

Dodds and Rothman [1999; 2000a; 2000b; 2000c; 2000d] showed that all the ratios and exponents above are determined by m and h, the meandering and Hack's exponents; see [Hack 1957; Dodds and Rothman 1999]. The origin of the meandering exponent m has recently been explained [Birnir et al. 2007b] but in this paper we discuss how it and Hack's exponent are determined by the scaling exponent of turbulent one-dimensional flow. Specifically, m and h are determined by the scaling exponent of the second structure function [Frisch 1995] in the statistical theory of the one-dimensional turbulent flow.

The breakthrough that initiated the theoretical advances discussed above was the proof of existence of turbulent solutions of the full Navier–Stokes equation driven by uniform flow, in dimensions one, two and three. These solutions turned out to have a finite velocity and velocity gradient but they are not smooth instead the velocity is Hölder continuous with a Hölder exponent depending on the dimension; see [Birnir 2007a; 2007b]. These solutions scale with the Kolmogorov scaling in three dimensions and the Batchelor–Kraichnan scaling in two dimensions. In one dimensions they scale with the exponent $\frac{3}{4}$, that is related to Hack's law [1957] of river basins; see [Birnir et al. 2001; 2007a]. The existence of these turbulent solutions is then used to proof the existence of an invariant measure in dimensions one, two and three; see [Birnir 2007a; 2007b]. The invariant measure characterizes the statistically stationary state of turbulence and it can be used to compute the statistically stationary quantities. These include all the deterministic properties of turbulence and everything that can be computed and measured. In particular, the invariant measure determines the probability density of the turbulent solutions and this can be used to develop accurate subgrid modeling in computations of turbulence, bypassing the problem that three-dimensional turbulence cannot be fully resolved with currently existing computer technology.

2. The initial value problem

Consider the Navier–Stokes equation

$$w_t + w \cdot \nabla w = \nu \Delta w - \nabla p \tag{2-1}$$
$$w(x, 0) = w_o,$$

where $\nu = \nu_0 / VL$, V being a typical velocity, L the length of a segment of the river and ν_0 the kinematic viscosity of water, with the incompressibility condition

$$\nabla \cdot w = 0. \tag{2-2}$$

Eliminating the pressure p using (2-2) gives the equation

$$w_t + w \cdot \nabla w = \nu \Delta w + \nabla \{ \Delta^{-1} [\text{trace}(\nabla w)^2] \} \tag{2-3}$$

We want to consider turbulent flow in the center of a wide and deep river and to do that we consider the flow to be in a box and impose periodic boundary conditions on the box. Since we are mostly interested in what happens in the direction along the river we take our x axis to be in that direction.

We will assume that the river flows fast and pick an initial condition of the form

$$w(0) = U_o e_1 \tag{2-4}$$

where U_o is a large constant and e_1 is a unit vector in the x direction. Clearly this initial condition is not sufficient because the fast flow may be unstable and the white noise ubiquitous in nature will grow into small velocity and pressure oscillations; see for example [Betchov and Criminale 1967]. But we perform a thought experiment where white noise is introduced into the fast flow at $t = 0$. This experiment may be hard to perform in nature but it is easily done numerically. It means that we should look for a solution of the form

$$w(x, t) = U_o e_1 + u(x, t) \tag{2-5}$$

where $u(x, t)$ is smaller that U_o but not necessarily small. However, in a small initial interval $[0, t_o]$ u is small and satisfies the equation (2-3) linearized about the fast flow U_o

$$u_t + U_o \partial_x u = \Delta u + f \tag{2-6}$$
$$u(x, 0) = 0$$

driven by the noise

$$f = \sum_{k \neq 0} h_k^{1/2} d\beta_t^k e_k$$

The $e_k = e^{2\pi i k \cdot x}$ are (three-dimensional) Fourier components and each comes with its own independent Brownian motion β_t^k. None of the coefficients of the vectors $h_k^{1/2} = (h_1^{1/2}, h_2^{1/2}, h_3^{1/2})$ vanish because the turbulent noise was seeded by truly white noise (white both is space and in time). f is not white in space because the coefficients $h_k^{1/2}$ must have some decay in k so that the noise term in (2-6) makes sense. However to determine the decay of the $h_k^{1/2}$s will now be part of the problem. The form of the turbulent noise f expresses the fact that in turbulent flow there is a continuous sources of small white noise that grows and saturates into turbulent noise that drives the fluid flow. The decay of the coefficients $h_k^{1/2}$ expresses the spatial coloring of this larger noise in turbulent flow. We have set the kinematic viscosity ν equal to one for computational convenience, but it can easily be restored in the formulas.

This modeling of the noise is the key idea that make everything else work. The physical reasoning is that the white noise ubiquitous in nature grows into the noise f that is characteristic for turbulence and the differentiability properties of the turbulent velocity u are the same as those of the turbulent noise.

The justification for considering the initial value problem (2-6) is that for a short time interval $[0, t_o]$ we can ignore the nonlinear terms

$$-u \cdot \nabla u + \nabla\{\Delta^{-1}[\text{trace}(\nabla u)^2]\}$$

in the equation (2-3). But this is only true for a short time t_o, after this time we have to start with the solution of (2-6)

$$u_o(x, t) = \sum_{k \neq 0} h_k^{1/2} \int_0^t e^{(-4\pi^2|k|^2 + 2\pi i U_o k_1)(t-s)} d\beta_s^k e_k(x) \qquad (2\text{-}7)$$

as the first iterate in the integral equation

$$u(x, t) = u_o(x, t) + \int_{t_o}^t K(t-s) * [-u \cdot \nabla u + \nabla \Delta^{-1}(\text{trace}(\nabla u)^2)] ds \quad (2\text{-}8)$$

where K is the (oscillatory heat) kernel in (2-7). In other words to get the turbulent solution we must take the solution of the linear equation (2-6) and use it as the first term in (2-8). It will also be the first guess in a Picard iteration. The solution of (2-6) can be written in the form

$$u_o(x, t) = \sum_{k \neq 0} h_k^{1/2} A_t^k e_k(x)$$

where the

$$A_t^k = \int_0^t e^{(-4\pi^2|k|^2 + 2\pi i U_o k_1)(t-s)} d\beta_s^k \qquad (2\text{-}9)$$

are independent Ornstein–Uhlenbeck processes with mean zero; see for example [Da Prato and Zabczyk 1996].

Now it is easy to see that the solution of the integral equation (2-8) $u(x,t)$ satisfies the driven Navier–Stokes equation

$$u_t + U_o \partial_x u = \Delta u - u \cdot \nabla u + \nabla \Delta^{-1}(\text{trace}(\nabla u)^2)$$
$$+ \sum_{k \neq 0} h_k^{1/2} d\beta_t^k e_k, \quad t > t_0,$$

(2-10)

$$u_t + U_o \partial_x u = \Delta u + \sum_{k \neq 0} h_k^{1/2} d\beta_t^k e_k, \qquad u(x,0) = 0, \quad t \leq t_0,$$

and the argument above is the justification for studying the initial value problem (2-10). We will do so from here on. The solution u of (2-10) still satisfies the periodic boundary conditions and the incompressibility condition

$$\nabla \cdot u = 0 \tag{2-11}$$

The mean of the solution u_o of the linear equation (2-6) is zero by the formula (2-7) and this implies that the solution u of (2-10) also has mean zero

$$\bar{u}(t) = \int_{\mathbb{T}^3} u(x,t)\, dx = 0 \tag{2-12}$$

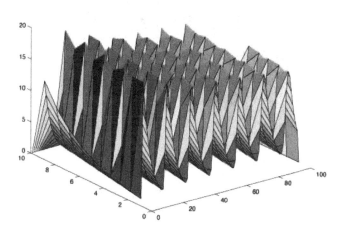

Figure 1. The traveling wave solution of the heat equation for the flow velocity $U_o = 85$. The perturbations are frozen in the flow. The x axis is space, the y axis time and the z axis velocity u.

Stability. The uniform flow $w = U_o e_1$ seem to be a stable solution of (2-6) judging from the solution (2-7). Namely, all the Fourier coefficients are decaying. However, this is deceiving, first the Brownian motion β_k is going to make the amplitude of the k-th Fourier coefficient large in due time with probability one. More importantly if U_o is large then (2-6) has traveling wave solutions that are perturbations "frozen in the flow", and for U_o even larger these traveling waves are unstable and start growing. For U_o large enough this happens after a very short initial time interval and makes the flow immediately become fully turbulent. The role of the white noise is then not to cause enough growth eventually for the nonlinearities to become important, but rather to immediately pick up (large) perturbations that grow exponentially. These are the large fluctuations that are observed in most turbulent flows. In Figure 1, we show the traveling wave solution of the transported heat equation (2-6), with $U_o = 85$. In Figure 2, where the flow has increased to $U_o = 94$, the traveling wave has become unstable and grows exponentially. Notice the difference in vertical scale between the figures.

Thus the white noise grows into a traveling wave that grows exponentially. This exponential growth is saturated by the nonlinearities and subsequently the flow becomes turbulent. This is the mechanism of explosive growth of turbulence of a uniform stream and describes what happens in our thought experiment described in Section 2.

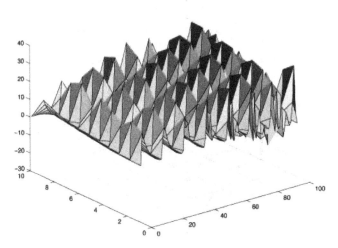

Figure 2. The traveling wave solution of the heat equation for the flow velocity $U_o = 94$. The perturbations are growing exponentially. The x axis is space, the y axis time and the z axis velocity u.

3. One-dimensional turbulence

In a deep and wide river it is reasonable to think that the directions transverse to the main flow, y the direction across the river, and z the horizontal direction, play a secondary role in the generation of turbulence. As a first approximation to the flow in the center of a deep and wide, fast-flowing river we will now drop these directions. Of course y and z play a role in the motion of the large eddies in the river but their motion is relatively slow compared to the smaller scale turbulence. Thus our initial value problem (2-10) becomes

$$u_t + U_o u_x$$
$$= u_{xx} - u u_x + \partial_x^{-1}((u_x)^2) - \int_0^1 \partial_x^{-1}((u_x)^2)\, dx + \sum_{k \neq 0} h_k^{1/2} d\beta_t^k e_k, \quad (3\text{-}1)$$

We still have periodic boundary condition on the unit interval but the incompressibility condition can be dropped at the price of subtracting the term

$$b = \int_0^1 \partial_x^{-1}((u_x)^2)\, dx$$

from the right hand side of the Navier–Stokes equation. This term keeps the mean of u, $\bar{u} = \int_0^1 u\, dx = 0$, equal to zero, see Equation (2-12). This equation (3-1) now describes the turbulent flow in the center of relative straight section of a fast river. The full three-dimensional flow will be treated in [Birnir 2007b].

The following theorem and corollaries are proved in [Birnir 2007a]. It states the existence of turbulent solutions in one dimension. First we write the initial value problem (3-1) as an integral equation

$$u(x,t) = u_0(x,t) + \int_{t_o}^t K(t-s) * \left(-\tfrac{1}{2}(u^2)_x + \partial_x^{-1}(u_x)^2 - b\right) ds. \quad (3\text{-}2)$$

Here K is the oscillatory heat kernel (2-7) in one dimension and

$$u_0(x,t) = \sum_{k \neq 0} h_k^{1/2} A_t^k e_k(x)$$

the A_t^ks being the Ornstein–Uhlenbeck processes from Equation (2-9).

If q/p is a rational number let $(q/p)^+$ denote any real number greater than q/p, and let E the expectation with a probability measure P on a set of events Ω.

THEOREM 3.1. *If the solution of the linear equation (2-6) satisfies the condition*

$$E(\|u_0\|_{(\frac{5}{4}^+,2)}^2) = \sum_{k \neq 0}(1 + (2\pi k)^{(5/2)^+})h_k E(|A_t^k|^2) \quad (3\text{-}3)$$

$$\leq \frac{1}{2}\sum_{k \neq 0} \frac{(1 + (2\pi i k)^{(5/2)^+})}{(2\pi k)^2} h_k < \infty$$

and U_0 is sufficiently large, then the integral equation (3-2) has a unique solution in the space $C\big([0, \infty); \mathcal{L}^2_{(\frac{5}{4}^+, 2)}\big)$ of stochastic processes with

$$\|u\|^2_{\mathcal{L}^2_{(\frac{5}{4}^+, 2)}} < \infty.$$

COROLLARY 3.1. *The solution of the linearized equation (2-6) uniquely determines the solution of the integral equation (3-2).*

COROLLARY 3.2 (ONSAGER'S CONJECTURE). *The solutions of the integral equation (3-2) are Hölder continuous with exponent $\frac{3}{4}$.*

REMARK. Hypothesis (3-3) is the answer to the question posed in Section 2 about how fast the coefficients $h_k^{1/2}$ have to decay in Fourier space. They have to decay sufficiently fast for the supremum in t of the expectation of the

$$H^{\frac{5}{4}^+} = W^{(\frac{5}{4}^+, 2)}$$

Sobolev norm of the initial function u_o, to be finite. In other words the sup in t of the $\mathcal{L}^2_{(\frac{5}{4}^+, 2)}$ norm has to be finite.

4. The existence and uniqueness of the invariant measure

We can define the invariant measure $d\mu$ for a stochastic partial differential equation (SPDE) by the limit

$$\lim_{t \to \infty} E(\phi(u(t))) = \int_{L^2(\mathbb{T}^n)} \phi(u)\, d\mu(u) \tag{4-1}$$

where E denotes the expectation, $u(t)$ is the solution of the SPDE, parametrized by time, and ϕ is any bounded function on $L^2(\mathbb{T}^n)$. $L^2(\mathbb{T}^n)$ is the space of square integrable functions on a torus $\mathbb{T}^n$ which means that we are imposing periodic boundary conditions on an interval, rectangle or a box, respectively $n = 1, 2, 3$ dimensions. However, the theory also carries over to other boundary conditions. One first uses the law $\mathcal{L}$ of the solution $u(t)$

$$P_t(w, \Gamma) = \mathcal{L}(u(w, t))(\Gamma), \quad \Gamma \subset \mathcal{E},$$

where $w = u_o$ is the initial condition for the SPDE and $\mathcal{E}$ is the σ algebra generated by the Borel subsets Γ of $L^2(\mathbb{T}^n)$, to define transition probabilities $P_t(w, \Gamma)$ on $L^2(\mathbb{T}^n)$. A stochastically continuous Markovian semigroup is called a Feller semigroup [Da Prato and Zabczyk 1996], and for such Feller semigroups

$$\frac{1}{T} \int_0^T P_t(w, \cdot)\, dt$$

defines a probability measure. This is how one forms probability measures on $L^2(\mathbb{T}^n)$ by taking these time averages of the transition probabilities and then one uses the Krylov–Bogoliubov Theorem [Da Prato and Zabczyk 1996] to show that the sequence of the resulting probability measures, indexed by time T, is tight. This is the first step, then the invariant measure exists and is the (weak) limit

$$d\mu(\cdot) = \lim_{T \to \infty} \frac{1}{T} \int_0^T P_t(w, \cdot)\, dt$$

Once the existence of the invariant measure has been established, one wishes to prove that it is unique. To prove this one first has to prove that P_t is in fact a strong Feller semigroup or that for all $T > 0$ there exists a constant $C > 0$, such that for all $\varphi \in B(L^2)$, the space of bounded functions on L^2, and $t \in [0, T]$

$$|P_t\varphi(x) - P_t\varphi(y)| \le C\|\varphi\|_\infty \|x - y\|, \quad x, y \in L^2.$$

Here $\|\cdot\|$ denotes the norm in L^2. Then one must prove the irreducibility of the P_t, namely that for any $\Gamma \subset L^2$ and $w \in \Gamma$

$$P_t(w, \Gamma) = P_t\chi_\Gamma(w) > 0,$$

where χ_Γ is the characteristic function of Γ. The strong Feller property and irreducibility are usually defined for a fixed t but by the semigroup property, if these hold at one t they also hold at any other t. Now if the transition semigroup P_t associated with the equation (5-1) below is a strong Feller semigroup and irreducible, then by Doob's theorem on invariant measures [Da Prato and Zabczyk 1996],

(i) The invariant measure μ associated with P_t is unique.
(ii) μ is strongly mixing and

$$\lim_{t \to \infty} P_t(w, \Gamma) = \mu(\Gamma),$$

for all $w \in L^2$ and $\Gamma \in \mathscr{E}$ where $\mathscr{E}(L^2)$ denotes the sigma field generated by the Borel subsets of L^2.
(iii) μ is equivalent to all measures $P_t(w, \cdot)$, for all $w \in L^2$ and all $t > 0$.

5. The statistical theory

The invariant measure can be used to compute statistical quantities characterizing the turbulent state. The mathematical model consists of the Navier–Stokes equation where we have used the incompressibility condition to eliminate the pressure,

$$\frac{\partial u}{\partial t} + u \cdot \nabla u = \nu \Delta u + \nabla \Delta^{-1}[\mathrm{trace}(\nabla u)^2] + f, \tag{5-1}$$

ν is the kinematic viscosity and f represents turbulent noise as in Equation (2-6). The velocity also satisfies the incompressibility condition

$$\nabla \cdot u = 0. \tag{5-2}$$

In one dimension, modeling a fast turbulent flow in a relatively narrow river, one can ignore the dimension transverse to the flow and the equation becomes,

$$u_t + uu_x = \nu u_{xx} + \partial_x^{-1}(u_x)^2 - b + f, \tag{5-3}$$

as discussed above. The existence of turbulent solutions of this equation and their associated invariant measures was established in [Birnir 2007a], following the method of McKean [2002]. The existence of invariant measures for the one-dimensional Navier–Stokes equation (dissipative Burger's equation) with stochastic forcing was established by Sinai [1996] (see also [Kuksin and Shirikyan 2001]) and McKean [2002]. The existence in the two-dimensional case was established by Flandoli [1994]; see also [Flandoli and Maslowski 1995; Mattingly 1999; Weinan et al. 2001; Bricmont et al. 2000; Hairer and Mattingly 2004; Hairer et al. 2004].

If one considers the second structure function

$$S_2(y) = E[|u(x + y) - u(x)|^2]$$

of the solution, one can show that it scales with the power $\frac{3}{2}$ in the lag variable y for the equation (5-3), in one dimension (see [Birnir et al. 2001; 2007a; Birnir 2007a], and $\frac{2}{3}$ for equation (5-1), in three dimensions, the latter is Kolmogorov's theory. The Kolmogorov scaling of the second structure function is usually written as

$$S_2(y) = C\varepsilon^{2/3}y^{2/3}.$$

where ε is the dissipation rate. In two dimensions the scaling is more complicated due to the existence of the inverse cascade (see [Kraichnan 1967]), and two scaling regimes may exist [Kraichnan 1967; Batchelor 1969; Kolmogorov 1941]. It is still an open problem to examine the higher moments for different scalings or multifractality [Frisch 1995; Lavallée et al. 1993], and the scalings at very small scales below the Kolmogorov scale. The latter is the scale below which dissipation and dissipative scaling is supposed to dominate. Finally, one needs to examine the scaling in time, that we have suppressed in the above formula, to see if one can characterize the transients to the stationary (fully developed turbulence) state.

If ϕ is a bounded function on $L^2(\mathbb{T}^1)$, then the invariant measure $d\mu$ for the SPDEs (3-1) is given by the limit

$$\lim_{t \to \infty} E(\phi(u(t))) = \int_{L^2(\mathbb{T}^1)} \phi(u) \, d\mu(u); \tag{5-4}$$

see (4-1). In [Birnir 2007a] we proved that this limit exists and is unique. We get the following theorem, as explained in Section 4,

THEOREM 5.1. *The integral equation (3-2) possesses a unique invariant measure.*

COROLLARY 5.1. *The invariant measure $d\mu$ is ergodic and strongly mixing.*

The corollary follows immediately from Doob's theorem for invariant measures; see for example [Da Prato and Zabczyk 1996].

The equations describing the erosion of a fluvial landsurface consist of a system of PDEs, one (u) equation describing the fluid flow, the other equation describing the sediment flow; see [Birnir et al. 2001]. Using these equations, Hack's law is proven in the following manner. In [Birnir et al. 2007a] the equations describing the sediment flow are linearized about convex (concave in the terminology of geomorphology) surface profiles describing mature surfaces. Then the colored noise generated by the turbulent flow (during big rainstorms) drives the linearized equations and the solutions obtain the same color (scaling), see Theorem 5.3 in [Birnir et al. 2007a]. The resulting variogram (second structure function) of the surfaces scales with the roughness exponent $\chi = \frac{3}{4}$, see Theorem 5.4 in the same reference. This determines the roughness coefficient χ of mature landsurfaces.

The final step is the following derivation of Hack's law is copied from [Birnir et al. 2001].

The origin of Hack's law. The preceding results allow us to derive some of the fundamental scaling results that are known to characterize fluvial landsurfaces. In particular, the avalanche dimension computed in [Birnir et al. 2001] and derived in [Birnir et al. 2007a], given the roughness coefficient χ, allows us to derive Hack's Law relating the length of a river l to the area A of the basin that it drains. This is the area of the river network that is given by the avalanche dimensions

$$A \sim l^D$$

and the avalanche dimensions is $D = 1 + \chi$. This relation says that if the length of the main river is l then the width of the basin in the direction, perpendicular to the main river, is l^χ. Stable scalings for the surface emerge together with the emergence of the separable solutions describing the mature surfaces; see [Birnir et al. 2001]. We note that in this case $\chi = \frac{3}{4}$, hence we obtain

$$l \sim A^{\frac{1}{1+\chi}} \approx A^{0.57}, \tag{5-5}$$

a number that is in excellent agreement with observed values of the exponent of Hack's law of 0.58; see [Gray 1961].

It still remains to explain how the roughness of the bottom and boundary of a river channel gets spread to the whole surface of the river basin over time. In [Birnir et al. 2007b] it is shown that the mechanism for this consists of the meanderings of the river. As the rivers meanders over time it sculpts a roughness of the surface with the roughness exponent $\frac{3}{4}$.

6. Invariant measures and turbulent mixing

Now how does the existence of the invariant measure help in determining the turbulent mixing properties on a small scale? First, the invariant measure is not only ergodic but in fact strongly mixing; see [Birnir 2007a]. Secondly, the invariant measure allows one to compute the statistical properties, in particular the mixing rates. This, of course sounds, a little too good to be true so what is the problem?

The main problem one has to tackle first is that no explicit formula exist for the invariant measure, such as the explicit formula one has for the Gaussian invariant measure of Brownian motion. Indeed no such formula can exist, no more than one can have an explicit formula for a general turbulent solution of the Navier–Stokes equation. However, since the invariant measure is both ergodic and weakly mixing, by Doob's theorem (see [Da Prato and Zabczyk 1996], for example) one can use the ergodic theorem and approximate the invariant measure by taking the long-time time average. In practice this means that we take the limit of the expectation of a computed solution or rather it substitute: an ensemble average of many computed solutions and the time average of this ensemble average, when time becomes large. Roughly speaking this means that we can approximate the invariant measure to the same accuracy as the computed solution. However, this means that we also have an approximation of the probability density and this can be used to make a subgrid model for (LES) computations.

It is desirable to go beyond the above approximation and develop approximations of the invariant measure that are independent of the computational accuracy. This requires one to find an approximations of the invariant measure by a sequence of measures that can be computed explicitly and an estimate of the error one makes by each approximation. There are some proposals for doing this that need to be explored. One also needs to investigate the properties of the invariant measure, what its continuity properties are with respect to other measures, etc. The discovery of these properties that now are completely unknown will help in determining good and efficient approximations to the invariant measure and the probability density.

If methods are found to efficiently approximate the invariant measure then there are no limits to the spatial and temporal scales that can be resolved except the theoretical one given by the Kolmogorov and dissipative scales. In other words with good methods to approximate the invariant measures the turbulent mixing problem can be solved and the mixing rates of the various components due to the turbulence computed. Furthermore, at least theoretically this can be done to any desired accuracy.

7. Approximations of the invariant measure

It is imperative for application to be able to approximate the invariant measure up a high order. This permits the computation of statistical quantities to within the desired accuracy in experiments or simulations. The first step in the approximation procedure is to use the same method that was used to construct the solutions to construct approximations of the invariant measure. If we linearize the Navier–Stokes equation (5-1) around a fast unidirectional flow $U_o e_1$ where e_1 is a unit vector in the x direction and include noise then we get a heat equation with a convective term that has the solution

$$u_o(x,t) = \sum_{k \neq 0} h_k^{1/2} \int_0^t e^{(-4\pi^2|k|^2 + 2\pi i U_o k_1)(t-s)} d\beta_s^k e_k(x) \qquad (7\text{-}1)$$

as explained in Section 2. The $\beta_t^k s$ are independent Brownian motions and the $e_k s$ are Fourier components. Then if we look for a solution of (5-1) of the form $U = U_o e_1 + u$ then u satisfies the integral equation

$$u(x,t) = u_o(x,t) + \int_{t_o}^t K(t-s) * [-u \cdot \nabla u + \nabla \Delta^{-1}(\text{trace}(\nabla u)^2)] ds \qquad (7\text{-}2)$$

where K is the (oscillatory heat) kernel in (2-7). The solution of the integral equation is constructed by substituting u_o as the first guess into the integral and then iterating the result. This produces a sequence of (Picard) iterates that one proves converges to the solution of the integral equation. No explicit formula can exist for the limit in general but one can iterate the integral equation as often as desired to produce an approximate solution. The formulas get more and more complicated but it is possible that one quickly gets a good approximation to the real solutions. This obviously depends on the rate of convergence. In any case the mth iterate u_m of the integral equation with $u_0 = u_o$ is an approximate solution that can be compared to a numerical solution of the equation (5-1).

It is conceivable that these approximations can be implemented by a symbolic or partially symbolic and partially numerical computation.

By the ergodic theorem the time average of the solution

$$\frac{1}{T} \int_0^T u(t) \, dt$$

converges to the invariant measure. In fact,

$$\lim_{T \to \infty} \frac{1}{T} \int_0^T \phi(u(t)) \, dt = \int_{L^2(\mathbb{T}^1)} \phi(u) \, d\mu(u) \tag{7-3}$$

where $\phi \in B(L^2)$ is any bounded function on L^2. Thus we can find approximations μ_m to the invariant measure μ by considering the sequence

$$\frac{1}{T_m} \int_0^{T_m} u_m(t) \, dt \sim \int_{L^2(\mathbb{T}^1)} u \, d\mu_m(u)$$

u_0 in these formulas is simply the solution of the linear equation (2-6) for uniform flows and the invariant measure μ_0 obtained in the limit is a weighted Gaussian; see [Birnir et al. 2007a]. The higher Picard iterates will give more complicated limits. Again, these approximations can probably be implemented by a symbolic or partially symbolic and partially numerical computation.

The problem is that this way of approximating the invariant measure may not be very inefficient. Thus it is important to seek more efficient ways of implementing these approximations first theoretically and then numerically.

8. RANS and LES models

The objective of RANS (Reynolds Averaged Navier Stokes) computations is to compute the spatial distribution of the mean velocity of the turbulent flow. To do this the velocity and pressure are decomposed into the mean $\bar{u}$ and the deviation from the mean $u = U - \bar{u}$ (or fluctuation)

$$U(x, t) = \bar{u}(x, t) + u(x, t)$$

The average denoted here by a bar is an ensemble average. Then, by definition, the mean of u is equal to zero. Similarly, the pressure is decomposed as

$$P(x, t) = \bar{p}(x, t) + p(x, t)$$

The divergence condition (5-2) gives that

$$\nabla \cdot \bar{u} = 0 = \nabla \cdot u$$

and averaging the Navier–Stokes equation (5-1) gives the equation for the mean velocity

$$\frac{\partial \bar{u}}{\partial t} + \bar{u} \cdot \nabla \bar{u} + \nabla \cdot \overline{u \otimes u} = \nu \Delta \bar{u} - \nabla \bar{p} \tag{8-1}$$

Thus the mean satisfies an equation similar to (2-1) except for an additional term due the Reynolds stress

$$\mathbf{R} = \overline{u \otimes u}$$

The additional term in (8-1) acts as an effective stress on the flow due to momentum transport cause by turbulent fluctuation. Until recently it has been impossible to determine this term from first principle and various approximations have been used. The simplest formulation is to set the Reynolds stress tensor to

$$\mathbf{R} = -\nu_T(x)\nabla\bar{u}$$

where $\nu_T(x)$ is called the turbulent eddy viscosity. This makes the additional term in the equation act as an additional (viscous) diffusion term. A better approximation is to develop an evolution equation for $\mathbf{R}$. This equation turns out to depend on the $\overline{u \otimes u \otimes u}$ and so on. Thus an infinite sequence of evolution equations for higher and higher moments is obtained and it must be closed at some level. This is done by approximating some higher moment by a formula depending only on lower moments. The closure problem is the problem of how to implement this moment truncation. A good recent exposition of the RANS models is contained in [Bernard and Wallace 2002].

The approximate invariant measure discussed above gives us a new insight into RANS models. In particular the mean is nothing but the expectation

$$\bar{u}(x,t) \sim \int_{L^2} u d\mu_m(u)$$

This obviously does not determine $\bar{u}$ since u is unknown but we can now work with the various closure approximations and improve them knowing what the spatial average actually means. This can be done and the result simulated. The hope is to develop RANS models that are less dependent on the available data and the parameter regions covered by that data.

In LES (see [Meneveau and Katz 2000]) the velocity is decomposed into Fourier modes and then the expansion truncated at some intermediate scale that are usually given by the grid resolution. Then one computes the large scales explicitly and models the effects of the small scales, smaller than the cutoff, on the large scales with a *subgrid model*. The cutoff is usually done with a smooth Gaussian filter. LES thus assumes that the small scale turbulence structures are not significantly dependent on the geometry of the flow and therefore can be represented by a general model. This method is able to handle transition to turbulence and the resulting turbulent regimes in the flow better than RANS that usually needs to be told explicitly where the transition occurs. Now if we let $\bar{u}$ and $\bar{u}$ denote resolved velocity and pressure then the Navier–Stokes equation for the resolve quantities can be written as

$$\frac{\partial\bar{u}}{\partial t} + \bar{u}\cdot\nabla\bar{u} + \nabla\cdot(\bar{u}\otimes\bar{u}) = \nu\Delta\bar{u} - \nabla\bar{p} - \nabla\cdot\tau \qquad (8\text{-}2)$$

where τ represents the subgrid stress tensor (SGS)

$$\tau = \overline{u \otimes u} - \bar{u} \otimes \bar{u}$$

and the resolved scales are divergence free

$$\nabla \cdot \bar{u} = 0$$

τ describes the effects of the subgrid scales on the resolved velocity.

The most common subgrid models use a relationship between SGS and e the resolved strain tensor

$$e = \tfrac{1}{2}(\nabla u + (\nabla u)^{\mathbf{T}})$$

where $(\nabla u)^{\mathbf{T}}$ denotes the transpose. The relationship between τ and e is

$$\tau - \tfrac{1}{3} \text{ trace } \tau \ \delta_{ij} = -2\nu_T e$$

Here δ_{ij} denotes Kronecker's delta and the eddy viscosity is

$$\nu_T = C\varepsilon^2 \sqrt{2e(e)^T}$$

The ε is a characteristic length scale for the subgrid. As it stands this subgrid model is purely dissipative and excessively so. If the constant in front of $|e| = \sqrt{2e(e)^T}$ is allowed to vary with time (see [Germano et al. 1991]) a much better result is obtained. Then the constant is computed dynamically during the simulation and with this modification the so-called Smagorinsky subgrid model does not produce excessive dissipation. However, it only work with situations where the flow is homogeneous in at least one direction and thus does not permit general geometries.

In general when modeling an experiment we want the subgrid model to reproduce the Kolmogorov $k^{-5/3}$ energy spectrum of homogeneous isotropic turbulence, and the statistics of turbulent channel flow. The advantage that we have with the approximate invariant measure is that we can base the cutoff on the approximately correct probability density function instead of a Gaussian that has nothing to do with the details of the small scale flow. This holds the promise that we can reproduce the correct scaling in the subgrid model. Ultimately this tests that the LES is producing the correct scaling down to the size of the computational grid.

9. Validation of the numerical methods

Turbulent fluids are highly unstable phenomena that are sensitive to noise and perturbation. Velocity trajectories depend sensitively on their initial conditions and it is not clear that they can be given a deterministic interpretation. This means that computations of such fluids are highly sensitive to truncation and even round-off errors. One must regards turbulent phenomena to be structurally unstable and stochastic. Statistical quantities associated to the turbulent fluids are deterministic and can be computed by taking appropriate statistical ensembles. However, one must be careful that the numerical methods one uses can be trusted to converge to the correct statistical quantity. It turns out that it is not enough to check that the conventional quantities such as energy or momentum and make sure that they converge. One must also consider the scalings of the statistical quantities and check that they show the correct scalings over a sufficiently large parameter range. In doing this one must choose the numerical methods carefully.

In a series of papers the author and his collaborators, [Smith et al. 1997b; Smith et al. 1997a; Birnir et al. 2001; Birnir et al. 2007a], showed that whereas explicit methods generally fail to produce the correct scalings over a large parameter interval, implicit methods do. This reason for this is that in an implicit method the time step is independent of the spatial discretization and does not go to zero as the spatial discretization decreases. Explicit methods obtain stability by inserting artificial viscosity into the problem and this artificial viscosity destroys the small scale scalings. Before the scaling of the small scales is obtained the time step goes to zero in the explicit method and the computation grinds to a halt. This makes implicit methods the methods of choice. Although the implicit methods also induce some viscosity, it is much smaller and does not interfere with the small scale scaling to the same extent as for explicit methods. The problem is that implicit methods are much slower than explicit and although this is not a serious obstacle in one dimension it is in two dimensions and makes the turbulence problem intractable in three dimensions. Thus it becomes imperative to compute correct closure approximations for RANS and subgrid models for LES in order to be able to solve these by implicit methods and produce numerically the correct scalings. One way of implementing this is to use the (approximate) invariant measure to develop tests on numerical methods to see if they produce correct scalings down to the size of the numerical grid.

Acknowledgments

This research for this paper was started during the author's sabbatical at the University of Granada, Spain. The author was supported by grants number DMS-0352563 from the National Science Foundation whose support is gratefully acknowledged. Some simulations are being done on a cluster of workstations, funded by a National Science Foundation SCREMS grant number DMS-0112388. The author wished to thank Professor Juan Soler of the Applied Mathematics Department at the University of Granada, for his support and the whole research group Delia Jiroveanu, José Luis López, Juanjo Nieto, Oscar Sánchez and María José Cáceres for their help and inspiration. His special thanks go to Professor José Martínez Aroza who cheerfully shared his office and the wonders of the fractal universe.

References

[Batchelor 1969] G. K. Batchelor, "Computation of the energy spectrum in homogeneous two-dimensional turbulence", *Phys. Fluids Suppl. II* **12** (1969), 233–239.

[Bernard and Wallace 2002] P. S. Bernard and J. M. Wallace, *Turbulent flow*, Wiley, Hoboken, NJ, 2002.

[Betchov and Criminale 1967] R. Betchov and W. O. Criminale, *Stability of parallel flows*, Academic Press, New York, 1967.

[Birnir 2007a] B. Birnir, "Turbulent rivers", preprint, 2007. Available at http://www.math.ucsb.edu/-birnir/papers. Submitted to *Quarterly Appl. Math.*

[Birnir 2007b] B. Birnir, "Uniqueness, an invariant measure and Kolmogorov's scaling for the stochastic Navier–Stokes equation", preprint, 2007. Available at http://www.math.ucsb.edu/-birnir/papers.

[Birnir et al. 2001] B. Birnir, T. R. Smith, and G. Merchant, "The scaling of fluvial landscapes", *Computers and Geosciences* **27** (2001), 1189–1216.

[Birnir et al. 2007a] B. Birnir, J. Hernández, and T. R. Smith, "The stochastic theory of fluvial landsurfaces", *J. Nonlinear Sci.* **17**:1 (2007), 13–57.

[Birnir et al. 2007b] B. Birnir, K. Mertens, V. Putkaradze, and P. Vorobieff, "Meandering of fluid streams on acrylic surface driven by external noise", preprint, 2007. Available at http://www.math.ucsb.edu/-birnir/papers.

[Bricmont et al. 2000] J. Bricmont, A. Kupiainen, and R. Lefevere, "Probabilistic estimates for the two-dimensional stochastic Navier–Stokes equations", *J. Statist. Phys.* **100**:3-4 (2000), 743–756.

[Da Prato and Zabczyk 1996] G. Da Prato and J. Zabczyk, *Ergodicity for infinite-dimensional systems*, London Mathematical Society Lecture Note Series **229**, Cambridge University Press, Cambridge, 1996.

[Dingman 1984] S. L. Dingman, *Fluvial hydrology*, W. H. Freeman, New York, 1984.

[Dodds and Rothman 1999] P. S. Dodds and D. Rothman, "Unified view of scaling laws for river networks", *Phys. Rev. E* **59**:5 (1999), 4865–4877.

[Dodds and Rothman 2000a] P. S. Dodds and D. Rothman, "Geometry of river networks, I: scaling, fluctuations and deviations", *Phys. Rev. E* **63** (2000), #016115.

[Dodds and Rothman 2000b] P. S. Dodds and D. Rothman, "Geometry of river networks, II: distributions of component size and number", *Phys. Rev. E* **63** (2000), #016116.

[Dodds and Rothman 2000c] P. S. Dodds and D. Rothman, "Geometry of river networks, III: characterization of component connectivity", *Phys. Rev. E* **63** (2000), #016117.

[Dodds and Rothman 2000d] P. S. Dodds and D. Rothman, "Scaling, universality and geomorphology", *Annu. Rev. Earth Planet. Sci.* **28** (2000), 571–610.

[Edwards and Wilkinson 1982] S. F. Edwards and D. R. Wilkinson, "The surface statistics of a granular aggregate", *Proc. Roy. Soc. London Ser. A* **381** (1982), 17–31.

[Flandoli 1994] F. Flandoli, "Dissipativity and invariant measures for stochastic Navier-Stokes equations", *Nonlinear Differential Equations Appl.* **1**:4 (1994), 403–423.

[Flandoli and Maslowski 1995] F. Flandoli and B. Maslowski, "Ergodicity of the 2-D Navier–Stokes equation under random perturbations", *Comm. Math. Phys.* **172**:1 (1995), 119–141.

[Frisch 1995] U. Frisch, *Turbulence: The legacy of A. N. Kolmogorov*, Cambridge University Press, Cambridge, 1995.

[Germano et al. 1991] M. Germano, U. Piomelli, P. Moin, and W. H. Cabot, "A dynamic subgrid-scale eddy viscosity model", *Phys. Fluids A* **3** (1991), 1760–1770.

[Gray 1961] D. M. Gray, "Interrelationships of watershed characteristics", *Journal of Geophysics Research* **66**:4 (1961), 1215–1223.

[Hack 1957] J. Hack, "Studies of longitudinal stream profiles in Virginia and Maryland", Professional Paper 294-B, 1957.

[Hairer and Mattingly 2004] M. Hairer and J. C. Mattingly, "Ergodic properties of highly degenerate 2D stochastic Navier–Stokes equations", *C. R. Math. Acad. Sci. Paris* **339**:12 (2004), 879–882.

[Hairer et al. 2004] M. Hairer, J. C. Mattingly, and É. Pardoux, "Malliavin calculus for highly degenerate 2D stochastic Navier–Stokes equations", *C. R. Math. Acad. Sci. Paris* **339**:11 (2004), 793–796.

[Holden et al. 1998] L. Holden, R. Hauge, O. Skare, and A. Skorstad, "Modeling of fluvial reservoirs with object models", *Math. Geol.* **30**:5 (1998), 473–496.

[Horton 1945] R. E. Horton, "Erosional development of streams and their drainage basins: a hydrophysical approach to quantitative morphology", *Geol. Soc. Am. Bull.* **56** (1945), 275–370.

[Kolmogorov 1941] A. N. Kolmogorov, "The local structure of turbulence in incompressible viscous fluid for very large Reynolds numbers", *Dokl. Akad. Nauk SSSR* **30**:4 (1941). In Russian; translated in *Proc. Roy. Soc. London Ser. A* **434** (1991), 9-13.

[Kraichnan 1967] R. H. Kraichnan, "Inertial ranges in two dimensional turbulence", *Phys. Fluids* **10** (1967), 1417–1423.

[Kuksin and Shirikyan 2001] S. Kuksin and A. Shirikyan, "A coupling approach to randomly forced nonlinear PDE's, I", *Comm. Math. Phys.* **221**:2 (2001), 351–366.

[Lavallée et al. 1993] D. Lavallée, S. Lovejoy, D. Schertzer, and P. Ladoy, "Nonlinear variability of landscape topography: Multifractal analysis and simulation", pp. 158–192 in *Fractals in geometry*, edited by N. S. Lam and L. De Cola, Prentice-Hall, Englewood Cliffs, NJ, 1993.

[Levi 1995] E. Levi, *The science of water*, ASCE Press, New York, 1995.

[Mattingly 1999] J. C. Mattingly, "Ergodicity of 2D Navier–Stokes equations with random forcing and large viscosity", *Comm. Math. Phys.* **206**:2 (1999), 273–288.

[McKean 2002] H. P. McKean, "Turbulence without pressure: existence of the invariant measure", *Methods Appl. Anal.* **9**:3 (2002), 463–467.

[Meneveau and Katz 2000] C. Meneveau and J. Katz, "Scale-invariance and turbulence models for large-eddy simulation", *Annu. Rev. Fluid Mech.* **32** (2000), 1–32.

[Rodriguez-Iturbe and Rinaldo 1997] I. Rodriguez-Iturbe and A. Rinaldo, *Fractal river basins: chance and self-organization*, Cambridge Univ. Press, Cambridge, 1997.

[Sidi and Dalaudier 1990] C. Sidi and F. Dalaudier, "Turbulence in the stratified atmosphere: recent theoretical developments and experimental results", *Advances in Space Research* **10** (1990), 25–36.

[Sinai 1996] Y. G. Sinai, "Burgers system driven by a periodic stochastic flow", pp. 347–353 in *Itô's stochastic calculus and probability theory*, edited by N. Ikeda et al., Springer, Tokyo, 1996.

[Smith et al. 1997a] T. R. Smith, B. Birnir, and G. E. Merchant, "Towards an elementary theory of drainage basin evolution, I: The theoretical basis", *Computers and Geosciences* **23**:8 (1997), 811–822.

[Smith et al. 1997b] T. R. Smith, G. E. Merchant, and B. Birnir, "Towards an elementary theory of drainage basin evolution, II: A computational evaluation", *Computers and Geosciences* **23**:8 (1997), 823–849.

[Smith et al. 2000] T. R. Smith, G. E. Merchant, and B. Birnir, "Transient attractors: towards a theory of the graded stream for alluvial and bedrock channels", *Computers and Geosciences* **26**:5 (2000), 531–541.

[Sølna 2002] K. Sølna, "Focusing of time-reversed reflections", *Waves Random Media* **12**:3 (2002), 365–385.

[Sølna 2003] K. Sølna, "Acoustic pulse spreading in a random fractal", *SIAM J. Appl. Math.* **63**:5 (2003), 1764–1788.

[Weinan et al. 2001] E. Weinan, J. C. Mattingly, and Y. Sinai, "Gibbsian dynamics and ergodicity for the stochastically forced Navier—Stokes equation", *Comm. Math. Phys.* **224**:1 (2001), 83–106.

[Welsh et al. 2006] E. Welsh, B. Birnir, and A. Bertozzi, "Shocks in the evolution of an eroding channel", *AMRX Appl. Math. Res. Express* (2006), Art. Id 71638.

BJÖRN BIRNIR
CENTER FOR COMPLEX AND NONLINEAR SCIENCE
AND
DEPARTMENT OF MATHEMATICS
UNIVERSITY OF CALIFORNIA, SANTA BARBARA
SANTA BARBARA, CA 93106
UNITED STATES
birnir@math.ucsb.edu

Probability, Geometry and Integrable Systems
MSRI Publications
Volume **55**, 2008

Riemann–Hilbert problem in the inverse scattering for the Camassa–Holm equation on the line

ANNE BOUTET DE MONVEL AND DMITRY SHEPELSKY

Dedicated to Henry McKean in deep admiration and friendship

ABSTRACT. We present a Riemann–Hilbert problem formalism for the initial value problem for the Camassa–Holm equation $u_t - u_{txx} + 2\omega u_x + 3uu_x = 2u_x u_{xx} + uu_{xxx}$ on the line (CH), where ω is a positive parameter. We show that, for all $\omega > 0$, the solution of this initial value problem can be obtained in a parametric form from the solution of some associated Riemann–Hilbert problem; that for large time, it develops into a train of smooth solitons; and that for small ω, this soliton train is close to a train of peakons, which are piecewise smooth solutions of the CH equation for $\omega = 0$.

1. Introduction

The main purpose of this paper is to develop an inverse scattering approach, based on an appropriate Riemann–Hilbert problem formulation, for the initial value problem for the Camassa–Holm (CH) equation [Camassa and Holm 1993] on the line, whose form is

$$u_t - u_{txx} + 2\omega u_x + 3uu_x = 2u_x u_{xx} + uu_{xxx}, \quad -\infty < x < \infty, \ t > 0, \quad \text{(1-1a)}$$

$$u(x, 0) = u_0(x), \quad \text{(1-1b)}$$

where ω is a positive parameter. The CH equation is a model equation describing the shallow-water approximation in inviscid hydrodynamics. In this equation $u = u(x, t)$ is a real-valued function that refers to the horizontal fluid velocity along the x direction (or equivalently, the height of the water's free surface above a flat bottom) as measured at time t. The constant ω is related to the critical shallow water wave speed $\sqrt{gh_0}$, where g is the acceleration of gravity and h_0 is the undisturbed water depth; hence, the case $\omega > 0$ is physically more relevant than the case $\omega = 0$, though the latter has attracted more attention in the mathematical studies due to interesting specific features such as the existence of peaked (nonanalytic) solitons.

In terms of the "momentum" variable

$$m := u - u_{xx}, \tag{1-2}$$

the CH equation (1-1a) reads

$$m_t + 2\omega u_x + u m_x + 2m u_x = 0. \tag{1-3}$$

Assuming further that $m + \omega \geq 0$, equation (1-1a) can be equivalently expressed as

$$\left(\sqrt{m+\omega}\right)_t = -\left(u\sqrt{m+\omega}\right)_x. \tag{1-4}$$

The function $u_0(x)$ in (1-1b), as well as

$$m_0(x) := u_0(x) - u_{0xx}(x)$$

is assumed to be sufficiently smooth and to decay fast as $|x| \to \infty$. It is known (see, e.g., [Constantin 2001]) that if $m_0(x) + \omega > 0$ for all x then the solution $m(x, t)$ to (1-3) exists for all t; moreover, $m(x, t) + \omega > 0$ for all $x \in \mathbb{R}$ and all $t > 0$. This justifies the form (1-4) of the CH equation, which will be used in our constructions below.

Our goal is to develop the inverse scattering approach to the CH equation, in view of its further application for studying the long-time asymptotics. The starting point of the approach is the Lax pair representation: the CH equation is indeed the compatibility condition of two linear equations [Camassa and Holm 1993]

$$\psi_{xx} = \tfrac{1}{4}\psi + \lambda(m + \omega)\psi, \tag{1-5a}$$

$$\psi_t = \left(\tfrac{1}{2\lambda} - u\right)\psi_x + \tfrac{1}{2}u_x\psi. \tag{1-5b}$$

Together with the fact that the x-equation of the Lax pair can be transformed to the spectral problem for the one-dimensional Schrödinger equation, by means of the Liouville transformation, this allows using the inverse scattering transform method to study the initial value problem for the CH equation with $\omega > 0$; see [Constantin 2001; Lenells 2002; Constantin and Lenells 2003; Johnson 2003; Constantin et al. 2006].

In the present paper, we propose a "scattering – inverse scattering" formalism, in which the Lax pair is used in the form of a system of first order matrix-valued linear equations. Then dedicated solutions of this system are defined and used to construct a multiplicative Riemann–Hilbert (RH) problem in the complex plane of the spectral parameter. The main advantage of the representation of a solution of the CH equation in terms of the solution of a RH problem is that it allows applying the nonlinear steepest descent method by Deift and Zhou [1993] in order to obtain rigorous results on the long-time asymptotic behavior of the solution.

An alternative inverse scattering method based on an additive RH problem formulation for the associated eigenfunctions is given in [Constantin et al. 2006].

In Section 2, we define appropriate eigenfunctions and spectral functions, which are used in Section 3 in the reformulation of the scattering problem as a Riemann–Hilbert problem of analytic conjugation in the complex plane of the spectral parameter. We also introduce a scale in which the Riemann–Hilbert problem becomes explicitly given. In Sections 4 and 5, we briefly discuss the soliton solutions and the soliton asymptotics of the solution of a general initial value problem. Finally, Section 6 deals with the small-ω analysis, the main result of which can be formulated as follows: for sufficiently small ω, the solution of the initial value problem for the Camassa–Holm equation is seen, for large time, as a train of "almost" peakons, which are piecewise smooth weak solutions of the Camassa–Holm equation with $\omega = 0$, the parameters of which are determined by the spectrum of the associated linear problem (1-5a) with $\omega = 0$.

The results of this paper were announced in [Boutet de Monvel and Shepelsky 2006b]. For their application to long-time asymptotics, see [Boutet de Monvel and Shepelsky 2007].

2. Eigenfunctions and spectral functions

2.1. Eigenfunctions. We present the general formalism scaling out the parameter ω in the CH equation and thus assuming that $\omega = 1$ and $m + 1 > 0$ (in Section 6 we will return to arbitrary positive ω when studying the small-ω limit of solutions).

Let $u(x, t)$ be a solution to (1-1a) with $\omega = 1$ such that $u(x, t) \to 0$ as $|x| \to \infty$ for all t. First, we rewrite the Lax pair in vector form. Let $\Phi_{\text{init}} := \left(\begin{smallmatrix} \psi \\ \psi_x \end{smallmatrix} \right)$; then (1-5) is equivalent to

$$(\Phi_{\text{init}})_x = \begin{pmatrix} 0 & 1 \\ \lambda(m+1) + \frac{1}{4} & 0 \end{pmatrix} \Phi_{\text{init}} =: U_{\text{init}} \Phi_{\text{init}}, \tag{2.1a}$$

$$(\Phi_{\text{init}})_t = \begin{pmatrix} \frac{1}{2}u_x & \frac{1}{2\lambda} - u \\ \frac{1}{8\lambda} + \frac{1}{2} + \frac{1}{4}u - \lambda u(m+1) & -\frac{1}{2}u_x \end{pmatrix} \Phi_{\text{init}} =: V_{\text{init}} \Phi_{\text{init}}. \tag{2.1b}$$

Our aim is to introduce special solutions to this system that are well-controlled as functions of the spectral parameter.

In order to control the behavior of solutions of this system for $\lambda \to \infty$, it is convenient to transform it in such a way that

(i) the principal terms for $\lambda \to \infty$ in the Lax equations be diagonal and the terms of order λ^0 be off-diagonal; and

(ii) all the lower order terms vanish as $|x| \to \infty$.

Writing U_{init} in the form

$$U_{\text{init}} = \begin{pmatrix} 0 & 1 \\ (\lambda + \frac{1}{4})(m+1) & 0 \end{pmatrix} - \frac{m}{4}\begin{pmatrix} 0 & 0 \\ 1 & 0 \end{pmatrix}$$

suggests introducing the spectral parameter k and the transformation matrix G_∞ by

$$k^2 = -\lambda - \frac{1}{4}, \quad G_\infty(x,t;k) := \frac{1}{2}\begin{pmatrix} 1 & -\frac{1}{ik} \\ 1 & \frac{1}{ik} \end{pmatrix}\begin{pmatrix} (m+1)^{1/4} & 0 \\ 0 & (m+1)^{-1/4} \end{pmatrix}.$$

Defining $\tilde{\Phi} := G_\infty \Phi_{\text{init}}$ transforms (2-1) into

$$\tilde{\Phi}_x + ik\sqrt{m+1}\sigma_3\tilde{\Phi} = U\tilde{\Phi}, \tag{2-2a}$$

$$\tilde{\Phi}_t + ik\left(\frac{1}{2\lambda} - u\sqrt{m+1}\right)\sigma_3\tilde{\Phi} = V\tilde{\Phi}, \tag{2-2b}$$

where $\sigma_3 = \begin{pmatrix} 1 & 0 \\ 0 & -1 \end{pmatrix}$,

$$U(x,t;k) = \frac{1}{4}\frac{m_x}{m+1}\begin{pmatrix} 0 & 1 \\ 1 & 0 \end{pmatrix} - \frac{1}{8ik}\frac{m}{\sqrt{m+1}}\begin{pmatrix} -1 & -1 \\ 1 & 1 \end{pmatrix} \tag{2-3}$$

and

$$V(x,t;k) = -\frac{u}{4}\frac{m_x}{m+1}\begin{pmatrix} 0 & 1 \\ 1 & 0 \end{pmatrix} + \frac{1}{8ik}\frac{u(m+2)}{\sqrt{m+1}}\begin{pmatrix} -1 & -1 \\ 1 & 1 \end{pmatrix}$$

$$+ \frac{ik}{4\lambda}\left\{\sqrt{m+1}\begin{pmatrix} -1 & 1 \\ -1 & 1 \end{pmatrix} + \frac{1}{\sqrt{m+1}}\begin{pmatrix} -1 & -1 \\ 1 & 1 \end{pmatrix}\right\} + \frac{ik}{2\lambda}\sigma_3. \tag{2-4}$$

It is clear that $U(x,t;k) \to 0$ as $|x| \to \infty$. As for the t-equation (2-2b), the term $\frac{ik}{2\lambda}\sigma_3$ has been introduced into the r.h.s. of (2-4) in order to provide $V(x,t;k) \to 0$ as $|x| \to \infty$.

Now the equations (2-2) suggest introducing a scalar function $p(x,t;k)$ in such a way that

$$p_x = \sqrt{m+1}, \quad p_t = \frac{1}{2\lambda} - u\sqrt{m+1}.$$

Indeed, due to (1-4), one can define such p (normalized by $p(x,0;k) \sim x$ as $x \to +\infty$) as follows:

$$p(x,t;k) := x - \int_x^\infty \left(\sqrt{m(\xi,t)+1} - 1\right) d\xi + \frac{t}{2\lambda(k)}$$

$$= x - \int_x^\infty \left(\sqrt{m(\xi,t)+1} - 1\right) d\xi - \frac{2}{1+4k^2}t. \tag{2-5}$$

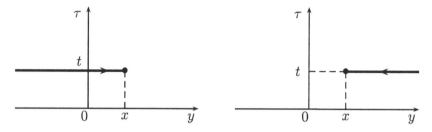

Figure 1. Paths of integration for Φ_- and Φ_+.

Finally, assuming $\tilde{\Phi}$ to be a 2×2 matrix-valued function, let $\Phi(x,t;k) := \tilde{\Phi} e^{ikp(x,t;k)\sigma_3}$. Then (2-2) becomes

$$\Phi_x + ikp_x[\sigma_3, \Phi] = U\Phi, \quad \Phi_t + ikp_t[\sigma_3, \Phi] = V\Phi, \tag{2-6}$$

where $[a, b] := ab - ba$.

The Lax pair in the form (2-6) is very convenient for defining dedicated solutions via integral Volterra equations by specifying the initial point of integration (x^*, t^*) in the (x, t)-plane:

$$\Phi(x,t;k) = I + \int_{(x^*,t^*)}^{(x,t)} e^{-ik(p(x,t;k)-p(y,\tau;k))\hat{\sigma}_3}$$
$$\times \{U(y,\tau;k)\Phi(y,\tau;k)dy + V(y,\tau;k)\Phi(y,\tau;k)d\tau\}, \tag{2-7}$$

where I is the 2×2 identity matrix, $e^{\hat{\sigma}_3} A := e^{\sigma_3} A e^{-\sigma_3}$ for any 2×2 matrix A, and the r.h.s. is independent of the integration path. Then the special structure of U and V (recall that their main terms as $k \to \infty$ are off-diagonal) provides a well-controlled behavior of solutions of (2-7) for large k.

Choosing the initial points of integration to be $(-\infty, t)$ and $(+\infty, t)$ and the paths of integration to be parallel to the x-axis (see Figure 1) leads to the integral equations for Φ_+ and Φ_-:

$$\Phi_-(x,t;k) = I + \int_{-\infty}^{x} e^{-ik \int_y^x \sqrt{m(\xi,t)+1}\, d\xi \hat{\sigma}_3} (U\Phi_-)(y,t;k)\, dy,$$
$$\Phi_+(x,t;k) = I - \int_{x}^{\infty} e^{ik \int_x^y \sqrt{m(\xi,t)+1}\, d\xi \hat{\sigma}_3} (U\Phi_+)(y,t;k)\, dy. \tag{2-8}$$

It is due to the condition $V \to 0$ as $|x| \to \infty$ as well as the compatibility of the two equations in (2-6) that the solutions of (2-8) satisfy the t-equation in (2-6). Since U and V are traceless, it follows that $\det \Phi_\pm \equiv 1$.

The structure of the integral equations (2-8) provides the following analytic properties of the eigenfunctions $\Phi_\pm$ as functions of k. We denote $\mu^{(1)}$ and $\mu^{(2)}$ the columns of a 2×2 matrix $\mu = (\mu^{(1)} \quad \mu^{(2)})$. Then, for all (x,t), the following conditions are satisfied:

(a) $\Phi_-^{(1)}$ and $\Phi_+^{(2)}$ are analytic in $\{k \mid \operatorname{Im} k > 0\}$ and continuous in $\{k \mid \operatorname{Im} k \geq 0,\ k \neq 0\}$;

(b) $\Phi_+^{(1)}$ and $\Phi_-^{(2)}$ are analytic in $\{k \mid \operatorname{Im} k < 0\}$ and continuous in $\{k \mid \operatorname{Im} k \leq 0,\ k \neq 0\}$;

(c) as $k \to \infty$ in $\{k \mid \operatorname{Im} k \geq 0\}$, $\big(\Phi_-^{(1)}\ \Phi_+^{(2)}\big) \to I$;

(d) as $k \to \infty$ in $\{k \mid \operatorname{Im} k \leq 0\}$, $\big(\Phi_+^{(1)}\ \Phi_-^{(2)}\big) \to I$;

(e) as $k \to 0$, $\Phi_\pm = \frac{\alpha_\pm(x,t)}{ik}\begin{pmatrix} -1 & -1 \\ 1 & 1 \end{pmatrix} + O(1)$ with $\alpha_\pm \in \mathbb{R}$.

REMARK 1. The Lax pair in the form (2-6) (see also (3-9) below) and the associated integral equations (2-7) turn out to be useful also in studying initial boundary value problems for the CH equation, see [Boutet de Monvel and Shepelsky 2006a]. Indeed, the structure of the t-equation is as "good" as the structure of the x-equation for controlling the properties of the appropriate eigenfunctions in the k-plane even in the case where the integration paths in (2-7) are not parallel to the x-axis; such paths (parallel to the t-axis) are needed in order to relate the eigenfunctions to the boundary values (i.e., at $x = 0$) of a solution of the nonlinear equation in question. These eigenfunctions can be viewed as coming from the simultaneous spectral analysis (of the x- and t-equations) of the Lax pair.

2.2. Spectral functions. For $k \in \mathbb{R}$, the eigenfunctions Φ_- and Φ_+, being the solutions of the system of differential equations (2-6), are related by a matrix independent of (x,t); this allows introducing the scattering matrix $s(k)$ by

$$\Phi_+(x,t;k) = \Phi_-(x,t;k)e^{-ikp(x,t;k)\hat{\sigma}_3}s(k), \quad k \in \mathbb{R},\ k \neq 0. \tag{2-9}$$

Since the matrix U satisfies the symmetry relations

$$\overline{U(\cdot,\cdot,\bar{k})} = U(\cdot,\cdot,-k) = \begin{pmatrix} 0 & 1 \\ 1 & 0 \end{pmatrix}U(\cdot,\cdot,k)\begin{pmatrix} 0 & 1 \\ 1 & 0 \end{pmatrix}, \tag{2-10}$$

the spectral matrix $s(k)$ can be written as

$$s(k) = \begin{pmatrix} \overline{a(k)} & b(k) \\ \overline{b(k)} & a(k) \end{pmatrix}, \quad k \in \mathbb{R}, \tag{2-11}$$

where $\overline{a(k)} = a(-k)$ and $\overline{b(k)} = b(-k)$.

Writing the spectral functions $a(k)$ and $b(k)$ in terms of determinants, namely

$$a(k) = \det\big(\Phi_-^{(1)}\ \Phi_+^{(2)}\big),$$
$$b(k) = e^{2ikp}\det\big(\Phi_+^{(2)}\ \Phi_-^{(2)}\big),$$

allows us to establish the following properties:

(i) $a(k)$ and $b(k)$ are determined by $\Phi_\pm$ for $t = 0$ and thus by $m(x,0)$ (or, equivalently, by $u(x,0)$).

(ii) $a(k)$ is analytic in $\{k \mid \operatorname{Im} k > 0\}$ and continuous in $\{k \mid \operatorname{Im} k \geq 0, k \neq 0\}$; moreover, $a(k) \to 1$ as $k \to \infty$.

(iii) $b(k)$ is continuous for $k \in \mathbb{R}, k \neq 0$ and $b(k) \to 0$ as $|k| \to \infty$.

(iv) as $k \to 0$, $a(k) = \alpha_0/(ik) + O(1)$ and $b(k) = -\alpha_0/(ik) + O(1)$ with $\alpha_0 \in \mathbb{R}$.

(v) $|a(k)|^2 - |b(k)|^2 = 1$ for $k \in \mathbb{R}, k \neq 0$.

(vi) Let $\{k_j\}_{j=1}^N$ be the set of zeros of $a(k)$: $a(k_j) = 0$. Then $N < \infty$ and the zeros are simple with $(da/dk)(k_j) \in i\mathbb{R}$; moreover, $k_j = i v_j$ with $0 < v_j < \frac{1}{2}$ for all $1 \leq j \leq N$; and the eigenvectors are related by

$$\Phi_-^{(1)}(x,t;iv_j) = \chi_j e^{-2v_j \, p(x,t;iv_j)} \Phi_+^{(2)}(x,t;iv_j) \tag{2-12}$$

with $\chi_j \in \mathbb{R}$.

The statements in the last item follow from the fact that $\lambda_j = v_j^2 - \frac{1}{4}$, $j = 1, 2, \ldots, N$ are the eigenvalues of the x-equation (1-5a) with $m = m(x,0)$: they are known (see, e.g., [Constantin 2001]) to be simple and to satisfy $-\frac{1}{4} < \lambda_j < 0$.

REMARK 2. It follows from the construction of the eigenfunctions $\Phi_\pm$ that $s(k)$ is the scattering matrix for the one-dimensional Schrödinger equation $-\varphi_{yy} + Q(y)\varphi = k^2\varphi$ associated to (1-5a) with $m = m(x,0)$ via the Liouville transformation:

$$y = x - \int_x^\infty \left(\sqrt{m(\xi,0) + 1} - 1\right) d\xi, \qquad q(y) = m(x,0) + 1,$$

$$\varphi(y,k) = \psi(x,k)q(y)^{1/4}, \qquad Q(y) = \frac{q_{yy}}{4q} - \frac{3(q_y)^2}{16q^2} + \frac{1-q}{4q}.$$

3. Riemann–Hilbert problem

3.1. Scattering problem as a Riemann–Hilbert problem in the (x,t) scale.

Regrouping the columns of the scattering relation (2-9) in accordance with their analyticity properties allows rewriting (2-9) in the form of a conjugation of piecewise meromorphic, matrix-valued functions. Let us define a 2×2 matrix function $M(x,t;k)$ by

$$M(x,t;k) = \begin{cases} \left(\dfrac{\Phi_-^{(1)}(x,t;k)}{a(k)} \quad \Phi_+^{(2)}(x,t;k) \right) & \text{if } \operatorname{Im} k > 0, \\[4mm] \left(\Phi_+^{(1)}(x,t;k) \quad \dfrac{\Phi_-^{(2)}(x,t;k)}{\overline{a(\bar{k})}} \right) & \text{if } \operatorname{Im} k < 0. \end{cases} \tag{3-1}$$

Then the limiting values $M_\pm(x,t;\zeta)$ of $M(x,t;k)$ as k approaches the real axis $(k = \zeta \pm i\varepsilon,\ \varepsilon > 0$ and $\varepsilon \to 0)$ are related as follows:

$$M_-(x,t;k) = M_+(x,t;k)e^{-ikp(x,t;k)\sigma_3}J_0(k)e^{ikp(x,t;k)\sigma_3},\quad k \in \mathbb{R},\quad (3\text{-}2)$$

where

$$J_0(k) = \begin{pmatrix} 1 & -r(k) \\ r(k) & 1-|r(k)|^2 \end{pmatrix} \qquad (3\text{-}3)$$

with $r(k) = b(k)/\overline{a(k)}$.

By construction, $M(x,t;k)$ also satisfies the following properties:

(i) $M \to I$ as $k \to \infty$.

(ii) $M = \dfrac{\alpha_+(x,t)}{ik}\begin{pmatrix} -c & -1 \\ c & 1 \end{pmatrix} + O(1)$ as $k \to 0$ in $\operatorname{Im} k \geq 0$, where $c = 0$ if $\lim\limits_{k \to 0} ka(k) \neq 0$.

(iii) Symmetry properties:

$$\overline{M(\cdot,\cdot,\bar{k})} = M(\cdot,\cdot,-k) = \begin{pmatrix} 0 & 1 \\ 1 & 0 \end{pmatrix}M(\cdot,\cdot,k)\begin{pmatrix} 0 & 1 \\ 1 & 0 \end{pmatrix}.$$

(iv) M has poles at the zeros $k_j = iv_j$ of $a(k)$ (in the upper half-plane $\operatorname{Im} k > 0$) and at $\bar{k}_j = -iv_j$ (in the lower half-plane $\operatorname{Im} k < 0$), $j = 1,2,\ldots,N$, with the following residue conditions:

$$\begin{aligned} \operatorname{Res}_{k=iv_j} M^{(1)}(x,t;k) &= i\gamma_j e^{-2v_j\,p(x,t;iv_j)}M^{(2)}(x,t;iv_j), \\ \operatorname{Res}_{k=-iv_j} M^{(2)}(x,t;k) &= -i\gamma_j e^{-2v_j\,p(x,t;iv_j)}M^{(1)}(x,t;-iv_j) \end{aligned} \qquad (3\text{-}4)$$

with $\gamma_j = -i\dfrac{\chi_j}{(da/dk)(iv_j)} \in \mathbb{R}$.

In the Riemann–Hilbert approach to nonlinear evolution equations, one tries to interpret a jump relation (of type (3-2)) across a contour in the k-plane, together with residue conditions (of type (3-4)) and certain normalization conditions, as a Riemann–Hilbert problem, the data for which are the jump matrix and the residue parameters (which can be obtained by solving the direct scattering problem for an operator with coefficients determined by the initial data for the nonlinear problem), and the solution of which gives the solution to the nonlinear equation in question.

In the case of the Camassa–Holm equation, the jump relation (3-2) cannot be used immediately for this purpose. In the construction of the jump matrix $e^{-ikp}J_0(k)e^{ikp}$ the factor $J_0(k)$ is indeed given in terms of the known initial data but $p(x,t;k)$ is not: it involves $m(x,t)$ which is unknown (and, in fact, is to be reconstructed) in the framework of the inverse problem.

To remedy this, we introduce the new (time-dependent) scale (cf. the Liouville transformation)

$$y(x,t) = x - \int_x^\infty \left(\sqrt{m(\xi,t)+1} - 1 \right) d\xi, \tag{3-5}$$

in terms of which the jump matrix and the residue conditions become explicit. The price to pay for this, however, is that the solution to the nonlinear problem can be given only implicitly, or parametrically: it will be given in terms of functions in the new scale, whereas the original scale will also be given in terms of functions in the new scale.

In order to achieve this program, we use a particular feature of the Lax pair for the Camassa–Holm equation, namely, the fact that the x-equation (1-5a) becomes trivial (independent of the "momentum" m) for $\lambda = 0$, which corresponds to $k = \pm\frac{1}{2}$. In order to translate this into the properties of the eigenfunctions involved in the construction of the analytic conjugation problem (3-2), it is convenient to transform the Lax pair equations in such a way that the main terms become diagonal as $\lambda \to 0$.

3.2. Eigenfunctions near $\lambda = 0$. Setting $\tilde{\Phi}^0 := \frac{1}{2}\begin{pmatrix} 1 & -\frac{1}{ik} \\ 1 & \frac{1}{ik} \end{pmatrix} \Phi_{\text{init}}$ transforms (2-1) into

$$\tilde{\Phi}_x^0 + ik\sigma_3\tilde{\Phi}^0 = U_0\tilde{\Phi}^0, \quad \tilde{\Phi}_t^0 + \frac{ik}{2\lambda}\sigma_3\tilde{\Phi}^0 = V_0\tilde{\Phi}^0, \tag{3-6}$$

where

$$U_0(x,t;k) = \frac{\lambda}{2ik}m(x,t)\begin{pmatrix} -1 & -1 \\ 1 & 1 \end{pmatrix}, \tag{3-7}$$

and

$$V_0(x,t;k) = \frac{u_x}{2}\begin{pmatrix} 0 & 1 \\ 1 & 0 \end{pmatrix}$$
$$+ \frac{u}{4ik}\begin{pmatrix} 0 & -1 \\ 1 & 0 \end{pmatrix} + \frac{\lambda u}{2ik}\left\{ 2\sigma_3 - m\begin{pmatrix} -1 & -1 \\ 1 & 1 \end{pmatrix} \right\}. \tag{3-8}$$

The eigenfunctions $\Phi_{\pm}^0$ are defined similarly to $\Phi_{\pm}$: setting

$$\Phi^0 := \tilde{\Phi}^0 \exp\left\{ \left(ikx + \frac{ik}{2\lambda}t \right)\sigma_3 \right\}$$

transforms (3-6) into

$$\Phi_x^0 + ik[\sigma_3, \Phi^0] = U_0\Phi^0, \quad \Phi_t^0 + \frac{ik}{2\lambda}[\sigma_3, \Phi^0] = V_0\Phi^0. \tag{3-9}$$

Now the eigenfunctions $\Phi_{\pm}^0$ are defined as solutions of the integral equations

$$\Phi_{\pm}^0(x,t;k) = I + \int_{\pm\infty}^{x} e^{-ik(x-y)\hat{\sigma}_3}(U_0\Phi_{\pm}^0)(y,t;k)\,dy \qquad (3\text{-}10)$$

(notice that the fact that $V_0 \to 0$ as $|x| \to \infty$ is again of importance here). Since $U_0(x,t;\pm\frac{i}{2}) \equiv 0$ in (3-10), we have that $\Phi_{\pm}^0(x,t;\pm\frac{i}{2}) \equiv I$.

Since $\Phi_{\pm}$ and $\Phi_{\pm}^0$ solve a system of differential equations which are transformations of the same system (2-1), they are related by matrices $C_{\pm}(k)$ independent of (x,t):

$$\Phi_{\pm}(x,t;k) = F(x,t)\Phi_{\pm}^0(x,t;k)e^{-ik(x+\frac{t}{2\lambda})\sigma_3}C_{\pm}(k)e^{ikp(x,t;k)\sigma_3}, \qquad (3\text{-}11)$$

where

$$F = \frac{1}{2}\begin{pmatrix} q+q^{-1} & q-q^{-1} \\ q-q^{-1} & q+q^{-1} \end{pmatrix}, \qquad q(x,t) = (m(x,t)+1)^{1/4},$$

and $C_{\pm}(k)$ are determined by the boundary conditions

$$\Phi_{\pm}(\pm\infty,t;k) = \Phi_{\pm}^0(\pm\infty,t;k) = I;$$

this gives $C_+(k) = I$ and $C_-(k) = e^{ik\varkappa\sigma_3}$ with $\varkappa = \int_{-\infty}^{\infty}(\sqrt{m(\xi,t)+1}-1)\,d\xi$ independent of t (conservation law).

In particular, evaluating (3-11) at $k = \pm\frac{i}{2}$ we have

$$\Phi_{+}^{(2)}(x,t;\tfrac{i}{2}) = F^{(2)}(x,t)e^{-\frac{1}{2}\int_x^\infty(\sqrt{m(\xi,t)+1}-1)\,d\xi},$$

$$\Phi_{-}^{(1)}(x,t;\tfrac{i}{2}) = F^{(1)}(x,t)e^{-\frac{1}{2}\int_{-\infty}^x(\sqrt{m(\xi,t)+1}-1)\,d\xi}.$$

Calculating $a(\frac{i}{2})$ using the determinant formula gives

$$a(\tfrac{i}{2}) = \det(\Phi_{-}^{(1)}\ \Phi_{+}^{(2)})\big|_{k=\frac{i}{2}} = e^{-\frac{1}{2}\int_{-\infty}^{\infty}(\sqrt{m(\xi,t)+1}-1)\,d\xi} = e^{-\varkappa/2},$$

which finally yields

$$M(x,t;\tfrac{i}{2}) = F(x,t)\begin{pmatrix} e^{\frac{1}{2}\int_x^\infty(\sqrt{m(\xi,t)+1}-1)\,d\xi} & 0 \\ 0 & e^{-\frac{1}{2}\int_x^\infty(\sqrt{m(\xi,t)+1}-1)\,d\xi} \end{pmatrix}. \qquad (3\text{-}12)$$

Equation (3-12) allows relating the original scale x and the new scale y, see (3-5), in terms of M evaluated at $k = \frac{i}{2}$. Indeed, let

$$\tilde{\mu}_1(x,t) := M_{11}(x,t;\tfrac{i}{2}) + M_{21}(x,t;\tfrac{i}{2}),$$

$$\tilde{\mu}_2(x,t) := M_{12}(x,t;\tfrac{i}{2}) + M_{22}(x,t;\tfrac{i}{2}).$$

Then from (3-12) and (3-5) we have

$$\frac{\tilde{\mu}_1(x,t)}{\tilde{\mu}_2(x,t)} = e^{\int_x^\infty(\sqrt{m(\xi,t)+1}-1)\,d\xi} = e^{x-y(x,t)} \qquad (3\text{-}13)$$

and

$$[\tilde{\mu}_1(x,t)\tilde{\mu}_2(x,t)]^2 = m(x,t) + 1. \tag{3-14}$$

3.3. Riemann–Hilbert problem in the (y,t) scale. We observe that the jump conditions (3-2) as well as the residue conditions (3-4) become explicit in the variables y and t. This, together with the considerations above concerning the relations involving the x and y scales, suggest introducing the vector Riemann–Hilbert problem, parametrized by (y,t), as follows (cf. (3-2), (3-4)):

RH-PROBLEM. Given $r(k)$ for $k \in \mathbb{R}$, $\{v_j\}_{j=1}^N$ $(0 < v_j < \frac{1}{2})$, and $\{\gamma_j\}_{j=1}^N$ $(\gamma_j > 0)$, find a row function $\mu(y,t;k) = (\mu_1(y,t;k) \quad \mu_2(y,t;k))$ such that:

(i) $\mu(\cdot, \cdot ; k)$ is analytic in $\{k \mid \operatorname{Im} k > 0\}$ and in $\{k \mid \operatorname{Im} k < 0\}$.
(ii) The limits $\mu_\pm(\cdot, \cdot, \zeta) = \lim_{\varepsilon \to +0} \mu(\cdot, \cdot ; \zeta \pm i\varepsilon)$, $\zeta \in \mathbb{R}$ are related by

$$\mu_-(y,t;\zeta) = \mu_+(y,t;\zeta)J(y,t;\zeta), \qquad \zeta \in \mathbb{R}, \tag{3-15}$$

where the jump matrix is

$$J(y,t;k) = e^{-ik\left(y-\frac{2}{1+4k^2}t\right)\sigma_3} J_0(k) e^{ik\left(y-\frac{2}{1+4k^2}t\right)\sigma_3} \tag{3-16}$$

with

$$J_0(k) = \begin{pmatrix} 1 & -r(k) \\ r(k) & 1 - |r(k)|^2 \end{pmatrix}.$$

(iii) Normalization at infinity:

$$\mu(y,t;k) \to (1 \quad 1) \quad \text{as } k \to \infty. \tag{3-17}$$

(iv) Residue conditions:

$$\operatorname{Res}_{k=iv_j} \mu_1(y,t;k) = i\gamma_j e^{-2v_j\left(y-\frac{2}{1-4v_j^2}t\right)} \mu_2(y,t;iv_j),$$
$$\operatorname{Res}_{k=-iv_j} \mu_2(y,t;k) = -i\gamma_j e^{-2v_j\left(y-\frac{2}{1-4v_j^2}t\right)} \mu_1(y,t;-iv_j). \tag{3-18}$$

REMARKS. The symmetry properties of the jump matrix imply that

$$\overline{\mu_1(\cdot, \cdot ; \bar{k})} = \mu_1(\cdot, \cdot ; -k) = \mu_2(\cdot, \cdot ; k). \tag{3-19}$$

Now we notice the following:

▷ The data for the Riemann–Hilbert problem (3-15)–(3-18) are determined in terms of the scattering data $r(k)$, $\{v_j\}_{j=1}^N$, $\{\gamma_j\}_{j=1}^N$, which, in turn, are determined by $m(x,0)$, the initial value of the solution of the Camassa–Holm equation, via the solutions $\Phi_\pm$ of the direct scattering problem at $t = 0$, see (2-8), (2-9), (2-12).

▷ This RH problem has the same structure as that for the Korteweg–de Vries equation (except for the k-dependence of the velocity in the phase factors, see (3-16) and (3-18)), which implies that there exists a vanishing lemma [Beals et al. 1988] guarantying that the RH problem has a unique solution $\mu(y, t; k)$ for all $y \in (-\infty, \infty)$ and $t > 0$.

Let us evaluate this solution at $k = \frac{i}{2}$. Then, the relations (3-13) and (3-14) allow us to:

(a) Determine the function $y = y(x, t)$ as the inverse to the function

$$x(y, t) = y + \ln \frac{\mu_1(y, t; \frac{i}{2})}{\mu_2(y, t; \frac{i}{2})}. \tag{3-20}$$

(b) Determine the momentum $m(x, t)$ of the Camassa–Holm equation by

$$m(x, t) = \left(\mu_1(y, t; \tfrac{i}{2}) \mu_2(y, t; \tfrac{i}{2}) \right)^2 \big|_{y=y(x,t)} - 1. \tag{3-21}$$

Now the solution $u(x, t)$ of the Camassa–Holm equation can be determined from $m(x, t)$ by

$$u(x, t) = \frac{1}{2} \left(\int_{-\infty}^{x} e^{y-x} m(y, t) \, dy + \int_{x}^{\infty} e^{x-y} m(y, t) \, dy \right). \tag{3-22}$$

Alternatively, and more directly, u can be determined in terms of μ_1 and μ_2 using the equality

$$\frac{\partial x}{\partial t}(y, t) = u(x, t), \tag{3-23}$$

which follows from the definition (3-5) of the function $y = y(x, t)$, in which m satisfies (1-4). In view of (3-20) one has

$$u(x, t) = \left(\frac{\partial}{\partial t} \ln \frac{\mu_1}{\mu_2}(y, t; \tfrac{i}{2}) \right) \Big|_{y=y(x,t)}. \tag{3-24}$$

Equation (3-12) provides also alternative (nonlocal) ways for determining $x = x(y, t)$. Indeed,

$$\frac{\partial x}{\partial y}(y, t) = (m(x(y, t), t) + 1)^{-\frac{1}{2}} = \frac{1}{\mu_1(y, t) \mu_2(y, t)} = \frac{e^{y-x}}{\mu_2^2(y, t)} = \frac{e^{x-y}}{\mu_1^2(y, t)},$$

and the integral formulae for $x = x(y, t)$ emerge.

The discussion above is summarized as follows:

PROPOSITION 1. *The solution $u(x, t)$ of the initial value problem for the Camassa–Holm equation (1-1) with $\omega = 1$, where the initial data $u_0(x)$ is rapidly decreasing as $|x| \to \infty$ and such that $u_{0xx}(x) - u_0(x) + 1 > 0$, can be expressed parametrically, by (3-20), (3-24), in terms of the solution of the Riemann–Hilbert problem (3-15)–(3-18).* □

4. Soliton solutions

Equations (3-20) and (3-24) give a parametric representation for the solution of the initial value problem for the Camassa–Holm equation for general initial data. They have the same structure as the parametric formulae representing pure multisoliton solutions [Matsuno 2005] in terms of two determinants (at the places of μ_1 and μ_2). Therefore, the multisoliton solutions [Matsuno 2005] are "embedded" into our scheme for the solution of the initial value problem: they correspond to reflectionless ($r(k) \equiv 0$) initial data, for which the solution of the RH problem is reduced to solving linear algebraic equations.

Notice also that the formulae by McKean [McKean 2003] for the solution of the Camassa–Holm equation with $\omega = 0$ in terms of the associated theta functions have a similar structure.

Since the algebraic structure of the RH problem is exactly the same as in the case of the KdV equation, its solution (for all k) can be obtained by solving the same linear algebraic equations. Then, comparing to the KdV, the difference in the construction of the solitons is threefold:

(i) the solution of the RH problem is to be evaluated at $k = \frac{i}{2}$ (rather than as $k \to \infty$ for the KdV);

(ii) the phases $yk + 4tk^3$ for $k = k_j$, $j = 1, 2, \ldots, N$ in the case of the KdV equation are to be replaced by $yk - t\frac{2k}{1+4k^2}$ for the Camassa–Holm equation;

(iii) the original scale x is to be related to the y-scale (again by using the solutions of the RH problem evaluated at $k = \frac{i}{2}$).

If $N = 1$, $k_1 = i v_1 \equiv i v$ then the solution of the corresponding RH problem (3-15) normalized by (3-17) and having the trivial jump $J(y, t; \zeta) \equiv I$ is a row-valued rational function with poles at $k = \pm i v$ and thus has the form

$$(\mu_1(y, t; k) \quad \mu_2(y, t; k)) = \left(\frac{k - B(y, t)}{k - i v} \quad \frac{k + B(y, t)}{k + i v} \right). \tag{4-1}$$

Here B can be calculated using the residue conditions (3-18); this gives $B = i v (1 - g)/(1 + g)$ with

$$g(y, t) = \begin{cases} \exp\{-2v(y - 4v^2 t - y_0)\} & \text{for KdV,} \\ \exp\left\{ -2v\left(y - \frac{2}{1 - 4v^2} t - y_0 \right) \right\} & \text{for CH,} \end{cases} \tag{4-2}$$

where y_0 is the phase shift determined by the norming constant $\gamma > 0$ in the residue relation: $y_0 = \frac{1}{2v} \ln \frac{\gamma}{2v}$. Then the 1-soliton solution for the KdV equation is given in terms of $\mu_1^0(y, t)$, where $\mu_1(y, t; k) = 1 + \mu_1^0(y, t)/k + o(1/k)$ as $k \to \infty$, by

$$u_{\mathrm{KdV}}(y, t) = -2i \frac{\partial}{\partial x} \mu_1^0(y, t) = -8v \frac{g_{\mathrm{KdV}}}{(1 + g_{\mathrm{KdV}})^2}(y, t), \tag{4-3}$$

whereas the 1-soliton solution for the CH equation (in the y scale) is given in terms of $\mu_j(y, t; \frac{i}{2})$, $j = 1, 2$ by (in a form comparable with that of (4-3))

$$u_{CH}(y, t) = \frac{\partial}{\partial t} \ln \frac{\mu_1}{\mu_2}(y, t; \frac{i}{2}) = \frac{32v^2}{(1 - 4v^2)^2} \frac{g_{CH}}{(1 + g_{CH})^2 + \frac{16v^2}{1 - 4v^2} g_{CH}}(y, t).$$

(4-4)

The associated relation between the scales (3-20) becomes

$$x(y, t) = y + \ln \frac{1 + g \frac{1 + 2v}{1 - 2v}}{1 + g \frac{1 - 2v}{1 + 2v}}.$$

(4-5)

Introducing

$$v := \frac{2}{1 - 4v^2}, \quad \phi := -2v(y - vt - y_0)$$

allows rewriting (4-4) as

$$u_{CH}(y, t) = \frac{16v^2}{1 - 4v^2} \times \frac{1}{1 + 4v^2 + (1 - 4v^2) \cosh \phi}.$$

(4-6)

Since $0 < v < 1/2$, it follows that the soliton velocity v is greater than 2. Similarly, the velocities of the N solitons appearing asymptotically, as $t \to \infty$, from the N-soliton solution (associated with N residue conditions of type (3-18) at the poles $k = \pm iv_j$, $j = 1, 2, \ldots, N$) are all greater than 2.

5. Long-time asymptotics

The representation of the solution of a nonlinear equation in terms of the solution of the associated Riemann–Hilbert problem has proved to be crucial in studying its long-time behavior using the nonlinear steepest descent method by Deift and Zhou [Deift and Zhou 1993]. The solution of the RH problem with poles (3-18) can be represented as

$$\mu(y, t; k) = \tilde{\mu}(y, t; k) M_r(y, t; k) D(k),$$

where

$$D = \begin{pmatrix} \prod\limits_{j=1}^{N} (k - iv_j)^{-1} & 0 \\ 0 & \prod\limits_{j=1}^{N} (k + iv_j)^{-1} \end{pmatrix},$$

M_r is the solution of the 2×2 regular (i.e., without residue conditions) RH problem with the jump matrix $\tilde{J} = DJD^{-1}$, and $\tilde{\mu}(y, t; k)$ is a row polynomial in k with coefficients determined by the residue conditions.

For the convenience of the reader, we present here a scheme for studying the large-t behavior of solutions of Riemann–Hilbert problems with rapidly oscillating jump data (cf. [Deift et al. 1993]).

PROPOSITION 2. *In the soliton region, $y > (2 + \delta)t$ with any $\delta > 0$, we have*

$$M_r(y, t; k) = I + o(1), \qquad t \to +\infty.$$

PROOF. In the spirit of the nonlinear steepest descent method, the RH problem (3-15) is to be deformed in a way that its jump matrix would approach the identity matrix. In the original setting, J in (3-16) is rapidly oscillating with t, with the exponential factors

$$J(y, t; k) = \begin{pmatrix} 1 & -r(k)e^{-2it\theta} \\ r(k)e^{2it\theta} & 1 - |r(k)|^2 \end{pmatrix},$$

where

$$\theta(y, t; k) = \frac{y}{t}k - \frac{2k}{1 + 4k^2}. \tag{5-1}$$

The deformation of the original contour (real axis) is guided by the "signature table" i.e., the decomposition of the k-plane into domains where Im θ keeps its sign. In the domain $y/t > 2$, the signature table is shown in Figure 2. Therefore, in this case the whole real axis is the boundary of the domains where Im $\theta > 0$ (for Im $k > 0$) and Im $\theta < 0$ (for Im $k < 0$). This suggests using the factorization of the jump matrix

$$J = \begin{pmatrix} 1 & 0 \\ r(k)e^{2it\theta} & 1 \end{pmatrix} \begin{pmatrix} 1 & -r(k)e^{-2it\theta} \\ 0 & 1 \end{pmatrix}, \qquad k \in \mathbb{R}.$$

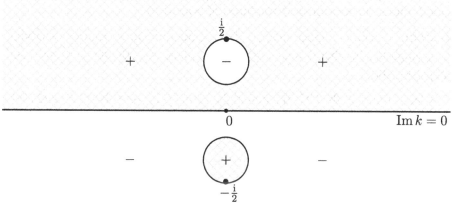

Figure 2. Signature table for Im $\theta(y, t; k)$ when $\dfrac{y}{t} > 2$.

Then appropriate rational approximations of $r(k)$ and $\overline{r(k)}$ are used in order to deform the contour into two lines $\operatorname{Im} k = \pm\varepsilon$ and to absorb the relevant triangular factors (near the real axis) into the new function

$$
\hat{M}_r = \begin{cases}
M_r & \text{if } |\operatorname{Im} k| > \varepsilon, \\[2mm]
M_r \begin{pmatrix} 1 & 0 \\ \dfrac{1}{r(k)} e^{2it\theta} & 1 \end{pmatrix} & \text{if } 0 < \operatorname{Im} k < \varepsilon, \\[4mm]
M_r \begin{pmatrix} 1 & r(k)e^{-2it\theta} \\ 0 & 1 \end{pmatrix} & \text{if } 0 > \operatorname{Im} k > -\varepsilon.
\end{cases}
$$

Now the Riemann–Hilbert problem for $\hat{M}_r$ becomes

$$
\hat{M}_{r-} = \hat{M}_{r+}\hat{J} \quad \text{for } |\operatorname{Im} k| = \varepsilon,
$$

where

$$
\hat{J} = \begin{cases}
\begin{pmatrix} 1 & 0 \\ \dfrac{1}{r(k)} e^{2it\theta} & 1 \end{pmatrix} & \text{if } \operatorname{Im} k = \varepsilon, \\[4mm]
\begin{pmatrix} 1 & -r(k)e^{-2it\theta} \\ 0 & 1 \end{pmatrix} & \text{if } \operatorname{Im} k = -\varepsilon.
\end{cases}
$$

Since the jump matrix $\hat{J}$ approaches I, as $t \to \infty$, exponentially fast, this implies $\hat{M}_r \to I$ and thus $M_r \to I$ for $\{k \mid |\operatorname{Im} k| > \varepsilon\}$. $\qquad\square$

As a consequence of Proposition 2 and the fact that $x - y = o(1)$ as $y \to +\infty$, see (3-5), we have that

$$
u(x,t) = u_{\text{solit}}(x,t) + o(1), \qquad t \to \infty, \tag{5-2}
$$

where $u_{\text{solit}}(x,t)$ is the pure N-soliton solution of the CH equation [Matsuno 2005], which corresponds to the Riemann–Hilbert problem with $r(k) \equiv 0$ and with residue parameters $\{v_j\}_{j=1}^N$ and $\{\gamma_j\}_{j=1}^N$. In turn, $u_{\text{solit}}(x,t)$ develops, for large t, into a superposition of 1-solitons of type (4-6).

REMARK 3. The nonlinear steepest descent method allows rigorous studying the asymptotics in other domains of the (y,t) plane; see [Boutet de Monvel and Shepelsky 2007].

Now let us return to the CH equation in the form (1-1a) (see also (1-3) and (1-4)) depending on the parameter $\omega > 0$. Observe that (1-4) and the x-equation of the Lax pair (1-5a) take their forms corresponding to $\omega = 1$ if we replace

$$
m \mapsto \frac{m}{\omega}, \qquad \lambda \mapsto \lambda\omega. \tag{5-3}
$$

Accordingly, the spectral parameter k is introduced by

$$
k^2 = -\lambda\omega - \frac{1}{4},
$$

(2-5) becomes

$$p(x,t;k) = x - \int_x^\infty \left(\sqrt{\frac{m(\xi,t)}{\omega} + 1} - 1 \right) d\xi - \frac{2\omega}{1+4k^2}t, \qquad (5\text{-}4)$$

and the scale

$$y(x,t) = x - \int_x^\infty \left(\sqrt{\frac{m(\xi,t)}{\omega} + 1} - 1 \right) d\xi \qquad (5\text{-}5)$$

is ω-dependent as well. The coefficient matrices in the Lax pair equations are also to be modified in accordance with (5-3).

6. The limit $\omega \to 0$

Now consider a family of initial value problems (1-1) parametrized by ω, where the initial function $u_0(x)$ in (1-1b) is the same for all the differential equations (1-1a). Then the spectral functions (2-11) as well as the parameters of the discrete spectrum $\{v_j^\omega\}$, $\{\gamma_j^\omega\}$ become ω-dependent.

The results of the Section 5 show that for every fixed ω, the observer will see a train of solitons, parameters of which are determined by $\{v_j^\omega\}_{j=1}^{N_\omega}$ and $\{\gamma_j^\omega\}_{j=1}^{N_\omega}$. Then an interesting question is as follows:

QUESTION. What happens with the solution of the initial value problem as $\omega \to 0$? More precisely, what will we see in the long-time asymptotics?

It has been observed (see, e.g., [Matsuno 2005; Parker 2004]) that if the parameters of an ω-soliton (i.e., a one-soliton solution of (1-1a)) are changing appropriately with ω, then this soliton approaches, as $\omega \to 0$, a peakon, which is a piecewise smooth (weak), stable (cf. [Beals et al. 1999; Constantin and Strauss 2000]) solution of (1-1a) with $\omega = 0$ having a peak at its maximum point:

$$u^0(x,t) = v^0 e^{-|x-v^0 t - x^0|}, \qquad v^0 > 0. \qquad (6\text{-}1)$$

We will show that the solutions of the initial value problem for the CH equations with varying ω but with the same initial data approach, as $\omega \to 0$, a train of peakons with parameters determined by the spectrum of (1-5a) for $\omega = 0$.

It is known (see [Constantin 2001]) that

- for any $\omega > 0$ fixed, the spectrum of (1-5a) consists of

  (i) a continuous part $\lambda \in (-\infty, -\frac{1}{4\omega})$ and

  (ii) a finite set of simple eigenvalues $\{\lambda_j^\omega\}_{j=1}^{N_\omega}$.

- For $\omega = 0$, the spectrum is discrete and consists of

▷ simple eigenvalues $\{\lambda_j^0\}_{j=1}^\infty$ which accumulate at $-\infty$ only: if they are enumerated in decreasing order $0 > \lambda_1^0 > \lambda_2^0 > \cdots$, then $\lambda_N^0 \to -\infty$ as $N \to \infty$.

Let $\Psi_\pm^\omega(x, \lambda_j^\omega)$, $\omega \geq 0$ be the eigenfunctions of (1-5a) associated with the eigenvalues λ_j^ω, and let $\hat{\chi}_j^\omega$ be the corresponding norming constants: $\Psi_\pm^\omega(x, \lambda_j^\omega)$ are normalized by the limit conditions

(a) $\Psi_\pm^\omega(x, \lambda_j^\omega) \to e^{\mp \nu_j^\omega x}$ as $x \to \pm\infty$ with $\nu_j^\omega = \sqrt{\omega \lambda_j^\omega + \frac{1}{4}}$ for $\omega > 0$,

(b) $\Psi_\pm^0(x, \lambda_j^0) \sim e^{\mp x/2}$ as $x \to \pm\infty$ for $\omega = 0$.

Then $\Psi_-^\omega(x, \lambda_j^\omega) = \hat{\chi}_j^\omega \Psi_+^\omega(x, \lambda_j^\omega)$. Passing from (1-5a) to the spectral problem for the operators $K^*(m + \omega)Kf = \frac{1}{\lambda}f$ with $(Kg)(x) = \int_x^\infty e^{(x-y)/2}g(y)\,dy$ [Constantin and McKean 1999; McKean 2003] and considering ω as the perturbation parameter, the following properties are not hard to obtain.

PROPOSITION 3. *As $\omega \to 0$, we have that $N_\omega \to \infty$, $\lambda_j^\omega \to \lambda_j^0$, and $\Psi_\pm^\omega(x, \lambda_j^\omega) \to \Psi_\pm^\omega(x, \lambda_j^0)$ in $L^2(-\infty, \infty)$, $j = 1, 2, \ldots$ in the sense that as $\omega \to 0$, new ω-eigenvalues are "escaping", one by one, from the continuous spectrum whereas the already existing ω-eigenvalues and the associated ω-eigenfunctions are approaching respectively the corresponding eigenvalues and eigenfunctions of (1-5a) with $\omega = 0$.* □

PROPOSITION 4. *Let $u_0(x)$ be a smooth, rapidly decreasing, as $|x| \to \infty$, function such that $m_0(x) := u_{0xx}(x) - u_0(x) > 0$ for all $x \in \mathbb{R}$. Let $\{\nu_j^\omega\}_{j=1}^{N_\omega}$, $\{\gamma_j^\omega\}_{j=1}^{N_\omega}$ be the residue parameters associated with $u_0(x)$ viewed as fixed initial data in the initial value problems (1-1) parametrized by $\omega > 0$.*

Then, as $\omega \to 0$, we have the following asymptotic behavior of these parameters:

$$\nu_j^\omega = \frac{1}{2} + \omega \lambda_j^0 + o(\omega), \tag{6-2}$$

$$\gamma_j^\omega = \omega \Gamma_j^0 + o(\omega), \tag{6-3}$$

where $\Gamma_j^0 = \dfrac{1}{\int_{-\infty}^\infty m_0(x)|\Psi_+^0(x, \lambda_j^0)|^2 dx}$.

PROOF. Taking into account the relation between ν_j^ω and λ_j^ω, (6-2) follows immediately from Proposition 3.

Differentiating (1-5a) with respect to k (the derivative with respect to k is denoted by dot) and combining the resulting equation with (1-5a), after some manipulations similar to those for the Sturm-Liouville equation [Marchenko 1986] leading to the expression relating the derivative of the spectral function $a(k)$ at the spectrum points with the norm of the corresponding eigenfunctions (the details for the CH equation are given in [Constantin 2001]) we arrive at the following expression

$$\frac{1}{\omega}\int_{-\infty}^{\infty}(m^0(x)+\omega)|\Psi_-^\omega(x,\lambda_j^\omega)|^2dx = \frac{\hat\chi_j^\omega}{2i\nu_j^\omega}\frac{\partial}{\partial k}W\{\Psi_+^\omega,\Psi_-^\omega\}\big|_{k=i\nu_j^\omega}, \quad (6\text{-}4)$$

where W is the Wronskian bilinear form $W\{f,g\}:=f'g-fg'$. Comparing the asymptotics of $\Psi_\pm^\omega(x,\lambda)$ with those of $\Phi_\pm(x,0;k)$ and taking into account that $x-y\to 0$ as $x\to+\infty$ and $x-y\to\exp\left\{\frac{1}{\omega}\int_{-\infty}^{\infty}\left(\sqrt{m+\omega}-\omega\right)dx\right\}=:\mathscr{E}(\omega)$ as $x\to-\infty$ we have that

$$a^\omega(k)=W\{\Psi_+^\omega,\Psi_-^\omega\}\cdot\mathscr{E}, \qquad \chi_j^\omega=\hat\chi_j^\omega\cdot\mathscr{E},$$

where χ_j^ω are the norming constants in (2-12). Therefore, (6-4) can be written as

$$\frac{1}{\omega}\int_{-\infty}^{\infty}(m^0(x)+\omega)|\Psi_-^\omega(x,\lambda_j^\omega)|^2dx$$

$$=\frac{1}{\mathscr{E}}\frac{\chi_j^\omega}{2i\nu_j^\omega}\frac{1}{\mathscr{E}}(-2\nu_j^\omega)\dot a^\omega(i\nu_j^\omega)=\frac{i\chi_j^\omega}{\mathscr{E}^2}\dot a^\omega(i\nu_j^\omega). \quad (6\text{-}5)$$

Now recall that γ_j^ω in the residue relations (3-18) are related to χ_j^ω by $i\gamma_j^\omega=\chi_j^\omega/\dot a^\omega(i\nu_j^\omega)$. Hence, (6-5) gives for γ_j^ω the expression

$$\gamma_j^\omega=\frac{-i\chi_j^\omega}{\dot a^\omega(i\nu_j^\omega)}=\omega\frac{(\hat\chi_j^\omega)^2}{\int_{-\infty}^{\infty}(m^0(x)+\omega)|\Psi_-^\omega(x,\lambda_j^\omega)|^2dx}$$

$$=\frac{\omega}{\int_{-\infty}^{\infty}(m^0(x)+\omega)|\Psi_+^\omega(x,\lambda_j^\omega)|^2dx} \qquad (6\text{-}6)$$

and, by Proposition 3, (6-3) follows.   □

Now, as we have established the behavior of the soliton parameters as $\omega\to 0$, we are able to study the limiting behavior of ω-solitons in the original scale; since, as $t\to+\infty$, a multisoliton solution behaves as a superposition of one-soliton solutions [Matsuno 2005], it is enough to see what happens with a one-soliton solution.

An ω-soliton (with parameters ν^ω and γ^ω) is given parametrically by equations of type (4-5), (4-6) appropriately modified in order to take into account the dependence on ω:

$$u(y,t)=\frac{16\omega(\nu^\omega)^2}{1-4(\nu^\omega)^2}\times\frac{1}{1+4(\nu^\omega)^2+(1-4(\nu^\omega)^2)\cosh\phi(y,t)}, \quad (6\text{-}7)$$

where

$$\phi(y,t) = -2v^\omega \left(y - \frac{2\omega}{1-4(v^\omega)^2} t - \frac{1}{2v^\omega} \ln \frac{\gamma^\omega}{2v^\omega} \right), \tag{6-8}$$

and

$$x(y,t) = y + \ln \frac{1 + g(y,t)\dfrac{1+2v^\omega}{1-2v^\omega}}{1 + g(y,t)\dfrac{1-2v^\omega}{1+2v^\omega}} \tag{6-9}$$

with $g = e^\phi$.

Rewrite (6-9) as

$$x(y,t) = y + \ln \frac{1+2v^\omega}{1-2v^\omega} + \ln \frac{1-2v^\omega + (1+2v^\omega)g}{1+2v^\omega + (1-2v^\omega)g} \tag{6-10}$$

and introduce the new (moving) variables

$$X = x - v^\omega t - x_0^\omega, \qquad Y = y - v^\omega t - y_0^\omega, \tag{6-11}$$

where

$$v^\omega = \frac{2\omega}{1-4(v^\omega)^2}, \quad x_0^\omega = \frac{1}{2v^\omega} \ln \frac{\gamma^\omega}{2v^\omega} + \ln \frac{1+2v^\omega}{1-2v^\omega}, \quad y_0^\omega = \frac{1}{2v^\omega} \ln \frac{\gamma^\omega}{2v^\omega}. \tag{6-12}$$

Then the one-soliton is given parametrically by

$$u(x,t) = U(Y(X))\big|_{X=x-v^\omega t-x_0^\omega} \equiv U^\omega(X)\big|_{X=x-v^\omega t-x_0^\omega}, \tag{6-13}$$

where

$$U(Y) = \frac{16\omega(v^\omega)^2}{1-4(v^\omega)^2} \times \frac{1}{1+4(v^\omega)^2 + (1-4(v^\omega)^2)\cosh(2v^\omega Y)} \tag{6-14}$$

and $Y(X)$ is inverse to

$$X(Y) = Y + \ln \frac{1-2v^\omega + (1+2v^\omega)e^{-2v^\omega Y}}{1+2v^\omega + (1-2v^\omega)e^{-2v^\omega Y}}. \tag{6-15}$$

Applying Proposition 4 to (6-12)–(6-15), we see that as $\omega \to 0$,

(i) The soliton velocity v^ω approaches the finite limit associated with the corresponding eigenvalue of (1-5a) with $\omega = 0$:

$$v^\omega \to -\frac{1}{2\lambda^0}.$$

(ii) The phase shift x_0^ω also approaches a finite value:

$$x_0^\omega \sim \ln \omega + \ln \Gamma^0 + \ln \frac{2}{-2\omega\lambda^0} \to \ln \frac{\Gamma^0}{-\lambda^0}.$$

(iii) For all $X > \omega^{\alpha} > 0$ with $\alpha \in (0, 1)$, $Y(X) \to +\infty$. Moreover,

$$X(Y) \sim Y + \ln \frac{2 - 2\omega\lambda^0 e^Y}{2e^Y} = \ln\left(1 - \omega\lambda^0 e^Y\right)$$

and thus

$$U(Y) \sim -\frac{1}{2\lambda^0} \frac{1}{1 - \omega\lambda^0 e^Y} \sim -\frac{1}{2\lambda^0} e^{-X}.$$

Since $X(-Y) = -X(Y)$ in (6-15) and $U(-Y) = U(Y)$ in (6-14), it finally follows that

$$U(Y(X)) \sim -\frac{1}{2\lambda^0} e^{-|X|}, \qquad |X| > \omega^{\alpha}. \qquad (6\text{-}16)$$

The right-hand side of (6-16) is nothing but the peakon solution (6-1) of (1-5a) with $\omega = 0$ having the velocity (= amplitude) $v^0 = -1/(2\lambda^0)$ associated with the corresponding eigenvalue λ^0. Now taking into account the phase shift when passing from a multisoliton solution to a superposition of one-solitons [Matsuno 2005], we arrive at the following proposition (see Figure 3).

PROPOSITION 5. *Let $u_0(x)$ be a smooth function, fast decreasing as $|x| \to \infty$, and such that $m_0(x) := u_{0xx}(x) - u_0(x) > 0$ for all $x \in \mathbb{R}$.*

▷ *Let $\{\lambda_j^0\}_{j=1}^{\infty}$ be the eigenvalues of the spectral problem (1-5a) with $\omega = 0$ and $m = m_0$.*

▷ *For $\omega > 0$, let $u^{\omega}(x,t)$ be the solution of the initial value problem for the Camassa–Holm equation (1-1).*

▷ *Fix $C > 0$, $\delta > 0$, and $\varepsilon > 0$.*

▷ *Let $\{\lambda_j^0\}_{j=1}^{N(C)}$ be those λ_j^0 satisfying $0 > \lambda_1^0 > \cdots > \lambda_{N(C)}^0 > -\frac{1}{2C}$.*

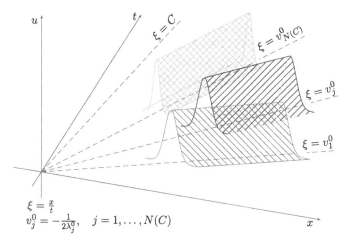

Figure 3. Long-time asymptotics of $u = u^{\omega}(x,t)$ for small ω.

Then there exists $\tilde{\omega} = \tilde{\omega}(C, \delta, \varepsilon)$ such that for all $0 < \omega < \tilde{\omega}$ the asymptotics of $u^{\omega}(x, t)$ in the domain $x > Ct$ is given by $N(C)$ one-solitons of type (6-13)–(6-15) with velocities and forms close to those of the corresponding peakons:

$$u^{\omega}(x, t) = U_j^{\omega}(X) + o(1) \text{ as } t \to \infty, \quad |X| = O(1) \text{ with } X = x - v_j^{\omega} t - x_{0j}^{\omega},$$

where

- o(1) *depends on* ω,
- $|v_j^{\omega} - v_j^0| < \varepsilon$ *with* $v_j^0 = -\dfrac{1}{2\lambda_j^0}$,
- $|x_{0j}^{\omega} - x_{0j}^0| < \varepsilon$ *with*

$$x_{0j}^0 = \ln\left(\frac{\Gamma_j^0}{-\lambda_j^0} \prod_{l=1}^{j-1} \left(\frac{\lambda_j^0}{\lambda_l^0} - 1 \right)^2 \right)$$

and $\left| U_j^{\omega}(X) - v_j^0 e^{-|X|} \right| < \varepsilon$ *for* $|X| > \delta$.

References

[Beals et al. 1988] R. Beals, P. Deift, and C. Tomei, *Direct and inverse scattering on the line*, Mathematical Surveys and Monographs **28**, American Mathematical Society, Providence, RI, 1988.

[Beals et al. 1999] R. Beals, D. H. Sattinger, and J. Szmigielski, "Multi-peakons and a theorem of Stieltjes", *Inverse Problems* **15**:1 (1999), L1–L4.

[Boutet de Monvel and Shepelsky 2006a] A. Boutet de Monvel and D. Shepelsky, "Initial boundary value problem for the Camassa–Holm equation on the half-line", Preprint, 2006.

[Boutet de Monvel and Shepelsky 2006b] A. Boutet de Monvel and D. Shepelsky, "Riemann–Hilbert approach for the Camassa–Holm equation on the line", *C. R. Math. Acad. Sci. Paris* **343**:10 (2006), 627–632.

[Boutet de Monvel and Shepelsky 2007] A. Boutet de Monvel and D. Shepelsky, "Long-time asymptotics of the Camassa–Holm equation on the line", Preprint, 2007. To appear in *Integrable systems, random matrices, and applications: A conference in honor of Percy Deift's 60th birthday*.

[Camassa and Holm 1993] R. Camassa and D. D. Holm, "An integrable shallow water equation with peaked solitons", *Phys. Rev. Lett.* **71**:11 (1993), 1661–1664.

[Constantin 2001] A. Constantin, "On the scattering problem for the Camassa–Holm equation", *R. Soc. Lond. Proc. Ser. A Math. Phys. Eng. Sci.* **457**:2008 (2001), 953–970.

[Constantin and Lenells 2003] A. Constantin and J. Lenells, "On the inverse scattering approach to the Camassa–Holm equation", *J. Nonlinear Math. Phys.* **10**:3 (2003), 252–255.

[Constantin and McKean 1999] A. Constantin and H. P. McKean, "A shallow water equation on the circle", *Comm. Pure Appl. Math.* **52**:8 (1999), 949–982.

[Constantin and Strauss 2000] A. Constantin and W. A. Strauss, "Stability of peakons", *Comm. Pure Appl. Math.* **53**:5 (2000), 603–610.

[Constantin et al. 2006] A. Constantin, V. S. Gerdjikov, and R. I. Ivanov, "Inverse scattering transform for the Camassa–Holm equation", *Inverse Problems* **22**:6 (2006), 2197–2207.

[Deift and Zhou 1993] P. Deift and X. Zhou, "A steepest descent method for oscillatory Riemann–Hilbert problems: Asymptotics for the MKdV equation", *Ann. of Math. (2)* **137**:2 (1993), 295–368.

[Deift et al. 1993] P. A. Deift, A. R. Its, and X. Zhou, "Long-time asymptotics for integrable nonlinear wave equations", pp. 181–204 in *Important developments in soliton theory*, Springer, Berlin, 1993.

[Johnson 2003] R. S. Johnson, "On solutions of the Camassa–Holm equation", *R. Soc. Lond. Proc. Ser. A Math. Phys. Eng. Sci.* **459**:2035 (2003), 1687–1708.

[Lenells 2002] J. Lenells, "The scattering approach for the Camassa–Holm equation", *J. Nonlinear Math. Phys.* **9**:4 (2002), 389–393.

[Marchenko 1986] V. A. Marchenko, *Sturm–Liouville operators and applications*, Operator Theory: Advances and Applications **22**, Birkhäuser, Basel, 1986.

[Matsuno 2005] Y. Matsuno, "Parametric representation for the multisoliton solution of the Camassa–Holm equation", *J. Phys. Soc. Japan* **74**:7 (2005), 1983–1987.

[McKean 2003] H. P. McKean, "Fredholm determinants and the Camassa–Holm hierarchy", *Comm. Pure Appl. Math.* **56**:5 (2003), 638–680.

[Parker 2004] A. Parker, "On the Camassa–Holm equation and a direct method of solution, I: Bilinear form and solitary waves", *Proc. R. Soc. Lond. Ser. A Math. Phys. Eng. Sci.* **460**:2050 (2004), 2929–2957.

ANNE BOUTET DE MONVEL
INSTITUT DE MATHÉMATIQUES DE JUSSIEU
CASE 7012
UNIVERSITÉ PARIS 7
2 PLACE JUSSIEU
75251 PARIS
FRANCE
    aboutet@math.jussieu.fr

DMITRY SHEPELSKY
MATHEMATICS DIVISION
INSTITUTE B. VERKIN
47 LENIN AVENUE
61103 KHARKIV
UKRAINE
    shepelsky@yahoo.com

Probability, Geometry and Integrable Systems
MSRI Publications
Volume **55**, 2008

The Riccati map in random Schrödinger and random matrix theory

SANTIAGO CAMBRONERO, JOSÉ RAMÍREZ, AND BRIAN RIDER

For H. P. McKean, who taught us this trick.

ABSTRACT. We discuss the relevance of the classical Riccati substitution to the spectral edge statistics in some fundamental models of one-dimensional random Schrödinger and random matrix theory.

1. Introduction

The Riccati map amounts to the observation that the Schrödinger eigenvalue problem $Q\psi = \lambda\psi$ for $Q = -d^2/dx^2 + q(x)$ is transformed into the first order relation

$$q(x) = \lambda + p'(x) + p^2(x) \tag{1-1}$$

upon setting $p(x) = \psi'(x)/\psi(x)$. That this simple fact has deep consequences for the problem of characterizing the spectrum of Q with a random potential q has been known for some time. It also turns out to be important for related efforts in random matrix theory (RMT). We will describe some of the recent progress on both fronts.

Random operators of type Q arise in the description of disordered systems. Their use goes back to Schmidt [1957], Lax and Phillips [1958], and Frisch and Lloyd [1960] in connection with disordered crystals, represented by potentials in the form of trains of signed random masses, randomly placed on the line. Consider instead the case of white noise potential, $q(x) = b'(x)$ with a standard brownian motion $x \mapsto b(x)$, which may be viewed as a simplifying caricature

Rider was supported in part by NSF grant DMS-0505680.

of the above. The problem $Q\psi = \lambda\psi$ then reads $d\psi'(x) = \psi(x)\,db(x) + \lambda\psi(x)\,dx$ and is solvable for $\psi \in C^{3/2}$.

A first order statistic of interest is the integrated density of states $N(\lambda) = \lim_{L\to\infty} L^{-1} \times \{$the number of eigenvalues $\leq \lambda\}$, in which we take Q on the interval $[0, L]$ with say Dirichlet boundary conditions. Build the sine-like solution $\psi_0(x, \lambda)$ of $Q\psi_0 = \lambda\psi_0$ with $\psi_0(0) = 0$ and $\psi_0'(0) = 1$. The pair $x \mapsto (\psi_0(x), \psi_0'(x))$ is clearly Markovian, as is the ratio $x \mapsto p(x) := \psi_0'(x)/\psi_0(x)$. Further, the latter solves a version of (1-1) which can only be interpreted as to say that p performs the diffusion with infinitesimal generator

$$\mathfrak{G} = (1/2)\partial^2/\partial p^2 - (\lambda + p^2)\partial/\partial p. \tag{1-2}$$

This motion begins at $p(0) = +\infty$, which is an entrance barrier, hits the exit barrier $-\infty$ at the first root $\mathfrak{m}_1$ of $\psi_0(x, \lambda) = 0$, then reappears at $+\infty$ whereupon everything starts afresh.

Now, to count the eigenvalues below a level λ is to count the number of roots of $\psi_0(x, \lambda)$ before $x = L$, and so the number of (independent) passages from $+\infty$ to $-\infty$ of the p motion. If this number is n, then L approximates $\mathfrak{s}_n = \mathfrak{m}_1 + \cdots + \mathfrak{m}_n$, the sum of the first n passage times, so that, by the law of large numbers

$$\frac{1}{N(\lambda)} = \lim_{n\to\infty} \mathfrak{s}_n/n = E[\mathfrak{m}_1] = \sqrt{2\pi} \int_0^\infty e^{-(p^3/6+\lambda p)} \frac{dp}{\sqrt{p}},$$

as may be worked out from the speed and scale associated with (1-2). This computation is due to Halperin [1965]; see also [Fukushima and Nakao 1976/77].

As for the fluctuations, McKean [1994] proved, via Riccati, that

$$\lim_{L\to\infty} P\left(\frac{L}{\pi}(-\Lambda_0(L))^{1/2} \exp\left[-\tfrac{8}{3}(-\Lambda_0(L))^{3/2}\right] > x\right) = \begin{cases} 1 & \text{for } x < 0, \\ e^{-x} & \text{for } x \geq 0, \end{cases} \tag{1-3}$$

where $\Lambda_0(L)$ pertains to the operator $-\frac{d^2}{dx^2} + b'(x)$ acting on $[0, L]$ with Dirichlet, Neumann, or periodic conditions. While a step forward, (1-3) is still thermodynamic in nature. More desirable is to use the Riccati trick to capture local spectral statistics in a fixed volume, and this is where the main part of our story begins.

Cambronero and McKean [1999] took the point of view that the Riccati map (1-1) represents a change of measure from potential, or q-path, space to the space of p-paths, resulting in an explicit functional integral formula for the probability density of Λ_0 under periodic conditions (Hill's equation). The method extends from white noise q, to any periodic diffusion potential of brownian motion type plus restoring drift. Section 2 describes all this. Given such integral expressions, the next natural task is to describe the shape of the ground state eigenvalue density. A summary of the results thus far makes up Section 3, with an emphasis on the differences between the white noise case, and the roughly universal nature of the shape for nice Gaussian potentials. Section 4 is devoted to the surprising recent discovery that a 1-d random Schrödinger operator and thus, via Riccati, the explosion probability of a certain diffusion figure into the celebrated Tracy–Widom laws of RMT along with their generalizations. We finish up with a collection of open questions.

Further background. As indicated, the Riccati substitution is a basic tool in the study of 1-d random Schrödinger, as may be gleaned from the comprehensive book [Carmona and Lacroix 1990]. Indeed, (1-3) is only one instance of a ground state limit theorem. For a large class of Markovian potentials it is understood that the spectrum is Poissonian and that the large volume limit of the edge eigenvalues follow standard i.i.d. extremal laws; see [Molčanov 1980/81] or [Grenkova et al. 1983]. The second reference also shows that the limit can be joint Gaussian (and so exhibit repulsion) when the Lyapunov exponent is degenerate at the spectral edge. In all these results the normalization depends on the smoothness of the potential, and this is one reason that (1-3) deserves to be set apart. Additionally, our shape results for the ground state density (Section 3) should be compared with the large body of work on the Lifschitz tails dating back to the 70's. Ideas connected to that work can in fact be used to obtain tail estimates on the *distribution function* in the case of continuous Gaussian potentials in a finite volume, including even multiple dimensions (exactly such bounds turn up in recent work on the parabolic Anderson model [Gärtner et al. 2000]). Finally, there is an extensive literature on the almost sure behavior of Λ_0 in the more physical $d > 1$ setting with Poisson-bump or Gibbsian type potentials; see [Merkl 2003; Sznitman 1998] and the many references therein. Our point here though is to focus on the ground state *density* and the approach inspired by McKean.

2. The Riccati map as a change of measure

Let $Q\psi = -\psi'' + q\psi = \lambda\psi$ be Hill's equation with standard white noise potential $q(x)$ on the circle $0 \leq x < 1 = S^1$. Bring in the sine and cosine-like solutions $\psi_0(x, \lambda)$ and $\psi_1(x, \lambda)$ satisfying $\psi_1(0) = 0, \psi_0'(0) = 1, \psi_1(0) = 1, \psi_1'(0) = 0$, and also the discriminant $\Delta(\lambda) = \frac{1}{2}[\psi_0(1, \lambda) + \psi_1'(1, \lambda)]$. The latter is an entire function of order $1/2$ and encodes the spectrum: $\Delta = \pm 1$ at the periodic/antiperiodic eigenvalues. In particular, if $\Lambda_0 = \Lambda_0(q)$ is the ground state eigenvalue for Q, $\Delta(\lambda)$ decreases from the left to its value $\Delta = 1$ at $\lambda = \Lambda_0$. Moreover, $Q\psi = \lambda\psi$ has a solution with multiplier m (a solution for which $\psi(x + 1) = m\psi(x)$), if and only if $m = \Delta(\lambda) \pm \sqrt{\Delta^2(\lambda) - 1}$. There is a positive solution of this type with $0 < m < \infty$ only when $\lambda \leq \Lambda_0$, in which case there are actually two such solutions with multipliers m_+ and $m_- = 1/m_+$; these fall together ($m_+ = m_- = 1$) at the periodic ground state when $\lambda = \Lambda_0$.

The corresponding Riccati equation,

$$q(x) = \lambda + p'(x) + p^2, \tag{2-1}$$

determines p as a diffusion on S^1 solving the stochastic differential equation

$$dp(x) = db(x) - (\lambda + p^2(x)) \, dx,$$

provided that $\Lambda_0(q) \geq \lambda$. In fact, if such a solution p exists and ϕ is a smooth periodic function with $\int_0^1 \phi^2(x) \, dx = 1$, then

$$\int_0^1 \left((\phi'(x))^2 + q(x)\phi^2(x) \right) dx \geq \lambda,$$

and therefore $\Lambda_0 \geq \lambda$. Conversely, if $\Lambda_0 \geq \lambda$ we have just explained that there is a positive solution $\psi(x)$ of $Q\psi = \lambda\psi$ with multiplier: $\psi(x+1) = m\psi(x)$ and $m \geq 1$. It follows that $p = \psi'/\psi$ solves (2-1) and satisfies the side condition $\int_0^1 p(x) \, dx = \log m \geq 0$.

This defines the Riccati map. In the $p \to q$ direction, it is one-to-one on $H = \left[\int_0^1 p = 0 \right]$, and also on $H^+ = \left[\int_0^1 p \geq 0 \right]$. The set H^+ is mapped onto $[\Lambda_0(q) \geq \lambda]$, while the mean-zero condition in p-space H coincides with $m = 1$ and so the event $[\Lambda_0(q) = \lambda]$.

Distribution of the ground state eigenvalue. Cambronero and McKean [1999] used the map above between $[\Lambda_0 \geq \lambda]$ and $\left[\int_0^1 p \geq 0 \right]$ to express the white noise measure of the former in terms of a circular brownian motion (CBM) integral over the latter. The CBM is formed by the standard brownian motion loop space with $p(0) = p(1)$, which is then distributed according to $P(p(0) \in da) = (1/\sqrt{2\pi}) \, da$. The result is,

$$Q_*[\Lambda_0(q) \geq \lambda] = \sqrt{\frac{2}{\pi}} \int_{H^+} e^{-\frac{1}{2} \int_0^1 (\lambda + p^2(x))^2 \, dx} \sinh\left(\int_0^1 p \right) dP_*(p), \tag{2-2}$$

where Q_* and P_* henceforth denote the white noise and CBM measures. By a more elaborate computation, considering the Riccati map on the product space of the potential and logarithmic multiplier $\log m$, [Cambronero and McKean 1999] also establishes a formula for the probability density $f(\lambda) = \frac{d}{d\lambda} Q_*[\Lambda_0 \leq \lambda]$. In particular,

$$f(\lambda) = \frac{1}{\sqrt{2\pi}} \int_H e^{-\frac{1}{2} \int_0^1 (\lambda + p^2(x))^2 \, dx} A(p) \, dP_0(p), \tag{2-3}$$

where $A(p) = \int_0^1 e^{2\int_0^x p} \times \int_0^1 e^{-2\int_0^x p}$ and P_0 is the CBM conditioned so that $\int_0^1 p = 0$. Unlike CBM which has infinite total mass, P_0 is a proper Gaussian probability measure on paths.

REMARK. The distribution (2-2) may be differentiated to produce the density in the form

$$f(\lambda) = \sqrt{\frac{2}{\pi}} \int_{H^+} \left(\lambda + \int_0^1 p \right) e^{-\frac{1}{2}\int_0^1 (\lambda + p^2(x))^2 \, dx} \sinh \left(\int_0^1 p \right) dP_*(p),$$

equating an integral over the half-space H^+ to an integral over its boundary H. One might suppose that the present is related to (2-3) by the appropriate function-space divergence theorem, and this in fact is verified in [Cambronero and McKean 1999].

Formally, the Riccati map relates the white noise measure to CBM via

$$dQ_* = \exp\left(-\frac{1}{2} \int_0^1 q^2 \right) \frac{d^\infty q}{(2\pi/0+)^{\infty/2}} = \exp\left(-\frac{1}{2} \int_0^1 (\lambda + p^2)^2 \right) |J| \, dP_*,$$

where

$$dP_* = \exp\left(-\frac{1}{2} \int_0^1 |p'|^2 \right) \frac{d^\infty p}{(2\pi 0+)^{\infty/2}}$$

is the CBM in symbols, and the Jacobian J is to be determined. One may be tempted to employ the Cameron–Martin formula and claim that

$$dQ_* = \exp\left(-\frac{1}{2} \int_0^1 (\lambda + p^2)^2 \right) \exp\left(\int_0^1 p \right) dP_*,$$

that is, $|J| = \exp(\int_0^1 p)$. But this does not apply here, the equation (2-1) being understood with periodic, and not initial, conditions.

The next section contains a sketch of the proper Jacobian calculation and so the verification of (2-2). This is followed by (the outline of) two proofs of the density formula (2-3). Last, it is explained how both types of expressions may be extended to a class of periodic diffusion potentials.

Jacobian of the Riccati map and distribution of Λ_0. The needed Jacobian is obtained by passing through the finite–dimensional distributions of Q_* and P_*. These spaces are furnished with a discrete version of the transformation (2-1) for which we can compute $|J|$ by hand. Afterward, limits may be performed to pin down the "infinite dimensional" Jacobian.

The appropriate discrete version of Riccati's transformation reads

$$q_i = \lambda + n^2(e^{hp_{i+1}} - 2 + e^{-hp_i}), \quad i = 0, \dots, n-1, \tag{2-4}$$

carrying $\mathbb{R}^n$ to $\mathbb{R}^n$, where $h = \frac{1}{n}$, and $q_n = q_0$ and $p_n = p_0$. Notice that, for hp_i small,

$$q_i \simeq \lambda + n(p_{i+1} - p_i) + \tfrac{1}{2}(p_{i+1}^2 + p_i^2)$$

provides an approximation to (2-1). Also, one easily computes that

$$|J| = \frac{2}{h^n}\left| \sinh\left(\sum_{i=0}^{n-1} p_i h \right) \right|$$

for the map (2-4). This expression vanishes only when $\left[\sum p_i h = 0 \right]$, and this discrete form of Riccati is actually one-to-one on the region $\left[\sum p_i h > 0 \right]$ onto $[\lambda^{(n)} \geq \lambda]$, $\lambda^{(n)}$ being the ground state of the discrete version of Hill's equation with potential vector $(q_0, \ldots, q_{n-1})$.

Next, bring in the discrete white noise

$$\bar{q}_i = n \int_{\frac{i}{n}}^{\frac{i+1}{n}} q = n(b_{i+1} - b_i),$$

$$\text{with } b_i = b\left(\frac{i}{n}\right) \text{ and a standard brownian motion } b(\cdot). \quad (2\text{-}5)$$

Assuming that $\Lambda_0(q) > \lambda$, it holds that $\lambda^{(n)}(q) > \lambda$ for all large values of n. Also, denoting by $\overline{p_0 \cdots p_{n-1}}$ the polygonal path determine by the points $p_0, \ldots, p_n$, and similarly for q, it may be checked that:

LEMMA 2.1. *For almost every white noise path q, with $\Lambda_0(q) > \lambda$, $\overline{p_0 \cdots p_{n-1}}$ converges uniformly to the solution $p(x) = \psi'(x, \lambda)/\psi(x, \lambda)$ of (2-1).*

As a consequence, if H_N denotes the set of white noise paths q for which $\lambda^{(n)}(q) > \lambda$ for all $n \geq N$, and $\max |p_i| \leq N$ for all $n \geq N$, then $Q_*(H_N) \to 1$, as $N \to \infty$. This allows one to further restrict the discrete transform to

$$D_N = H_N \cap \{q : \max_{i=0,\ldots,n-1} |b_{i+1} - b_i| \leq 2\sqrt{h \log n} \text{ for all } n \geq N\},$$

where the convergence may be controlled. (By Levy's modulus of continuity $Q_*(D_N)$ tends to 1, so this is enough.) Now, on D_N and taking $\lambda = 0$ for convenience, one has

$$-\frac{1}{2}\sum_{i=0}^{n-1} q_i^2 h = -\frac{1}{2h}\sum_{i=0}^{n-1}(p_{i+1} - p_i)^2 - \frac{1}{8}\sum_{i=0}^{n-1}(p_{i+1}^2 + p_i^2)^2 h + R_n,$$

with a remainder $R_n \to 0$ boundedly. The discrete white noise measure

$$\exp\left(-\frac{1}{2}\sum q_i^2 h\right) \frac{dq_0 \ldots dq_{n-1}}{(2\pi/h)^{n/2}}$$

may then be written as

$$\sqrt{\frac{2}{\pi}} \exp\left(-\frac{1}{8}\sum_{i=0}^{n-1}(p_{i+1}^2 + p_i^2)^2 h + R_n\right) \left|\sinh \sum_{i=0}^{n-1} p_i h\right| d\mu_n,$$

where

$$d\mu_n = \sqrt{2\pi} \exp\left(-\frac{1}{2h}\sum_{i=0}^{n-1}(p_{i+1} - p_i)^2\right)\frac{dp_0 \ldots dp_{n-1}}{(2\pi h)^{n/2}}.$$

Thus, for a bounded continuous function ϕ of the path q vanishing off D_N, it holds that

$$\int_{[\lambda_0 \geq 0]} \phi(q) \, dQ_*$$
$$= \lim_{n\to\infty} \int_{\mathbb{R}^n} \phi_n(q_0, \ldots, q_{n-1}) \exp\left(-\frac{1}{2}\sum_{i=0}^{n-1} q_i^2 h\right)\frac{dq_0 \ldots dq_{n-1}}{(2\pi/h)^{n/2}}$$
$$= \lim_{n\to\infty} \sqrt{\frac{2}{\pi}} \int_{\mathbb{R}^n} \hat{\phi}_n(p_0, \ldots, p_{n-1}) \, d\nu_n,$$

in which

$$d\nu_n = \exp\left(-\frac{1}{8}\sum_{i=0}^{n-1}(p_{i+1}^2 + p_i^2)^2 h + R_n\right) \sinh\left(\sum_{i=0}^{n-1} p_i h\right) d\mu_n.$$

ϕ_n denotes ϕ evaluated on the discrete q-path, and $\hat{\phi}_n(p) := \phi_n(q)$. Then, by dominated convergence we have the identity

$$\int_{[\Lambda_0 \geq 0]} \phi(q) \, dQ_* = \sqrt{\frac{2}{\pi}} \int_{H^+} \hat{\phi}(p) \, \exp\left(-\frac{1}{2}\int_0^1 p^4\right) \sinh\left(\int_0^1 p\right) dP_*,$$

where $\hat{\phi}(p)$ is defined through the Riccati correspondence; it is sensible along with $\phi(q)$. A standard argument will extend the picture to any bounded continuous ϕ and also to $\lambda \neq 0$. To summarize:

THEOREM 2.2. *If Q_* is the restriction of the white noise measure to the region $[\lambda_0(q) \geq \lambda]$, and if P_* is the restriction of circular brownian motion measure to H^+, then*

$$dQ_* = \sqrt{\frac{2}{\pi}} \exp\left(-\frac{1}{2}\int_0^1 (\lambda + p^2)^2\right) \sinh\left(\int_0^1 p\right) dP_*.$$

The formula (2-2) for the distribution of $\Lambda_0(q)$ follows immediately.

REMARK. As an entertaining aside one learns that

$$\lim_{\lambda\to-\infty}\int_{H+}\exp\left(-\frac{1}{2}\int_0^1(\lambda+p^2)^2\right)\sinh\left(\int_0^1 p\right)dP_* = \sqrt{\frac{\pi}{2}},$$

which is not at all obvious.

The measure induced by Q_* on $[\Lambda_0 = \lambda]$ and the density formula. Here is a way to understand (2-3) not reported in [Cambronero and McKean 1999]. To start, define Q_λ by

$$\int_{[\Lambda_0=\lambda]}\phi(q)\,dQ_\lambda = \lim_{h\to 0}\frac{1}{h}\int\phi(q)\chi_{[\lambda\le\Lambda_0\le\lambda+h]}\,dQ_*, \qquad (2\text{-}6)$$

for any bounded continuous ϕ.

Next, being analytic, $\Delta(\lambda)$ is locally bounded in both λ and $|b|$, and the same is true of $\dot\Delta(\lambda) = (d/d\lambda)\,\Delta(\lambda)$ and $\ddot\Delta(\lambda)$. So, $\Delta(\lambda) = 1 + (\Lambda_0 - \lambda)\,|\dot\Delta(\Lambda_0)| + O(h^2)$ with $\lambda \le \Lambda \le \lambda+h$. It follows that

$$m = \Delta + \sqrt{\Delta^2 - 1} = 1 + \sqrt{2\,(\Lambda_0 - \lambda)\,|\dot\Delta(\Lambda_0)|} + O(h),$$

and for $q = \lambda + p' + p^2$, we also conclude

$$\int_0^1 p = \log m = \sqrt{2\,(\Lambda_0 - \lambda)\,|\dot\Delta(\lambda_0)|} + O(h).$$

Coupled with the classical fact that

$$-2\dot\Delta(\Lambda_0) = \int_0^1\psi^2(t)\,dt\int_0^1\frac{dt}{\psi^2(t)}.$$

for ψ the periodic ground state, $2|\dot\Delta(\lambda_0)| = A(p_0)(1 + O(h))$ where $p_0 = p - \int_0^1 p$ and

$$A(p_0) = \int_0^1 e^{-2\int_0^x p_0}dx\int_0^1 e^{2\int_0^x p_0}dx.$$

Now introduce the identity

$$\int_H\phi(p)B^2(p)\,dP_0(p)$$
$$= \lim_{\varepsilon\downarrow 0}\frac{2}{\varepsilon^2}\int\phi(p)\sinh\left(\int_0^1 p\right)1_{[0\le\int_0^1 p\le B(p-\int_0^1 p)\varepsilon]}\,dP_*(p),$$

which is proved directly from the definition of P_0 as the conditional P_*; it holds for bounded continuous ϕ and a large class of $B: H\to\mathbb{R}^+$ including $B(\cdot) = \sqrt{A(\cdot)}$. With that choice, the previous estimates can be used to effectively replace $\{0 \le \int_0^1 p \le \left(A(p-\int p)\right)^{1/2}\varepsilon\}$ with $\{0 \le \Lambda_0 \le \varepsilon^2\}$. If that substitution is made, we understand at once that the measure Q_λ induced by Q_* on $[\Lambda_0 = \lambda]$ satisfies

$$dQ_\lambda = \frac{1}{\sqrt{2\pi}} \exp\left(-\frac{1}{2} \int_0^1 (\lambda + p^2)^2\right) A(p)\, dP_0$$

under the Riccati transformation, and this is equivalent to (2-3).

Joint distribution of $(q, \log m)$ and a second proof. Perhaps a more formulaic route to the density formula is available by way of the joint transformation

$$(q, \log m) \leftrightarrow (p, \lambda).$$

Given (p, λ) with p in the CBM space, we set $\log m = \int_0^1 p$ and $q = \lambda + p' + p^2$. Mapping back, given $(q, \log m)$ with q in the white noise space, we take $\lambda \le \Lambda_0(q)$ so that $\Delta(\lambda, q) = \frac{1}{2}(m + \frac{1}{m})$. This λ is unique since $\dot{\Delta}(\lambda, q) < 0$ for $\lambda < \Lambda_0(q)$. One may then choose ψ to be the positive Hill's solution with multiplier m and set $p = \psi'/\psi$. This (p, λ) pair is thus unique and will reproduce the original $(q, \log m)$, showing that the augmented Riccati map is one to one and onto.

To compute the joint distribution of q and $\log m$ in terms of p and λ, [Cambronero and McKean 1999] again considers the approximating discrete (one-to-one and onto) transformation

$$(p_0, \ldots, p_{n-1}, \lambda) \longrightarrow (q_0, \ldots, q_{n-1}, \log m),$$

from $\mathbb{R}^{n+1}$ to $\mathbb{R}^{n+1}$, defined by

$$q_i = \lambda + n^2(e^{hp_{i+1}} - 2 + e^{-hp_i}), \quad \log m = \sum_{i=0}^{n-1} p_i h, \qquad (2\text{-}7)$$

where $h = \dfrac{1}{n}$ and $p_n = p_0$. The corresponding Jacobian is now

$$h^n |J_n| = \sum_{i=0}^{n-1} \frac{h}{m\varphi_i^2} \sum_{k=i+1}^{i+n} \varphi_k^2 h + O(h) \quad \text{for } \varphi_i = \exp\left(\sum_{j=1}^{i} p_j h\right).$$

As before, the discrete white noise $\times d \log m$ measure may then be reexpressed as in

$$\exp\left(-\frac{1}{2} \sum q_i^2 h\right) \frac{dq_0 \ldots dq_{n-1}}{(2\pi/h)^{n/2}} \times d \log m$$

$$= \exp\left(-\frac{1}{2h} \sum_{i=0}^{n-1} (p_{i+1} - p_i)^2 - \frac{1}{8} \sum_{i=0}^{n-1} (p_{i+1}^2 + p_i^2)^2 h \right.$$

$$\left. -\frac{\lambda}{2} \sum_{i=0}^{n-1} (p_i^2 + p_{i+1}^2) h - \frac{\lambda^2}{2} + R_n\right) \frac{h^n |J_n|}{(2\pi h)^{n/2}} \, dp_0 \ldots dp_{n-1}\, d\lambda,$$

where again $R_n \to 0$ boundedly on certain sets of large measure. Thus, on any such set, we have

$$\lim_{n\to\infty} h^n |J_n| = A(p) = \int_0^1 \frac{dx}{m\varphi^2(x)} \int_x^{x+1} \varphi^2(y)\, dy,$$

for $\varphi(x) = \exp(\int_0^x p)$, and it is only a bit more effort to arrive at the following.

THEOREM 2.3. *For any bounded ϕ, compactly supported with respect to $a = \log m$, we have*

$$\int \phi(Q,a)\, dQ_*\, da = \int \phi(\lambda + p' + p^2, \int p)\, \Theta(p,\lambda)\, dP_*\, d\lambda,$$

where

$$\Theta(p,\lambda) = \frac{1}{\sqrt{2\pi}} \exp\left(-\frac{1}{2}\int_0^1 (\lambda + p^2)^2 - \int_0^1 p\right) A(p).$$

In brief, $dQ_*\, da = \Theta(p,\lambda)\, dP_*\, d\lambda$.

Now employ the relation between $dQ_*\, da$ and $dP_*\, d\lambda$ as follows. First,

$$Q_*[\lambda \leq \Lambda_0(q) \leq \lambda+\varepsilon] = \frac{1}{\delta}\int_0^\delta \int_{[\lambda\leq\Lambda_0(q)\leq\lambda+\varepsilon]} dQ_*\, da$$

$$= \frac{1}{\delta}\int_{[0\leq\int p\leq\delta]} \Theta(p,\lambda)\chi_{[\lambda\leq\Lambda_0(\lambda+p'+p^2)\leq\lambda+\varepsilon]}\, dP_*\, d\lambda.$$

The left-hand side is independent of δ, so for $\delta\to 0$ we find

$$Q_*[\lambda \leq \Lambda_0(q) \leq \lambda + \varepsilon] = \int_H \Theta(p,\lambda)\chi_{[\lambda\leq\Lambda_0(\lambda+p'+p^2)\leq\lambda+\varepsilon]}\, dP_0\, d\lambda.$$

Now $\int_0^1 p = 0$ implies $\Lambda_0(\lambda + p' + p^2) = \lambda$, and therefore

$$\varepsilon^{-1} Q_*[\lambda \leq \Lambda_0(q) \leq \lambda + \varepsilon] = \int_H \left(\varepsilon^{-1}\int_\lambda^{\lambda+\varepsilon} \Theta(p,\lambda)\, d\lambda\right) dP_0.$$

As $\varepsilon \to 0$, the left-hand side converges to $f(\lambda) = (d/d\lambda)\, Q_*[\Lambda_0 \leq \lambda]$, and the integrand on the right-hand side converges to $\Theta(p,\lambda)$. Moreover, there is the needed domination to prove that

$$f(\lambda) = \int_H \Theta(p,\lambda)\, dP_0 = \frac{1}{\sqrt{2\pi}}\int_H e^{-\frac{1}{2}\int(\lambda+p^2)^2} A(p)\, dP_0,$$

as advertised.

Ornstein–Uhlenbeck type potentials. The methods above extend from white noise potentials to a whole class of q's which perform a periodic diffusion. For example, let $\hat{Q}$ denote periodic Ornstein–Uhlenbeck (OU) measure (of mass m). This is the rotation invariant Gaussian process on S^1 arrived at by conditioning the OU paths so that $q(0) = q(1)$ and then distributing that common point according to the stationary measure for the full-line OU.

Similarly to white noise one gets:

THEOREM 2.4 [Cambronero and McKean 1999]. *Under the transformation $q = \lambda + p' + p^2$, the periodic OU measure $\hat{Q}$, restricted to $[\Lambda_0 \geq \lambda]$, is transformed into the measure $dP_0\, d\alpha$ according to*

$$\int_{[\Lambda_0 \geq \lambda]} \phi\, d\hat{Q}$$

$$= C \int_H \int_{\mathcal{I}(p')}^{\infty} \phi(\lambda + p' + p^2) e^{-\frac{1}{2}m^2 \int_0^1 (\lambda + p' + p^2)^2} G(\alpha, p')\, d\alpha\, dP_0(p'), \quad (2\text{-}8)$$

with $C = (4/\sqrt{2\pi})\, \sinh(m/2)$, $p = \alpha + \int_0^t p'$, $\mathcal{I}(p') = -\int_0^1 \int_0^t p'$, and

$$G(\alpha, p') = \exp\left(\int_0^1 (p'^3 - 2p^2 p' + p^2)\, dt \right) \sinh\left(\int_0^1 p \right).$$

In particular, the distribution is read off upon setting $\phi \equiv 1$ in (2-8), providing the analogue of (2-2). Further, one can move on to other potentials of type brownian motion plus drift,

$$dq(x) = db(x) - m(q)\, dx,$$

where it is assumed that m is an odd function with $m(q) > 0$ for $q > 0$ to avoid explosion. The periodic versions of these processes are built in the same way as for OU; the added condition

$$\int_{-\infty}^{\infty} e^{\frac{1}{2}(m'(q) - m^2(q))}\, dq < \infty \quad (2\text{-}9)$$

being required to ensure the periodic measure has finite total mass.

THEOREM 2.5 [Cambronero 1996]. *Let Q_* be a periodic diffusion with odd drift $m(q)$ subject to $m(q) > 0$ for $q > 0$ and (2-9). Then*

$$Q_*[\Lambda_0 \geq \lambda]$$

$$= 2C_0 \int_H \int_{\mathcal{I}(p')}^{\infty} \exp\left(-\frac{1}{2} \int_0^1 F(\lambda + p'(x) + p^2(x))\, dx \right) G(\alpha, p')\, d\alpha\, dP_0,$$

where $F = -m' + m^2$, and $C_0^{-1} = \int \exp(-\frac{1}{2}\int_0^1 F(q))\, dP_*$ is a normalizing constant.

And, again by considering joint distributions of q and the multiplier, there is also a formula for the density.

THEOREM 2.6 [Cambronero 1996]. *The density of Λ_0 under Q_* is given by*

$$f(\lambda) = C_0 \int_H e^{-\frac{1}{2}\int F(\lambda + p' + p^2)}\mathcal{E}(p') \, dP_0(p')$$

where $p = \Im(p') + \int_0^t p'$ and $\mathcal{E}(p') = \exp\left(\int_0^1 (p'^3 - 2p^2 p' + p^2)\right) A(p)$.

After this parade of formulae, it is probably helpful to write out the linear (OU or $m(q) = mq$) case in full:

$$f_{OU}(\lambda) = \sqrt{\frac{2}{\pi}} \sinh\frac{m}{2} \int_H e^{-\frac{1}{2}m^2 \int_0^1 (\lambda + p' + p^2)^2} e^{\int_0^1 (p'^3 - 2p^2 p' + p^2)} A(p) \, dP_0(p'). \quad (2\text{-}10)$$

It is now p' that is locally brownian. Starting with white noise, p is CBM under the Riccati map. Starting with an additional derivative in potential space results in an additional derivative in p-space. The added dependence in the field makes integrals like (2-10) harder to analyze than their white noise counterparts. This is the subject of the next section.

3. Ground state energy asymptotics

As an application of the above integral expressions we consider the shape of the ground state energy density for various random potentials. We begin again in the white noise case, for which detailed asymptotics are available:

THEOREM 3.1 [Cambronero et al. 2006]. *Let $f_{WN}(\lambda)$ denote the density function for $\Lambda_0(q)$, the minimal eigenvalue for Hill's operator on the circle of perimeter one with white noise potential. Then*

$$f_{WN}(\lambda) = \sqrt{\frac{\lambda}{\pi}} \exp\left(-\frac{1}{2}\lambda^2 - \frac{1}{\sqrt{2}}\lambda^{1/2}\right)(1 + o(1)),$$

as $\lambda \to +\infty$ and,

$$f_{WN}(\lambda) = \frac{4}{3\pi}|\lambda| \exp\left(-\frac{8}{3}|\lambda|^{3/2} - \frac{1}{2}|\lambda|^{1/2}\right)(1 + o(1)),$$

as $\lambda \to -\infty$.

The overall asymmetry has an intuitive explanation: level-repulsion holds down the right tail, while a large negative deviation can be affected by a single excursion of the potential. The 3/2-exponent in $\lambda \to -\infty$ direction is shared by the allied tail in the Tracy–Widom laws of RMT, but more on this later.

The above result stems from the second version of the density:

$$f_{WN}(\lambda) = \frac{1}{\sqrt{2\pi}} \int_H e^{-\frac{1}{2}\int_0^1(\lambda+p^2)^2} A(p)\,dP_0(p),$$

where P_0 is the CBM conditioned to be mean-zero. In either the $\lambda \to +\infty$ or $\lambda \to -\infty$ direction, the leading order, or logarithmic scale, asymptotics of f_{WN} are governed by those of the infimum of

$$I_\lambda(p) := \frac{1}{2}\int_0^1 \left(\lambda + p^2(x)\right)^2 dx + \frac{1}{2}\int_0^1 \left(p'(x)\right)^2 dx, \qquad (3\text{-}1)$$

over $p \in H$. When $\lambda \to +\infty$ it is plain that it is most advantageous for the path p to sit in a vicinity of the origin, which already accounts for the appraisal $f_{WN}(\lambda) \sim e^{-\lambda^2/2}$. For a more complete picture, $\int_0^1(\lambda+p^2)^2$ may be expanded, and both $A(p)$ and $e^{-1/2\int_0^1 p^4}$ are seen to be unimportant in comparison with $e^{-\lambda\int_0^1 p^2}$. That is, $E_0[e^{-\lambda\int_0^1 p^2 - 1/2\int_0^1 p^4}A(p)] \simeq E_0[e^{-\lambda\int_0^1 p^2}]$, and the computation is finished with aid of the explicit formula

$$\int_H e^{-\lambda\int_0^1 p^2}\,dP_0(p) = \frac{\sqrt{\lambda/2}}{\sinh\sqrt{\lambda/2}}.$$

All this had already been noticed in [Cambronero and McKean 1999].

The behavior as $\lambda \to -\infty$ is far less transparent. Now there is the possibility of cancellation in the first part of the variational formula $\int_0^1(|\lambda| - p^2)^2$, compelling the path to live near $\pm\sqrt{-\lambda}$. However, the mean-zero condition ($p \in H$) dictates that p must its time between these two levels, while sharp transitions from $-\sqrt{-\lambda}$ to $+\sqrt{-\lambda}$ or back are penalized by the energy $\int_0^1 p'^2$. The heavier left tail is the outcome of this competition.

Getting started, the Euler–Lagrange equation for any $\lambda < 0$ minimizer p_λ of (3-1) may be computed,

$$p_\lambda'' = 2p_\lambda^3 - 2p_\lambda^2, \qquad (3\text{-}2)$$

and solved explicitly in terms of the Jacobi elliptic function sin-amp,

$$p_\lambda(x) = k\sqrt{|\lambda|} \times \operatorname{sn}(\sqrt{|\lambda|}x, k), \qquad (3\text{-}3)$$

with modulus satisfying $k^2 \simeq 1 - 16e^{-\sqrt{|\lambda|}/2}$ to fix the period at one.[1] Substituting back yields $I_\lambda(p_\lambda) \sim \frac{8}{3}|\lambda|^{3/2}$, and there follows the first-order large-deviation type estimate

$$f_{WN}(\lambda) \simeq \exp\left(-\frac{8}{3}|\lambda|^{3/2}\right) \quad \text{for } \lambda \to -\infty.$$

---

[1] Technical aside: the equation (3-2) reported in [Cambronero et al. 2006] includes an additive constant, but this was later understood to vanish in [Ramírez and Rider 2006].

Toward more exact asymptotics, there are various degeneracy problems that need to be addressed. First is the obvious lack of uniqueness: any translation $p_\lambda^a(x) = p_\lambda(x + a)$ of (3-3) also minimizes I_λ. Second, and more obscure, is an asymptotic degeneracy in the direction of the low lying eigenfunctions of the Hessian of I_λ.

The translational issue is dealt with by conditioning: the minimizing path is pinned at zero at some predetermined point. Then, by a change of measure computation, we arrive at the following Rice-type formula. With $\{p_\lambda^a\}$ the one-parameter family of minimizers, $d(\cdot, \{p_\lambda^a\})$ the sup-norm distance to that family, and any $\varepsilon > 0$, we have

$$E_0\left[e^{-\frac{1}{2}\int_0^1 (|\lambda|-p^2)^2} A(p), d(p, \{p_\lambda^a\}) \le \varepsilon\sqrt{|\lambda|}\right]$$
$$= E_0^0\left[e^{-\frac{1}{2}\int_0^1 (|\lambda|-p^2)^2} A(p)R(p), d(p, \{p_\lambda^a\}) \le \varepsilon\sqrt{|\lambda|}\right] P_0\left(\int_0^1 \phi_1^\lambda p = 0\right).$$

Here, ϕ_1^λ is the $L^2(S^1)$-normalized derivative of p_λ (the derivative generating all translations), E_0^0 is now the CBM conditioned so that both $\int_0^1 p = 0$ and $\int_0^1 \phi_1^\lambda p = 0$, and $R(p)$ is a Radon–Nikodym factor which we will not make explicit. On the left-hand side, note that the integral is localized about the full family of minimizers. On the right-hand side, it is easy to see that the intersection of a small tube about $\{p_\lambda^a\}$ and the plane $\left[p : \int_0^1 p\phi_1^\lambda = 0\right]$ may be replaced with a similarly small neighborhood about $p_\lambda^0 = p_\lambda$. In this way the expectation has in fact been localized about a fixed path.

Next, the obvious shift $p \to p + p_\lambda$ results in

$$f_{WN}(\lambda) \simeq$$
$$e^{-I_\lambda(p_\lambda)} E_0^0\left[e^{-\frac{1}{2}\int_0^1 (q_\lambda+2\lambda)p^2} S(p, p_\lambda), \|p\|_\infty \le \varepsilon\sqrt{|\lambda|}\right] P_0\left(\int_0^1 \phi_1^\lambda p = 0\right),$$

where

$$S(p, p_\lambda) = e^{-2\int_0^1 p_\lambda p^3 - \frac{1}{2}\int_0^1 p^4} A(p + p_\lambda)R(p + p_\lambda),$$
$$q_\lambda(x) = 6|\lambda|k^2 \operatorname{sn}^2(\sqrt{|\lambda|}x, k).$$

One expects the Gaussian measure tied to the quadratic form

$$Q_\lambda = -\frac{d^2}{dx^2} + q_\lambda(x) + 2\lambda \tag{3-4}$$

to dominate the higher order nonlinearities in $S(\cdot, p_\lambda)$ and focus the path at $p = 0$. This deterministic Hill's operator Q_λ is of course the Hessian of I_λ, and it is no small piece of good fortune that it coincides with one of Lamé's finite-gap operators for which simple spectrum and corresponding eigenfunctions are explicitly computable [Ince 1940]. Information about the rest of the spectrum is obtained from a beautiful formula of Hochstadt [1961] for the discriminant.

HOCHSTADT'S FORMULA. *Let Q be finite-gap with $2g + 1$ simple eigenvalues. Then $\Delta(\lambda) = 2 \cos \psi(\lambda)$ with*

$$\psi(\lambda) = \frac{\sqrt{-1}}{2} \int_{\lambda_0}^{\lambda} \frac{(s - \lambda_1') \cdots (s - \lambda_g')}{\sqrt{-(s - \lambda_0) \cdots (s - \lambda_{2g})}} \, ds, \tag{3-5}$$

in which $\lambda_1' < \cdots < \lambda_g'$ are the points $\lambda_{2\ell-1} < \lambda_\ell' < \lambda_{2\ell}$, where $\Delta'(\lambda) = 0$. They are determined from the simple spectrum through the requirement $\psi(\lambda_{2\ell}) - \psi(\lambda_{2\ell-1}) = 0$ for $\ell = 1, 2, \ldots g$.

In the case of Q_λ for example we have $g = 2$.

Moving on, as alluded to just above we claim that, for $\lambda \to -\infty$ and all $\varepsilon > 0$ sufficiently small:

$$E_0^0 \left[e^{2 \int_0^1 p_\lambda p^3 - \frac{1}{2} \int_0^1 p^4} A(p + p_\lambda) R(p + p_\lambda) e^{-\frac{1}{2} \int_0^1 (q_\lambda(x) + 2\lambda) p^2}, \|p\|_\infty \le \varepsilon \sqrt{|\lambda|} \right]$$
$$= A(p_\lambda) R(p_\lambda) Z(\lambda)(1 + o(1)), \tag{3-6}$$

where

$$Z(\lambda) = E_0^0 \left[e^{-\frac{1}{2} \int_0^1 (q_\lambda(x) + 2\lambda) p^2(x) \, dx} \right] P_0 \left(\int_0^1 \phi_1^\lambda(x) p(x) \, dx = 0 \right).$$

This rests on the coercive properties of the measure $e^{-\frac{1}{2} \int_0^1 (q_\lambda + 2\lambda) p^2} d\,\mathrm{CBM}(p)$ restricted to $\int_0^1 p = 0$ and $\int_0^1 p \phi_1^\lambda = 0$, which is to say, on the spectral gap of Q_λ restricted to the same space. Here lies the second degeneracy in the problem. This gap actually goes to zero as $\lambda \to -\infty$, making the estimate (3-6) rather laborious and hard to imagine without having the Q_λ spectrum explicitly at hand.

Taking the last appraisal for granted, it remains to find a closed expression for $Z(\lambda)$. This plays the role of the usual (though now infinite-dimensional) Gaussian correction in any Laplace-type analysis, and the fact is

$$Z(\lambda) \simeq C(\lambda_0, \ldots \lambda_4; c_0, \ldots, c_4) \times \frac{1}{\sqrt{\Delta^2(2|\lambda|) - 4}}. \tag{3-7}$$

The prefactor $C(\cdot)$ is a rational function of the (explicitly known) simple spectrum of Q_λ (eigenvalues λ_k and corresponding norming-constants c_k, $k = 0, \ldots, 4$). Hochstadt's formula now comes to the rescue, expressing the discriminant Δ back in terms of the same $\lambda_0, \ldots, \lambda_4$. Putting together the asymptotics of $A(p_\lambda)$, $R(\lambda)$, and those for $Z(\lambda)$ via the above expression will complete the proof for the left tail.

REMARK. It is enlightening to run the Riccati correspondence in reverse, the concentration of p about p_λ resulting in an optimal potential of the form

$$q(x; \lambda) = \lambda + p_\lambda'(x) + p_\lambda^2(x) \simeq -2|\lambda| \operatorname{sech}^2\left(\sqrt{|\lambda|}\,(x - 1/2)\right).$$

That is, the white noise "path" must perform a single excursion of depth $O(\lambda)$ in an $O(|\lambda|^{-1/2})$ span to produce a large negative eigenvalue.

Nice gaussian potentials. Ay this point is natural to ask: To what extent is the white noise result universal for some class of potentials? The general case remains a question for the future; we describe here what is known for a class of *nice* stationary Gaussian potentials q.

The periodic diffusion setting is not the appropriate theater to explore questions of universality; certainly the details of the force F in Theorem 2.6 will play out in the shape of the density. Instead we consider the case that q is a stationary Gaussian process of periodicity one, continuous (and so, nice) such that

$$E[q(x)] = 0, \quad E[q(x)q(y)] = K(x - y), \tag{3-8}$$

with K satisfying the technical condition $\bar{K} = \int_0^1 K(x)\,dx > 0$. There is of course a point in common with the previously discussed potentials, namely periodic OU with mass m, in which case $K(z) = \frac{1}{2m}\left(\frac{e^{mz}}{e^m - 1} - \frac{e^{-mz}}{e^{-m} - 1}\right)$. Generally, however, the Cambronero–McKean formulas do not carry over to this Gaussian potential framework. Because the Riccati map is nonlinear, it is not always the case that q, under the $(\,\cdot\,, K^{-1}, \cdot\,)$ Gaussian measure, and p, under the $(\,\cdot\,, DK^{-1}D\,\cdot\,)$ Gaussian measure, are absolutely continuous. Take for example the situation when only a finite number of modes in the spectral expansion of q are charged.

For these reasons we rely on a yet another formula for the density, the idea behind which is to carry out the Riccati map on only part of the space. Denote by P the measure of q and let $\hat{P}$ be the measure induced on $\hat{q} = q - \int_0^1 q$. Then, this new formula for the density, established in [Ramírez and Rider 2006], is

$$f_K(\lambda) = \frac{1}{\sqrt{2\pi}} \int_H \exp\left(-\frac{1}{2\bar{K}}(\lambda + \Phi(\hat{q}))^2\right) d\hat{P}(\hat{q}). \tag{3-9}$$

Here, Φ is some implicitly defined nonlinear functional of the path, expressible through the Riccati map. When available, the Cambronero–McKean formula is certainly more powerful, being so explicit. On the other hand, (3-9) suffices to uncover the asymptotic shape of the density.

THEOREM 3.2 [Ramírez and Rider 2006]. *The probability density function f_K for $\Lambda_0(q)$ corresponding to any Gaussian random potential as above is C^∞ and* *satisfies*

$$\lim_{\lambda \to +\infty} \frac{1}{\lambda^2} \log f_K(\lambda) = -\frac{1}{2\bar{K}}, \quad \lim_{\lambda \to -\infty} \frac{1}{\lambda^2} \log f_K(\lambda) = -\frac{1}{2K(0)}.$$

As in the white noise case, the computation for the right tail is relatively simple, stemming from an optimal potential $q \simeq \lambda$ (or $p \simeq 0$). Further, the estimate for the left tail is connected with q-paths concentrating around $q(x) \simeq \lambda K(x)/K(0)$. That is, the covariance structure provides just enough freedom for the path to oscillate in accordance with K itself. Very loosely speaking, this falls in line with the white noise result where, after rescaling, the minimizing potential approaches a Dirac delta, which is the right kernel K for that process.

Lastly, we should reiterate the connection between Theorem 3.2 and well known Lifschitz tail results. For example, Pastur [1972] proved that, with $Q_L = -\Delta + q(x)$ on the cube of side-length L in $\mathbb{R}^d$ and q stationary Gaussian with covariance K satisfying a Hölder estimate: $\lim_{\lambda \to -\infty} \lambda^{-2} \log N(\lambda) = -1/(2K(0))$ where $N(\lambda)$ equals the $L \uparrow \infty$ density of states. Moreover, the basic method employed will provide tail bounds on the distribution function of the ground state eigenvalue for a large class of continuous q and $L < \infty$. From here, our own result could very well be anticipated. On the other hand, we know of no way to access the density function directly other than through the Riccati-as-a-change-of-measure idea.

4. General Tracy–Widom laws

The study of detailed limit theorems at the spectral edge is far more highly developed in RMT than in random Schrödinger. This is easiest to describe for the Gaussian Unitary Ensemble (GUE). GUE is an $n \times n$ Hermitian matrix ensemble M comprised of independent complex Gaussians: $M_{ij} = \overline{M_{ji}} \sim \mathcal{N}_\mathbb{C}(0, 1/4)$, while $M_{ii} \sim \mathcal{N}(0, 1/2)$. Equivalently, it is drawn from the distribution with increment $dP(M) = \frac{1}{Z} e^{-\mathrm{tr}M^2} dM$; dM denoting Lebesgue measure on the space of n-dimensional Hermitian matrices and $Z < \infty$ a normalizing factor.

Regarding spectral properties, GUE is integrable in so far as the full joint density of eigenvalues $\lambda_1, \lambda_2, \dots, \lambda_n$ is known:

$$P_{GUE}(\lambda_1, \lambda_2, \dots, \lambda_n) = \frac{1}{Z_n} e^{-\sum_{k=1}^n \lambda_k^2} \prod_{k<j} |\lambda_j - \lambda_k|^2 \tag{4-1}$$

$$= \frac{1}{n!} \det\big(K_n(\lambda_j, \lambda_k)\big)_{1 \le j, k \le n}.$$

On the second line, $K_n(\lambda, \mu)$ is the kernel for the projection onto the span of the first n Hermite polynomials in $L^2(\mathbb{R}, e^{-\mu^2})$; it follows from line one by simple row operations in the square Vandermonde component of the density. In fact, all finite dimensional correlations are expressed in terms of determinants of the same kernel. As a consequence there is the explicit formula at the spectral edge,[2]

$$P(\lambda_{max} \le \lambda) = \det\big(I - K_n \mathbf{1}_{(\lambda,\infty)}\big),$$

the right-hand side denoting the Fredholm determinant of the integral operator associated with K_n restricted to (λ, ∞). The classical Plancherel–Rotach asymptotics for Hermite polynomials and a marvelous identity from [Tracy and Widom 1994] now provide the distributional limit as $n \to \infty$:

$$P_{GUE}\Big(\frac{1}{\sqrt{2}}n^{1/6}(\lambda_{max} - \sqrt{2n}) \le \lambda\Big)$$

$$\to \exp\Big(-\int_{\lambda}^{\infty}(s - \lambda)u^2(s)\,ds\Big) =: F_{GUE}(\lambda). \quad (4\text{-}2)$$

Here $u(s)$ is the solution of $u'' = su + 2u^3$ (Painlevé II) subject to $u(s) \sim Ai(s)$ (the standard Airy function) as $s \to +\infty$.

Associated with GUE are the Gaussian Orthogonal and Symplectic Ensembles (GOE and GSE) of real symmetric or self-dual quaternion Gaussian matrices. These are again integrable, with joint eigenvalue densities of a similar shape to line one of (4-1), the power two on the absolute Vandermonde interaction term being replaced by a 1 or 4. While not determinantal in the same way, there are again closed expressions for the largest eigenvalue distribution, and, at the same basic scalings, limit laws due to Tracy and Widom [1996]:

$$F_{G(O/S)E}(\lambda) = \begin{cases} \exp\big(-\frac{1}{2}\int_{\lambda}^{\infty}(s - \lambda)u^2(s)\,ds\big)\exp\big(-\frac{1}{2}\int_{\lambda}^{\infty}u(s)\,ds\big), \\ \exp\big(-\frac{1}{2}\int_{\lambda'}^{\infty}(s - \lambda')u^2(s)\,ds\big)\cosh\big(\int_{\lambda'}^{\infty}u(s)\,ds\big). \end{cases} \quad (4\text{-}3)$$

with $\lambda' = 2^{2/3}\lambda$ and u is the same solution of Painlevé II. For each of these three special ensembles there are also Painlevé expressions for the limiting distribution of the scaled second and higher largest eigenvalues, see again [Tracy and Widom 1994] and [Dieng 2005].

While striking in and of themselves, these results of Tracy–Widom have surprising importance in physics, combinatorics, multivariate statistics, engineering, and applied probability. A few highlights include [Johansson 2000; Baik et al. 1999; Johnstone 2001; Baryshnikov 2001]. From a probabilist's perspective, the laws (4-2) and (4-3) should be regarded as important new points in the space of distributions. In particular, one would like to understand $F_{G(O/U/S)E}$ in the same way that we do the Normal or Poisson distribution, being able to set down a few characterizing conditions. As it stands, the Tracy–Widom laws

---

[2]In RMT it is customary here to look at largest, rather than smallest, eigenvalues as is the case in random Schrödinger.

seem to live in the realm of integrable systems: we know of many interesting examples in which they arise, but that is about all.

One avenue to a deeper understanding of the Tracy–Widom laws would be some explanation of how to interpolate between them. For any $\beta > 0$, consider the following measure on n real points

$$P_\beta(\lambda_1, \lambda_2, \ldots, \lambda_n) = \frac{1}{Z_{n,\beta}} e^{-\beta \sum_{k=1}^n \lambda_k^2} \prod_{k<j} |\lambda_j - \lambda_k|^\beta. \tag{4-4}$$

G(O/U/S)E occur for $\beta = 1, 2, 4$; on physical grounds β plays the role of inverse temperature in a 1-d "coulomb" gas. The limiting distribution of the largest P_β-point would give us a general Tracy–Widom law. While there appears to be no hope of integrating out a correlation for general beta, Dumitriu and Edelman [2002] have discovered a matrix model for all $\beta > 0$. The fact is: with $g_1, g_2, \ldots$ i.i.d. unit Gaussians and each χ_r an independent chi random variables of parameter r, the symmetric tridiagonal ensemble

$$H_n^\beta = \frac{1}{\sqrt{2}} \begin{bmatrix} \sqrt{2}g_1 & \chi_{\beta(n-1)} & & & & \\ \chi_{\beta(n-1)} & \sqrt{2}g_2 & \chi_{\beta(n-2)} & & & \\ & \ddots & \ddots & \ddots & & \\ & & \chi_{\beta 2} & \sqrt{2}g_{n-1} & \chi_\beta & \\ & & & \chi_\beta & \sqrt{2}g_n \end{bmatrix} \tag{4-5}$$

has joint eigenvalue law with density (4-4).

The simplicity of (4-5) opens up the possibility of scaling the operator itself rather than dealing with the eigenvalue law. Formally invoking the central limit theorem in the form $\chi_{(n-k)\beta} \approx \sqrt{n\beta} - (\sqrt{\beta}k/2n) + (1/\sqrt{2})\mathcal{N}(0,1)$ in the off-diagonal entries, one can readily understand the conjecture of Sutton and Edelman [Edelman and Sutton 2007] that the rescaled matrices

$$-\tilde{H}_n^\beta = -\frac{\sqrt{2}}{\sqrt{\beta}} n^{1/6}(H_n^\beta - \sqrt{2\beta n}I),$$

should go over into

$$\mathcal{H}_\beta = -\frac{d^2}{dx^2} + x + \frac{2}{\sqrt{\beta}} b'(x) \tag{4-6}$$

in the $n \to \infty$, or continuum, limit. As before, b' indicates a white noise, and the scaling in (4) corresponds to that for the spectral edge in the known $\beta = 1, 2, 4$ cases. Thus, were it to hold, the above correspondence would entail that the low-lying eigenvalues of $\mathcal{H}_\beta$ agree in distribution with the limiting largest eigenvalues of H_n^β. Recently, the second two authors and B. Virág [Ramírez et al. 2006] have proved this conjecture.

THEOREM 4.1. ([*Ramírez et al. 2006*]) *Let* $\lambda_{\beta,1} \geq \lambda_{\beta,2} \geq \cdots$ *be the ordered eigenvalues of the β-ensemble H_n^β, and $\Lambda_0(\beta) \leq \Lambda_1(\beta) \leq \cdots$ the spectral points of $\mathcal{H}_\beta$ in $L^2(\mathbb{R}_+)$ with Dirichlet conditions at $x = 0$. Then, for any finite k, the family*

$$\left\{ \sqrt{\frac{2}{\beta}} n^{1/6} (\lambda_{\beta,\ell} - \sqrt{2\beta n}) \right\}_{\ell=1,\ldots,k}$$

converges in distribution as $n \to \infty$ to $\{-\Lambda_0(\beta), -\Lambda_1(\beta), \cdots, -\Lambda_{k-1}(\beta)\}$.

Part of this result is the fact that the Schrödinger operator $\mathcal{H}_\beta$, referred to as the Stochastic Airy operator for obvious reasons, has an almost surely finite ground state eigenvalue Λ_0, as well as well defined higher eigenvalues Λ_1, and so on. Though no longer on a finite volume, the compactifying linear restoring force proves enough to tame the white noise at infinity. It is also remarked that the proof of Theorem 4.1 is actually made almost surely — eigenvalue by eigenvalue — after coupling the noise in the matrix model H_n^β to the brownian motion $b(x)$ in the limiting $\mathcal{H}_\beta$.

Next recall that the densities f_β of the $\beta = 1, 2, 4$ Tracy–Widom laws satisfy

$$f_\beta(\lambda) \sim e^{-\frac{1}{24}\beta|\lambda|^3}$$

for $\lambda \to -\infty$ and

$$f_\beta(\lambda) \sim e^{-\frac{2}{3}\beta\lambda^{3/2}}$$

for $\lambda \to +\infty$. Coupled with Theorem 4.1 this sheds new light on the results just discussed for the shape of the ground state eigenvalue density of the simple Hill operator $-d^2/dx^2 + b'(x)$. Moving into the spectrum, white noise on S^1 and white noise plus linear force on $\mathbb{R}^+$ certainly should give rise to different phenomena. On the other hand, when pulling far away from the spectrum, it is intuitive that these potentials would have roughly the same effect.

That said, the reader will anticipate what comes next. The Riccati map immediately gives a second description of the limiting distribution of the largest β-ensemble eigenvalues in terms of the explosion question for the one dimensional diffusion $x \mapsto p(x)$

$$dp(x) = \frac{2}{\sqrt{\beta}} \, db(x) + (x - \lambda - p^2(x)) \, dx. \tag{4-7}$$

To make things precise, return to the eigenvalue problem,

$$d\psi'(x) = \frac{2}{\sqrt{\beta}} \psi(x) \, db(x) + (x - \lambda)\psi(x) \, dx,$$

restricted to $[0, L]$ subject to $\psi(L) = 0$ as well as $\psi(0) = 0$. Denote by $\Lambda_0(L)$ the minimal Dirichlet eigenvalue, and take $\psi_0(x, \lambda)$ the solution of the initial value problem with $\psi_0(0) = 0$ and $\psi_0'(0) = 1$. As already mentioned, the event that

$\Lambda_0(L) \geq \lambda$ is the event that ψ_0 does not vanish before $x \leq L$. This is the classical "shooting method". Now make the Riccati move: $p(x, \lambda) = \psi_0'(x, \lambda)/\psi_0(x, \lambda)$ is the diffusion (4-7), and the event that $\psi_0(x, \lambda)$ has no root before $x = L$ is the event that the p motion, begun from $p(0, \lambda) = \psi'(0, \lambda)/\psi(0, \lambda) = +\infty$ at $x = 0$, fails to explode down to $-\infty$ before $x = L$. (While it is not customary to use the entrance/exit terminology for inhomogeneous motions, comparison with the homogeneous case will explain why $p(x, \lambda)$ may be started at $+\infty$ and leaves its domain only at $-\infty$.)

Granted Theorem 4.1, Λ_0, the ground state eigenvalue of the full line problem, exists, and it is obvious that $\Lambda_0(L)$ converges almost surely to that variable as $L \to \infty$. In other words,

$$P(\Lambda_0 > \lambda) = P(\psi(\cdot, \lambda) \text{ never vanishes}) = P_{+\infty}(p(\cdot, \lambda) \text{ does not explode}).$$

A description of $P(\Lambda_k > \lambda)$ is similar for all k. The probability that the second eigenvalue exceeds λ is the $P_{+\infty}$ probability that p explodes *at most* once to $-\infty$, and so on. All this, with its implications for the limiting largest eigenvalues in the β-ensembles is summarized in the next statement.

THEOREM 4.2. (*[Ramírez et al. 2006]*) *With* $x \mapsto p(x) = p(x, \lambda)$ *the motion* (4-7), *let* $P_\bullet$ *denote the measure on paths induced by* p *begun at* $p(0) = \bullet$ *and let* $m(\lambda, \beta)$ *denote the passage time of* p *to* $-\infty$. *Then,*

$$\lim_{n \to \infty} P\left(\sqrt{\frac{2}{\beta}} n^{1/6} (\lambda_{\beta,1} - \sqrt{2\beta n}) \leq \lambda \right) = P_{+\infty}(m(-\lambda, \beta) = +\infty),$$

and also

$$\lim_{n \to \infty} P\left(\sqrt{\frac{2}{\beta}} n^{1/6} (\lambda_{\beta,k} - \sqrt{2\beta n}) \leq \lambda \right)$$
$$= \sum_{\ell=1}^{k} \int_0^\infty \cdots \int_0^\infty P_{+\infty}(m(-\lambda, \beta) \in dx_1) P_{+\infty}(m(-\lambda + x_1, \beta) \in dx_2) \cdots$$
$$\cdots P_{+\infty}(m(-\lambda + x_1 + \cdots + x_{\ell-1}, \beta) = +\infty),$$

for any fixed k.

Even at $\beta = 1, 2$, and 4, Theorems 4.1 and 4.2 provide yet another vantage point on the Tracy–Widom laws. Not only are these laws now tied to a much simpler mechanical model (1-d Schrödinger), the Riccati map has introduced a Markovian structure where none appeared to exist.

5. Questions for the future

Shape of Hill's ground state density. This is still in its infancy. In particular, the exact regularity of the potential at which one sees a transition between the white noise 3/2-heavy tail (Theorem 3.1) and the Gaussian tail (Theorem 3.2) is an interesting question.

Non-i.i.d. matrix ensembles. Little is known about the limiting scaled distribution of $\lambda_{\max}$ for Hermitian matrix ensembles with entries exhibiting correlations which do not vanish in the $n \uparrow \infty$ limit. For the sake of discussion, consider such a non-i.i.d. Gaussian matrix M. Given Theorem 4.1, it is believable that M has some random differential operator as its continuum limit. Further, if the correlations in M are strong enough, one might imagine that the white noise type potential of Stochastic Airy is replaced by a smoother Gaussian potential, and then Theorem 3.2 would become relevant.

Sample covariance ensembles. Of importance in statistics are ensembles of the form $X^T C X$ where X is comprised of say independent identically distributed real or complex Gaussians and C may be assumed diagonal. If $C = Id$, these are the classical null-Wishart or Laguerre ensembles at $\beta = 1$ (real) and $\beta = 2$ (complex), and the corresponding Tracy–Widom laws turn up at the spectral edge. In fact, Edelman and Dumitriu also have general $\beta > 0$ tridiagonal versions of these null ensembles to which the results of Section 4 apply. On the other hand, if C is not the identity the picture is rather murky. The possibility of phase transition away from Tracy–Widom if C is sufficiently "spiked" is proved in [Baik et al. 2005], while [El Karoui 2007] provides some conditions on C which will result in Tracy–Widom for $\lambda_{\max}(X^T C X)$. Both results however pertain only to the $\beta = 2$ case as the rely on the special structure of the eigenvalue density at that value of the parameter. Perhaps the strategy outlined above — scaling directly in the operator rather than in the spectral distribution — can be successfully employed in this direction.

Painlevé expressions. One hopes that either the random Airy operator or the associated diffusion will lead to explicit formulas in terms of Painlevé II for the limiting largest eigenvalue distributions at all $\beta > 0$. While we appear to be far from realizing this goal, here perhaps is a hint. By the Cameron–Martin formula: with $F_\beta(\lambda)$ the distribution function of $-\Lambda_0(\beta)$,

$$
\begin{aligned}
F_\beta(\lambda) \\
= \lim_{L \to \infty} \lim_{a \to \infty} \int_{p(0)=a} e^{-\frac{\beta}{8} \int_0^L (\lambda + x - p^2(x))\, dp(x)} e^{-\frac{\beta}{8} \int_0^L (\lambda + x - p^2(x))^2 dx} \\
\times \frac{e^{-\frac{\beta}{2} \int_0^L (p'(x))^2 dx}}{(2\pi 0^+)^{\infty/2}}\, dp^\infty.
\end{aligned}
$$

The Itô factor

$$
\int_0^L ((\lambda + x - p^2(x))\, dp(x)
$$

only contributes boundary terms, leaving the integral to concentrate on minimizers of the functional

$$p \mapsto \int_0^L \left([\lambda + x - p^2(x)]^2 + (p'(x))^2 \right) dx.$$

The associated Euler–Lagrange equation is Painlevé II.

Acknowledgements

We thank the referees for pointing out several important references.

References

[Baik et al. 1999] J. Baik, P. Deift, and K. Johansson, "On the distribution of the length of the longest increasing subsequence of random permutations", *J. Amer. Math. Soc.* **12**:4 (1999), 1119–1178.

[Baik et al. 2005] J. Baik, G. Ben Arous, and S. Péché, "Phase transition of the largest eigenvalue for nonnull complex sample covariance matrices", *Ann. Probab.* **33**:5 (2005), 1643–1697.

[Baryshnikov 2001] Y. Baryshnikov, "GUEs and queues", *Probab. Theory Related Fields* **119**:2 (2001), 256–274.

[Cambronero 1996] S. Cambronero, *The ground state of Hill's equation with random potential*, Ph.D. thesis, New York University, 1996.

[Cambronero and McKean 1999] S. Cambronero and H. P. McKean, "The ground state eigenvalue of Hill's equation with white noise potential", *Comm. Pure Appl. Math.* **52**:10 (1999), 1277–1294.

[Cambronero et al. 2006] S. Cambronero, B. Rider, and J. Ramírez, "On the shape of the ground state eigenvalue density of a random Hill's equation", *Comm. Pure Appl. Math.* **59**:7 (2006), 935–976.

[Carmona and Lacroix 1990] R. Carmona and J. Lacroix, *Spectral theory of random Schrödinger operators*, Birkhäuser, Boston, 1990.

[Dieng 2005] M. Dieng, "Distribution functions for edge eigenvalues in orthogonal and symplectic ensembles: Painlevé representations", *Int. Math. Res. Not.* no. 37 (2005), 2263–2287.

[Dumitriu and Edelman 2002] I. Dumitriu and A. Edelman, "Matrix models for beta ensembles", *J. Math. Phys.* **43**:11 (2002), 5830–5847.

[Edelman and Sutton 2007] A. Edelman and B. D. Sutton, "From random matrices to stochastic operators", *J. Stat. Phys.* **127**:6 (2007), 1121–1165.

[El Karoui 2007] N. El Karoui, "Tracy-Widom limit for the largest eigenvalue of a large class of complex sample covariance matrices.", *Ann. Probab.* **35**:2 (2007), 663–714.

[Frisch and Lloyd 1960] H. L. Frisch and S. P. Lloyd, "Electron levels in a one-dimensional lattice", *Phys. Rev.* **120**:4 (1960), 1175–1189.

[Fukushima and Nakao 1976/77] M. Fukushima and S. Nakao, "On spectra of the Schrödinger operator with a white Gaussian noise potential", *Z. Wahrscheinlichkeitstheorie und Verw. Gebiete* **37**:3 (1976/77), 267–274.

[Gärtner et al. 2000] J. Gärtner, W. König, and S. A. Molchanov, "Almost sure asymptotics for the continuous parabolic Anderson model", *Probab. Theory Related Fields* **118**:4 (2000), 547–573.

[Grenkova et al. 1983] L. N. Grenkova, S. A. Molčanov, and J. N. Sudarev, "On the basic states of one-dimensional disordered structures", *Comm. Math. Phys.* **90**:1 (1983), 101–123.

[Halperin 1965] B. I. Halperin, "Green's functions for a particle in a one-dimensional random potential", *Phys. Rev.* (2) **139** (1965), A104–A117.

[Hochstadt 1961] H. Hochstadt, "Asymptotic estimates for the Sturm-Liouville spectrum", *Comm. Pure Appl. Math.* **14** (1961), 749–764.

[Ince 1940] E. L. Ince, "The periodic Lamé functions", *Proc. Roy. Soc. Edinburgh* **60** (1940), 47–63.

[Johansson 2000] K. Johansson, "Shape fluctuations and random matrices", *Comm. Math. Phys.* **209**:2 (2000), 437–476.

[Johnstone 2001] I. M. Johnstone, "On the distribution of the largest eigenvalue in principal components analysis", *Ann. Statist.* **29**:2 (2001), 295–327.

[Lax and Pillips 1958] M. Lax and J. C. Pillips, "One-dimensional impurity bands", *Phys. Rev.* (2) **110** (1958), 41–49.

[McKean 1994] H. P. McKean, "A limit law for the ground state of Hill's equation", *J. Statist. Phys.* **74**:5-6 (1994), 1227–1232.

[Merkl 2003] F. Merkl, "Quenched asymptotics of the ground state energy of random Schrödinger operators with scaled Gibbsian potentials", *Probab. Theory Related Fields* **126**:3 (2003), 307–338.

[Molčanov 1980/81] S. A. Molčanov, "The local structure of the spectrum of the one-dimensional Schrödinger operator", *Comm. Math. Phys.* **78**:3 (1980/81), 429–446.

[Pastur 1972] L. A. Pastur, "The distribution of eigenvalues of the Schrödinger equation with a random potential", *Funkcional. Anal. i Priložen.* **6**:2 (1972), 93–94. In Russian; translated in *Funct. Anal. Appl.* **6** (1972), 163–165.

[Ramírez and Rider 2006] J. Ramírez and B. Rider, "On the ground state energy of Hill's equation with nice Gaussian potential", Preprint, 2006. Available at http://arxiv.org/abs/math.PR/0611555.

[Ramírez et al. 2006] J. Ramírez, B. Rider, and B. Virág, "Beta ensembles, stochastic Airy spectrum, and a diffusion", Preprint, 2006. Available at http://arxiv.org/abs/math.PR/0607331.

[Schmidt 1957] H. Schmidt, "Disordered one-dimensional crystals", *Phys. Rev.* (2) **105** (1957), 425–441.

[Sznitman 1998] A.-S. Sznitman, *Brownian motion, obstacles and random media*, Springer, Berlin, 1998.

[Tracy and Widom 1994] C. A. Tracy and H. Widom, "Level-spacing distributions and the Airy kernel", *Comm. Math. Phys.* **159**:1 (1994), 151–174.

[Tracy and Widom 1996] C. A. Tracy and H. Widom, "On orthogonal and symplectic matrix ensembles", *Comm. Math. Phys.* **177**:3 (1996), 727–754.

SANTIAGO CAMBRONERO
DEPARTMENT OF MATHEMATICS
UNIVERSIDAD DE COSTA RICA
SAN JOSE 2060
COSTA RICA
  sambro@emate.ucr.ac.cr

JOSÉ RAMÍREZ
DEPARTMENT OF MATHEMATICS
UNIVERSIDAD DE COSTA RICA
SAN JOSE 2060
COSTA RICA
  jaramirez@cariari.ucr.ac.cr

BRIAN RIDER
DEPARTMENT OF MATHEMATICS
UNIVERSITY OF COLORADO
UCB 395
BOULDER, CO 80309
UNITED STATES
  brider@euclid.colorado.edu

Probability, Geometry and Integrable Systems
MSRI Publications
Volume **55**, 2008

SLE_6 and CLE_6 from critical percolation

FEDERICO CAMIA AND CHARLES M. NEWMAN

ABSTRACT. We review some of the recent progress on the scaling limit of two-dimensional critical percolation; in particular, the convergence of the exploration path to chordal SLE_6 and the full scaling limit of cluster interface loops. The results given here on the full scaling limit and its conformal invariance extend those presented previously. For site percolation on the triangular lattice, the results are fully rigorous. We explain some of the main ideas, skipping most technical details.

1. Introduction

In the theory of critical phenomena it is usually assumed that a physical system near a continuous phase transition is characterized by a single length scale (the correlation length) in terms of which all other lengths should be measured. When combined with the experimental observation that the correlation length diverges at the phase transition, this simple but strong assumption, known as the scaling hypothesis, leads to the belief that at criticality the system has no characteristic length, and is therefore invariant under scale transformations. This suggests that all thermodynamic functions at criticality are homogeneous functions, and predicts the appearance of power laws.

It also implies that if one rescales appropriately a critical lattice model, shrinking the lattice spacing to zero, it should be possible to obtain a continuum model, known as the scaling limit. The scaling limit is not restricted to a lattice and may possess more symmetries than the original model. Indeed, the scaling limits

Keywords: continuum scaling limit, percolation, critical behavior, triangular lattice, conformal invariance, SLE.

AMS 2000 Subject Classification: 82B27, 60K35, 82B43, 60D05, 30C35.

Research supported in part by the NSF under grant PHY99-07949 (preprint no. NSF-KITP-06-76). Camia was supported in part by a Veni grant of the NWO (Dutch Organization for Scientific Research). Newman was supported in part by the NSF under grants DMS-01-04278 and DMS-06-06696.

of many critical lattice models are believed to be conformally invariant and to correspond to Conformal Field Theories (CFTs). But until recently, such a correspondence was at most heuristic, and was assumed as a starting point by physicists working in CFT. The methods of CFT themselves proved hard to put into a rigorous mathematical formulation.

The introduction by Oded Schramm [2000] of Stochastic/Schramm Loewner Evolution (SLE) has provided a new powerful and mathematically rigorous tool to study scaling limits of critical lattice models. Thanks to this, in recent years tremendous progress has been made in understanding the conformally invariant nature of the scaling limits of several such models.

While CFT focuses on correlation functions of local operators (e.g., spin variables in the Ising model), SLE describes the behavior of macroscopic random curves present in these models, such as percolation cluster boundaries. In the scaling limit, the distribution of such random curves can be uniquely identified thanks to their conformal invariance and a certain "Markovian" property. There is a one-parameter family of SLEs, indexed by a positive real number κ, and they appear to be essentially the only possible candidates for the scaling limits of interfaces of two-dimensional critical systems that are believed to be conformally invariant.

The main power of SLE stems from the fact that it allows to compute different quantities; for example, percolation crossing probabilities and various percolation critical exponents. Therefore, relating the scaling limit of a critical lattice model to SLE allows for a rigorous determination of some aspects of the large scale behavior of the lattice model.

In the context of the Ising, Potts and $O(n)$ models, an SLE curve is believed to describe the scaling limit of a single interface, which can be obtained by imposing special boundary conditions. A single SLE curve is therefore not in itself sufficient to immediately describe the scaling limit of the unconstrained model without boundary conditions in the whole plane (or in domains with boundary conditions that do not determine a single interface), and contains only limited information concerning the connectivity properties of the model.

A more complete description can be obtained in terms of loops, corresponding to the scaling limits of cluster boundaries. Such loops should also be random and have a conformally invariant distribution. This approach led Wendelin Werner [2005b; 2005a] (see also [Werner 2003]) to the definition of Conformal Loop Ensembles (CLEs), which are, roughly speaking, random collections of fractal loops with a certain conformal restriction/renewal property.

For percolation, a complete proof of the connection with SLE, first conjectured in [Schramm 2000], has recently been given in [Camia and Newman 2007]. The proof relies heavily on the ground breaking result of Stas Smirnov [2001] about the existence and conformal invariance of the scaling limit of crossing probabilities (see [Cardy 1992]). The last section of this paper explains the main ideas of that proof, highlighting the role of conformal invariance, but without dwelling on the heavy technical details.

As for the Ising, Potts and $O(n)$ models, the scaling limit of percolation in the whole plane should be described by a measure on loops, where the loops are closely related to SLE curves. Such a description in the case of percolation was presented in [Camia and Newman 2004], where the authors of the present paper constructed a probability measure on collections of fractal conformally invariant loops in the plane (closely related to a CLE), arguing that it corresponds to the full scaling limit of critical two-dimensional percolation. A proof of that statement was subsequently provided in [Camia and Newman 2006].

Here, we will briefly explain how to go from a single SLE curve to the full scaling limit, again skipping the technical details, for the case of a Jordan domain with monochromatic boundary conditions (see Theorem 2). This extends the results presented in [Camia and Newman 2006], where the scaling limit was first taken in the unit disc and then an infinite volume limit was taken in order to obtain the full scaling limit in the whole plane. Moving from the unit disc (or any convex domain) to a general Jordan domain introduces extra complications that are dealt with using a new argument, developed in [Camia and Newman 2007], that exploits the continuity of Cardy's formula [1992] with respect to changes in the shape of the domain (see the discussion in Section 5). Taking scaling limits in general Jordan domains is a necessary step in order to consider conformal restriction/renewal properties as in Theorem 4 below.

Using the full scaling limit, one can attempt to understand the geometry of the near-critical scaling limit, where the percolation density tends to the critical one in an appropriate way as the lattice spacing tends to zero. A heuristic analysis [Camia et al. 2006a; 2006b] based on a natural ansatz leads to a one-parameter family of loop models (i.e., probability measures on random collections of loops), with the critical full scaling limit corresponding to a particular choice of the parameter. Except for the latter case, these measures are not scale invariant, but are mapped into one another by scale transformations. This framework can be used to define a renormalization group flow (under the action of dilations), and to describe the scaling limit of related models, such as invasion and dynamical percolation and the minimal spanning tree. In particular, this analysis helps explain why the scaling limit of the minimal spanning tree may be scale invariant but *not* conformally invariant, as first observed numerically by Wilson [2004].

2. SLE and CLE

The Stochastic/Schramm Loewner Evolution with parameter $\kappa > 0$ (SLE$_\kappa$) was introduced by Schramm [2000] as a tool for studying the scaling limit of two-dimensional discrete (defined on a lattice) probabilistic models whose scaling limits are expected to be conformally invariant. In this section we define the chordal version of SLE$_\kappa$; for more on the subject, the interested reader can consult Schramm's paper as well as the fine reviews by Lawler [2004], Kager and Nienhuis [2004], and Werner [2004], and Lawler's book [2005].

Let $\mathbb{H}$ denote the upper half-plane. For a given continuous real function U_t with $U_0 = 0$, define, for each $z \in \overline{\mathbb{H}}$, the function $g_t(z)$ as the solution to the ODE

$$\partial_t g_t(z) = \frac{2}{g_t(z) - U_t}, \tag{2-1}$$

with $g_0(z) = z$. This is well defined as long as $g_t(z) - U_t \neq 0$, i.e., for all $t < T(z)$, where

$$T(z) \equiv \sup\{t \geq 0 : \min_{s \in [0,t]} |g_s(z) - U_s| > 0\}. \tag{2-2}$$

Let $K_t \equiv \{z \in \overline{\mathbb{H}} : T(z) \leq t\}$ and let $\mathbb{H}_t$ be the unbounded component of $\mathbb{H} \setminus K_t$; it can be shown that K_t is bounded and that g_t is a conformal map from $\mathbb{H}_t$ onto $\mathbb{H}$. For each t, it is possible to write $g_t(z)$ as

$$g_t(z) = z + \frac{2t}{z} + O\left(\frac{1}{z^2}\right), \tag{2-3}$$

when $z \to \infty$. The family $(K_t, t \geq 0)$ is called the *Loewner chain* associated to the driving function $(U_t, t \geq 0)$.

DEFINITION 2.1. *Chordal SLE$_\kappa$* is the Loewner chain $(K_t, t \geq 0)$ that is obtained when the driving function $U_t = \sqrt{\kappa} B_t$ is $\sqrt{\kappa}$ times a standard real-valued Brownian motion $(B_t, t \geq 0)$ with $B_0 = 0$.

For all $\kappa \geq 0$, chordal SLE$_\kappa$ is almost surely generated by a continuous random curve γ in the sense that, for all $t \geq 0$, $\mathbb{H}_t \equiv \mathbb{H} \setminus K_t$ is the unbounded connected component of $\mathbb{H} \setminus \gamma[0, t]$; γ is called the *trace* of chordal SLE$_\kappa$.

It is not hard to see, as argued by Schramm, that any continuous random curve γ in the upper half-plane starting at the origin and going to infinity must be an SLE curve if it possesses the following *conformal Markov property*. For any fixed $T \in \mathbb{R}$, conditioning on $\gamma[0, T]$, the image under g_T of $\gamma(T, \infty)$ is distributed like an independent copy of γ, up to a time reparametrization. This implies that the driving function U_t in the Loewner chain associated to the curve γ is continuous and has stationary and independent increments. If the time parametrization implicit in Definition 2.1 and the discussion preceding it is chosen for γ, then scale invariance also implies that the law of U_t is the same as the law of $\lambda^{-1/2}U_{\lambda t}$ when $\lambda > 0$. These properties together imply that U_t must be a constant multiple of standard Brownian motion.

Now let $D \subset \mathbb{C}$ $(D \neq \mathbb{C})$ be a simply connected domain whose boundary is a continuous curve. By Riemann's mapping theorem, there are (many) conformal maps from the upper half-plane $\mathbb{H}$ onto D. In particular, given two distinct points $a, b \in \partial D$ (or more accurately, two distinct prime ends), there exists a conformal map f from $\mathbb{H}$ onto D such that $f(0) = a$ and $f(\infty) \equiv \lim_{|z| \to \infty} f(z) = b$. In fact, the choice of the points a and b on the boundary of D only characterizes $f(\cdot)$ up to a scaling factor $\lambda > 0$, since $f(\lambda \cdot)$ would also do.

Suppose that $(K_t, t \geq 0)$ is a chordal SLE$_\kappa$ in $\mathbb{H}$ as defined above; we define chordal SLE$_\kappa$ $(\tilde{K}_t, t \geq 0)$ in D from a to b as the image of the Loewner chain $(K_t, t \geq 0)$ under f. It is possible to show, using scaling properties of SLE$_\kappa$, that the law of $(\tilde{K}_t, t \geq 0)$ is unchanged, up to a linear time-change, if we replace $f(\cdot)$ by $f(\lambda \cdot)$. This makes it natural to consider $(\tilde{K}_t, t \geq 0)$ as a process from a to b in D, ignoring the role of f. The trace of chordal SLE in D from a to b will be denoted by $\gamma_{D,a,b}$.

We now move from the conformally invariant random curves of SLE to collections of conformally invariant random loops and introduce the concept of Conformal Loop Ensemble (CLE — see [Werner 2003; 2005b; 2005a; Sheffield 2006]). The key feature of a CLE is a sort of conformal restriction/renewal property. Roughly speaking, a CLE in D is a random collection $\mathcal{L}_D$ of loops such that if all the loops intersecting a (closed) subset D' of D or of its boundary are removed, the loops in any one of the various remaining (disjoint) subdomains of D form a random collection of loops distributed as an independent copy of $\mathcal{L}_D$ conformally mapped to that subdomain (see Theorem 4). We will not attempt to be more precise here since somewhat different definitions (although, in the end, substantially equivalent) have appeared in the literature, but the meaning of the conformal restriction/renewal property should be clear from Theorem 4.

For formal definitions and more discussion on the properties of a CLE, see the original literature on the subject [Werner 2005b; 2005a; Sheffield 2006], where it is shown that there is a one-parameter family CLE$_\kappa$ of conformal loop ensembles with the above conformal restriction/renewal property and that for $\kappa \in (8/3, 8]$, the CLE$_\kappa$ loops locally look like SLE$_\kappa$ curves.

There are numerous lattice models that can be described in terms of random curves and whose scaling limits are assumed (and in a few cases proved) to be conformally invariant. These include the Loop Erased Random Walk, the Self-Avoiding Walk and the Harmonic Explorer, all of which can be defined as polygonal paths along the edges of a lattice. The Ising, Potts and percolation models instead are naturally defined in terms of clusters, and the interfaces between different clusters form random loops. In the $O(n)$ model, configurations of loops along the edges of the hexagonal lattice are weighted according to the total number and length of the loops. All of these models are supposed to have scaling limits described by SLE_κ or CLE_κ for some value of κ between 2 and 8. For more information on these lattice models and their scaling limits, the interested reader can consult [Cardy 2001; 2005; Kager and Nienhuis 2004; Werner 2005b; Sheffield 2006].

In the rest of the paper we will restrict attention to percolation, where the connection with SLE_6 and CLE_6 has been made rigorous [Smirnov 2001; Camia and Newman 2006; 2007].

3. Conformal invariance of critical percolation

In this section we will consider critical site percolation on the triangular lattice, for which conformal invariance in the scaling limit was rigorously proved [Smirnov 2001]. A precise formulation of conformal invariance, attributed to Michael Aizenman, is that the probability that a percolation cluster crosses between two disjoint segments of the boundary of some simply connected domain should converge to a conformally invariant function of the domain and the two segments of the boundary. This conjecture is connected with the extensive numerical investigations reported in [Langlands et al. 1994]. A formula for the purposed limit was then derived by John Cardy [1992] using (nonrigorous) field theoretical methods. The interest of mathematicians was already evident in [Langlands et al. 1994], but a proof of the conjecture [Smirnov 2001] (and of Cardy's formula) did not come until 2001.

We will denote by $\mathcal{T}$ the two-dimensional triangular lattice, whose sites are identified with the elementary cells of a regular hexagonal lattice $\mathcal{H}$ embedded in the plane as in Figure 1. We say that two hexagons are neighbors (or that they are adjacent) if they have a common edge. A sequence $(\xi_0, \dots, \xi_n)$ of hexagons of $\mathcal{H}$ such that ξ_{i-1} and ξ_i are neighbors for all $i = 1, \dots, n$ and $\xi_i \neq \xi_j$ whenever $i \neq j$ will be called a $\mathcal{T}$-path. If the first and last hexagons of the path are neighbors, the path will be called a $\mathcal{T}$-loop.

Let D be a bounded simply connected domain containing the origin whose boundary ∂D is a continuous curve. Let $\phi : \overline{\mathbb{D}} \to D$ be the (unique) continuous function that maps $\mathbb{D}$ onto D conformally and such that $\phi(0) = 0$ and $\phi'(0) > 0$. Let z_1, z_2, z_3, z_4 be four points of ∂D in counterclockwise order — i.e., such that $z_j = \phi(w_j)$, $j = 1, 2, 3, 4$, with $w_1, \dots, w_4$ in counterclockwise order. Also, let

$$\eta = \frac{(w_1 - w_2)(w_3 - w_4)}{(w_1 - w_3)(w_2 - w_4)}.$$

Cardy's formula [1992] for the probability $\Phi_D(z_1, z_2; z_3, z_4)$ of a crossing inside D from the counterclockwise arc $\overline{z_1 z_2}$ to the counterclockwise arc $\overline{z_3 z_4}$ is

$$\Phi_D(z_1, z_2; z_3, z_4) = \frac{\Gamma(2/3)}{\Gamma(4/3)\Gamma(1/3)} \eta^{1/3} {}_2F_1(1/3, 2/3; 4/3; \eta), \qquad (3\text{-}1)$$

where ${}_2F_1$ is a hypergeometric function.

For a given mesh $\delta > 0$, the probability of a blue crossing inside D from the counterclockwise arc $\overline{z_1 z_2}$ to the counterclockwise arc $\overline{z_3 z_4}$ is the probability of the existence of a blue $\mathcal{T}$-path $(\xi_0, \ldots, \xi_n)$ such that ξ_0 intersects the counterclockwise arc $\overline{z_1 z_2}$, ξ_n intersects the counterclockwise arc $\overline{z_3 z_4}$, and $\xi_1, \ldots, \xi_{n-1}$ are all contained in D. Smirnov [2001] proved that crossing probabilities converge in the scaling limit to conformally invariant functions of the domain and the four points on its boundary, and identified the limit with Cardy's formula (3-1).

The proof of Smirnov's theorem is based on the identification of certain generalized crossing probabilities that are almost discrete harmonic functions and whose scaling limits converge to harmonic functions. The behavior on the boundary of such functions is easy to determine and is sufficient to specify them uniquely. The relevant crossing probabilities can be expressed in terms of the boundary values of such harmonic functions, and as a consequence are invariant under conformal transformations of the domain and the two segments of its boundary.

The presence of a blue crossing in D from the counterclockwise boundary arc $\overline{z_1 z_2}$ to the counterclockwise boundary arc $\overline{z_3 z_4}$ can be determined using a clever algorithm that explores the percolation configuration inside D starting at, say, z_1 and assumes that the hexagons just outside $\overline{z_1 z_2}$ are all blue and those just outside $\overline{z_4 z_1}$ are all yellow. The exploration proceeds following the interface between the blue cluster adjacent to $\overline{z_1 z_2}$ and the yellow cluster adjacent to $\overline{z_4 z_1}$. A blue crossing is present if the exploration process reaches $\overline{z_3 z_4}$ before $\overline{z_2 z_3}$. This *exploration process* and the *exploration path* (see Figure 1) associated to it were introduced in [Schramm 2000].

The exploration process can be carried out in $\mathbb{H} \cap \mathcal{H}$, where the hexagons in the lowest row and to the left of a chosen hexagon have been colored yellow and the remaining hexagons in the lowest row have been colored blue. This produces an infinite exploration path, whose scaling limit was conjectured [Schramm 2000] by Schramm to converge to SLE$_6$.

It is easy to see that the exploration process is Markovian in the sense that, conditioned on the exploration up to a certain (stopping) time, the future of the exploration evolves in the same way as the past except that it is now performed in a different domain, where some of the explored hexagons have become part of the boundary (see, e.g., Figure 1).

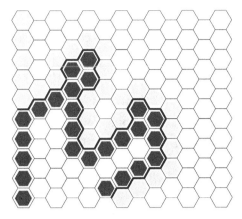

Figure 1. Percolation exploration path in a portion of the hexagonal lattice with blue/yellow boundary conditions on the first column, corresponding to the boundary of the region where the exploration is carried out. The colored hexagons that do not belong to the first column have been explored during the exploration process. The heavy line between yellow (light) and blue (dark) hexagons is the exploration path produced by the exploration process.

This observation, together with the connection between the exploration process and crossing probabilities, Smirnov's theorem about the conformal invariance of crossing probabilities in the scaling limit, and Schramm's characterization of SLE via the conformal Markov property discussed in Section 2, strongly support the above conjecture.

As we now explain, the natural setting to define the exploration process is that of *lattice domains*, i.e., sets D^δ of hexagons of $\delta\mathcal{H}$ that are *connected* in the sense that any two hexagons in D^δ can be joined by a $(\delta\mathcal{T})$-path contained in D^δ. We say that a bounded lattice domain D^δ is *simply connected* if both D^δ and $\delta\mathcal{T} \setminus D^\delta$ are connected. A *lattice-Jordan* domain D^δ is a bounded simply connected lattice domain such that the set of hexagons adjacent to D^δ is a $(\delta\mathcal{T})$-loop.

Given a lattice-Jordan domain D^δ, the set of hexagons adjacent to D^δ can be partitioned into two (lattice-)connected sets. If those two sets of hexagons are assigned different colors, for any coloring of the hexagons inside D^δ, there is an interface between two clusters of different colors starting and ending at two boundary points, a^δ and b^δ, corresponding to the locations on the boundary of D^δ where the color changes. If one performs an exploration process in D^δ starting at a^δ, one ends at b^δ, producing an exploration path γ^δ that traces the entire interface from a^δ to b^δ.

Given a planar domain D, we denote by ∂D its topological boundary. Let ∂D be locally connected (i.e., a continuous curve), and assume that D contains the origin. Then one can parametrize ∂D by $\varphi : S^1 \to \partial D$, where φ is the restriction to the unit circle S^1 of the continuous map $\phi : \overline{\mathbb{D}} \to \overline{D}$ that is conformal in $\mathbb{D}$ and satisfies $\phi(0) = 0$, $\phi'(0) > 0$. With this notation, we say that D^δ converges to D as $\delta \to 0$ if

$$\liminf_{\delta \to 0} \sup_{h} \sup_{z \in S^1} |\varphi(z) - \varphi^\delta(h(z))| = 0, \tag{3-2}$$

where the infimum is over monotonic functions $h : S^1 \to S^1$ (and the objects with the superscript δ refer to D^δ — for simplicity we are assuming that all domains contain the origin). If moreover two points, $a^\delta, b^\delta \in \partial D^\delta$, converge respectively to $a, b \in \partial D$ as $\delta \to 0$, we write $(D^\delta, a^\delta, b^\delta) \to (D, a, b)$. In the following theorem the topology on curves is that induced by the supremum norm, but with monotonic reparametrizations of the curves allowed (see [Aizenman and Burchard 1999; Camia and Newman 2006; 2007]), i.e., the distance between curves is

$$d(\gamma, \gamma^\delta) = \inf_{h} \sup_{t \in [0,\infty)} |\gamma(t) - \gamma^\delta(h(t))|, \tag{3-3}$$

where $\gamma(t), \gamma^\delta(t), t \in [0, \infty)$, are parametrizations of $\gamma_{D,a,b}$ and $\gamma^\delta_{D,a,b}$ respectively, and the infimum is over monotonic functions $h : [0, \infty) \to [0, \infty)$. A proof of the theorem can be found in [Camia and Newman 2007] and a detailed sketch is presented in Section 6 below.

THEOREM 1. *Let (D, a, b) be a Jordan domain with two distinct selected points on its boundary ∂D. Then, for lattice-Jordan domains D^δ from $\delta\mathcal{H}$ with $a^\delta, b^\delta \in \partial D^\delta$ such that $(D^\delta, a^\delta, b^\delta) \to (D, a, b)$ as $\delta \to 0$, the percolation exploration path $\gamma^\delta_{D,a,b}$ in D^δ from a^δ to b^δ converges in distribution to the trace $\gamma_{D,a,b}$ of chordal SLE₆ in D from a to b, as $\delta \to 0$.*

4. The full scaling limit in a Jordan domain

In this section we define the *Continuum Nonsimple Loop (CNL) process* in a Jordan domain D, a random collection of countably many nonsimple fractal loops in D which corresponds to the full scaling limit of percolation in D with monochromatic boundary conditions. This refers to the collection of all cluster boundaries of percolation configurations in D with the hexagons at the boundary of D all blue (obviously, one could as well choose yellow boundary conditions). The algorithmic construction that we present below is analogous to that of [Camia and Newman 2004; 2006] for the unit disc $\mathbb{D}$, but here we perform it in a general Jordan domain.

The CNL process on the full plane can be obtained by taking a sequence of domains D tending to $\mathbb{C}$. This was done in the two works just cited, and for that purpose, discs of radius R with $R \to \infty$ suffice. This full plane CNL process is the scaling limit of the collection of all cluster boundaries in the full lattice (without boundary conditions). In order to consider conformal restriction/renewal properties (as we do in Theorem 4 below), one needs to consider the CNL process in fairly general bounded domains D. There are extra complications in taking the scaling limit when D is nonconvex, as discussed in Section 5.

The basic ingredient in our algorithmic construction consists of a chordal SLE_6 path between two points on the boundary of a Jordan domain. As we will explain soon, sometimes the two boundary points are naturally determined as a product of the construction itself, and sometimes they are given as an input to the construction. In the second case, there are various procedures which would yield the "correct" distribution for the resulting CNL process; one possibility is as follows. Given a domain D, choose a and b so that, of all points in ∂D, they have maximal x-distance or maximal y-distance, whichever is greater. It is important to stress that in the end, the CNL process will turn out to be independent of the actual choice of boundary points, as is evident in Theorem 2. (One caveat is that one should avoid malicious choices of the boundary points for which the entire original domain would not be explored asymptotically.)

The first step of our construction is a chordal SLE_6, $\gamma \equiv \gamma_{D,a,b}$, between two boundary points $a, b \in \partial D$ chosen according to the above rule (see Figure 2). The set $D \setminus \gamma_{D,a,b}[0, \infty)$ is a countable union of its connected components, which are open and simply connected. If z is a deterministic point in D, then with probability one, z is not touched by γ [Rohde and Schramm 2005] and so belongs to a unique one of these, that we denote $D_{a,b}(z)$. There are four kinds of components which may be usefully thought of in terms of how a point z in the interior of the component was first trapped at some time t_1 by $\gamma[0, t_1]$ perhaps together with either the counterclockwise arc $\partial_{a,b} D$ of ∂D between a and b or the counterclockwise arc $\partial_{b,a} D$ of ∂D between b and a: (1) those components whose boundary contains a segment of $\partial_{b,a} D$ between two successive visits at $\gamma(t_0)$ and $\gamma(t_1)$ to $\partial_{b,a} D$ (where here and below $t_0 < t_1$), (2) the analogous components with $\partial_{b,a} D$ replaced by the other part of the boundary $\partial_{a,b} D$, (3) those components formed when $\gamma(t_0) = \gamma(t_1)$ with γ winding about z in a counterclockwise direction between t_0 and t_1, and finally (4) the analogous clockwise components.

To conclude the first step, we consider all domains of type (1), corresponding to excursions of the SLE_6 path from the portion $\partial_{b,a} D$ of ∂D. For each such domain D', the points a' and b' on its boundary are chosen to be respectively those points where the excursion ends and where it begins, that is, for $D_{a,b}(z)$ we set $a' = \gamma((t_1(z))$ and $b' = \gamma(t_0(z))$. We then run a chordal SLE_6 from a' to b'. The loop obtained by pasting together the excursion from b' to a' followed

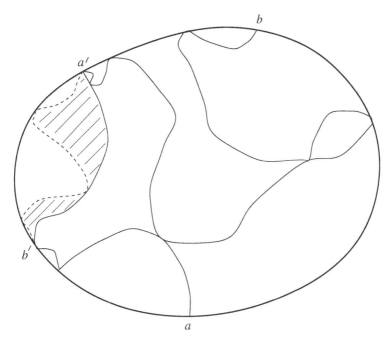

Figure 2. Schematic drawing of the construction of continuum nonsimple loops inside a Jordan domain D. The construction starts with a chordal SLE$_6$ (full line) between two points, a and b, on ∂D. To obtain loops, other chordal SLE$_6$s (e.g., dashed line) are run (e.g., from a' to b') between where an excursion from the counterclockwise arc $\partial_{b,a} D$ of ∂D of the first SLE$_6$ respectively ends and starts. The inside of one such loop is shaded.

by the new SLE$_6$ path from a' to b' is one of our continuum loops (see Figure 2). At the end of the first step, then, the procedure has generated countably many loops that touch $\partial_{b,a} D$; each of these loops touches $\partial_{b,a} D$ but may or may not touch $\partial_{a,b} D$.

The last part of the first step also produces new domains, corresponding to the connected components of $D' \setminus \gamma_{D',a',b'}[0, \infty)$ for all domains D' of type (1). Each one of these components, together with all the domains of type (2), (3) and (4) previously generated, is to be used in the next step of the construction, playing the role of the original domain D. For each one of these domains, we choose the new a and new b on the boundary as explained before, and then continue with the construction. Note that the new a and new b are chosen according to the rule explained at the beginning of this section also for domains of type (2), even though they are generated by excursions like the domains of type (1).

This iterative procedure produces at each step a countable set of loops. The limiting object, corresponding to the collection of all such loops, is our basic process. (Technically speaking, we should include also trivial loops fixed at each $z \in D$ so that the collection of loops is closed in an appropriate sense [Aizenman and Burchard 1999].)

As explained, the construction is carried out iteratively and can be performed simultaneously on all the domains that are generated at each step. We wish to emphasize, though, that the obvious monotonicity of the procedure, where at each step new paths are added independently in different domains, and new domains are formed from the existing ones, implies that any other choice of the order in which the domains are used would give the same result (i.e., produce the same limiting distribution), provided that every domain that is formed during the construction is eventually used.

The main interest of the loop process defined above is in the following theorem, where the topology on collections of loops is that of [Aizenman and Burchard 1999] (see also [Camia and Newman 2006]).

THEOREM 2. *In the scaling limit, $\delta \to 0$, the collection of all cluster boundaries of critical site percolation on the triangular lattice in a Jordan domain D with monochromatic boundary conditions converges in distribution to the Continuum Nonsimple Loop process in D.*

A key property of the CNL process is conformal invariance.

THEOREM 3. *Let D, D' be Jordan domains and $f : \overline{D} \to \overline{D}'$ a continuous function that maps D conformally onto D'. Then the CNL process in D' is distributed like the image under f of the CNL process in D.*

Moreover, as shown in the next theorem, the outermost loops of the CNL process in a Jordan domain satisfy a conformal restriction/renewal property, as in the definitions of the Conformal Loop Ensembles of Werner [2005b] and Sheffield [2006].

THEOREM 4. *Let D be a Jordan domain and $\mathcal{L}_D$ be the collection of CN loops in $\overline{D}$ that are not surrounded by any other loop. Consider an arc Γ of ∂D and let $\mathcal{L}_{D,\Gamma}$ be the set of loops of $\mathcal{L}_D$ that touch Γ. Then, conditioned on $\mathcal{L}_{D,\Gamma}$, for any connected component D' of $D \setminus \overline{\bigcup \{L : L \in \mathcal{L}_{D,\Gamma}\}}$, the loops in $\overline{D}'$ form a random collection of loops distributed as an independent copy of $\mathcal{L}_D$ conformally mapped to D'.*

Yet another form of conformal invariance is illustrated by showing how to obtain a (conformally invariant) SLE_6 curve from the CNL process. Given a Jordan domain D and two points $a, b \in \partial D$, let $\Gamma = \overline{ba}$ be the counterclockwise closed arc $\overline{ba}$ of ∂D. Define $\mathcal{L}_D$ and $\mathcal{L}_{D,\Gamma}$ as in Theorem 4. For each $L \in \mathcal{L}_{D,\Gamma}$, going from a to b clockwise, there are a first and a last point, x and y respectively, where L intersects Γ. We call the counterclockwise arc $\overline{xy}(L)$ of L between

x and y a (counterclockwise) *excursion* from $\overline{ba}$. We call such an $\overline{xy}(L)$ a *maximal* excursion if there is no other excursion from $\overline{ba}$ in (the closure of) the domain created by $\overline{xy}(L)$ and the counterclockwise arc $\overline{yx}$ of ∂D. The random curve obtained by pasting together (in the order in which they are encountered going from a to b clockwise) all such maximal excursions from $\overline{ba}$ is distributed like a chordal SLE$_6$ in D from a to b.

The procedure described above obviously requires some care, since there are countably many such excursions and there is no such thing as the first excursion encountered from a, or the next excursion. What this means is that in order to properly define the curve, one needs to use a limiting procedure. Since it is quite obvious how to do it but tedious to explain, we leave the details to the interested reader; see [Camia and Newman 2006].

5. Convergence and conformal invariance of the full scaling limit

SKETCH OF THE PROOF OF THEOREM 2. It follows directly from [Aizenman and Burchard 1999] that the family of distributions of the collections of cluster boundaries in D with monochromatic boundary conditions is tight, as $\delta \to 0$, in the sense of the induced Hausdorff metric on closed sets of curves based on the metric (3-3) for single curves (see [Aizenman and Burchard 1999] and [Camia and Newman 2006]), and so there is convergence along subsequences $\delta_k \to 0$. What needs to be proved is that the limiting distribution is that of the CNL process, independently of the subsequence δ_k.

The key to the proof is an algorithmic construction on the lattice which parallels the continuum construction of Section 4 used to define the CNL process in D. The construction takes place in a lattice-domain $D_k \equiv D^{\delta_k}$ that converges to D in the sense of (3-2) as $k \to \infty$ ($\delta_k \to 0$) and is essentially the same as the continuum one but with exploration paths instead of the SLE$_6$ curves.

This raises the question of how to define an exploration process and obtain an exploration path in a lattice-domain with monochromatic boundary conditions. The basic idea is that away from the boundary, the exploration process does not know the boundary conditions. For two given points x and y on the boundary of a lattice-domain with, say, blue boundary conditions, split the boundary into two arcs, the counterclockwise arc $\overline{xy}$ and the counterclockwise arc $\overline{yx}$. Then, one can run an exploration process from x to y with the usual rule inside the domain and on the counterclockwise arc $\overline{xy}$, while pretending that the counterclockwise arc $\overline{yx}$ is colored yellow (see Figure 3).

If we run such an exploration process in D_k and then look at the hexagons that have not yet been explored, we will see several disjoint lattice subdomains, all of which are lattice-Jordan. This amounts to removing the fattened exploration

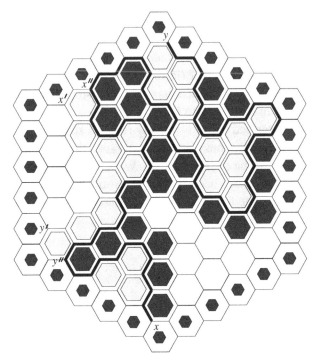

Figure 3. First step of the construction of the outer contour of a cluster of yellow (light in the figure) hexagons consisting of an exploration (heavy line) from x to y. The outer layer of hexagons does not belong to the domain where the explorations are carried out, but represents its monochromatic blue external boundary. x'' and y'' are the ending and starting points of an excursion that determines a new domain D', and x' and y' are the vertices where the edges that separate the yellow and blue portions of the external boundary of D' intersect $\partial D'$. The second step will consist of an exploration process in D' from x' to y'.

path consisting of the exploration path $\gamma_k \equiv \gamma_{D_k,x,y}^{\delta_k}$ itself and the hexagons immediately to its right and to its left.

The resulting lattice-Jordan subdomains are of four types, which may be usefully thought of in terms of their external boundaries: (1) those components whose boundary contains both sites in the fattened exploration path and in ∂_{yx}^k, the counterclockwise portion between y and x of the boundary of D_k, (2) the analogous components with ∂_{yx}^k replaced by the other boundary portion ∂_{xy}^k, (3) those components whose boundary only contains yellow hexagons from the fattened exploration path and finally (4) the analogous components whose boundary only contains blue hexagons from the fattened exploration path.

Notice that the components of type 1 are the only ones with mixed (partly blue and partly yellow) boundary conditions, while all other components have

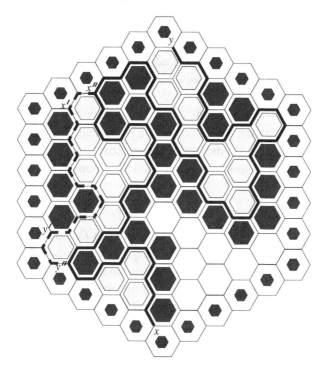

Figure 4. Second step of the construction of the outer contour of a cluster of yellow (light in the figure) hexagons consisting of an exploration from x' to y' whose resulting path (heavy broken line) is pasted to a portion of the previous exploration path with the help of the edges (indicated again by a heavy broken line) between x' and x'' and between y' and y'' in such a way as to obtain a loop around a yellow cluster (light in the figure) touching the boundary portion ∂^k_{yx}.

monochromatic (blue or yellow) boundary conditions; type 1 components are special because we have taken blue boundary conditions on D_k while the exploration path has yellow on its left and blue on its right. Because of the mixed boundary conditions, each lattice subdomain of type 1 must contain an interface between the two boundary points where the color changes. It is also clear that to find such an interface one has to start an exploration process at one of the two boundary points where the color changes (the two choices give the same exploration path).

If we run such an exploration process inside a lattice subdomain D'_k of type 1 and paste it to a portion of γ_k as in Figure 4, we obtain a loop corresponding to the interface surrounding a yellow cluster that touches ∂^k_{yx}. If we then again remove the fattened exploration path, D'_k is split into various components, but this time those lattice subdomains all have monochromatic boundary conditions.

If we do the same in each subdomain of type 1, we obtain a collection of loops. Moreover, all the lattice subdomains of D_k of nonexplored hexagons then have monochromatic boundary conditions. Thus we can iterate the whole procedure inside each of those lattice subdomains, until we have found all the interfaces contained in D_k.

The similarity between this construction and the continuum one of the CNL process should be apparent. To continue the proof one needs first to show that the exploration paths used in the lattice construction converge to chordal SLE$_6$ curves. The first step is a simple application of Theorem 1 to the first exploration path

$$\gamma_k = \gamma_{D_k,x_k,y_k}^{\delta_k},$$

where D_k, x_k, y_k are chosen so that D_k converges to D and x_k and y_k converge to the a and b of the continuum construction. However, in order to iterate this step and apply Theorem 1 again, we need to also show that the subdomains of the lattice construction converge to those of the continuum construction.

The convergence in distribution of γ_k to $\gamma = \gamma_{D,a,b}$ implies that we can find versions of γ_k and γ on some probability space $(\Omega, \mathcal{B}, \mathbb{P})$ such that $\gamma_k(\omega)$ converges to $\gamma(\omega)$ for all $\omega \in \Omega$. Using the coupling, γ_k and γ, for δ_k small, are close in the sense of (3-3). This is, however, not sufficient. If we want to conclude convergence of the subdomains, we need that wherever γ touches the boundary of D, γ_k touches the boundary of D_k nearby. Closeness in the sense of (3-3) does not ensure this but only that γ_k gets close to the boundary ∂D_k.

Note that, if γ_k gets within distance R_1 of some point z on ∂D_k without touching ∂D_k within distance R_2 of z, with $R_2 > R_1 > \delta_k$, considering the fattened version of γ_k shows the existence of two $(\delta_k \mathcal{T})$-paths of one color, say yellow, and one $(\delta_k \mathcal{T})$-path of the other color, blue, crossing the annulus of inner radius R_1 and outer radius R_2 centered at z.

In [Camia and Newman 2006], where the construction of the CN loops is carried out in the unit disc $\mathbb{D}$, the problem is solved by using the fact that $\mathbb{D}$ is convex and resorting to an upper bound (see, e.g., [Lawler et al. 2002]) on the probability that three disjoint monochromatic $\mathcal{T}$-paths cross a semiannulus in a half-plane. More precisely, the probability that the upper half-plane $\mathbb{H}$ contains three disjoint monochromatic $(\delta\mathcal{T})$-paths crossing the annulus of inner radius R_1 and outer radius R_2 centered at a point z of the real axis is bounded above by a constant times $(R_1/R_2)^{1+\varepsilon}$ for some $\varepsilon > 0$ (for all $\delta < R_1 < R_2$). Since $\varepsilon > 0$, if we let $\delta, R_1 \to 0$ and cover any finite part of $\partial \mathbb{H}$ by $O(R_1)$ such annuli, the bound shows that such three-arm events with $R_1 \to 0$ do not occur in the scaling limit $\delta \to 0$ near $\partial \mathbb{H}$. For a domain D with a locally flat boundary or for a convex domain, this implies that, as $k \to \infty$ ($\delta_k \to 0$), the (lim sup of the) probability that γ_k gets within distance R_1 of *any* $z \in \partial D_k$ without touching the boundary within distance R_2 of z goes to zero as $R_1 \to 0$ for all (fixed) $R_2 > 0$. (In the case of a convex D this follows from the fact that the intersection of an annulus centered at the origin of the real axis with an appropriate translation and rotation of D is smaller than the intersection of the same annulus with $\mathbb{H}$, thus making the probability of three arms even smaller than in the case of the upper half-plane.)

We cannot use that bound here, since D is not necessarily convex (and even if it were, the D' domains of Theorems 3 and 4 will not generally be convex). Instead, we will use the continuity of Cardy's formula with respect to small changes in the shape of the domain. We postpone this issue until later and proceed with the sketch of the proof assuming that γ_k does not get close to the boundary of the domain without touching it nearby (probably).

Then the boundaries of the lattice/continuum subdomains obtained after running the first (coupled) exploration path/SLE$_6$ curve are close to each other in the metric (3-3). I.e., we can match lattice and continuum subdomains, at least for those whose diameter is larger than some ε_k which depends on δ_k. It is important that, as $k \to \infty$ (and $\delta_k \to 0$), we can let $\varepsilon_k \to 0$.

If we run an exploration process inside a (large) lattice subdomain D'_k converging to a continuum subdomain D', Theorem 1 allows us to conclude that the exploration path γ'_k in D'_k converges to the SLE$_6$ curve γ' in D' from a' to b', provided that the starting and ending points x'_k and y'_k of the exploration process are chosen so that they converge to a' and b' respectively as $k \to \infty$. We can now work with coupled versions of γ'_k and γ' and repeat the above argument with the new subdomains that they produce, obtaining again a match (with high probability).

This allows us to keep the lattice and continuum constructions coupled, which ensures in particular that the $(\delta_k \mathcal{T})$-loops obtained in the lattice construction converge, as $\delta_k \to 0$, to the loops obtained in the continuum construction.

For any fixed δ_k, it is clear that the lattice construction eventually finds all the boundary loops. However, to conclude that the CNL process is indeed the scaling limit of the collection of all interfaces, we need to show that, for any $\varepsilon > 0$, the number of steps of the discrete construction needed to find all the loops of diameter at least ε does not diverge as $k \to \infty$ (otherwise some loops would never be found in the scaling limit).

In [Camia and Newman 2006], this is resolved using percolation arguments (that make use of the RSW theorem [Russo 1978; Seymour and Welsh 1978] and FKG inequalities) to show that the size of the subdomains has a bounded away from zero probability of decreasing significantly at each iteration. We point out that the argument used in [Camia and Newman 2006], where the construction of the CN loops in carried out in the unit disc, is independent of the actual shape of

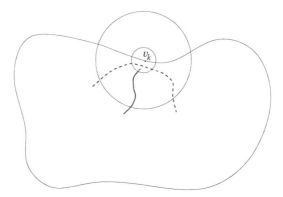

Figure 5. The figure shows a blue $(\delta_k \mathcal{T})$-path (heavy full line) crossing the partial annulus $D_k \cap \{B(v_k, R) \setminus B(v_k, r)\}$ that fails to connect to ∂D_k near v_k because it is blocked by a yellow $(\delta_k \mathcal{T})$-path (heavy dashed line) that twice crosses the annulus $B(v_k, R) \setminus B(v_k, r)$.

the domain so that it can be applied to the present situation. Since that argument is long, we will not repeat it here.

Returning to the problem of close encounters of γ_k with ∂D_k, we will try to provide the intuition on which the proof of touching is based. Suppose, by contradiction, that γ_k enters the disc $B(v_k, \varepsilon_k)$ of radius ε_k centered at $v_k \in \partial D_k$ without touching ∂D_k inside the disc $B(v_k, r)$ of radius r, and that $\varepsilon_k \to 0$. As $k \to \infty$, $D_k \to D$ and we can assume by compactness that v_k converges to some $v \in \partial D$. Considering the fattened version of γ_k shows the existence of two $(\delta_k \mathcal{T})$-paths of one color, say yellow, and one $(\delta_k \mathcal{T})$-path of the other color, blue, crossing the annulus $B(v_k, r) \setminus B(v_k, \varepsilon_k)$ (see Figure 5).

Assume for simplicity that v is far enough from a and b so that $a, b \notin \overline{B(v, R)}$ for some $R > r$, and consequently $x_k, y_k \notin B(v_k, R)$ for k large enough. Then, in the domain $D_k \cap \{B(v_k, R) \setminus B(v_k, r)\}$ there is a blue crossing between a certain portion J_k of the circle of radius R centered at v_k and a certain portion J'_k of the circle of radius r centered at v_k. If we consider instead the domain $D_k \cap B(v_k, R)$, there is no blue crossing between J_k and the portion of $\partial D_k \cap B(v_k, r)$ containing v_k (see Figure 5). If this discrepancy persists as $k \to \infty$, it must show up in the scaling limit of crossing probabilities for the domains $D \cap \{B(v, R) \setminus B(v, r)\}$ and $D \cap B(v, R)$. On the other hand, since $\varepsilon_k \to 0$, we can take r very small, and so $D \cap \{B(v, R) \setminus B(v, r)\}$ is very close to $D \cap B(v, R)$ so that the crossing probabilities in the two domains between the corresponding arcs, given in the continuum by Cardy's formula, should be very close. This follows from the continuity of Cardy's formula with respect to the shape of the domain and the positions of the boundary arcs (see, e.g., Lemma A.2 of [Camia and Newman 2007]).

Using this idea, one can show that the assumption that γ_k comes close to ∂D_k without touching it nearby produces a contradiction. Although the idea outlined above is relatively simple, the arguments needed to obtain a contradiction are rather involved (see Lemmas 7.1, 7.2, 7.3 and 7.4 of [Camia and Newman 2007]), so we will not present them here, except for a brief discussion in the proof of Lemma 6.2 below. □

SKETCH OF THE PROOF OF THEOREM 3. In order to prove the claim, we will define a lattice construction inside D' coupled to the continuum construction inside D, by means of the conformal map f from D to D'. Roughly speaking, this new lattice construction for D' is one in which the (x, y) pairs at each step are chosen to be close to the $(f(a), f(b))$ points in D' mapped from D via f, where the pairs (a, b) are those that appear at the corresponding steps of the continuum construction inside D.

More precisely, let $\gamma_{(1)}$ be the first SLE$_6$ curve in D from $a_{(1)}$ to $b_{(1)}$. Because of the conformal invariance of SLE$_6$, the image $f(\gamma_{(1)})$ of $\gamma_{(1)}$ under f is a curve distributed as the trace of chordal SLE$_6$ in D' from $f(a_{(1)})$ to $f(b_{(1)})$. Therefore, the exploration path $\gamma_{(1)}^\delta$ inside D' from $x_{(1)}$ to $y_{(1)}$, chosen so that they converge to $f(a_{(1)})$ and $f(b_{(1)})$ respectively as $\delta \to 0$, converges in distribution to $f(\gamma_{(1)})$, as $\delta \to 0$, which means that there exists a coupling between $\gamma_{(1)}^\delta$ and $f(\gamma_{(1)})$ such that the curves stay close for δ small.

We see that one can use the same strategy as in the sketch of the proof of Theorem 1, and obtain a lattice construction whose exploration paths are coupled to the SLE$_6$ curves in D' that are the images under f of the SLE$_6$ curves in D. Then, for this discrete construction, the scaling limits of the exploration paths will be distributed as the images of the SLE$_6$ curves in D.

To conclude the proof, we should show that the lattice construction inside D' defined above finds all the boundaries in a number of steps that is bounded in probability as $\delta \to 0$. But this is essentially equivalent to the analogous claim in the sketch of the proof of Theorem 1. Thus the scaling limit, as $\delta \to 0$, of this new lattice construction for D' gives the CNL process in D', which by construction is distributed like the image under f of the CNL process in D. □

SKETCH OF THE PROOF OF THEOREM 4. Let $a, b \in \partial D$ be the endpoints of Γ in clockwise order, i.e., $\Gamma = \overline{ba}$ is the counterclockwise arc of ∂D from b to a. As explained at the end of Section 4, the random curve γ obtained by pasting together the maximal excursions $\overline{xy}(L)$ from $\overline{ba}$, for $L \in \mathcal{L}_{D,\Gamma}$, is distributed like chordal SLE$_6$ in D from a to b. Indeed, removing γ from D is equivalent (in distribution) to the first step of the algorithmic construction presented in Section 4 to produce a realization of the CNL process, if we choose

a and b with $\overline{ba} = \Gamma$ as starting and ending points of the first SLE$_6$ curve of the construction.

Note that γ is in $\mathcal{L}^*_{D,\Gamma} \equiv \overline{\bigcup\{L : L \in \mathcal{L}_{D,\Gamma}\}}$, and the remaining pieces of $\mathcal{L}^*_{D,\Gamma}$ are all in (the closures of) subdomains of $D \setminus \gamma$ of type 1. If we condition on γ and run the algorithmic construction described in Section 4 inside a subdomain of $D \setminus \gamma$ of type 2, 3 or 4, we get an independent CNL process or, by Theorem 3, an independent copy of $\mathcal{L}_D$ conformally mapped to that domain. This already proves part of the claim.

Consider now a subdomain D' of $D \setminus \gamma$ of type 1 and let a', b' be the endpoints of the excursion that generated D'. Part of $\partial D'$ is in ∂D and we choose a', b' so that the counterclockwise arc $\Gamma' = \overline{b'a'} \subset \partial D$ is that part of $\partial D'$. The excursion that generated D' is part of a loop L' whose other "half" is in D' and runs from b' to a'. We know from the construction of Section 4 that if we trace the "half" of L' contained in D' from b' to a' we get a curve γ' distributed like chordal SLE$_6$ in D' from b' to a'. Note that γ' is contained in $\mathcal{L}^*_{D,\Gamma}$.

The subdomains of $D' \setminus \gamma'$ are of two types: (I) those whose boundary does not contain a portion of ∂D and (II) those whose boundary does contain a portion, $\Gamma'' = \overline{b''a''} \subset \partial D$, of Γ. If we condition on γ and γ' and run the algorithmic construction described in Section 4 inside a subdomain of $D' \setminus \gamma'$ of type I, we get an independent CNL process or, by Theorem 3, an independent copy of $\mathcal{L}_D$ conformally mapped to that domain.

The remaining pieces of $\mathcal{L}^*_{D,\Gamma}$ are all contained inside the (closures of) domains of type II (for all the subdomains of $D \setminus \gamma$ of type 1). Inside each subdomain D'' of type II, the CN loops that touch Γ'' are contained in $\mathcal{L}^*_{D,\Gamma}$ and can be used to obtain a curve γ'' distributed like chordal SLE$_6$ in D'' from a'' to b'' by pasting together maximal excursions as above (and at the end of Section 4). It should now be clear how to complete the argument by iterating the steps described above inside each subdomain D''.   □

6. Convergence of exploration path to SLE$_6$

SKETCH OF THE PROOF OF THEOREM 1. We begin discussing the proof of Theorem 1 by noting, as in the proof of Theorem 2 discussed in Section 5, that it follows from [Aizenman and Burchard 1999] that the family of distributions of $\gamma^\delta_{D,a,b}$ is tight (as $\delta \to 0$, in the sense of the metric (3-3)) and so there is convergence along subsequences $\delta_k \to 0$. We write, in simplified notation, $\gamma_k \to \tilde{\gamma}$ along such a convergent subsequence. What needs to be proved is that the distribution $\tilde{\mu}$ of $\tilde{\gamma}$ is that of γ^{SLE_6}, the trace of chordal SLE$_6$ in D from a to b.

We next discuss how much information about $\tilde{\mu}$ can be extracted from Cardy's formula for crossing probabilities. We note that there are versions of Smirnov's

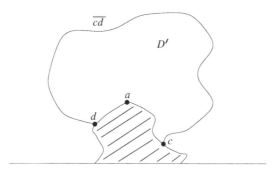

Figure 6. D is the upper half-plane $\mathbb{H}$ with the shaded portion removed, $b = \infty$, C' is an unbounded subdomain, and $D' = D \setminus C'$ is indicated in the figure. The counterclockwise arc $\overline{cd}$ indicated in the figure belongs to $\partial D'$.

result on convergence of crossing probabilities to Cardy's formula that allow the domains being crossed and the target boundary arcs to vary as $\delta \to 0$. Theorem 3 of [Camia and Newman 2007] is such a version that suffices for our purposes. Let $D_t \equiv D \setminus \tilde{K}_t$ denote the (unique) connected component of $D \setminus \tilde{\gamma}[0, t]$ whose closure contains b, where $\tilde{K}_t$, the *filling* of $\tilde{\gamma}[0, t]$, is a closed connected subset of $\overline{D}$. $\tilde{K}_t$ is called a *hull* if it satisfies the condition

$$\overline{\tilde{K}_t \cap D} = \tilde{K}_t. \tag{6-1}$$

We will consider curves $\tilde{\gamma}$ such that $\tilde{K}_t$ is a hull for each t, although here we only consider $\tilde{K}_T$ at certain stopping times T.

Let $C' \subset D$ be a closed subset of $\overline{D}$ such that $a \notin C'$, $b \in C'$, and $D' = D \setminus C'$ is a bounded simply connected domain whose boundary contains the counterclockwise arc $\overline{cd}$ that does not belong to ∂D (except for its endpoints c and d – see Figure 6).

Let $T' = \inf\{t : \tilde{K}_t \cap C' \neq \varnothing\}$ be the first time that $\tilde{\gamma}(t)$ hits C' and assume that the filling $\tilde{K}_{T'}$ of $\tilde{\gamma}[0, T']$ is a hull. We say that the hitting distribution of $\tilde{\gamma}(t)$ at the stopping time T' is determined by Cardy's formula (see (3-1)) if, for any C' and any counterclockwise arc $\overline{xy}$ of $\overline{cd}$, the probability that $\tilde{\gamma}$ hits C' at time T' on $\overline{xy}$ is given by

$$\mathbb{P}(\tilde{\gamma}(T') \in \overline{xy}) = \Phi_{D'}(a, c; x, d) - \Phi_{D'}(a, c; y, d). \tag{6-2}$$

We want to relate the distribution of $\tilde{K}_{T'}$ to the distribution of hitting *locations* for a family of C'''s related to C'. To explain, consider the set $\tilde{\mathcal{A}}$ of closed subsets $\tilde{A}$ of $\overline{D'}$ that do not contain a and such that $\partial \tilde{A} \setminus \partial D'$ is a simple (continuous) curve contained in D' except for its endpoints, one of which is on $\partial D' \cap D$ and the other is on ∂D (see Figure 7). Let $\mathcal{A}$ be the set of closed subsets of $\overline{D'}$ of the form $\tilde{A}_1 \cup \tilde{A}_2$, where $\tilde{A}_1, \tilde{A}_2 \in \tilde{\mathcal{A}}$ and $\tilde{A}_1 \cap \tilde{A}_2 = \varnothing$.

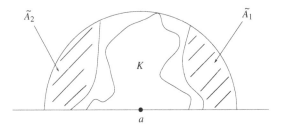

Figure 7. Example of a hull K and a set $\tilde{A}_1 \cup \tilde{A}_2$ (shaded regions) in $\mathcal{A}$. Here, $D = \mathbb{H}$ and D' is the semidisc centered at a.

It is easy to see that if the hitting distribution of $\tilde{\gamma}(T')$ is determined by Cardy's formula, then the probabilities of events of the form $\{\tilde{K}_{T'} \cap A = \varnothing\}$ for $A \in \mathcal{A}$ are also determined by Cardy's formula in the following way. Let $A \in \mathcal{A}$ be the union of $\tilde{A}_1, \tilde{A}_2 \in \tilde{A}$, with $\partial \tilde{A}_1 \setminus \partial D'$ given by a curve from $u_1 \in \partial D' \cap D$ to $v_1 \in \partial D$ and $\partial \tilde{A}_2 \setminus \partial D'$ given by a curve from $u_2 \in \partial D' \cap D$ to $v_2 \in \partial D$; then, assuming that a, v_1, u_1, u_2, v_2 are ordered counterclockwise around $\partial D'$,

$$\mathbb{P}(\tilde{K}_{T'} \cap A = \varnothing) = \Phi_{D' \setminus A}(a, v_1; u_1, v_2,) - \Phi_{D' \setminus A}(a, v_1; u_2, v_2). \quad (6\text{-}3)$$

The probabilities of such events determine uniquely the distribution of the hull (for more detail, see Section 5 of [Camia and Newman 2007]). Thus we have the following useful lemma, since the hitting distribution for SLE_6 is determined by Cardy's formula [Lawler et al. 2001].

LEMMA 6.1. *If $\tilde{K}_{T'}$ is a hull and the hitting distribution of $\tilde{\gamma}$ at the stopping time T' is determined by Cardy's formula, then $\tilde{K}_{T'}$ is distributed like the corresponding hull of γ^{SLE_6}.*

We next define the sequence of hitting times for $\tilde{\gamma}$ that will be used to compare it to γ^{SLE_6}. They involve conformal maps of semiballs (i.e., half-disks) in the upper half-plane. Let $\tilde{f}_0$ be a conformal map from the upper half-plane $\mathbb{H}$ to D such that $\tilde{f}_0^{-1}(a) = 0$ and $\tilde{f}_0^{-1}(b) = \infty$. (Since ∂D is a continuous curve, the map $\tilde{f}_0^{-1}$ has a continuous extension from D to $D \cup \partial D$ and, by a slight abuse of notation, we do not distinguish between $\tilde{f}_0^{-1}$ and its extension; the same applies to $\tilde{f}_0$.) These two conditions determine $\tilde{f}_0$ only up to a scaling factor. For $\varepsilon > 0$ fixed, let $C(u, \varepsilon) = \{z : |u - z| < \varepsilon\} \cap \mathbb{H}$ denote the semiball of radius ε centered at u on the real line and let $\tilde{T}_1 = \tilde{T}_1(\varepsilon)$ denote the first time $\tilde{\gamma}(t)$ hits $D \setminus \tilde{G}_1$, where $\tilde{G}_1 \equiv \tilde{f}_0(C(0, \varepsilon))$. Define recursively $\tilde{T}_{j+1}$ as the first time $\tilde{\gamma}[\tilde{T}_j, \infty)$ hits $\tilde{D}_{\tilde{T}_j} \setminus \tilde{G}_{j+1}$, where $\tilde{D}_{\tilde{T}_j} \equiv D \setminus \tilde{K}_{\tilde{T}_j}$, $\tilde{G}_{j+1} \equiv \tilde{f}_{\tilde{T}_j}(C(0, \varepsilon))$, and $\tilde{f}_{\tilde{T}_j}$ is a conformal map from $\mathbb{H}$ to $\tilde{D}_{\tilde{T}_j}$ whose inverse maps $\tilde{\gamma}(\tilde{T}_j)$ to 0 and b to ∞. We also define $\tilde{\tau}_{j+1} \equiv \tilde{T}_{j+1} - \tilde{T}_j$, so that $\tilde{T}_j = \tilde{\tau}_1 + \ldots + \tilde{\tau}_j$. We choose

$\tilde{f}_{\tilde{T}_j}$ so that its inverse is the composition of the restriction of $\tilde{f}_0^{-1}$ to $\tilde{D}_{\tilde{T}_j}$ with $\tilde{\varphi}_{\tilde{T}_j}$, where $\tilde{\varphi}_{\tilde{T}_j}$ is the unique conformal transformation from $\mathbb{H} \setminus \tilde{f}_0^{-1}(\tilde{K}_{\tilde{T}_j})$ to $\mathbb{H}$ that maps ∞ to ∞ and $\tilde{f}_0^{-1}(\tilde{\gamma}(\tilde{T}_j))$ to the origin of the real axis, and has derivative at ∞ equal to 1.

Notice that $\tilde{G}_{j+1}$ is a bounded simply connected domain chosen so that the conformal transformation which maps $\tilde{D}_{\tilde{T}_j}$ to $\mathbb{H}$ maps $\tilde{G}_{j+1}$ to the semiball $C(0, \varepsilon)$ centered at the origin on the real line. With these definitions, we consider the (discrete-time) stochastic process $\tilde{X}_j \equiv (\tilde{K}_{\tilde{T}_j}, \tilde{\gamma}(\tilde{T}_j))$ for $j = 1, 2, \ldots$. Analogous quantities can be defined for the trace of chordal SLE$_6$. They are indicated by the superscript SLE$_6$; we choose $f_0^{\mathrm{SLE}_6} = \tilde{f}_0$, so that $G_1^{\mathrm{SLE}_6} = \tilde{G}_1$. Our aim is to prove that the variables $\tilde{X}_1, \tilde{X}_2, \ldots$ are (jointly) equidistributed with the corresponding SLE$_6$ hull and tip variables $X_1^{\mathrm{SLE}_6}, X_2^{\mathrm{SLE}_6}, \ldots$. By letting $\varepsilon \to 0$, this will directly yield that $\tilde{\gamma}$ is equidistributed with γ^{SLE_6} as desired. Since γ_k converges in distribution to $\tilde{\gamma}$, we can find coupled versions of γ_k and $\tilde{\gamma}$ on some probability space $(\Omega, \mathcal{B}, \mathbb{P})$ such that γ_k converges to $\tilde{\gamma}$ for all $\omega \in \Omega$; in the rest of the proof we work with these new versions which, with a slight abuse of notation, we denote with the same names as the original ones.

For each k, let K_t^k denote the filling (or *lattice hull*) at time t of γ_k, i.e., the set of hexagons that at time t have been explored or have been disconnected from b by the exploration path. Let now f_0^k be a conformal transformation that maps $\mathbb{H}$ to $D_k \equiv D^{\delta_k}$ such that $(f_0^k)^{-1}(a_k) = 0$ and $(f_0^k)^{-1}(b_k) = \infty$ and let $T_1^k = T_1^k(\varepsilon)$ denote the first exit time of $\gamma_k^{\delta_k}(t)$ from $G_1^k \equiv f_0^k(C(0, \varepsilon))$ defined as the first time that γ_k intersects the image under f_0^k of the semicircle $\{z : |z| = \varepsilon\} \cap \mathbb{H}$. Define recursively T_{j+1}^k as the first exit time of $\gamma_k^{\delta_k}[T_j^k, \infty)$ from $G_{j+1}^k \equiv f_{T_j^k}^k(C(0, \varepsilon))$, where $f_{T_j^k}^k$ is a conformal map from $\mathbb{H}$ to $D_k \setminus K_{T_j^k}^k$ whose inverse maps $\gamma_k(T_j^k)$ to 0 and b_k to ∞. Each of the maps $f_{T_j^k}^k$, where $j \geq 1$, is defined only up to a scaling factor. We also set $\tau_{j+1}^k \equiv T_{j+1}^k - T_j^k$, so $T_j^k = \tau_1^k + \ldots + \tau_j^k$, and define the (discrete-time) stochastic process

$$X_j^k \equiv (K_{T_j^k}^k, \gamma_k^{\delta_k}(T_j^k)) \quad \text{for } j = 1, 2, \ldots.$$

We want to show recursively that, for any j, as $k \to \infty$, $\{X_1^k, \ldots, X_j^k\}$ converge jointly in distribution to $\{\tilde{X}_1, \ldots, \tilde{X}_j\}$. By recursively applying the convergence of crossing probabilities to Cardy's formula (i.e., Theorem 3 of [Camia and Newman 2007]) and Lemma 6.1, we will then be able to conclude, as explained in more detail below, that $\{\tilde{X}_1, \tilde{X}_2, \dots\}$ are jointly equidistributed with the corresponding SLE$_6$ hull variables (at the corresponding stopping times) $\{X_1^{\text{SLE}_6}, X_2^{\text{SLE}_6}, \dots\}$.

The zeroth step consists in noticing that the convergence of (D_k, a_k, b_k) to (D, a, b) as $k \to \infty$ allows us to select a sequence of conformal maps f_0^k that converge to $f_0^{\text{SLE}_6} = \tilde{f}_0$ uniformly in $\overline{\mathbb{H}}$ as $k \to \infty$, which implies that the boundary ∂G_1^k of $G_1^k = f_0^k(C(0, \varepsilon))$ converges to the boundary $\partial \tilde{G}_1$ of $\tilde{G}_1 = \tilde{f}_0(C(0, \varepsilon))$ in the uniform metric on continuous curves (see Corollary A.2 of [Camia and Newman 2007]).

The next lemma is the technical heart of the proof. It basically allows us to interchange the scaling limit $\delta \to 0$ and the process of filling (which generates hulls) by declaring that the hull of the limiting curve is the limit of the (lattice) hulls. The proof of the lemma involves extensive use of nontrivial results from percolation theory. Although the lemma is stated here in the framework of the first step of the proof where we are analyzing convergence of X_1^k to $\tilde{X}_1$, essentially the same lemma can be applied sequentially to the convergence of X_j^k conditioned on $\{X_1^k, \dots, X_{j-1}^k\}$.

LEMMA 6.2. $(\gamma_k, K_{T_1^k}^k)$ *converges in distribution to* $(\tilde{\gamma}, \tilde{K}_{\tilde{T}_1})$ *as* $k \to \infty$. *Furthermore* $\tilde{K}_{\tilde{T}_1}$ *is almost surely a hull equidistributed with the hull* $K_{T_1}^{\text{SLE}_6}$ *of* SLE$_6$ *at the corresponding stopping time* T_1.

PROOF. Proving the first claim, that for the exploration path γ_k in G_1^k one can interchange the limit $k \to \infty$ ($\delta_k \to 0$) with the process of filling, requires showing two things about the exploration path: (1) the return of a (macroscopic) segment of the path close to an earlier segment (and away from ∂G_1^k) without nearby (microscopic) touching does not occur (probably), and (2) the close approach of a (macroscopic) segment of the path to ∂G_1^k without nearby (microscopic) touching either of ∂G_1^k itself or else of another segment of the path that touches ∂G_1^k does not occur (probably). If G_1^k (or more accurately, its limit $\tilde{G}_1$) were replaced by a convex domain like the unit disk, these could be controlled by known estimates on probabilities of six-arm events in the full plane for (1) and of three-arm events in the half-plane for (2). But $\tilde{G}_1$ is not in general convex and then the three-arm event argument for (2) appears to break down. The replacement in [Camia and Newman 2007] is the use of several lemmas in Section 7 there. Basically, these control (2) by a novel argument about "mushroom events" in $\tilde{G}_1$, which is based on continuity of Cardy's formula with respect to changes in $\partial \tilde{G}_1$. Roughly speaking, mushroom events are ones where (in the limit $k \to \infty$) there is a macroscopic monochromatic path in $\tilde{G}_1$ just reaching to $\partial \tilde{G}_1$, but blocked from it by a macroscopic path in $\tilde{G}_1$ of the other color (see Figure 5). It is shown in [Camia and Newman 2007] (see Lemma 7.4 there) that mushroom events cannot occur with positive probability while on the other hand they would occur if (2) were not the case. The second claim of Lemma 6.2 now follows from Smirnov's result [2001] on convergence to Cardy's formula (see also Theorem 3 of [Camia and Newman 2007]) and Lemma 6.1.   □

Using Lemma 6.2, the first step of our recursion argument is organized as follows, where all limits and equalities are in distribution:

(i) $K^k_{T^k_1} \to \tilde{K}_{\tilde{T}_1} = K^{\mathrm{SLE}_6}_{T_1}$ by Lemma 6.2.

(ii) By (i), $D_k \setminus K^k_{T^k_1} \to D \setminus \tilde{K}_{\tilde{T}_1} = D \setminus K^{\mathrm{SLE}_6}_{T_1}$.

(iii) By (ii), $f^{\mathrm{SLE}_6}_{T_1} = \tilde{f}_{\tilde{T}_1}$, and we can select a sequence $f^k_{T^k_1} \to \tilde{f}_{\tilde{T}_1} = f^{\mathrm{SLE}_6}_{T_1}$.

(iv) By (iii), $G^k_2 \to \tilde{G}_2 = G^{\mathrm{SLE}_6}_2$.

At this point, we are in the same situation as at the zeroth step, but with G^k_1, $\tilde{G}_1$ and $G^{\mathrm{SLE}_6}_1$ replaced by G^k_2, $\tilde{G}_2$ and $G^{\mathrm{SLE}_6}_2$, and we proceed by induction, as follows.

The next step consists in proving that

$$\left((K^k_{T^k_1}, \gamma^{\delta_k}_k(T^k_1)), (K^k_{T^k_2}, \gamma^{\delta_k}_k(T^k_2)) \right)$$

converges in distribution to $\left((\tilde{K}_{\tilde{T}_1}, \tilde{\gamma}(\tilde{T}_1)), (\tilde{K}_{\tilde{T}_2}, \tilde{\gamma}(\tilde{T}_2)) \right)$. Since we have already proved the convergence of $(K^k_{T^k_1}, \gamma^{\delta_k}_k(T^k_1))$ to $(\tilde{K}_{\tilde{T}_1}, \tilde{\gamma}(\tilde{T}_1))$, all we need to prove is the convergence of $(K^k_{T^k_2} \setminus K^k_{T^k_1}, \gamma^{\delta_k}_k(T^k_2))$ to $(\tilde{K}_{\tilde{T}_2} \setminus \tilde{K}_{\tilde{T}_1}, \tilde{\gamma}(\tilde{T}_2))$. To do this, notice that $K^k_{T^k_2} \setminus K^k_{T^k_1}$ is distributed like the lattice hull of a percolation exploration path inside $D_k \setminus K^k_{T^k_1}$. Besides, the convergence in distribution of $(K^k_{T^k_1}, \gamma^{\delta_k}_k(T^k_1))$ to $(\tilde{K}_{\tilde{T}_1}, \tilde{\gamma}(\tilde{T}_1))$ implies that we can find versions of $(\gamma^{\delta_k}_k, K^k_{T^k_1})$ and $(\tilde{\gamma}, \tilde{K}_{\tilde{T}_1})$ on some probability space $(\Omega, \mathcal{B}, \mathbb{P})$ such that $\gamma^{\delta_k}_k(\omega)$ converges to $\tilde{\gamma}(\omega)$ and $(K^k_{T^k_1}, \gamma^{\delta_k}_k(T^k_1))$ converges to $(\tilde{K}_{\tilde{T}_1}, \tilde{\gamma}(\tilde{T}_1))$ for all $\omega \in \Omega$. These two observations imply that, if we work with the coupled versions of $(\gamma^{\delta_k}_k, K^k_{T^k_1})$ and $(\tilde{\gamma}, \tilde{K}_{\tilde{T}_1})$, we are in the same situation as before, but with D_k and D replaced by $D_k \setminus K^k_{T^k_1}$ and $D \setminus \tilde{K}_{\tilde{T}_1}$, and a_k and a replaced by $\gamma^{\delta_k}_k(T^k_1)$ and $\tilde{\gamma}(\tilde{T}_1)$, respectively. Then, the conclusion that $(K^k_{T^k_2} \setminus K^k_{T^k_1}, \gamma^{\delta_k}_k(T^k_2))$ converges in distribution to $(\tilde{K}_{\tilde{T}_2} \setminus \tilde{K}_{\tilde{T}_1}, \tilde{\gamma}(\tilde{T}_2))$ follows, as before, by arguments like those used for Lemma 6.2. We can now iterate the above arguments j times, for any $j > 1$. If we keep track at each step of the previous ones, this provides the *joint* convergence of all the curves and lattice hulls involved at each step.

The proof of Theorem 1 is concluded by letting $\varepsilon \to 0$. We note that in this paper we circumvent the use of a "spatial Markov property" that played a role in [Camia and Newman 2007] in the $\varepsilon \to 0$ limit. The point is that that property was proved as a consequence of the equidistribution of $\tilde{X}_1, \tilde{X}_2, \ldots$ with $X_1^{\text{SLE}_6}, X_2^{\text{SLE}_6}, \ldots$ and here we apply the equidistribution directly. It should be noted however that there needs to be some a priori information about $\tilde{\gamma}$ to insure that this equidistribution for each $\varepsilon > 0$ implies equidistribution of $\tilde{\gamma}$ with γ^{SLE_6}. For example, one could create by hand a process $\hat{\gamma}$ which behaved like γ^{SLE_6} except that at random times it retraced back and forth part of its previous path. Such a $\hat{\gamma}$ would have its $\hat{X}_j$ variables equidistributed with those of SLE_6 but as a random curve (modulo monotonic reparametrizations) would not be equidistributed with γ^{SLE_6}; it would also not be describable by a Loewner chain. Such possibilities can be ruled out by the same arguments as those used in proving Lemma 6.2; see Lemma 6.4 of [Camia and Newman 2007]. □

Acknowledgements

The authors thank the Kavli Institute for Theoretical Physics for its hospitality in 2006, when this paper was mostly written. Newman thanks the Clay Mathematics Institute for partial support of his visit at KITP, the Mathematical Sciences Research Institute, Berkeley for organizing the 2005 workshop in honor of Henry McKean and he thanks Henry for many interesting conversations over many years.

References

[Aizenman and Burchard 1999] M. Aizenman and A. Burchard, "Hölder regularity and dimension bounds for random curves", *Duke Math. J.* **99**:3 (1999), 419–453.

[Camia and Newman 2004] F. Camia and C. M. Newman, "Continuum nonsimple loops and 2D critical percolation", *J. Statist. Phys.* **116**:1-4 (2004), 157–173.

[Camia and Newman 2006] F. Camia and C. M. Newman, "Two-dimensional critical percolation: the full scaling limit", *Comm. Math. Phys.* **268**:1 (2006), 1–38.

[Camia and Newman 2007] F. Camia and C. M. Newman, "Critical percolation exploration path and SLE$_6$: a proof of convergence", *Probab. Theory Related Fields* **139**:3-4 (2007), 473–519.

[Camia et al. 2006a] F. Camia, L. R. G. Fontes, and C. M. Newman, "The scaling limit geometry of near-critical 2D percolation", *J. Stat. Phys.* **125**:5-6 (2006), 1159–1175.

[Camia et al. 2006b] F. Camia, L. R. G. Fontes, and C. M. Newman, "Two-dimensional scaling limits via marked nonsimple loops", *Bull. Braz. Math. Soc. (N.S.)* **37**:4 (2006), 537–559.

[Cardy 1992] J. L. Cardy, "Critical percolation in finite geometries", *J. Phys. A* **25**:4 (1992), L201–L206.

[Cardy 2001] J. Cardy, "Lectures on conformal invariance and percolation", preprint, 2001. Available at http://www.arxiv.org/abs/math-ph/0103018.

[Cardy 2005] J. Cardy, "SLE for theoretical physicists", *Ann. Physics* **318**:1 (2005), 81–118.

[Kager and Nienhuis 2004] W. Kager and B. Nienhuis, "A guide to stochastic Löwner evolution and its applications", *J. Statist. Phys.* **115**:5-6 (2004), 1149–1229.

[Langlands et al. 1994] R. Langlands, P. Pouliot, and Y. Saint-Aubin, "Conformal invariance in two-dimensional percolation", *Bull. Amer. Math. Soc. (N.S.)* **30**:1 (1994), 1–61.

[Lawler 2004] G. F. Lawler, "Conformally invariant processes in the plane", pp. 305–351 in *School and Conference on Probability Theory*, ICTP Lecture Notes **17**, Abdus Salam Int. Cent. Theoret. Phys., Trieste, 2004.

[Lawler 2005] G. F. Lawler, *Conformally invariant processes in the plane*, Mathematical Surveys and Monographs **114**, American Mathematical Society, Providence, RI, 2005.

[Lawler et al. 2001] G. F. Lawler, O. Schramm, and W. Werner, "Values of Brownian intersection exponents. I. Half-plane exponents", *Acta Math.* **187**:2 (2001), 237–273.

[Lawler et al. 2002] G. F. Lawler, O. Schramm, and W. Werner, "One-arm exponent for critical 2D percolation", *Electron. J. Probab.* **7** (2002), paper no. 2.

[Rohde and Schramm 2005] S. Rohde and O. Schramm, "Basic properties of SLE", *Ann. of Math.* (2) **161**:2 (2005), 883–924.

[Russo 1978] L. Russo, "A note on percolation", *Z. Wahrscheinlichkeitstheorie und Verw. Gebiete* **43**:1 (1978), 39–48.

[Schramm 2000] O. Schramm, "Scaling limits of loop-erased random walks and uniform spanning trees", *Israel J. Math.* **118** (2000), 221–288.

[Seymour and Welsh 1978] P. D. Seymour and D. J. A. Welsh, "Percolation probabilities on the square lattice", pp. 227–245 in *Advances in graph theory* (Cambridge, 1977), edited by B. Bollobás, Ann. Discrete Math. **3**, 1978.

[Sheffield 2006] S. Sheffield, "Exploration trees and conformal loop ensembles", preprint, 2006. Available at arXiv:math.PR/0609167.

[Smirnov 2001] S. Smirnov, "Critical percolation in the plane: conformal invariance, Cardy's formula, scaling limits", *C. R. Acad. Sci. Paris Sér. I Math.* **333**:3 (2001), 239–244. A longer version, dated Nov. 15, 2001, is available at http://www.math.kth.se/~stas/papers/index.html.

[Werner 2003] W. Werner, "SLEs as boundaries of clusters of Brownian loops", *C. R. Math. Acad. Sci. Paris* **337**:7 (2003), 481–486.

[Werner 2004] W. Werner, "Random planar curves and Schramm-Loewner evolutions", pp. 107–195 in *Lectures on probability theory and statistics*, Lecture Notes in Math. **1840**, Springer, Berlin, 2004.

[Werner 2005a] W. Werner, "The conformally invariant measure on self-avoiding loops", preprint, 2005. Available at arXiv:math.PR/0511605.

[Werner 2005b] W. Werner, "Some recent aspects of random conformally invariant systems", preprint, 2005. Available at arXiv:math.PR/0511268.

[Wilson 2004] D. B. Wilson, "Red-green-blue model", *Phys. Rev. E* **69** (2004), paper #037105.

FEDERICO CAMIA
DEPARTMENT OF MATHEMATICS
VRIJE UNIVERSITEIT AMSTERDAM
DE BOELELAAN 1081A
1081 HV AMSTERDAM
THE NETHERLANDS
fede@few.vu.nl

CHARLES M. NEWMAN
MATHEMATICS DEPARTMENT
COURANT INSTITUTE OF MATHEMATICAL SCIENCES
NEW YORK UNIVERSITY
251 MERCER STREET
NEW YORK, NY 10012
UNITED STATES
newman@courant.nyu.edu

Probability, Geometry and Integrable Systems
MSRI Publications
Volume **55**, 2008

Global optimization, the Gaussian ensemble, and universal ensemble equivalence

MARIUS COSTENIUC, RICHARD S. ELLIS, HUGO TOUCHETTE, AND BRUCE TURKINGTON

With great affection this paper is dedicated to Henry McKean
on the occasion of his 75th birthday.

ABSTRACT. Given a constrained minimization problem, under what conditions does there exist a related, unconstrained problem having the same minimum points? This basic question in global optimization motivates this paper, which answers it from the viewpoint of statistical mechanics. In this context, it reduces to the fundamental question of the equivalence and nonequivalence of ensembles, which is analyzed using the theory of large deviations and the theory of convex functions.

In a 2000 paper appearing in the *Journal of Statistical Physics*, we gave necessary and sufficient conditions for ensemble equivalence and nonequivalence in terms of support and concavity properties of the microcanonical entropy. In later research we significantly extended those results by introducing a class of Gaussian ensembles, which are obtained from the canonical ensemble by adding an exponential factor involving a quadratic function of the Hamiltonian. The present paper is an overview of our work on this topic. Our most important discovery is that even when the microcanonical and canonical ensembles are not equivalent, one can often find a Gaussian ensemble that satisfies a strong form of equivalence with the microcanonical ensemble known as universal equivalence. When translated back into optimization theory, this implies that an unconstrained minimization problem involving a Lagrange multiplier and a quadratic penalty function has the same minimum points as the original constrained problem.

The results on ensemble equivalence discussed in this paper are illustrated in the context of the Curie–Weiss–Potts lattice-spin model.

Keywords: Equivalence of ensembles, Gaussian ensemble, microcanonical entropy, large deviation principle, Curie–Weiss–Potts model.

131

1. Introduction

Oscar Lanford, at the beginning of his groundbreaking paper [Lanford 1973], describes the underlying program of statistical mechanics:

> The objective of statistical mechanics is to explain the macroscopic properties of matter on the basis of the behavior of the atoms and molecules of which it is composed. One of the most striking facts about macroscopic matter is that in spite of being fantastically complicated on the atomic level — to specify the positions and velocities of all molecules in a glass of water would mean specifying something of the order of 10^{25} parameters — its macroscopic behavior is describable in terms of a very small number of parameters; e.g., the temperature and density for a system containing only one kind of molecule.

Lanford shows how the theory of large deviations enables this objective to be realized. In statistical mechanics one determines the macroscopic behavior of physical systems not from the deterministic laws of Newtonian mechanics, but from a probability distribution that expresses both the behavior of the system on the microscopic level and the intrinsic inability to describe precisely what is happening on that level. Using the theory of large deviations, one shows that, with probability converging to 1 exponentially fast as the number of particles tends to ∞, the macroscopic behavior is describable in terms of a very small number of parameters

The success of this program depends on the correct choice of probability distribution, also known as an ensemble. One starts with a prior measure on configuration space, which, as an expression of the lack of information concerning the behavior of the system on the atomic level, is often taken to be the uniform measure. As Boltzmann recognized, the most natural choice of ensemble is the microcanonical ensemble, obtained by conditioning the prior measure on the set of configurations for which the Hamiltonian per particle equals a constant energy u. Boltzmann also introduced a mathematically more tractable probability distribution known as the canonical ensemble, in which the conditioning that defines the microcanonical ensemble is replaced by an exponential factor involving the Hamiltonian and the inverse temperature β, a parameter dual to the energy parameter u [Gibbs 1902].

Among other reasons, the canonical ensemble was introduced in the hope that in the limit $n \to \infty$ the two ensembles are equivalent; i.e., all macroscopic properties of the model obtained via the microcanonical ensemble could be realized as macroscopic properties obtained via the canonical ensemble. While ensemble equivalence is valid for many standard and important models, ensemble equivalence does not hold in general, as numerous studies cited later in this introduction show. There are many examples of statistical mechanical models for which nonequivalence of ensembles holds over a wide range of model parameters and for which physically interesting microcanonical equilibria are often omitted by the canonical ensemble.

The present paper is an overview of our work on this topic. One of the beautiful aspects of the theory is that it elucidates a fundamental issue in global optimization, which in fact motivated our work on the Gaussian ensemble. Given a constrained minimization problem, under what conditions does there exist a related, unconstrained minimization problem having the same minimum points?

In order to explain the connection between ensemble equivalence and global optimization and in order to outline the contributions of this paper, we introduce some notation. Let X be a space, I a function mapping X into $[0, \infty]$, and $\tilde{H}$ a function mapping X into $\mathbb{R}$. For $u \in \mathbb{R}$ we consider the following constrained minimization problem:

$$\text{minimize } I(x) \text{ over } x \in X \text{ subject to the contraint } \tilde{H}(x) = u. \qquad (1\text{-}1)$$

A partial answer to the question posed at the end of the preceding paragraph can be found by introducing the following related, unconstrained minimization problem for $\beta \in \mathbb{R}$:

$$\text{minimize } I(x) + \beta \tilde{H}(x) \text{ over } x \in X. \qquad (1\text{-}2)$$

The theory of Lagrange multipliers outlines suitable conditions under which the solutions of the constrained problem (1-1) lie among the critical points of $I + \beta \tilde{H}$. However, it does not give, as we will do in Theorems 3.1 and 3.3, necessary and sufficient conditions for the solutions of (1-1) to coincide with the solutions of the unconstrained minimization problem (1-2) and with the solutions of the unconstrained minimization problem appearing in (1-5).

We denote by $\mathcal{E}^u$ and $\mathcal{E}_\beta$ the respective sets of solutions of the minimization problems (1-1) and (1-2). These problems arise in a natural way in the context of equilibrium statistical mechanics [Ellis et al. 2000], where u denotes the energy and β the inverse temperature. As we will outline in Section 2, the theory of large deviations allows one to identify the solutions of these problems as the respective sets of equilibrium macrostates for the microcanonical ensemble and the canonical ensemble.

The paper [Ellis et al. 2000] analyzes equivalence of ensembles in terms of relationships between $\mathcal{E}^u$ and $\mathcal{E}_\beta$. In turn, these relationships are expressed in terms of support and concavity properties of the microcanonical entropy

$$s(u) = -\inf\{I(x) : x \in X, \tilde{H}(x) = u\}. \qquad (1\text{-}3)$$

The main results in [Ellis et al. 2000] are summarized in Theorem 3.1. Part (a) of that theorem states that if s has a strictly supporting line at an energy value u, then full equivalence of ensembles holds in the sense that there exists a β such that $\mathcal{E}^u = \mathcal{E}_\beta$. In particular, if s is strictly concave on dom s, then s has a strictly supporting line at all $u \in$ dom s except possibly boundary points [Theorem 3.2(a)] and thus full equivalence of ensembles holds at all such u. In this case we say that the microcanonical and canonical ensembles are universally equivalent.

The most surprising result, given in part (c), is that if s does not have a supporting line at u, then nonequivalence of ensembles holds in the strong sense that $\mathcal{E}^u \cap \mathcal{E}_\beta = \varnothing$ for all $\beta \in \mathbb{R}^\sigma$. That is, if s does not have a supporting line at u — equivalently, if s is not concave at u — then microcanonical equilibrium macrostates cannot be realized canonically. This is to be contrasted with part (d), which states that for any $x \in \mathcal{E}_\beta$ there exists u such that $x \in \mathcal{E}^u$; i.e., canonical equilibrium macrostates can always be realized microcanonically. Thus of the two ensembles, in general the microcanonical is the richer.

The paper [Costeniuc et al. 2005b] addresses the natural question suggested by part (c) of Theorem 3.1. If the microcanonical ensemble is not equivalent with the canonical ensemble on a subset of energy values u, then is it possible to replace the canonical ensemble with another ensemble that is universally equivalent with the microcanonical ensemble? We answered this question by introducing a penalty function $\gamma[\tilde{H}(x) - u]^2$ into the unconstrained minimization problem (1-2), obtaining the following:

$$\text{minimize } I(x) + \beta \tilde{H}(x) + \gamma[\tilde{H}(x) - u]^2 \text{ over } x \in \mathcal{X}. \qquad (1\text{-}4)$$

Since for each $x \in \mathcal{X}$

$$\lim_{\gamma \to \infty} \gamma[\tilde{H}(x) - u]^2 = \begin{cases} 0 & \text{if } \tilde{H}(x) = u \\ \infty & \text{if } \tilde{H}(x) \neq u, \end{cases}$$

it is plausible that for all sufficiently large γ minimum points of the penalized problem (1-4) are also minimum points of the constrained problem (1-1). Since β can be adjusted, (1-4) is equivalent to the following:

$$\text{minimize } I(x) + \beta \tilde{H}(x) + \gamma[\tilde{H}(x)]^2 \text{ over } x \in \mathcal{X}. \qquad (1\text{-}5)$$

The theory of large deviations allows one to identify the solution of this problem as the set of equilibrium macrostates for the so-called Gaussian ensemble. It is obtained from the canonical ensemble by adding an exponential factor involving γh_n^2, where h_n denotes the Hamiltonian energy per particle. The utility of the Gaussian ensemble rests on the simplicity with which the quadratic function γu^2 defining this ensemble enters the formulation of ensemble equivalence. Essentially all the results in [Ellis et al. 2000] concerning ensemble equivalence, including Theorem 3.1, generalize to the setting of the Gaussian ensemble by replacing the microcanonical entropy $s(u)$ by the generalized microcanonical entropy

$$s_\gamma(u) = s(u) - \gamma u^2. \tag{1-6}$$

The generalization of Theorem 3.1 is stated in Theorem 3.3, which gives all possible relationships between the set $\mathcal{E}^u$ of equilibrium macrostates for the microcanonical ensemble and the set $\mathcal{E}_{\beta,\gamma}$ of equilibrium macrostates for the Gaussian ensemble. These relationships are expressed in terms of support and concavity properties of s_γ.

For the purpose of applications the most important consequence of Theorem 3.3 is given in part (a), which states that if s_γ has a strictly supporting line at an energy value u, then full equivalence of ensembles holds in the sense that there exists a β such that $\mathcal{E}^u = \mathcal{E}_{\beta,\gamma}$. In particular, if s_γ is strictly concave on dom s, then s_γ has a strictly supporting line at all $u \in$ dom s except possibly boundary points [Theorem 3.4(a)] and thus full equivalence of ensembles holds at all such u. In this case we say that the microcanonical and Gaussian ensembles are universally equivalent.

In the case in which s is C^2 and s'' is bounded above on the interior of dom s, then the strict concavity of s_γ is easy to show. In fact, the strict concavity is a consequence of

$$s_\gamma''(u) = s''(u) - 2\gamma < 0 \quad \text{for all } u \in \text{int(dom } s),$$

and this in turn is valid for all sufficiently large γ [Theorem 4.2]. For such γ it follows, therefore, that the microcanonical and Gaussian ensembles are universally equivalent.

Defined in (2.6), the Gaussian ensemble is mathematically much more tractable than the microcanonical ensemble, which is defined in terms of conditioning. The simpler form of the Gaussian ensemble is reflected in the simpler form of the unconstrained minimization problem (1-5) defining the set $\mathcal{E}_{\beta,\gamma}$ of Gaussian equilibrium macrostates. In (1-5) the constraint appearing in the minimization problem (1-1) defining the set $\mathcal{E}^u$ of microcanonical equilibrium macrostates is replaced by the linear and quadratic terms involving $\tilde{H}(x)$. The virtue of the Gaussian formulation should be clear. When the microcanonical and Gaussian ensembles are universally equivalent, then from a numerical point of view, it is better to use the Gaussian ensemble because in contrast to the microcanonical one, the Gaussian ensemble does not involve an equality constraint, which is difficult to implement numerically. Furthermore, within the context of the Gaussian ensemble, it is possible to use Monte Carlo techniques without any constraint on the sampling [Challa and Hetherington 1988a; Challa and Hetherington 1988b].

By giving necessary and sufficient conditions for the equivalence of the three ensembles in Theorems 3.1 and 3.3, we make contact with the duality theory of global optimization and the method of augmented Lagrangians [Bertsekas 1982, §2.2], [Minoux 1986, §6.4]. In the context of global optimization the primal function and the dual function play the same roles that the microcanonical entropy (resp., generalized microcanonical entropy) and the canonical free energy (resp., Gaussian free energy) play in statistical mechanics. Similarly, the replacement of the Lagrangian by the augmented Lagrangian in global optimization is paralleled by our replacement of the canonical ensemble by the Gaussian ensemble.

The Gaussian ensemble is a special case of the generalized canonical ensemble, which is obtained from the canonical ensemble by adding an exponential factor involving $g(h_n)$, where g is a continuous function that is bounded below. Our paper [Costeniuc et al. 2005b] gives all possible relationships between the sets of equilibrium macrostates for the microcanonical and generalized canonical ensembles in terms of support and concavity properties of an appropriate entropy function. Our paper [Touchette et al. 2006] shows that the generalized canonical ensemble can be used to transform metastable or unstable nonequilibrium macrostates for the standard canonical ensemble into stable equilibrium macrostates for the generalized canonical ensemble.

Equivalence and nonequivalence of ensembles is the subject of a large literature. An overview is given in the introduction of [Lewis et al. 1995]. A number of theoretical papers on this topic, including [Deuschel et al. 1991; Ellis et al. 2000; Eyink and Spohn 1993; Georgii 1993; Lewis et al. 1994; Lewis et al. 1995; Roelly and Zessin 1993], investigate equivalence of ensembles using the theory of large deviations. In [Lewis et al. 1994, §7] and [Lewis et al. 1995, §7.3] there is a discussion of nonequivalence of ensembles for the simplest mean-field model in statistical mechanics; namely, the Curie–Weiss model of a ferromagnet. However, despite the mathematical sophistication of these and other studies, none of them except for our papers [Costeniuc et al. 2005b; Ellis et al. 2000] explicitly addresses the general issue of the nonequivalence of ensembles.

Nonequivalence of ensembles has been observed in a wide range of systems that involve long-range interactions and that can be studied by the methods of [Costeniuc et al. 2005b; Ellis et al. 2000]. In all of these cases the microcanonical formulation gives rise to a richer set of equilibrium macrostates. For example, it has been shown computationally that the strongly reversing zonal-jet structures on Jupiter as well as the Great Red Spot fall into the nonequivalent range of an appropriate microcanonical ensemble [Turkington et al. 2001]. Other models for which ensemble nonequivalence has been observed include a number of long-range, mean-field spin models including the Hamiltonian mean-field model [Dauxois et al. 2002], the mean-field X-Y model [Dauxois et al. 2000], and the mean-field Blume–Emery–Griffiths model [Barré et al. 2002; 2001; Ellis et al. 2004b]. For a mean-field version of the Potts model called the Curie–Weiss–Potts model, equivalence and nonequivalence of ensembles is analyzed in detail in [Costeniuc et al. 2005a; Costeniuc et al. 2006a]. Ensemble nonequivalence has also been observed in models of turbulent vorticity dynamics [DiBattista et al. 2001; Dibattista et al. 1998; Ellis et al. 2002a; Eyink and Spohn 1993; Kiessling and Lebowitz 1997; Robert and Sommeria 1991], models of plasmas [Kiessling and Neukirch 2003; Smith and O'Neil 1990], gravitational systems [Gross 1997; Hertel and Thirring 1971; Lynden-Bell and Wood 1968; Thirring 1970], and models of the Lennard–Jones gas [Borges and Tsallis 2002; Kiessling and Percus 1995]. A detailed discussion of ensemble nonequivalence for models of coherent structures in two dimensional turbulence is given in [Ellis et al. 2000, § 1.4].

Gaussian ensembles were introduced in [Hetherington 1987] and studied further in [Challa and Hetherington 1988a; Challa and Hetherington 1988b; Hetherington and Stump 1987; Johal et al. 2003; Stump and Hetherington 1987]. As these papers discuss, an important feature of Gaussian ensembles is that they allow one to account for ensemble-dependent effects in finite systems. Although not referred to by name, the Gaussian ensemble also plays a key role in [Kiessling and Lebowitz 1997], where it is used to address equivalence-of-ensemble questions for a point-vortex model of fluid turbulence.

Another seed out of which the research summarized in the present paper germinated is the paper [Ellis et al. 2002a]. There we study the equivalence of the microcanonical and canonical ensembles for statistical equilibrium models of coherent structures in two-dimensional and quasigeostrophic turbulence. Numerical computations demonstrate that, as in other cases, nonequivalence of ensembles occurs over a wide range of model parameters and that physically interesting microcanonical equilibria are often omitted by the canonical ensemble. In addition, in Section 5 of [Ellis et al. 2002a], we establish the nonlinear stability of the steady mean flows corresponding to microcanonical equilibria via a new Lyapunov argument. The associated stability theorem refines the well-known Arnold stability theorems, which do not apply when the microcanonical and canonical ensembles are not equivalent. The Lyapunov functional appearing in this new stability theorem is defined in terms of a generalized thermodynamic potential similar in form to $I(x) + \beta \tilde{H}(x) + \gamma [\tilde{H}(x)]^2$, the minimum points of which define the set of equilibrium macrostates for the Gaussian ensemble [see (2.14)].

Our goal in this paper is to give an overview of our theoretical work on ensemble equivalence presented in [Costeniuc et al. 2005b; Ellis et al. 2000]. The paper [Costeniuc et al. 2006b] investigates the physical principles underlying this theory. In Section 2 of the present paper, we first state the assumptions on the statistical mechanical models to which the theory of the present paper applies. We then define the three ensembles — microcanonical, canonical, and Gaussian — and specify the three associated sets of equilibrium macrostates in terms of large deviation principles. In Section 3 we state two sets of results on ensemble equivalence. The first involves the equivalence of the microcanonical and canonical ensembles, necessary and sufficient conditions for which are given in terms of support properties of the microcanonical entropy s defined in (1-3). The second involves the equivalence of the microcanonical and Gaussian ensembles, necessary and sufficient conditions for which are given in terms of support properties of the generalized microcanonical entropy s_γ defined in (1-6). Section 4 addresses a basic foundational issue in statistical mechanics. There we show that when the canonical ensemble is nonequivalent to the microcanonical ensemble on a subset of energy values u, it can often be replaced by a Gaussian ensemble that is universally equivalent to the microcanonical ensemble. In Section 5 the results on ensemble equivalence discussed in this paper are illustrated in the context of the Curie–Weiss–Potts lattice-spin model, a mean-field approximation to the nearest-neighbor Potts model. Several of the results presented near the end of this section are new.

2. Definitions of models and ensembles

One of the objectives of this paper is to show that when the canonical ensemble is nonequivalent to the microcanonical ensemble on a subset of energy values u, it can often be replaced by a Gaussian ensemble that is equivalent to the microcanonical ensemble for all u. Before introducing the various ensembles as well as the methodology for proving this result, we first specify the class of statistical mechanical models under consideration. The models are defined in terms of the following quantities.

1. A sequence of probability spaces $(\Omega_n, \mathcal{F}_n, P_n)$ indexed by $n \in \mathbb{N}$, which typically represents a sequence of finite dimensional systems. The Ω_n are the configuration spaces, $\omega \in \Omega_n$ are the microstates, and the P_n are the prior measures on the σ fields $\mathcal{F}_n$.

2. A sequence of positive scaling constant $a_n \to \infty$ as $n \to \infty$. In general a_n equals the total number of degrees of freedom in the model. In many cases a_n equals the number of particles.

3. For each $n \in \mathbb{N}$ a measurable functions H_n mapping Ω_n into $\mathbb{R}$. For $\omega \in \Omega_n$ we define the energy per degree of freedom by

$$h_n(\omega) = \frac{1}{a_n} H_n(\omega).$$

Typically, H_n in item 3 equals the Hamiltonian, which is associated with energy conservation in the model. The theory is easily generalized by replacing H_n by a vector of appropriate functions representing additional dynamical invariants associated with the model [Costeniuc et al. 2005b; Ellis et al. 2000].

A large deviation analysis of the general model is possible provided that there exist a space of macrostates, macroscopic variables, and an interaction representation function and provided that the macroscopic variables satisfy the large deviation principle (LDP) on the space of macrostates. These concepts are explained next.

4. **Space of macrostates.** This is a complete, separable metric space $\mathcal{X}$, which represents the set of all possible macrostates.

5. **Macroscopic variables.** These are a sequence of random variables Y_n mapping Ω_n into $\mathcal{X}$. These functions associate a macrostate in $\mathcal{X}$ with each microstate $\omega \in \Omega_n$.

6. **Interaction representation function.** This is a bounded, continuous functions $\tilde{H}$ mapping $\mathcal{X}$ into $\mathbb{R}$ such that as $n \to \infty$

$$h_n(\omega) = \tilde{H}(Y_n(\omega)) + o(1) \quad \text{uniformly for } \omega \in \Omega_n; \qquad (2.1)$$

i.e.,

$$\lim_{n \to \infty} \sup_{\omega \in \Omega_n} |h_n(\omega) - \tilde{H}(Y_n(\omega))| = 0.$$

The function $\tilde{H}$ enable us to write h_n, either exactly or asymptotically, as a function of the macrostate via the macroscopic variables Y_n.

7. **LDP for the macroscopic variables.** There exists a function I mapping $\mathcal{X}$ into $[0, \infty]$ and having compact level sets such that with respect to P_n the sequence Y_n satisfies the LDP on $\mathcal{X}$ with rate function I and scaling constants a_n. In other words, for any closed subset F of $\mathcal{X}$

$$\limsup_{n \to \infty} \frac{1}{a_n} \log P_n\{Y_n \in F\} \leq - \inf_{x \in F} I(x),$$

and for any open subset G of $\mathcal{X}$

$$\liminf_{n \to \infty} \frac{1}{a_n} \log P_n\{Y_n \in G\} \geq - \inf_{x \in G} I(x).$$

It is helpful to summarize the LDP by the formal notation $P_n\{Y_n \in dx\} \asymp \exp[-a_n I(x)]$. This notation expresses the fact that, to a first degree of approximation, $P_n\{Y_n \in dx\}$ behaves like an exponential that decays to 0 whenever $I(x) > 0$.

The assumptions on the statistical mechanical models just stated, as well as a number of definitions to appear later, follow the presentation in [Costeniuc et al. 2005b], which is adapted for applications to lattice spin systems and related models. These assumptions and definitions differ slightly from those in [Ellis et al. 2000], where they are adapted for applications to statistical mechanical models of coherent structures in turbulence. The major difference is that in the asymptotic relationship (2.1) and in the definition (2.3) of the microcanonical ensemble $P_n^{u,r}$, h_n is replaced by H_n in [Ellis et al. 2000]. In addition, in the definition (2.4) of the canonical ensemble $P_{n,\beta}$, $a_n h_n$ is replaced by H_n in [Ellis et al. 2000]. Similarly, in the definition (2.6) of the Gaussian ensemble $P_{n,\beta,\gamma}$, $a_n h_n$ and $a_n h_n^2$ are replaced by H_n and H_n^2 to yield the Gaussian ensemble used to study models of coherent structures in turbulence. Finally, in the present paper the LDP for Y_n is derived with respect to $P_{n,\beta}$ and $P_{n,\beta,\gamma}$ while in models of coherent structures in turbulence the LDP for Y_n is derived with respect to $P_{n,a_n\beta}$ and $P_{n,a_n\beta,a_n\gamma}$, in which β and γ are both scaled by a_n. With only these minor changes in notation, all the results stated here are applicable to models of coherent structures in turbulence and in turn, all the results derived in [Ellis et al. 2000] for models of coherent structures in turbulence are applicable here.

A wide variety of statistical mechanical models satisfy the assumptions listed in items 1–7 or the modifications just discussed. Hence they can be studied by the methods of [Costeniuc et al. 2005b; Ellis et al. 2000]. We next give six examples. The first two are long-range spin systems, the third a class of short-range spin systems, the fourth a model of two-dimensional turbulence, the fifth a model of quasigeostrophic turbulence, and the sixth a model of dispersive wave turbulence.

1. The mean-field Blume–Emery–Griffiths model [Blume et al. 1971] is one of the simplest lattice-spin models known to exhibit both a continuous, second-order phase transition and a discontinuous, first-order phase transition. The space of macrostates for this model is the set of probability measures on a certain finite set, the macroscopic variables are the empirical measures associated with the spin configurations, and the associated LDP is Sanov's Theorem, for which the rate function is a relative entropy. Various features of this model are studied in [Barré et al. 2002; Barré et al. 2001; Ellis et al. 2005; Ellis et al. 2004b].

2. The Curie–Weiss–Potts model is a mean-field approximation to the nearest-neighbor Potts model [Wu 1982]. For the Curie–Weiss–Potts model, the space of macrostates, the macroscopic variables, and the associated LDP are similar to those in the mean-field Blume–Emery–Griffiths model. The

Curie–Weiss–Potts model nicely illustrates the general results on ensemble equivalence discussed in this paper and is discussed in Section 5.

3. Short-range spin systems such as the Ising model on $\mathbb{Z}^d$ and numerous generalizations can also be handled by the methods of this paper. The large deviation techniques required to analyze these models are much more subtle than in the case of the long-range, mean-field models considered in items 1 and 2. For the Ising model the space of macrostates is the space of translation-invariant probability measures on $\mathbb{Z}^d$, the macroscopic variables are the empirical processes associated with the spin configurations, and the rate function in the associated LDP is the mean relative entropy [Ellis 1985; Föllmer and Orey 1988; Olla 1988].

4. The Miller–Robert model is a model of coherent structures in an ideal, two-dimensional fluid that includes all the exact invariants of the vorticity transport equation [Miller 1990; Robert 1991]. The space of macrostates is the space of Young measures on the vorticity field. The large deviation analysis of this model developed first in [Robert 1991] and more recently in [Boucher et al. 2000] gives a rigorous derivation of maximum entropy principles governing the equilibrium behavior of the ideal fluid.

5. In geophysical applications, another version of the model in item 4 is preferred, in which the enstrophy integrals are treated canonically and the energy and circulation are treated microcanonically [Ellis et al. 2002a]. In those formulations, the space of macrostates is $L^2(\Lambda)$ or $L^\infty(\Lambda)$ depending on the constraints on the vorticity field. The large deviation analysis is carried out in [Ellis et al. 2002b]. The paper [Ellis et al. 2002a] shows how the nonlinear stability of the steady mean flows arising as equilibrium macrostates can be established by utilizing the appropriate generalized thermodynamic potentials.

6. A statistical equilibrium model of solitary wave structures in dispersive wave turbulence governed by a nonlinear Schrödinger equation is studied in [Ellis et al. 2004a]. The large deviation analysis given in [Ellis et al. 2004a] derives rigorously the concentration phenomenon observed in long-time numerical simulations and predicted by mean-field approximations [Jordan et al. 2000; Lebowitz et al. 1989]. The space of macrostates is $L^2(\Lambda)$, where Λ is a bounded interval or more generally a bounded domain in $\mathbb{R}^d$. The macroscopic variables are certain Gaussian processes.

We now return to the general theory, first introducing the function whose support and concavity properties completely determine all aspects of ensemble equivalence and nonequivalence. This function is the microcanonical entropy, defined for $u \in \mathbb{R}$ by

$$s(u) = -\inf\{I(x) : x \in \mathcal{X}, \tilde{H}(x) = u\}. \qquad (2.2)$$

Since I maps $\mathfrak{X}$ into $[0, \infty]$, s maps $\mathbb{R}^\sigma$ into $[-\infty, 0]$. Moreover, since I is lower semicontinuous and $\tilde{H}$ is continuous on $\mathfrak{X}$, s is upper semicontinuous on $\mathbb{R}^\sigma$. We define dom s to be the set of $u \in \mathbb{R}^\sigma$ for which $s(u) > -\infty$. In general, dom s is nonempty since $-s$ is a rate function [Ellis et al. 2000, Prop. 3.1(a)]. For each $u \in$ dom s, $r > 0$, $n \in \mathbb{N}$, and set $B \in \mathcal{F}_n$ the microcanonical ensemble is defined to be the conditioned measure

$$P_n^{u,r}\{B\} = P_n\{B \mid h_n \in [u - r, u + r]\}. \tag{2.3}$$

As shown in [Ellis et al. 2000, p. 1027], if $u \in$ dom s, then for all sufficiently large n, $P_n\{h_n \in [u - r, u + r]\} > 0$; thus the conditioned measures $P_n^{u,r}$ are well defined.

A mathematically more tractable probability measure is the canonical ensemble. For each $n \in \mathbb{N}$, $\beta \in \mathbb{R}$, and set $B \in \mathcal{F}_n$ we define the partition function

$$Z_n(\beta) = \int_{\Omega_n} \exp[-a_n \beta h_n]\, dP_n,$$

which is well defined and finite, and the probability measure

$$P_{n,\beta}\{B\} = \frac{1}{Z_n(\beta)} \cdot \int_B \exp[-a_n \beta h_n]\, dP_n. \tag{2.4}$$

The measures $P_{n,\beta}$ are Gibbs states that define the canonical ensemble for the given model.

The Gaussian ensemble is a natural perturbation of the canonical ensemble. For each $n \in \mathbb{N}$, $\beta \in \mathbb{R}$, and $\gamma \in [0, \infty)$ we define the Gaussian partition function

$$Z_n(\beta, \gamma) = \int_{\Omega_n} \exp[-a_n \beta h_n - a_n \gamma h_n^2]\, dP_n. \tag{2.5}$$

This is well defined and finite because the h_n are bounded. For $B \in \mathcal{F}_n$ we also define the probability measure

$$P_{n,\beta,\gamma}\{B\} = \frac{1}{Z_n(\beta, \gamma)} \cdot \int_B \exp[-a_n \beta h_n - a_n \gamma h_n^2]\, dP_n, \tag{2.6}$$

which we call the Gaussian ensemble. One can generalize this by replacing the quadratic function by a continuous function g that is bounded below. This gives rise to the generalized canonical ensemble, which the theory developed in [Costeniuc et al. 2005b] allows one to treat.

Using the theory of large deviations, one introduces the sets of equilibrium macrostates for each ensemble. It is proved in [Ellis et al. 2000, Theorem 3.2] that with respect to the microcanonical ensemble $P_n^{u,r}$, Y_n satisfies the LDP on $\mathcal{X}$, in the double limit $n \to \infty$ and $r \to 0$, with rate function

$$I^u(x) = \begin{cases} I(x) + s(u) & \text{if } \tilde{H}(x) = u \\ \infty & \text{otherwise .} \end{cases} \tag{2.7}$$

I^u is nonnegative on $\mathcal{X}$, and for $u \in \text{dom } s$, I^u attains its infimum of 0 on the set

$$\mathcal{E}^u = \{x \in \mathcal{X} : I^u(x) = 0\} \tag{2.8}$$
$$= \{x \in \mathcal{X} : I(x) \text{ is minimized subject to } \tilde{H}(x) = u\}.$$

This set is precisely the set of solutions of the constrained minimization problem (1-1).

In order to state the LDPs for the other two ensembles, we bring in the canonical free energy, defined for $\beta \in \mathbb{R}$ by

$$\varphi(\beta) = - \lim_{n \to \infty} \frac{1}{a_n} \log Z_n(\beta),$$

and the Gaussian free energy, defined for $\beta \in \mathbb{R}$ and $\gamma \geq 0$ by

$$\varphi(\beta, \gamma) = - \lim_{n \to \infty} \frac{1}{a_n} \log Z_n(\beta, \gamma).$$

It is proved in [Ellis et al. 2000, Theorem 2.4] that the limit defining $\varphi(\beta)$ exists and is given by

$$\varphi(\beta) = \inf_{y \in \mathcal{X}} \{I(y) + \beta \tilde{H}(y)\} \tag{2.9}$$

and that with respect to $P_{n,\beta}$, Y_n satisfies the LDP on $\mathcal{X}$ with rate function

$$I_\beta(x) = I(x) + \beta \tilde{H}(x) - \varphi(\beta). \tag{2.10}$$

I_β is nonnegative on $\mathcal{X}$ and attains its infimum of 0 on the set

$$\mathcal{E}_\beta = \{x \in \mathcal{X} : I_\beta(x) = 0\} \tag{2.11}$$
$$= \{x \in \mathcal{X} : I(x) + \langle \beta, \tilde{H}(x) \rangle \text{ is minimized}\}.$$

This set is precisely the set of solutions of the unconstrained minimization problem (1-2).

A straightforward extension of these results shows that the limit defining $\varphi(\beta, \gamma)$ exists and is given by

$$\varphi(\beta, \gamma) = \inf_{y \in \mathcal{X}} \{I(y) + \beta \tilde{H}(y) + \gamma [\tilde{H}(y)]^2\} \tag{2.12}$$

and that with respect to $P_{n,\beta,g}$, Y_n satisfies the LDP on $\mathcal{X}$ with rate function

$$I_{\beta,\gamma}(x) = I(x) + \beta \tilde{H}(x) + \gamma [\tilde{H}(x)]^2 - \varphi(\beta, \gamma). \tag{2.13}$$

$I_{\beta,\gamma}$ is nonnegative on $\mathcal{X}$ and attains its infimum of 0 on the set

$$\mathcal{E}_{\beta,\gamma} = \{x \in \mathcal{X} : I_{\beta,\gamma}(x) = 0\} \tag{2.14}$$
$$= \{x \in \mathcal{X} : I(x) + \langle \beta, \tilde{H}(x) \rangle + \gamma[\tilde{H}(x)]^2 \text{ is minimized}\}.$$

This set is precisely the set of solutions of the penalized minimization problem (1-5).

For $u \in \text{dom}\, s$, let x be any element of $\mathcal{X}$ satisfying $I^u(x) > 0$. The formal notation

$$P_n^{u,r}\{Y_n \in dx\} \asymp e^{-a_n I^u(x)}$$

suggests that x has an exponentially small probability of being observed in the limit $n \to \infty$, $r \to 0$. Hence it makes sense to identify $\mathcal{E}^u$ with the set of microcanonical equilibrium macrostates. In the same way we identify with $\mathcal{E}_\beta$ the set of canonical equilibrium macrostates and with $\mathcal{E}_{\beta,\gamma}$ the set of generalized canonical equilibrium macrostates. A rigorous justification is given in [Ellis et al. 2000, Theorem 2.4(d)].

3. Equivalence and nonequivalence of the three ensembles

Having defined the sets of equilibrium macrostates $\mathcal{E}^u$, $\mathcal{E}_\beta$, and $\mathcal{E}_{\beta,\gamma}$ for the microcanonical, canonical and Gaussian ensembles, we now show how these sets are related to one another. In Theorem 3.1 we state the results proved in [Ellis et al. 2000] concerning equivalence and nonequivalence of the microcanonical and canonical ensembles. Then in Theorem 3.3 we extend these results to the Gaussian ensemble [Costeniuc et al. 2005b].

Parts (a)–(c) of Theorem 3.1 give necessary and sufficient conditions, in terms of support properties of s, for equivalence and nonequivalence of $\mathcal{E}^u$ and $\mathcal{E}_\beta$. These assertions are proved in Theorems 4.4 and 4.8 in [Ellis et al. 2000]. Part (a) states that s has a strictly supporting line at u if and only if full equivalence of ensembles holds; i.e., if and only if there exists a β such that $\mathcal{E}^u = \mathcal{E}_\beta$. The most surprising result, given in part (c), is that s has no supporting line at u if and only if nonequivalence of ensembles holds in the strong sense that $\mathcal{E}^u \cap \mathcal{E}_\beta = \varnothing$ for all β. Part (c) is to be contrasted with part (d), which states that for any β canonical equilibrium macrostates can always be realized microcanonically. Part (d) is proved in Theorem 4.6 in [Ellis et al. 2000]. Thus one conclusion of this theorem is that at the level of equilibrium macrostates, in general the microcanonical ensemble is the richer of the two ensembles.

THEOREM 3.1. *In parts* (a), (b), *and* (c), *u denotes any point in* dom *s*.

(a) **Full equivalence**. *There exists β such that $\mathcal{E}^u = \mathcal{E}_\beta$ if and only if s has a strictly supporting line at u with slope β; i.e.,*

$$s(v) < s(u) + \beta(v - u) \text{ for all } v \neq u .$$

(b) **Partial equivalence**. *There exists β such that $\mathcal{E}^u \subset \mathcal{E}_\beta$ but $\mathcal{E}^u \neq \mathcal{E}_\beta$ if and only if s has a nonstrictly supporting line at u with slope β; i.e.,*

$$s(v) \leq s(u) + \beta(v - u) \text{ for all } v \text{ with equality for some } v \neq u.$$

(c) **Nonequivalence**. *For all β, $\mathcal{E}^u \cap \mathcal{E}_\beta = \varnothing$ if and only if s has no supporting line at u; i.e.,*

$$\text{for all } \beta \text{ there exists } v \text{ such that } s(v) > s(u) + \beta(v - u).$$

(d) **Canonical is always realized microcanonically**. *For any $\beta \in \mathbb{R}$ we have $\tilde{H}(\mathcal{E}_\beta) \subset \text{dom } s$ and*

$$\mathcal{E}_\beta = \bigcup_{u \in \tilde{H}(\mathcal{E}_\beta)} \mathcal{E}^u.$$

We highlight several features of the theorem in order to illuminate their physical content. In part (a) let us add the assumption that for a given $u \in \text{dom } s$ there exists a unique β such that $\mathcal{E}^u = \mathcal{E}_\beta$. If s is differentiable at u and s and the double Legendre–Fenchel transform s^{**} are equal in a neighborhood of u, then β is given by the standard thermodynamic formula $\beta = s'(u)$ [Costeniuc et al. 2005b, Theorem A.4(b)]. The inverse relationship can be obtained from part (d) of the theorem under the added assumption that $\mathcal{E}_\beta$ consists of a unique macrostate or more generally that for all $x \in \mathcal{E}_\beta$ the values $\tilde{H}(x)$ are equal. Then $\mathcal{E}_\beta = \mathcal{E}^{u(\beta)}$, where $u(\beta) = \tilde{H}(x)$ for any $x \in \mathcal{E}_\beta$; $u(\beta)$ denotes the mean energy realized at equilibrium in the canonical ensemble. The relationship $u = u(\beta)$ inverts the relationship $\beta = s'(u)$. Partial ensemble equivalence can be seen in part (d) under the added assumption that for a given β, $\mathcal{E}_\beta$ can be partitioned into at least two sets $\mathcal{E}_{\beta,i}$ such that for all $x \in \mathcal{E}_{\beta,i}$ the values $\tilde{H}(x)$ are equal but $\tilde{H}(x) \neq \tilde{H}(y)$ whenever $x \in \mathcal{E}_{\beta,i}$ and $y \in \mathcal{E}_{\beta,j}$ for $i \neq j$. Then $\mathcal{E}_\beta = \bigcup_i \mathcal{E}^{u_i(\beta)}$, where $u_i(\beta) = \tilde{H}(x)$, $x \in \mathcal{E}_{\beta,i}$. Clearly, for each i, $\mathcal{E}^{u_i(\beta)} \subset \mathcal{E}_\beta$ but $\mathcal{E}^{u_i(\beta)} \neq \mathcal{E}_\beta$. Physically, this corresponds to a situation of coexisting phases that normally takes place at a first-order phase transition [Touchette et al. 2004].

Before continuing with our analysis of ensemble equivalence, we make a number of basic definitions. A function f on $\mathbb{R}$ is said to be concave on $\mathbb{R}$ if f maps $\mathbb{R}$ into $\mathbb{R} \cup \{-\infty\}$, $f \not\equiv -\infty$, and for all u and v in $\mathbb{R}$ and all $\lambda \in (0, 1)$

$$f(\lambda u + (1 - \lambda)v) \geq \lambda f(u) + (1 - \lambda) f(v).$$

Let $f \not\equiv -\infty$ be a function mapping $\mathbb{R}$ into $\mathbb{R} \cup \{-\infty\}$. We define dom f to be the set of u for which $f(u) > -\infty$. For β and u in $\mathbb{R}$ the Legendre–Fenchel transforms f^* and f^{**} are defined by

$$f^*(\beta) = \inf_{u \in \mathbb{R}} \{\langle \beta, u \rangle - f(u)\} \text{ and } f^{**}(u) = \inf_{\beta \in \mathbb{R}} \{\langle \beta, u \rangle - f^*(\beta)\}.$$

The function f^* is concave and upper semicontinuous on $\mathbb{R}$ and for all u we have $f^{**}(u) = f(u)$ if and only if f is concave and upper semicontinuous on $\mathbb{R}$ [Ellis 1985, Theorem VI.5.3]. When f is not concave and upper semicontinuous, then f^{**} is the smallest concave, upper semicontinuous function on $\mathbb{R}$ that satisfies $f^{**}(u) \geq f(u)$ for all u [Costeniuc et al. 2005b, Prop. A.2]. In particular, if for some u, $f(u) \neq f^{**}(u)$, then $f(u) < f^{**}(u)$.

Let $f \not\equiv -\infty$ be a function mapping $\mathbb{R}$ into $\mathbb{R} \cup \{-\infty\}$, u a point in dom f, and K a convex subset of dom f. We have the following four additional definitions: f is concave at u if $f(u) = f^{**}(u)$; f is not concave at u if $f(u) < f^{**}(u)$; f is concave on K if f is concave at all $u \in K$; and f is strictly concave on K if for all $u \neq v$ in K and all $\lambda \in (0, 1)$

$$f(\lambda u + (1 - \lambda)v) > \lambda f(u) + (1 - \lambda) f(v).$$

We also introduce two sets that play a central role in the theory. Let f be a concave function on $\mathbb{R}$ whose domain is an interval having nonempty interior. For $u \in \mathbb{R}$ the superdifferential of f at u, denoted by $\partial f(u)$, is defined to be the set of β such that β is the slope of a supporting line of f at u. Any such β is called a supergradient of f at u. Thus, if f is differentiable at $u \in \text{int}(\text{dom } f)$, then $\partial f(u)$ consists of the unique point $\beta = f'(u)$. If f is not differentiable at $u \in \text{int}(\text{dom } f)$, then dom ∂f consists of all β satisfying the inequalities

$$(f')^+(u) \leq \beta \leq (f')^-(u),$$

where $(f')^-(u)$ and $(f')^+(u)$ denote the left-hand and right-hand derivatives of f at u. The domain of ∂f, denoted by dom ∂f, is then defined to be the set of u for which $\partial f(u) \neq \varnothing$.

Complications arise because dom ∂f can be a proper subset of dom f, as simple examples clearly show. Let b be a boundary point of dom f for which $f(b) > -\infty$. Then b is in dom ∂f if and only if the one-sided derivative of f at b is finite. For example, if b is a left hand boundary point of dom f and $(f')^+(b)$ is finite, then $\partial f(b) = [(f')^+(b), \infty)$; any $\beta \in \partial f(b)$ is the slope of a supporting line at b. The possible discrepancy between dom ∂f and dom f introduces unavoidable technicalities in the statements of several results concerning the existence of supporting lines.

One of our goals is to find concavity and support conditions on the micro-canonical entropy guaranteeing that the microcanonical and canonical ensembles are fully equivalent at all points $u \in$ dom s except possibly boundary points. If this is the case, then we say that the ensembles are universally equivalent. Here is a basic result in that direction. The universal equivalence stated in part (b) follows from part (a) and from part (a) of Theorem 3.1. The rest of the theorem depends on facts concerning concave functions [Costeniuc et al. 2005b, p. 1305].

THEOREM 3.2. *Assume that* dom s *is an interval having nonempty interior and that* s *is strictly concave on* int(dom s) *and continuous on* dom s. *The following conclusions hold.*

(a) s *has a strictly supporting line at all* $u \in$ dom s *except possibly boundary points.*

(b) *The microcanonical and canonical ensembles are universally equivalent*; *i.e., fully equivalent at all* $u \in$ dom s *except possibly boundary points.*

(c) s *is concave on* $\mathbb{R}$, *and for each* u *in part* (b) *the corresponding* β *in the statement of full equivalence is any element of* $\partial s(u)$.

(d) *If* s *is differentiable at some* $u \in$ dom s, *then the corresponding* β *in part* (b) *is unique and is given by the standard thermodynamic formula* $\beta = s'(u)$.

The next theorem extends Theorem 3.1 by giving equivalence and nonequivalence results involving $\mathcal{E}^u$ and $\mathcal{E}_{\beta,\gamma}$, the sets of equilibrium macrostates with respect to the microcanonical and Gaussian ensembles. The chief innovation is that $s(u)$ in Theorem 3.1 is replaced here by the generalized microcanonical entropy $s(u) - \gamma u^2$. As we point out after the statement of Theorem 3.3, for the purpose of applications part (a) is its most important contribution. The usefulness of Theorem 3.3 is matched by the simplicity with which it follows from Theorem 3.1. Theorem 3.3 is a special case of Theorem 3.4 in [Costeniuc et al. 2005b], obtained by specializing the generalized canonical ensemble and the associated set of equilibrium macrostates to the Gaussian ensemble and the set $\mathcal{E}_{\beta,\gamma}$ of Gaussian equilibrium macrostates.

THEOREM 3.3. *Given* $\gamma \geq 0$, *define* $s_\gamma(u) = s(u) - \gamma u^2$. *In parts* (a), (b), *and* (c), u *denotes any point in* dom s.

(a) **Full equivalence.** *There exists* β *such that* $\mathcal{E}^u = \mathcal{E}_{\beta,\gamma}$ *if and only if* s_γ *has a strictly supporting line at* u *with slope* β.

(b) **Partial equivalence.** *There exists* β *such that* $\mathcal{E}^u \subset \mathcal{E}_{\beta,\gamma}$ *but* $\mathcal{E}^u \neq \mathcal{E}_{\beta,\gamma}$ *if and only if* s_γ *has a nonstrictly supporting line at* u *with slope* β.

(c) **Nonequivalence.** *For all* β, $\mathcal{E}^u \cap \mathcal{E}_{\beta,\gamma} = \varnothing$ *if and only if* s_γ *has no supporting line at* u.

(d) *Gaussian is always realized microcanonically.* For any β we have

$$\tilde{H}(\mathcal{E}_{\beta,\gamma}) \subset \text{dom } s, \quad \mathcal{E}_{\beta,\gamma} = \bigcup_{u \in \tilde{H}(\mathcal{E}_{\beta,\gamma})} \mathcal{E}^u.$$

PROOF. For $\gamma \geq 0$ and $B \in \mathcal{F}_n$ we define a new probability measure

$$P_{n,\gamma}\{B\} = \frac{1}{\int_{\Omega_n} \exp[-a_n\gamma h_n^2] \, dP_n} \cdot \int_B \exp[-a_n\gamma h_n^2] \, dP_n.$$

With respect to $P_{n,\gamma}$, Y_n satisfies the LDP on $\mathcal{X}$ with rate function

$$I_\gamma(x) = I(x) + \gamma[\tilde{H}(x)]^2 - \psi(\gamma),$$

where $\psi(\gamma) = \inf_{y \in \mathcal{X}}\{I(y) + \gamma[\tilde{H}(y)]^2\}$. Replacing the prior measure P_n in the canonical ensemble with $P_{n,\gamma}$ gives the Gaussian ensemble $P_{n,\beta,\gamma}$, which has $\mathcal{E}_{\beta,\gamma}$ as the associated set of equilibrium macrostates. On the other hand, replacing the prior measure P_n in the microcanonical ensemble with $P_{n,\gamma}$ gives

$$P_{n,\gamma}^{u,r}\{B\} = P_{n,\gamma}\{B \mid h_n \in [u - r, u + r]\},$$

By continuity, for ω satisfying $h_n(\omega) \in [u - r, u + r]$, $[h_n(\omega)]^2$ converges to u^2 uniformly in ω and n as $r \to 0$. It follows that with respect to $P_{n,\gamma}^{u,r}$, Y_n satisfies the LDP on $\mathcal{X}$, in the double limit $n \to \infty$ and $r \to 0$, with the same rate function I^u as in the LDP for Y_n with respect to $P_n^{u,r}$. As a result, the set of equilibrium macrostates corresponding to $P_{n,\gamma}^{u,r}$ coincides with the set $\mathcal{E}^u$ of microcanonical equilibrium macrostates.

It follows from parts (a)–(c) of Theorem 3.1 that all equivalence and non-equivalence relationships between $\mathcal{E}^u$ and $\mathcal{E}_{\beta,\gamma}$ are expressed in terms of support properties of the function $\tilde{s}_\gamma$ obtained from s by replacing the rate function I by the new rate function I_γ. The function $\tilde{s}_\gamma$ is given by

$$\begin{aligned}
\tilde{s}_\gamma(u) &= -\inf\{I_\gamma(x) : x \in \mathcal{X}, \tilde{H}(x) = u\} \\
&= -\inf\{I(x) + \gamma\tilde{H}(x)^2 : x \in \mathcal{X}, \tilde{H}(x) = u\} + \psi(\gamma) \\
&= s(u) - \gamma u^2 + \psi(\gamma).
\end{aligned}$$

Since $\tilde{s}_\gamma(u)$ differs from $s_\gamma(u) = s(u) - \gamma u^2$ by the constant $\psi(\gamma)$, we conclude that all equivalence and nonequivalence relationships between $\mathcal{E}^u$ and $\mathcal{E}_{\beta,\gamma}$ are expressed in terms of the same support properties of s_γ. This completes the derivation of parts (a)–(c) of Theorem 3.3 from parts (a)–(c) of Theorem 3.1. Similarly, part (d) of Theorem 3.3 follows from part (d) of Theorem 3.1.   □

The importance of part (a) of Theorem 3.3 in applications is emphasized by the following theorem, which will be applied in the sequel. This theorem is the analogue of Theorem 3.2 for the Gaussian ensemble, s in that theorem being replaced by s_γ. The functions s and s_γ have the same domains. The universal equivalence stated in part (b) of the next theorem follows from part (a) and from part (a) of Theorem 3.3.

THEOREM 3.4. *For $\gamma \geq 0$, define $s_\gamma(u) = s(u) - \gamma u^2$. Assume that* dom s *is an interval having nonempty interior and that s_γ is strictly concave on* int(dom s) *and continuous on* dom s. *The following conclusions hold.*

(a) *s_γ has a strictly supporting line at all $u \in$ dom s except possibly boundary points.*

(b) *The microcanonical ensemble and the Gaussian ensemble defined in terms of this γ are universally equivalent; i.e., fully equivalent at all $u \in$ dom s except possibly boundary points.*

(c) *s_γ is concave on $\mathbb{R}$, and for each u in part (b) the corresponding β in the statement of full equivalence is any element of $\partial s_\gamma(u)$.*

(d) *If s_γ is differentiable at some $u \in$ dom s, then the corresponding β in part (b) is unique and is given by the thermodynamic formula $\beta = s'_\gamma(u)$.*

The most important repercussion of Theorem 3.4 is the ease with which one can prove that the microcanonical and Gaussian ensembles are universally equivalent in those cases in which the microcanonical and canonical ensembles are not fully or partially equivalent. This rests mainly on part (b) of Theorem 3.4, which states that universal equivalence of ensembles holds if there exists a $\gamma \geq 0$ such that s_γ is strictly concave on int(dom s). The existence of such a γ follows from a natural set of hypotheses on s stated in Theorem 4.2 in the next section.

4. Universal equivalence via the Gaussian ensemble

This section addresses a basic foundational issue in statistical mechanics. Under the assumption that the microcanonical entropy is C^2 and s'' is bounded above, we show in Theorem 4.2 that when the canonical ensemble is nonequivalent to the microcanonical ensemble on a subset of energy values u, it can often be replaced by a Gaussian ensemble that is universally equivalent to the microcanonical ensemble; i.e., fully equivalent at all $u \in$ dom s except possibly boundary points. Theorem 4.3 is a weaker version that can often be applied when s'' is not bounded above. In the last section of the paper, these results will be illustrated in the context of the Curie–Weiss–Potts lattice-spin model.

In Theorem 4.2 the strategy is to find a quadratic function γu^2 such that $s_\gamma(u) = s(u) - \gamma u^2$ is strictly concave on int(dom s) and continuous on dom s. Parts (a) and (b) of Theorem 3.4 then yields the universal equivalence. As the next proposition shows, an advantage of working with quadratic functions is that support properties of s_γ involving a supporting line are equivalent to support properties of s involving a supporting parabola defined in terms of γ. This observation gives a geometrically intuitive way to find a quadratic function guaranteeing universal ensemble equivalence.

In order to state the proposition, we need a definition. Let f be a function mapping $\mathbb{R}$ into $\mathbb{R} \cup \{-\infty\}$, u and β points in $\mathbb{R}$, and $\gamma \geq 0$. We say that f has a supporting parabola at u with parameters (β, γ) if

$$f(v) \leq f(u) + \beta(v - u) + \gamma(v - u)^2 \quad \text{for all } v. \tag{4.1}$$

The parabola is said to be strictly supporting if the inequality is strict for all $v \neq u$.

PROPOSITION 4.1. f has a (strictly) supporting parabola at u with parameters (β, γ) if and only if $f - \gamma(\cdot)^2$ has a (strictly) supporting line at u with slope $\tilde{\beta}$. The quantities β and $\tilde{\beta}$ are related by $\tilde{\beta} = \beta - 2\gamma u$.

PROOF. We use the identity $(v - u)^2 = v^2 - 2u(v - u) - u^2$. If f has a strictly supporting parabola at u with parameters (β, γ), then for all $v \neq u$

$$f(v) - \gamma v^2 < f(u) - \gamma u^2 + \tilde{\beta}(v - u),$$

where $\tilde{\beta} = \beta - 2\gamma u$. Thus $f - \gamma(\cdot)^2$ has a strictly supporting line at u with slope $\tilde{\beta}$. The converse is proved similarly, as is the case in which the supporting line or parabola is supporting but not strictly supporting.   □

The first application of Theorem 3.4 is Theorem 4.2, which gives a criterion guaranteeing the existence of a quadratic function γu^2 such that $s_\gamma(u) = s(u) - \gamma u^2$ is strictly concave on $\text{dom}\, s$. The criterion — that s'' is bounded above on the interior of $\text{dom}\, s$ — is essentially optimal for the existence of a fixed quadratic function γu^2 guaranteeing the strict concavity of s_γ. The situation in which s'' is not bounded above on the interior of $\text{dom}\, s$ can often be handled by Theorem 4.3, which is a local version of Theorem 4.2.

THEOREM 4.2. Assume that $\text{dom}\, s$ is an interval having nonempty interior. Assume also that s is continuous on $\text{dom}\, s$, s is twice continuously differentiable on $\text{int}(\text{dom}\, s)$, and s'' is bounded above on $\text{int}(\text{dom}\, s)$. Then for all sufficiently large $\gamma \geq 0$, conclusions (a)–(c) hold. Specifically, if s is strictly concave on $\text{dom}\, s$, then we choose any $\gamma \geq 0$, and otherwise we choose

$$\gamma > \gamma_0 = \tfrac{1}{2} \cdot \sup_{u \in \text{int}(\text{dom}\, s)} s''(u). \tag{4.2}$$

(a) $s_\gamma(u) = s(u) - \gamma u^2$ is strictly concave and continuous on $\text{dom}\, s$.

(b) s_γ *has a strictly supporting line, and s has a strictly supporting parabola, at all* $u \in \mathrm{dom}\, s$ *except possibly boundary points. At a boundary point* s_γ *has a strictly supporting line, and s has a strictly supporting parabola, if and only if the one-sided derivative of* s_γ *is finite at that boundary point.*

(c) *The microcanonical ensemble and the Gaussian ensemble defined in terms of this* γ *are universally equivalent; i.e., fully equivalent at all* $u \in \mathrm{dom}\, s$ *except possibly boundary points. For all* $u \in \mathrm{int}(\mathrm{dom}\, s)$ *the value of* β *defining the universally equivalent Gaussian ensemble is unique and equals* $\beta = s'(u) - 2\gamma u$.

PROOF. (a) If s is strictly concave on $\mathrm{dom}\, s$, then s_γ is also strictly concave on this set for any $\gamma \geq 0$. We now consider the case in which s is not strictly concave on $\mathrm{dom}\, s$. For any $\gamma \geq 0$, s_γ is continuous on $\mathrm{dom}\, s$. If, in addition, we choose $\gamma > \gamma_0$ in accordance with (4.2), then for all $u \in \mathrm{int}(\mathrm{dom}\, s)$

$$s''_\gamma(u) = s''(u) - 2\gamma < 0.$$

A straightforward extension of the proof of Theorem 4.4 in [Rockafellar 1970], in which the inequalities in the first two displays are replaced by strict inequalities, shows that $-s_\gamma$ is strictly convex on $\mathrm{int}(\mathrm{dom}\, s)$ and thus that s_γ is strictly concave on $\mathrm{int}(\mathrm{dom}\, s)$. If s_γ is not strictly concave on $\mathrm{dom}\, s$, then s_γ must be affine on an interval. Since this violates the strict concavity on $\mathrm{int}(\mathrm{dom}\, s)$, part (a) is proved.

(b) The first assertion follows from part (a) of the present theorem, part (a) of Theorem 3.4, and Proposition 4.1. Concerning the second assertion about boundary points, the reader is referred to the discussion before Theorem 3.2.

(c) The universal equivalence of the two ensembles is a consequence of part (a) of the present theorem and part (b) of Theorem 3.4. The full equivalence of the ensembles at all $u \in \mathrm{int}(\mathrm{dom}\, s)$ is equivalent to the existence of a strictly supporting line at each $u \in \mathrm{int}(\mathrm{dom}\, s)$ [Theorem 3.3(a)]. Since $s_\gamma(u)$ is differentiable at all $u \in \mathrm{int}(\mathrm{dom}\, s)$, for each u the slope of the strictly supporting line at u is unique and equals $s'_\gamma(u)$ [Costeniuc et al. 2005b, Theorem A.1(b)]. □

Suppose that s is C^2 on the interior of $\mathrm{dom}\, s$ but the second-order partial derivatives of s are not bounded above. This arises, for example, in the Curie–Weiss–Potts model, in which $\mathrm{dom}\, s$ is a closed, bounded interval of $\mathbb{R}$ and $s''(u) \to \infty$ as u approaches the right hand endpoint of $\mathrm{dom}\, s$ [see §5]. In such cases one cannot expect that the conclusions of Theorems 4.2 will be satisfied; in particular, that there exists $\gamma \geq 0$ such that $s_\gamma(u) = s(u) - \gamma u^2$ has a strictly supporting line at each point of the interior of $\mathrm{dom}\, s$ and thus that the ensembles are universally equivalent.

In order to overcome this difficulty, we introduce Theorem 4.3, a local version of Theorem 4.2. Theorem 4.3 handles the case in which s is C^2 on an open set K but either K is not all of $\mathrm{int}(\mathrm{dom}\, s)$ or $K = \mathrm{int}(\mathrm{dom}\, s)$ and s'' is not bounded above on K. In neither of these situations are the hypotheses of Theorem 4.2 satisfied.

In Theorem 4.3 other hypotheses are given guaranteeing that for each $u \in K$ there exists γ such that s_γ has a strictly supporting line at u; in general, γ depends on u. However, with the same γ, s_γ might also have a strictly supporting line at other values of u. In general, as one increases γ, the set of u at which s_γ has a strictly supporting line cannot decrease. Because of part (a) of Theorem 3.3, this can be restated in terms of ensemble equivalence involving the set $\mathcal{E}_{\beta,\gamma}$ of Gaussian equilibrium macrostates. Defining

$$F_\gamma = \{u \in K : \text{there exists } \beta \text{ such that } \mathcal{E}_{\beta,\gamma} = \mathcal{E}^u\},$$

we have $F_{\gamma_1} \subset F_{\gamma_2}$ whenever $\gamma_2 > \gamma_1$ and because of Theorem 4.3, $\bigcup_{\gamma>0} F_\gamma = K$. This phenomenon is investigated in Section 5 for the Curie–Weiss–Potts model.

In order to state Theorem 4.3, we define for $u \in K$ and $\lambda \geq 0$

$$D(u, s'(u), \lambda) = \{v \in \text{dom } s : s(v) \geq s(u) + s'(u)(v-u) + \lambda(v-u)^2\}.$$

Geometrically, this set contains all points for which the parabola with parameters $(s'(u), \lambda)$ passing through $(u, s(u))$ lies below the graph of s. Clearly, since $\lambda \geq 0$, we have $D(u, s'(u), \lambda) \subset D(u, s'(u), 0)$; the set $D(u, s'(u), 0)$ contains all points for which the graph of the line with slope $s'(u)$ passing through $(u, s(u))$ lies below the graph of s. Thus, in the next theorem the hypothesis that for each $u \in K$ the set $D(u, s'(u), \lambda)$ is bounded for some $\lambda \geq 0$ is satisfied if dom s is bounded or, more generally, if $D(u, s'(u), 0)$ is bounded. The latter set is bounded if, for example, $-s$ is superlinear; i.e.,

$$\lim_{|v| \to \infty} s(v)/|v| = -\infty.$$

The quantity $\gamma_0(u)$ appearing in the next theorem is defined in equation (5.7) in [Costeniuc et al. 2005b].

THEOREM 4.3. *Let K an open subset of* dom s *and assume that s is twice continuously differentiable on K. Assume also that* dom s *is bounded or, more generally, that for every $u \in$ int K there exists $\lambda \geq 0$ such that $D(u, s'(u), \lambda)$ is bounded. Then for each $u \in K$ there exists $\gamma_0(u) \geq 0$ with the following properties.*

(a) *For each $u \in K$ and any $\gamma > \gamma_0(u)$, s has a strictly supporting parabola at u with parameters $(s'(u), \gamma)$.*

(b) *For each $u \in K$ and any $\gamma > \gamma_0(u)$, $s_\gamma = s - \gamma(\cdot)^2$ has a strictly supporting line at u with slope $s'(u) - 2\gamma u$.*

(c) *For each $u \in K$ and any $\gamma > \gamma_0(u)$, the microcanonical ensemble and the Gaussian ensemble defined in terms of this γ are fully equivalent at u. The value of β defining the Gaussian ensemble is unique and is given by $\beta = s'(u) - 2\gamma u$.*

COMMENTS ON THE PROOF. (a) We first choose a parabola that is strictly supporting in a neighborhood of u and then adjust γ so that the parabola becomes strictly supporting on all $\mathbb{R}$. Proposition 4.1 guarantees that $s - \gamma(\cdot)^2$ has a strictly supporting line at u. Details are given in [Costeniuc et al. 2005b, pp. 1319–1321].

(b) This follows from part (a) of the present theorem and Proposition 4.1.

(c) For $u \in K$ the full equivalence of the ensembles follows from part (b) of the present theorem and part (a) of Theorem 3.3. The value of β defining the fully equivalent Gaussian ensemble is determined by a routine argument given in [Costeniuc et al. 2005b, p. 1321]. □

Theorem 4.3 suggests an extended form of the notion of universal equivalence of ensembles. In Theorem 4.2 we are able to achieve full equivalence of ensembles for all $u \in \mathrm{dom}\, s$, except possibly boundary points, by choosing an appropriate γ that is valid for all u. This leads to the observation that the microcanonical ensemble and the Gaussian ensemble defined in terms of this γ are universally equivalent. In Theorem 4.3 we can also achieve full equivalence of ensembles for all $u \in K$. However, in contrast to Theorem 4.2, the choice of γ for which the two ensembles are fully equivalent depends on u. We summarize the ensemble equivalence property articulated in part (c) of Theorem 4.3 by saying that relative to the set of quadratic functions, the microcanonical and Gaussian ensembles are universally equivalent on the open set K of energy values.

We complete our discussion of the generalized canonical ensemble and its equivalence with the microcanonical ensemble by noting that the smoothness hypothesis on s in Theorem 4.3 is essentially satisfied whenever the microcanonical ensemble exhibits no phase transition at any $u \in K$. In order to see this, we recall that a point u_c at which s is not differentiable represents a first-order, microcanonical phase transition [Ellis et al. 2004b, Figure 3]. In addition, a point u_c at which s is differentiable but not twice differentiable represents a second-order, microcanonical phase transition [Ellis et al. 2004b, Figure 4]. It follows that s is smooth on any open set K not containing such phase-transition points. Hence, if the other hypotheses in Theorem 4.3 are valid, then the microcanonical and Gaussian ensembles are universally equivalent on K relative to the set of quadratic functions. In particular, if the microcanonical ensemble exhibits no phase transitions, then s is smooth on all of $\mathrm{int}(\mathrm{dom}\, s)$. This implies the universal equivalence of the two ensembles provided that the other conditions are valid in Theorem 4.2.

In the next section we apply the results in this paper to the Curie–Weiss–Potts model.

5. Applications to the Curie–Weiss–Potts model

The Curie–Weiss–Potts model is a mean-field approximation to the nearest-neighbor Potts model, which takes its place next to the Ising model as one of the most versatile models in equilibrium statistical mechanics [Wu 1982]. Although the Curie–Weiss–Potts model is considerably simpler to analyze, it is an excellent model to illustrate the general theory presented in this paper, lying at the boundary of the set of models for which a complete analysis involving explicit formulas is available. As we will see, there exists an interval N such that for any $u \in N$ the microcanonical ensemble is nonequivalent to the canonical ensemble. The main result, stated in Theorem 5.2, is that for any $u \in N$ there exists $\gamma \geq 0$ such that the microcanonical ensemble and the Gaussian ensemble defined in terms of this γ are fully equivalent for all $v \leq u$. While not as strong as universal equivalence, the ensemble equivalence proved in Theorem 5.2 is considerably stronger than the local equivalence stated in Theorem 4.3.

Let $q \geq 3$ be a fixed integer and define $\Lambda = \{\theta^1, \theta^2, \ldots, \theta^q\}$, where the θ^i are any q distinct vectors in $\mathbb{R}^q$. In the definition of the Curie–Weiss–Potts model, the precise values of these vectors is immaterial. For each $n \in \mathbb{N}$ the model is defined by spin random variables $\omega_1, \omega_2, \ldots, \omega_n$ that take values in Λ. The ensembles for the model are defined in terms of probability measures on the configuration spaces Λ^n, which consist of the microstates $\omega = (\omega_1, \omega_2, \ldots, \omega_n)$. We also introduce the n-fold product measure P_n on Λ^n with identical one-dimensional marginals

$$\bar{\rho} = \frac{1}{q} \sum_{i=1}^{q} \delta_{\theta^i}.$$

Thus for all $\omega \in \Lambda^n$, $P_n(\omega) = \frac{1}{q^n}$. For $n \in \mathbb{N}$ and $\omega \in \Lambda^n$ the Hamiltonian for the q-state Curie–Weiss–Potts model is defined by

$$H_n(\omega) = -\frac{1}{2n} \sum_{j,k=1}^{n} \delta(\omega_j, \omega_k),$$

where $\delta(\omega_j, \omega_k)$ equals 1 if $\omega_j = \omega_k$ and equals 0 otherwise. The energy per particle is defined by $h_n(\omega) = \frac{1}{n} H_n(\omega)$.

With this choice of h_n and with $a_n = n$, the microcanonical, canonical, and Gaussian ensembles for the model are the probability measures on Λ^n defined as in (2.3), (2.4), and (2.6). The key to our analysis of the Curie–Weiss–Potts model is to express h_n in terms of the macroscopic variables

$$L_n = L_n(\omega) = (L_{n,1}(\omega), L_{n,2}(\omega), \ldots, L_{n,q}(\omega)),$$

the ith component of which is defined by

$$L_{n,i}(\omega) = \frac{1}{n} \sum_{j=1}^{n} \delta(\omega_j, \theta^i).$$

This quantity equals the relative frequency with which ω_j, $j \in \{1, \ldots, n\}$, equals θ^i. The empirical vectors L_n take values in the set of probability vectors

$$\mathcal{P} = \left\{ v \in \mathbb{R}^q : v = (v_1, v_2, \ldots, v_q), \text{ each } v_i \geq 0, \sum_{i=1}^{q} v_i = 1 \right\}.$$

Each probability vector in $\mathcal{P}$ represents a possible equilibrium macrostate for the model.

There is a one-to-one correspondence between $\mathcal{P}$ and the set $\mathcal{P}(\Lambda)$ of probability measures on Λ, $v \in \mathcal{P}$ corresponding to the probability measure $\sum_{i=1}^{q} v_i \delta_{\theta^i}$. The element $\rho \in \mathcal{P}$ corresponding to the one-dimensional marginal $\tilde{\rho}$ of the prior measures P_n is the uniform vector having equal components $\frac{1}{q}$. For $\omega \in \Lambda^n$ the element of $\mathcal{P}$ corresponding to the empirical vector $L_n(\omega)$ is the empirical measure of the spin random variables $\omega_1, \omega_2, \ldots, \omega_n$.

We denote by $\langle \cdot, \cdot \rangle$ the inner product on $\mathbb{R}^q$. Since

$$\sum_{i=1}^{q} \sum_{j=1}^{n} \delta(\omega_j, \xi^i) \cdot \sum_{k=1}^{n} \delta(\omega_k, \xi^i) = \sum_{j,k=1}^{n} \delta(\omega_j, \omega_k),$$

it follows that the energy per particle can be rewritten as

$$h_n(\omega) = -\frac{1}{2n^2} \sum_{j,k=1}^{n} \delta(\omega_j, \omega_k) = -\tfrac{1}{2} \langle L_n(\omega), L_n(\omega) \rangle,$$

i.e.,

$$h_n(\omega) = \tilde{H}(L_n(\omega)), \text{ where } \tilde{H}(v) = -\tfrac{1}{2} \langle v, v \rangle \text{ for } v \in \mathcal{P}.$$

$\tilde{H}$ is the energy representation function for the model.

In order to define the sets of equilibrium macrostates with respect to the three ensembles, we appeal to Sanov's Theorem. This states that with respect to the product measures P_n, the empirical vectors L_n satisfy the LDP on $\mathcal{P}$ with rate function given by the relative entropy $R(\cdot \mid \rho)$ [Ellis 1985, Theorem VIII.2.1]. For $v \in \mathcal{P}$ this is defined by

$$R(v|\rho) = \sum_{i=1}^{q} v_i \log(q v_i).$$

With the choices $I = R(\cdot|\rho)$, $\tilde{H} = -\frac{1}{2}\langle\cdot,\cdot\rangle$, and $a_n = n$, L_n satisfies the LDP on $\mathcal{P}$ with respect to each of the three ensembles with the rate functions given by (2.7), (2.10), and (2.13). In turn, the corresponding sets of equilibrium macrostates are given by

$$\mathcal{E}^u = \{v \in \mathcal{P} : R(v|\rho) \text{ is minimized subject to } \tilde{H}(v) = u\},$$

$$\mathcal{E}_\beta = \{v \in \mathcal{P} : R(v|\rho) + \beta\tilde{H}(v) \text{ is minimized }\},$$

$$\mathcal{E}_{\beta,\gamma} = \{v \in \mathcal{P} : R(v|\rho) + \beta\tilde{H}(v) + \gamma[\tilde{H}(v)]^2 \text{ is minimized }\}.$$

Each element v in $\mathcal{E}^u$, $\mathcal{E}_\beta$, and $\mathcal{E}_{\beta,\gamma}$ describes an equilibrium configuration of the model with respect to the corresponding ensemble in the thermodynamic limit. The ith component v_i gives the asymptotic relative frequency of spins taking the value θ^i.

As in (2.2), the microcanonical entropy is defined by

$$s(u) = -\inf\{R(v|\rho) : v \in \mathcal{P}, \tilde{H}(v) = u\}.$$

Since $R(v|\rho) < \infty$ for all $v \in \mathcal{P}$, dom s equals the range of $\tilde{H}(v) = -\frac{1}{2}\langle v, v\rangle$ on $\mathcal{P}$, which is the closed interval $[-\frac{1}{2}, -\frac{1}{2q}]$. The set $\mathcal{E}^u$ of microcanonical equilibrium macrostates is nonempty precisely for $u \in$ dom s. For $q = 3$, the microcanonical entropy can be determined explicitly. For all $q \geq 4$ the micro-canonical entropy can also be determined explicitly provided Conjecture 4.1 in [Costeniuc et al. 2005a] is valid; this conjecture has been verified numerically for all $q \in \{4, 5, \ldots, 10^4\}$. The formulas for the microcanonical entropy are given in Theorem 4.3 in [Costeniuc et al. 2005a].

We first consider the relationships between $\mathcal{E}^u$ and $\mathcal{E}_\beta$, which according to Theorem 3.1 are determined by support properties of s. These properties can be seen in Figure 1. The quantity u_0 appearing in this figure equals $[-q^2 + 3q - 3]/[2q(q-1)]$ [Costeniuc et al. 2005a, Lem. 6.1]. Figure 1 is not the actual graph of s but a schematic graph that accentuates the shape of the graph of s together with the intervals of strict concavity and nonconcavity of this function.

These and other details of the graph of s are also crucial in analyzing the relationships between $\mathcal{E}^u$ and $\mathcal{E}_{\beta,\gamma}$. Denote dom s by $[u_\ell, u_r]$, where $u_\ell = -\frac{1}{2}$ and $u_r = -\frac{1}{2q}$. These details include the observation that there exists $w_0 \in (u_0, u_r)$ such that s is a concave-convex function with break point w_0; i.e., the restriction of s to (u_ℓ, w_0) is strictly concave and the restriction of s to (w_0, u_r) is strictly convex. A difficulty in validating this observation is that for certain values of q, including $q = 3$, the intervals of strict concavity and strict convexity are shallow and therefore difficult to discern. Furthermore, what seem to be strictly concave and strictly convex portions of this function on the scale of the entire graph might reveal themselves to be much less regular on a finer scale. Conjecture 5.1 gives a set of properties of s implying there exists $w_0 \in (u_0, u_r)$

such that s is a concave-convex function with break point w_0. In particular, this property of s guarantees that s has the support properties stated in the three items appearing in the next paragraph. Conjecture 5.1 has been verified numerically for all $q \in \{4, 5, \ldots, 10^4\}$.

We define the sets

$$F = (u_\ell, u_0) \cup \{u_r\}, \quad P = \{u_0\}, \quad and \quad N = (u_0, u_r).$$

Figure 1 and Theorem 3.1 then show that these sets are respectively the sets of full equivalence, partial equivalence, and nonequivalence of the microcanonical and canonical ensembles. The details are given in the next three items. In Theorem 6.2 in [Costeniuc et al. 2005a] all these conclusions concerning ensemble equivalence and nonequivalence are proved analytically without reference to the form of s given in Figure 1.

1. s is strictly concave on the interval (u_ℓ, u_0) and has a strictly supporting line at each $u \in (u_\ell, u_0)$ and at u_r. Hence for $u \in F = (u_\ell, u_0) \cup \{u_r\}$ the ensembles are fully equivalent in the sense that there exists β such that $\mathcal{E}^u = \mathcal{E}_\beta$ [Theorem 3.1(a)].

2. s is concave but not strictly concave at u_0 and has a nonstrictly supporting line at u_0 that also touches the graph of s over the right hand endpoint u_r. Hence for $u \in P = \{u_0\}$ the ensembles are partially equivalent in the sense that there exists β such that $\mathcal{E}^u \subset \mathcal{E}_\beta$ but $\mathcal{E}^u \neq \mathcal{E}_\beta$ [Theorem 3.1(b)].

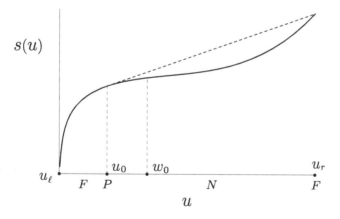

Figure 1. Schematic graph of $s(u)$, showing the set $F = (u_\ell, u_0) \cup \{u_r\}$ of full ensemble equivalence, the singleton set $P = \{u_0\}$ of partial equivalence, and the set $N = (u_0, u_r)$ of nonequivalence, where $u_\ell = -\frac{1}{2}$ and $u_r = -\frac{1}{2q}$. For $u \in F \cup P = (u_\ell, u_0] \cup \{u_r\}$, $s(u) = s^{**}(u)$; for $u \in N$, $s(u) < s^{**}(u)$ and the graph of s^{**} consists of the dotted line segment with slope β_c. The slope of s at u_ℓ is ∞. The quantity w_0 is discussed after Conjecture 5.1.

3. s is not concave on $N = (u_0, u_r)$ and has no supporting line at any $u \in N$. Hence for $u \in N$ the ensembles are nonequivalent in the sense that for all β, $\mathcal{E}^u \cap \mathcal{E}_\beta = \varnothing$ [Theorem 3.1(c)].

The explicit calculation of the elements of $\mathcal{E}_\beta$ and $\mathcal{E}^u$ given in [Costeniuc et al. 2005a] shows different continuity properties of these two sets. $\mathcal{E}_\beta$ undergoes a discontinuous phase transition as β increases through the critical inverse temperature $\beta_c = \frac{2(q-1)}{q-2} \log(q-1)$, the unique macrostate ρ for $\beta < \beta_c$ bifurcating discontinuously into the q distinct macrostates for $\beta > \beta_c$. By contrast, $\mathcal{E}^u$ undergoes a continuous phase transition as u decreases from the maximum value $u_r = -\frac{1}{2q}$, the unique macrostate ρ for $u = u_r$ bifurcating continuously into the q distinct macrostates for $u < u_r$. The different continuity properties of these phase transitions shows already that the canonical and microcanonical ensembles are nonequivalent.

For u in the interval N of ensemble nonequivalence, the graph of s^{**} is affine; this is depicted by the dotted line segment in Figure 1. One can show that the slope of the affine portion of the graph of s^{**} equals the critical inverse temperature β_c.

This completes the discussion of the equivalence and nonequivalence of the microcanonical and canonical ensembles. The equivalence and nonequivalence of the microcanonical and Gaussian ensembles depends on the relationships between the sets $\mathcal{E}^u$ and $\mathcal{E}_{\beta,\gamma}$ of corresponding equilibrium macrostates, which in turn are determined by support properties of the generalized microcanonical entropy $s_\gamma(u) = s(u) - \gamma u^2$. As we just saw, for each $u \in N = (u_0, u_r)$, the microcanonical and canonical ensembles are nonequivalent. For $u \in N$ we would like to recover equivalence by replacing the canonical ensemble by an appropriate Gaussian ensemble.

Theorem 4.2 is not applicable. Although the first three of the hypotheses are valid, unfortunately s'' is not bounded above on the interior of dom s. Indeed, using the explicit formula for s given in Theorem 4.3 in [Costeniuc et al. 2005a], one verifies that $\lim_{u \to (u_r)^-} s''(u) = \infty$. However, we can appeal to Theorem 4.3 in the present paper, which is applicable since s is twice continuously differentiable on N. We conclude that for each $u \in N$ and all sufficiently large γ there exists a corresponding Gaussian ensemble that is equivalent to the microcanonical ensemble for that u.

By using other conjectured properties of the microcanonical entropy, we are able to deduce the stronger result on the equivalence of the microcanonical and Gaussian ensembles stated in Theorem 5.2. As before, we denote dom s by $[u_\ell, u_r]$, where $u_\ell = -\frac{1}{2}$ and $u_r = -\frac{1}{2q}$, and write

$$s'(u_\ell) = \lim_{u \to (u_\ell)^+} s'(u) \quad \text{and} \quad s'(u_r) = \lim_{u \to (u_r)^-} s'(u)$$

with a similar notation for $s''(u_\ell)$ and $s''(u_r)$. Using the explicit but complicated formula for s given in Theorem 4.2 in [Costeniuc et al. 2005a], the following conjecture was verified numerically for all $q \in \{4, 5, \ldots, 10^4\}$ and all $u \in (u_\ell, u_r)$ of the form $u = u_\ell + 0.02k$, where k is a positive integer.

CONJECTURE 5.1. *For all $q \geq 3$ the microcanonical entropy s has the following two properties.*
(a) $s'''(u) > 0$ *for all* $u \in (u_\ell, u_r)$.
(b) $s'(u_\ell) = \infty$, $0 < s'(u_r) < \infty$, $s''(u_\ell) = -\infty$, *and* $s''(u_r) = \infty$.

The conjecture implies that s'' is an increasing bijection of (u_ℓ, u_r) onto $\mathbb{R}$. Therefore, there exists a unique point $w_0 \in (u_\ell, u_r)$ such that $s''(u) < 0$ for all $u \in (u_\ell, w_0)$, $s''(w_0) = 0$, and $s''(u) > 0$ for all $u \in (w_0, u_r)$. It follows that the restriction of s to $[u_\ell, w_0]$ is strictly concave and the restriction of s to $[w_0, u_r]$ is strictly convex. These properties, which can be seen in Figure 1, are summarized by saying that s is a concave-convex function with break point w_0.

The interval $N = (u_0, u_r)$ exhibited in Figure 1 contains all energy values u for which there exists no canonical ensemble that is equivalent with the microcanonical ensemble. Assuming the truth of Conjecture 5.1, we now show that for each $u \in N$ there exists $\gamma \geq 0$ and an associated Gaussian ensemble that is equivalent with the microcanonical ensemble for all $v \leq u$. In order to do this, for $\gamma \geq 0$ we bring in the generalized microcanonical entropy

$$s_\gamma(u) = s(u) - \gamma u^2$$

and note that the properties of s stated in Conjecture 5.1 are invariant under the addition of the quadratic $-\gamma u^2$. Hence, if Conjecture 5.1 is valid, then s_γ satisfies the same properties as s. In particular, s_γ must be a concave-convex function with some break point w_γ, which is the unique point in (u_ℓ, u_r) such that $s_\gamma''(u) < 0$ for all $u \in (u_\ell, w_\gamma)$, $s_\gamma''(w_\gamma) = 0$, and $s_\gamma''(u) > 0$ for all $u \in (w_\gamma, u_r)$. A straightforward argument, which we omit, and an appeal to Theorem 3.3 show that there exists a unique point $u_\gamma \in (u_\ell, w_\gamma)$ having the properties listed in the next three items. These properties show that u_γ plays the same role for ensemble equivalence involving the Gaussian ensemble that the point u_0 plays for ensemble equivalence involving the canonical ensemble.

1. For $\gamma \geq 0$, s_γ is strictly concave on the interval (u_ℓ, u_γ) and has a strictly supporting line at each $u \in (u_\ell, u_\gamma)$ and at u_r. Hence for $u \in F_\gamma = (u_\ell, u_\gamma) \cup \{u_r\}$ the ensembles are fully equivalent in the sense that there exists β such that $\mathcal{E}^u = \mathcal{E}_{\beta,\gamma}$ [Theorem 3.3(a)].
2. For $\gamma \geq 0$, s_γ is concave but not strictly concave at u_γ and has a nonstrictly supporting line at u_γ that also touches the graph of s over the right hand endpoint u_r. Hence for $u \in P_\gamma = \{u_\gamma\}$ the ensembles are partially equivalent

in the sense that there exists β such that $\mathcal{E}^u \subset \mathcal{E}_{\beta,\gamma}$ but $\mathcal{E}^u \neq \mathcal{E}_{\beta,\gamma}$ [Theorem 3.3(b)].

3. For $\gamma \geq 0$, s_γ is not concave on the interval $N = (u_\gamma, u_r)$ and has no supporting line at any $u \in N$. Hence for $u \in N_\gamma$ the ensembles are nonequivalent in the sense that for all β, $\mathcal{E}^u \cap \mathcal{E}_{\beta,\gamma} = \varnothing$ [Theorem 3.3(c)].

We now state our main result.

THEOREM 5.2. *We assume that Conjecture 5.1 is valid. Then as a function of $\gamma \geq 0$, $F_\gamma = (u_\ell, u_\gamma) \cup \{u_r\}$ is strictly increasing, and as $\gamma \to \infty$, $F_\gamma \uparrow (u_\ell, u_r]$. It follows that for any $u \in N = (u_0, u_r)$, there exists $\gamma \geq 0$ such that the microcanonical ensemble and the Gaussian ensemble defined in terms of this γ are fully equivalent for all $v \in (u_\ell, u_r)$ satisfying $v \leq u$. The value of β defining the Gaussian ensemble is unique and is given by $\beta = s'(v) - 2\gamma v$.*

The proof of the theorem relies on the next lemma, part (a) of which uses Proposition 4.1. When applied to s_γ, this proposition states that s_γ has a strictly supporting line at a point if and only if s has a strictly supporting parabola at that point. Proposition 4.1 illustrates why one can achieve full equivalence with the Gaussian ensemble when full equivalence with the canonical ensemble fails. Namely, even when s does not have a supporting line at a point, it might have a supporting parabola at that point; in this case the supporting parabola can be made strictly supporting by increasing γ. The proofs of parts (b)–(d) of the next lemma rely on Theorem 4.3 and on the properties of the sets F_γ, P_γ, and N_γ stated in the three items appearing just before Theorem 5.2.

LEMMA 5.3. *We assume that Conjecture 5.1 is valid. Then:*

(a) *If for some $\gamma \geq 0$, s_γ has a supporting line at a point u, then for any $\tilde{\gamma} > \gamma$, $s_{\tilde{\gamma}}$ has a strictly supporting line at u.*

(b) *For any $0 \leq \gamma < \tilde{\gamma}$, $F_\gamma \cup P_\gamma \subset F_{\tilde{\gamma}}$.*

(c) *u_γ is a strictly increasing function of $\gamma \geq 0$ and $\lim_{\gamma \to \infty} u_\gamma = u_r$.*

(d) *As a function of $\gamma \geq 0$, F_γ is strictly increasing.*

PROOF. (a) Suppose that s_γ has a supporting line at u with slope $\bar{\beta}$. Then by Proposition 4.1 s has a supporting parabola at u with parameters (β, γ), where $\beta = \bar{\beta} + 2\gamma u$. As the definition (4.1) makes clear, replacing γ by any $\tilde{\gamma} > \gamma$ makes the supporting parabola at u strictly supporting. Again by Proposition 4.1 $s_{\tilde{\gamma}}$ has a strictly supporting line at u.

(b) If $u \in F_\gamma \cup P_\gamma$, then s_γ has a supporting line at u. Since $0 \leq \gamma < \tilde{\gamma}$, part (a) implies that $s_{\tilde{\gamma}}$ has a strictly supporting line at u. Hence u must be an element of $F_{\tilde{\gamma}}$.

(c) If $0 \leq \gamma < \tilde{\gamma}$, then by part (a) of the present lemma $u_\gamma \in P_\gamma \subset F_{\tilde{\gamma}}$. Since $F_{\tilde{\gamma}} = (u_\ell, u_{\tilde{\gamma}}) \cup \{u_r\}$ and since $u_\gamma < u_r$, it follows that $u_\gamma < u_{\tilde{\gamma}}$. Thus u_γ is a strictly increasing function of $\gamma \geq 0$. We now prove that $\lim_{\gamma \to \infty} u_\gamma = u_r$. For any $u \in (u_\ell, u_r)$, part (b) of Theorem 4.3 states that there exists $\gamma_u > 0$ such that $s_{\gamma_u}(u)$ has a strictly supporting line at u. It follows that $u \in F_{\gamma_u} = (u_\ell, u_{\gamma_u}) \cup \{u_r\}$ and thus that $u < u_{\gamma_u} < u_r$. Since u_γ is a strictly increasing function of γ, it follows that for all $\gamma > \gamma_u$, we have $u_\gamma > u_{\gamma_u} > u$. We have shown that for any $u \in (u_\ell, u_r)$, there exists $\gamma_u > 0$ such that for all $\gamma > \gamma_u$, we have $u_\gamma > u$. This completes the proof that $\lim_{\gamma \to \infty} u_\gamma = u_r$.

(d) Since $F_\gamma = (u_\ell, u_\gamma) \cup \{u_r\}$, this follows immediately from the first property of u_γ in part (c). The proof of the lemma is complete. □

We are now ready to prove Theorem 5.2. The properties of F_γ stated there follow immediately from Lemma 5.3. Indeed, since u_γ is a strictly increasing function of $\gamma \geq 0$, F_γ is also strictly increasing. In addition, since $\lim_{\gamma \to \infty} u_\gamma = u_r$ it follows that as $\gamma \to \infty$, $F_\gamma \uparrow (u_\ell, u_r]$. Since F_γ is the set of full ensemble equivalence, we conclude that for any $u \in N = (u_0, u_r)$, there exists $\gamma > 0$ such that the microcanonical ensemble and the Gaussian ensemble defined in terms of this γ are fully equivalent for all $v \in (u_\ell, u_r)$ satisfying $v \leq u$. The last statement concerning β is a consequence of part (c) of Theorem 4.3. The proof of Theorem 5.2 is complete.

Acknowledgements

The research of Marius Costeniuc was supported by a grant from the National Science Foundation (DMS-0202309), the research of Richard S. Ellis was supported by grants from the National Science Foundation (DMS-0202309 and DMS-0604071), the research of Bruce Turkington was supported by grants from the National Science Foundation (DMS-0207064 and DMS-0604071), and the research of Hugo Touchette was supported by the Natural Sciences and Engineering Research Council of Canada and the Royal Society of London (Canada–UK Millennium Fellowship).

References

[Barré et al. 2001] J. Barré, D. Mukamel, and S. Ruffo, "Inequivalence of ensembles in a system with long-range interactions", *Phys. Rev. Lett.* **87** (2001), #030601.

[Barré et al. 2002] J. Barré, D. Mukamel, and S. Ruffo, "Ensemble inequivalence in mean-field models of magnetism", pp. 45–67 in *Dynamics and thermodynamics of systems with long-range interactions* (Les Houches, 2002), edited by T. Dauxois et al., Lecture Notes in Phys. **602**, Springer, Berlin, 2002.

[Bertsekas 1982] D. P. Bertsekas, *Constrained optimization and Lagrange multiplier methods*, Academic Press, New York, 1982.

[Blume et al. 1971] M. Blume, V. J. Emery, and R. B. Griffiths, "Ising model for the λ transition and phase separation in ^{3}He-^{4}He mixtures", *Phys. Rev. A* **4** (1971), 1071–1077.

[Borges and Tsallis 2002] E. P. Borges and C. Tsallis, "Negative specific heat in a Lennard–Jones-like gas with long-range interactions", *Physica A* **305** (2002), 148–151.

[Boucher et al. 2000] C. Boucher, R. S. Ellis, and B. Turkington, "Derivation of maximum entropy principles in two-dimensional turbulence via large deviations", *J. Statist. Phys.* **98**:5-6 (2000), 1235–1278.

[Challa and Hetherington 1988a] M. S. S. Challa and J. H. Hetherington, "Gaussian ensemble: an alternate Monte-Carlo scheme", *Phys. Rev. A* **38** (1988), 6324–6337.

[Challa and Hetherington 1988b] M. S. S. Challa and J. H. Hetherington, "Gaussian ensemble as an interpolating ensemble", *Phys. Rev. Lett.* **60** (1988), 77–80.

[Costeniuc et al. 2005a] M. Costeniuc, R. S. Ellis, and H. Touchette, "Complete analysis of phase transitions and ensemble equivalence for the Curie–Weiss–Potts model", *J. Math. Phys.* **46**:6 (2005), #063301.

[Costeniuc et al. 2005b] M. Costeniuc, R. S. Ellis, H. Touchette, and B. Turkington, "The generalized canonical ensemble and its universal equivalence with the microcanonical ensemble", *J. Statist. Phys.* **119**:5-6 (2005), 1283–1329.

[Costeniuc et al. 2006a] M. Costeniuc, R. S. Ellis, and H. Touchette, "Nonconcave entropies from generalized canonical ensembles", *Phys. Rev. E* **74**:1 (2006), #010105.

[Costeniuc et al. 2006b] M. Costeniuc, R. S. Ellis, H. Touchette, and B. Turkington, "Generalized canonical ensembles and ensemble equivalence", *Phys. Rev. E* **73**:2 (2006), #026105.

[Dauxois et al. 2000] T. Dauxois, P. Holdsworth, and S. Ruffo, "Violation of ensemble equivalence in the antiferromagnetic mean-field XY model", *Eur. Phys. J. B* **16** (2000), 659–667.

[Dauxois et al. 2002] T. Dauxois, V. Latora, A. Rapisarda, S. Ruffo, and A. Torcini, "Dynamics and thermodynamics of systems with long-range interactions", pp. 458–487 in *Dynamics and thermodynamics of systems with long-range interactions* (Les Houches), edited by T. Dauxois et al., Lecture Notes in Physics **602**, Springer, Berlin, 2002.

[Deuschel et al. 1991] J.-D. Deuschel, D. W. Stroock, and H. Zessin, "Microcanonical distributions for lattice gases", *Comm. Math. Phys.* **139**:1 (1991), 83–101.

[Dibattista et al. 1998] M. T. Dibattista, A. J. Majda, and B. Turkington, "Prototype geophysical vortex structures via large-scale statistical theory", *Geophys. Astrophys. Fluid Dynam.* **89**:3-4 (1998), 235–283.

[DiBattista et al. 2001] M. T. DiBattista, A. J. Majda, and M. J. Grote, "Meta-stability of equilibrium statistical structures for prototype geophysical flows with damping and driving", *Physica D* **151**:2-4 (2001), 271–304.

[Ellis 1985] R. S. Ellis, *Entropy, large deviations, and statistical mechanics*, Grundlehren der math. Wissenschaften **271**, Springer, Berlin, 1985. Reprinted 2006.

[Ellis et al. 2000] R. S. Ellis, K. Haven, and B. Turkington, "Large deviation principles and complete equivalence and nonequivalence results for pure and mixed ensembles", *J. Statist. Phys.* **101**:5-6 (2000), 999–1064.

[Ellis et al. 2002a] R. S. Ellis, K. Haven, and B. Turkington, "Nonequivalent statistical equilibrium ensembles and refined stability theorems for most probable flows", *Nonlinearity* **15**:2 (2002), 239–255.

[Ellis et al. 2002b] R. S. Ellis, B. Turkington, and K. Haven, "Analysis of statistical equilibrium models of geostrophic turbulence", *J. Appl. Math. Stochastic Anal.* **15**:4 (2002), 341–361.

[Ellis et al. 2004a] R. S. Ellis, R. Jordan, P. Otto, and B. Turkington, "A statistical approach to the asymptotic behavior of a class of generalized nonlinear Schrödinger equations", *Comm. Math. Phys.* **244**:1 (2004), 187–208.

[Ellis et al. 2004b] R. S. Ellis, H. Touchette, and B. Turkington, "Thermodynamic versus statistical nonequivalence of ensembles for the mean-field Blume–Emery–Griffiths model", *Physica A* **335**:3-4 (2004), 518–538.

[Ellis et al. 2005] R. S. Ellis, P. T. Otto, and H. Touchette, "Analysis of phase transitions in the mean-field Blume–Emery–Griffiths model", *Ann. Appl. Probab.* **15**:3 (2005), 2203–2254.

[Eyink and Spohn 1993] G. L. Eyink and H. Spohn, "Negative-temperature states and large-scale, long-lived vortices in two-dimensional turbulence", *J. Statist. Phys.* **70**:3-4 (1993), 833–886.

[Föllmer and Orey 1988] H. Föllmer and S. Orey, "Large deviations for the empirical field of a Gibbs measure", *Ann. Probab.* **16**:3 (1988), 961–977.

[Georgii 1993] H.-O. Georgii, "Large deviations and maximum entropy principle for interacting random fields on $\mathbf{Z}^d$", *Ann. Probab.* **21**:4 (1993), 1845–1875.

[Gibbs 1902] J. W. Gibbs, *Elementary principles in statistical mechanics: developed with especial reference to the rational foundation of thermodynamics*, Yale University Press, New Haven, 1902. Reprinted Dover, New York, 1960.

[Gross 1997] D. H. E. Gross, "Microcanonical thermodynamics and statistical fragmentation of dissipative systems: the topological structure of the n-body phase space", *Phys. Rep.* **279** (1997), 119–202.

[Hertel and Thirring 1971] P. Hertel and W. Thirring, "A soluble model for a system with negative specific heat", *Ann. Phys. (NY)* **63** (1971), 520.

[Hetherington 1987] J. H. Hetherington, "Solid ^{3}He magnetism in the classical approximation", *J. Low Temp. Phys.* **66** (1987), 145–154.

[Hetherington and Stump 1987] J. H. Hetherington and D. R. Stump, "Sampling a Gaussian energy distribution to study phase transitions of the $Z(2)$ and $U(1)$ lattice gauge theories", *Phys. Rev. D* **35** (1987), 1972–1978.

[Johal et al. 2003] R. S. Johal, A. Planes, and E. Vives, "Statistical mechanics in the extended Gaussian ensemble", *Phys. Rev. E* **68** (2003), #056113.

[Jordan et al. 2000] R. Jordan, B. Turkington, and C. L. Zirbel, "A mean-field statistical theory for the nonlinear Schrödinger equation", *Physica D* **137**:3-4 (2000), 353–378.

[Kiessling and Lebowitz 1997] M. K.-H. Kiessling and J. L. Lebowitz, "The microcanonical point vortex ensemble: beyond equivalence", *Lett. Math. Phys.* **42** (1997), 43–58.

[Kiessling and Neukirch 2003] M. K.-H. Kiessling and T. Neukirch, "Negative specific heat of a magnetically self-confined plasma torus", *Proc. Natl. Acad. Sci. USA* **100**:4 (2003), 1510–1514.

[Kiessling and Percus 1995] M. K.-H. Kiessling and J. K. Percus, "Nonuniform van der Waals theory", *J. Statist. Phys.* **78**:5-6 (1995), 1337–1376.

[Lanford 1973] O. E. Lanford, "Entropy and equilibrium states in classical statistical mechanics", pp. 1–113 in *Statistical mechanics and mathematical problems: Battelle Rencontres* (Seattle, WA, 1971), edited by A. Lenard, Lecture Notes in Physics **20**, Springer, Berlin, 1973.

[Lebowitz et al. 1989] J. L. Lebowitz, H. A. Rose, and E. R. Speer, "Statistical mechanics of the nonlinear Schrödinger equation. II. Mean field approximation", *J. Statist. Phys.* **54**:1-2 (1989), 17–56.

[Lewis et al. 1994] J. T. Lewis, C.-E. Pfister, and W. G. Sullivan, "The equivalence of ensembles for lattice systems: some examples and a counterexample", *J. Statist. Phys.* **77**:1-2 (1994), 397–419.

[Lewis et al. 1995] J. T. Lewis, C.-E. Pfister, and W. G. Sullivan, "Entropy, concentration of probability and conditional limit theorems", *Markov Process. Related Fields* **1**:3 (1995), 319–386.

[Lynden-Bell and Wood 1968] D. Lynden-Bell and R. Wood, "The gravo-thermal catastrophe in isothermal spheres and the onset of red-giant structure for stellar systems", *Mon. Notic. Roy. Astron. Soc.* **138** (1968), 495.

[Miller 1990] J. Miller, "Statistical mechanics of Euler equations in two dimensions", *Phys. Rev. Lett.* **65**:17 (1990), 2137–2140.

[Minoux 1986] M. Minoux, *Mathematical programming: theory and algorithms*, Wiley, Chichester, 1986.

[Olla 1988] S. Olla, "Large deviations for Gibbs random fields", *Probab. Theory Related Fields* **77**:3 (1988), 343–357.

[Robert 1991] R. Robert, "A maximum-entropy principle for two-dimensional perfect fluid dynamics", *J. Statist. Phys.* **65**:3-4 (1991), 531–553.

[Robert and Sommeria 1991] R. Robert and J. Sommeria, "Statistical equilibrium states for two-dimensional flows", *J. Fluid Mech.* **229** (1991), 291–310.

[Rockafellar 1970] R. T. Rockafellar, *Convex analysis*, Princeton Mathematical Series **28**, Princeton Univ. Press, Princeton, NJ, 1970.

[Roelly and Zessin 1993] S. Roelly and H. Zessin, "The equivalence of equilibrium principles in statistical mechanics and some applications to large particle systems", *Exposition. Math.* **11**:5 (1993), 385–405.

[Smith and O'Neil 1990] R. A. Smith and T. M. O'Neil, "Nonaxisymmetric thermal equilibria of a cylindrically bounded guiding center plasma or discrete vortex system", *Phys. Fluids B* **2** (1990), 2961–2975.

[Stump and Hetherington 1987] D. R. Stump and J. H. Hetherington, "Remarks on the use of a microcanonical ensemble to study phase transitions in the lattice gauge theory", *Phys. Lett. B* **188** (1987), 359–363.

[Thirring 1970] W. Thirring, "Systems with negative specific heat", *Z. Physik* **235** (1970), 339–352.

[Touchette et al. 2004] H. Touchette, R. S. Ellis, and B. Turkington, "An introduction to the thermodynamic and macrostate levels of nonequivalent ensembles", *Physica A* **340**:1-3 (2004), 138–146.

[Touchette et al. 2006] H. Touchette, M. Costeniuc, R. S. Ellis, , and B. Turkington, "Metastability within the generalized canonical ensemble", *Physica A* **365** (2006), 132–137.

[Turkington et al. 2001] B. Turkington, A. Majda, K. Haven, and M. DiBattista, "Statistical equilibrium predictions of jets and spots on Jupiter", *Proc. Natl. Acad. Sci. USA* **98**:22 (2001), 12346–12350.

[Wu 1982] F. Y. Wu, "The Potts model", *Rev. Modern Phys.* **54**:1 (1982), 235–268.

MARIUS COSTENIUC
MAX PLANCK INSTITUTE FOR MATHEMATICS IN THE SCIENCES
INSELSTRASSE 22–26
D-04103 LEIPZIG
GERMANY
    marius.costeniuc@gmail.com

RICHARD S. ELLIS
DEPARTMENT OF MATHEMATICS AND STATISTICS
UNIVERSITY OF MASSACHUSETTS
AMHERST, MA 01003
UNITED STATES
    rsellis@math.umass.edu

HUGO TOUCHETTE
SCHOOL OF MATHEMATICAL SCIENCES
QUEEN MARY, UNIVERSITY OF LONDON
LONDON E1 4NS
UNITED KINGDOM
    htouchet@alum.mit.edu

BRUCE TURKINGTON
DEPARTMENT OF MATHEMATICS AND STATISTICS
UNIVERSITY OF MASSACHUSETTS
AMHERST, MA 01003
UNITED STATES
    turk@math.umass.edu

Probability, Geometry and Integrable Systems
MSRI Publications
Volume **55**, 2008

Stochastic evolution of inviscid Burgers fluid

ANA BELA CRUZEIRO AND PAUL MALLIAVIN

Dedicated to Henry McKean with admiration

ABSTRACT. We study a stochastic Burgers equation using the geometric point of view initiated by Arnold for the incompressible Euler flow evolution. The geometry is developed as a Cartan-type geometry, using a frame bundle approach (stochastic, in this case) with respect to the infinite-dimensional Lie group where the evolution takes place. The existence of the stochastic Burgers flow is a consequence of the control in the mean of the energy transfer from low modes to high modes during the evolution, together with the use of a Girsanov transformation.

Introduction

Many distinguished authors have made notable contributions to the stochastic Burgers equation, of which a small sample appears in our very short bibliography. It is not our purpose to review those contributions; it is perhaps appropriate that we underline here that which seems to us the novelty of our approach.

We start from the viewpoint of geometrization of inertial evolution initiated in [Arnold 1966] and systematically developed in [Ebin and Marsden 1970; Brenier 2003; Constantin and Kolev 2002], based on infinite-dimensional Riemannian geometry; the classical approach of [Ebin and Marsden 1970] is to use Banach-modeled manifold theory; inherent difficulties appear in the construction of exponential charts and in the introduction of appropriate function spaces. We circumvent these difficulties by using the viewpoint [Malliavin 2007] of *Itô charts, Itô atlas*; in short Itô calculus makes it possible to compute any derivative of a smooth function f on the path p of a diffusion from the unique knowledge

Cruzeiro developed her work in the framework of the research project POCI/MAT/55977/2004.

of its restriction $f_{|p}$. Then no more function spaces are a priori introduced: the path of the diffusion constructs dynamically its canonical tangent space, built from the evolution of the system.

How do we make explicit computations without local coordinates? We take the viewpoint of [Arnold 1966; Cruzeiro and Malliavin 1996; Airault and Malliavin 2006; Cruzeiro et al. 2007], using the parallelism defined by the infinite-dimensional Lie group structure.

In fluid dynamics the escape of the energy from low modes to higher modes induces a lack of compactness which ruins the advantage of energy conservation for inertial evolution. The key point of our approach is the control of this ultraviolet divergence. We control the ultraviolet divergence in the case of the stochastic Burgers equation with vanishing initial value. Then symmetries appear which, as in [Airault and Malliavin 2006; Cruzeiro et al. 2007], make it possible to compute exactly the expectation of the energy transfer by the exponentiation of a numerical symmetric matrix.

Then we have solved our stochastic Burgers equation for vanishing initial data: we reduce, as in [Cruzeiro et al. 2007], the nonvanishing initial data case to this trivial case by a symmetry breaking expressed at the level of probability space by a Girsanov functional.

We emphasize that the noise that we use is neither an external force nor a damping. This important point is made explicit in the next section.

1. Random regularization of nonlinear evolution

In order to clarify our objectives, we shall proceed in this section at a conceptual level, which has the disadvantage that we cannot produce at this level of generality a single mathematical statement: the considered objects will not be exactly defined; the reader will have to wait until Section 2 before getting into mathematics.

Numerical integration of an evolution equation through a time discretization scheme introduces at each step a numerical error; if the scheme is "well chosen", it will be unbiased: therefore the cumulative effect of numerical errors will converge *locally* to a Brownian motion.

Let us axiomatize the previous empirical situation. Denote by $\mathcal{S}$ the infinitesimal generator of an evolution equation, which is not assumed to be linear; the operator $\mathcal{S}$ is operating on Cauchy data; then consider the Stratonovitch SDE

$$d_t u_t^\varepsilon = \mathcal{S}\left(u_t^\varepsilon\, dt + \varepsilon dx(t)\right), \quad u_0^\varepsilon \text{ deterministic and independent of } \varepsilon, \quad (1.1)_a$$

where x is a suitable Brownian motion modeling *the instantaneous discretization error* and where $\varepsilon > 0$. We call the solution of $(1.1)_a$ the *random regular-*

ization of the evolution equation

$$d_t u_t = \mathcal{S}(u_t \, dt), \quad u_0 \text{ given.} \tag{1.1}_b$$

The disadvantage of $(1.1)_a$ versus $(1.1)_b$ is to replace an ODE by an SDE; this disadvantage is balanced by the advantage that the introduction of a small noise can smooth out resonances leading to the system explosion.

The terminology used, *random regularization*, is parallel to the classical terminology *elliptic regularization*. This choice of terminology can be justified by the fact that dealing with the Brownian motion x is equivalent to dealing with some infinite-dimensional elliptic operator defined on the path space of $x(*)$.

2. The Burgers equation as a geodesic flow

Consider the group G of C^∞ diffeomorphisms of the circle S^1, denote by $\mathcal{G}$ its Lie algebra of right invariant first order differential operators on G; we identify $\mathcal{G}$ to vector fields on S^1; define on $\mathcal{G}$ the pre-Hilbertian metric

$$\|u\|^2 = \frac{1}{\pi} \int_0^{2\pi} |u|^2(\theta) \, d\theta; \tag{2.1}$$

then G becomes an "infinite-dimensional Riemannian manifold".

THEOREM [Arnold 1966; Constantin and Kolev 2003]. *Let $v_t(\theta)$ a be smooth vector field defined on S^1, depending smoothly on time t, which is assumed to satisfy the Burgers equation*

$$\frac{\partial v}{\partial t} = v \times \frac{\partial v}{\partial \theta}. \tag{2.2}$$

Let g_t be the time dependent diffeomorphism of S^1 defined by the family of ODEs

$$\frac{d}{dt} g_t(\theta) = v_t(g_t(\theta)); \quad g_0(\theta) = \theta. \tag{2.3}$$

Then

$$t \mapsto g_t \text{ is a geodesic of the Riemannian manifold } G. \tag{2.4}$$

3. Structure constants of $\mathcal{G}$

The vector fields

$$A_k = \cos k\theta, \quad B_k = \sin k\theta, \quad k > 0, \quad A_0 = \frac{1}{\sqrt{2}} \tag{3.1}$$

constitute an orthonormal basis of $\mathcal{G}$. In this basis, the Lie brackets are as follows:

$$[A_0, A_k] = -(k/\sqrt{2})\, B_k,$$
$$[A_0, B_k] = (k/\sqrt{2})\, A_k, \quad k > 0,$$
$$[A_s, A_k] = \tfrac{1}{2}((s-k)B_{k+s} + (s+k)B_{s-k}),$$
$$[B_s, B_k] = \tfrac{1}{2}((k-s)B_{k+s} + (s+k)B_{s-k}),$$
$$[A_s, B_k] = \tfrac{1}{2}((k-s)A_{k+s} + (s+k)A_{k-s}), \quad s \neq k,$$
$$[B_k, A_s] = \tfrac{1}{2}((s-k)A_{k+s} - (s+k)A_{s-k}), \quad s \neq k,$$
$$[A_k, B_k] = \sqrt{2}k\, A_0.$$

PROOF.

$$[A_s, A_k] = -kA_s \times B_k + sA_k \times B_s = \tfrac{1}{2}(-k(B_{k+s} + B_{k-s}) + s(B_{k+s} + B_{s-k}))$$
$$= \tfrac{1}{2}((s-k)B_{k+s} + (s+k)B_{s-k}),$$
$$[B_s, B_k] = kB_s \times A_k - sB_k \times A_s = \tfrac{1}{2}(k(B_{k+s} + B_{s-k}) - s(B_{k+s} + B_{k-s}))$$
$$= \tfrac{1}{2}((k-s)B_{k+s} + (s+k)B_{s-k}),$$
$$[A_s, B_k] = kA_s \times A_k + sB_k \times B_s = \tfrac{1}{2}(k(A_{k+s} + A_{k-s}) + s(-A_{k+s} + A_{s-k}))$$
$$= \tfrac{1}{2}((k-s)A_{k+s} + (s+k)A_{k-s}),$$

Analogously,

$$[B_k, A_s] = \tfrac{1}{2}((s-k)A_{k+s} - (s+k)A_{s-k}). \qquad \square$$

4. The Christoffel tensor

We have on G two connections:

(i) the algebraic connection defined by the right invariant parallelism on G;
(ii) the Riemannian connection defined by the Levi-Civita parallel transport.

The difference of two connections defines a tensor field $\Gamma^{*}_{*,*}$.
We have the key general lemma:

LEMMA [Arnold 1966; Cruzeiro and Malliavin 1996; Airault and Malliavin 2006]. *Let G be a group with a right-invariant Hilbertian metric, and let $\{e_k\}$ be an orthonormal basis of its Lie algebra $\mathcal{G}$. Then*

$$\Gamma^{l}_{s,k} = \tfrac{1}{2}(c^{l}_{s,k} - c^{s}_{k,l} + c^{k}_{l,s}), \quad \text{where } [e_s, e_k] = \sum_{l} c^{l}_{s,k} e_l. \qquad (4.1)$$

We deduce immediately from the structural constants the identities

$$2\Gamma^{A_l}_{A_s A_k} = ([A_s, A_k]|A_l) - ([A_k, A_l]|A_s) + ([A_l, A_s]|A_k) = 0,$$

$$2\Gamma^{B_l}_{A_s B_k} = ([A_s, B_k]|B_l) - ([B_k, B_l]|A_s) + ([B_l, A_s]|B_k) = 0,$$

$$2\Gamma^{A_l}_{B_s B_k} = ([B_s, B_k]|A_l) - ([B_k, A_l]|B_s) + ([A_l, B_s]|B_k) = 0,$$

$$2\Gamma^{B_l}_{B_s A_k} = -2\Gamma^{A_k}_{B_s B_l} = 0.$$

It remains to compute

$$\Gamma^{B_l}_{A_s A_k}, \quad \Gamma^{A_l}_{B_s A_k}, \quad \Gamma^{B_l}_{B_s B_k}, \quad \Gamma^{A_l}_{A_s B_k}.$$

THEOREM.

- *Assume $0 < s < k$. Then*

$$\Gamma_{A_s A_k} = -\left(k - \tfrac{1}{2}s\right)B_{k-s} - \left(k + \tfrac{1}{2}s\right)B_{k+s},$$

$$\Gamma_{A_s B_k} = \left(k - \tfrac{1}{2}s\right)A_{k-s} + \left(k + \tfrac{1}{2}s\right)A_{k+s},$$

$$\Gamma_{B_s A_k} = -\left(k - \tfrac{1}{2}s\right)A_{k-s} + \left(k + \tfrac{1}{2}s\right)A_{k+s},$$

$$\Gamma_{B_s B_k} = -\left(k - \tfrac{1}{2}s\right)B_{k-s} + \left(k + \tfrac{1}{2}s\right)B_{k+s}.$$

- *Assume $0 < k < s$. Then*

$$\Gamma_{A_s A_k} = \left(k - \tfrac{1}{2}s\right)B_{s-k} - \left(k + \tfrac{1}{2}s\right)B_{k+s},$$

$$\Gamma_{A_s B_k} = \left(k - \tfrac{1}{2}s\right)A_{s-k} + \left(k + \tfrac{1}{2}s\right)A_{k+s},$$

$$\Gamma_{B_s A_k} = -\left(k - \tfrac{1}{2}s\right)A_{s-k} + \left(k + \tfrac{1}{2}s\right)A_{k+s},$$

$$\Gamma_{B_s B_k} = \left(k - \tfrac{1}{2}s\right)B_{s-k} + \left(k + \tfrac{1}{2}s\right)B_{k+s},$$

In each case the two first lines define an antisymmetric operator $\Gamma(A_s)$ and the two last lines define an operator $\Gamma(B_s)$.

- *For $k > 0$,*

$$\Gamma_{A_k A_k} = -\Gamma_{B_k B_k} = -\tfrac{3}{2}k B_{2k},$$

$$\Gamma_{A_k B_k} = \tfrac{3}{2}k A_{2k} + \tfrac{\sqrt{2}}{2}k A_0, \quad \Gamma_{B_k A_k} = \tfrac{3}{2}k A_{2k} - \tfrac{\sqrt{2}}{2}k A_0,$$

$$\Gamma_{A_0 A_k} = -\sqrt{2}k B_k, \qquad\qquad \Gamma_{A_k A_0} = -\tfrac{\sqrt{2}}{2}k B_k,$$

$$\Gamma_{A_0 B_k} = \tfrac{\sqrt{2}}{2}k A_k + \tfrac{\sqrt{2}}{2}k A_0, \quad \Gamma_{B_k A_0} = \tfrac{\sqrt{2}}{2}k A_0,$$

- *Finally, $\Gamma_{A_0 A_0} = 0$.*

PROOF.   Consider the case $0 < s < k$. We have $4\Gamma^{B_l}_{A_s A_k} = I - II - III$, with

$$I = 2([A_s, A_k]|B_l), \quad II = 2([A_k, B_l]|A_s), \quad III = -2([B_l, A_s]|A_k).$$

The term I is equal to $s - k$ when $l = k + s$ and to $-(s + k)$ when $l = k - s$.

Other contributions to the component B_{k+s} are $s + 2k$ from II in the case $k < l$ and $-(2s + k)$ from III corresponding to the case $s < l$. Concerning the component B_{k-s} we have to consider the contribution $2k - s$ from II when $l < k$ and the contribution from III in the case $s < l$, which is equal to $2s - k$. Summing up all the terms gives the result.

In more detail, introduce for $s > 0$ the new Kronecker symbol

$$\varepsilon_p^s = \delta_p^s, \; p > 0, \quad \varepsilon_p^s = -\delta_{-p}^s, \; p < 0, \quad \varepsilon_0^s = 0.$$

Take $s, k, l > 0$; then $4\Gamma_{A_s A_k}^{B_l}$ equals

$$(s-k)\delta_{k+s}^l + (s+k)\varepsilon_{s-k}^l + (k-l)\delta_{k+l}^s - (l+k)\delta_{|l-k|}^s + (s-l)\delta_{l+s}^k - (s+l)\delta_{|s-l|}^k.$$

Consider first the case $0 < s < k$; then $4\Gamma_{A_s A_k}^{B_l}$ equals

$$(s-k)\delta_{k+s}^l - (s+k)\delta_{k-s}^l + (k-l)\delta_{k+l}^s - (l+k)\delta_{|l-k|}^s + (s-l)\delta_{l+s}^k - (s+l)\delta_{|l-s|}^k.$$

(1) Take the subcase $0 < s < k < l$. Then $4\Gamma_{A_s A_k}^{B_l}$ equals

$$(s-k)\delta_{k+s}^l - (s+k)\delta_{k-s}^l + (k-l)\delta_{k+l}^s - (l+k)\delta_{l-k}^s + (s-l)\delta_{l+s}^k - (s+l)\delta_{l-s}^k;$$

expressing the δ functions relatively to l, this expression becomes

$$(s-k)\delta_{k+s}^l - (s+k)\delta_{k-s}^l + (k-l)\delta_{s-k}^l - (l+k)\delta_{k+s}^l + (s-l)\delta_{k-s}^l - (s+l)\delta_{k+s}^l,$$

so

$$4\Gamma_{A_s A_k}^{B_l} = \big((s-k) - (l+k) - (s+l)\big)\delta_{k+s}^l = -2(k+l)\delta_{k+s}^l.$$

(2) In the subcase $0 < s < l < k$, we obtain for $4\Gamma_{A_s A_k}^{B_l}$ successively

$$(s-k)\delta_{k+s}^l - (s+k)\delta_{k-s}^l + (k-l)\delta_{k+l}^s - (l+k)\delta_{k-l}^s + (s-l)\delta_{l+s}^k - (s+l)\delta_{l-s}^k =$$
$$(s-k)\delta_{k+s}^l - (s+k)\delta_{k-s}^l + (k-l)\delta_{s-k}^l - (l+k)\delta_{k-s}^l + (s-l)\delta_{k-s}^l - (s+l)\delta_{k+s}^l =$$
$$\big(-(s+k) - (l+k) + (s-l)\big)\delta_{k-s}^l = -2(k+l)\delta_{k-s}^l = -2(2k-s)\delta_{k-s}^l.$$

(3) In the subcase $0 < l < s < k$, we obtain for $4\Gamma_{A_s A_k}^{B_l}$

$$(s-k)\delta_{k+s}^l - (s+k)\delta_{k-s}^l + (k-l)\delta_{k+l}^s - (l+k)\delta_{k-l}^s + (s-l)\delta_{l+s}^k - (s+l)\delta_{s-l}^k =$$
$$(s-k)\delta_{k+s}^l - (s+k)\delta_{k-s}^l + (k-l)\delta_{s-k}^l - (l+k)\delta_{k-s}^l + (s-l)\delta_{k-s}^l - (s+l)\delta_{s-k}^l =$$
$$\big(-(s+k) - (l+k) + (s-l)\big)\delta_{k-s}^l = -2(k+l)\delta_{k-s}^l.$$

Finally, still for $0 < s < k$ we have

$$\Gamma_{A_s A_k} = -\big(k - \tfrac{1}{2}s\big)B_{k-s} - \big(k + \tfrac{1}{2}s\big)B_{k+s}.$$

We now consider a rotation of angle φ. Define

$$A_k^\varphi = A_k \cos k\varphi - B_k \sin k\varphi, \quad B_q^\varphi = B_q \cos q\varphi + A_q \sin q\varphi.$$

The metric on $\mathcal{G}$ is invariant under translation by φ. Therefore the Christoffel symbols commute with this translation:

$$-\left(k - \tfrac{1}{2}s\right) B_{k-s}^\varphi - \left(k + \tfrac{1}{2}s\right) B_{k+s}^\varphi = \Gamma_{A_s^\varphi A_k^\varphi}$$

$$= \Gamma_{A_s A_k} \cos s\varphi \cos k\varphi + \Gamma_{B_s B_k} \sin s\varphi \sin k\varphi$$
$$- \Gamma_{A_s B_k} \cos s\varphi \sin k\varphi - \Gamma_{B_s A_k} \sin s\varphi \cos k\varphi.$$

On the other hand,

$$-\left(k - \tfrac{1}{2}s\right) B_{k-s}^\varphi - \left(k + \tfrac{1}{2}s\right) B_{k+s}^\varphi$$
$$= -\left(k - \tfrac{1}{2}s\right)\left(B_{k-s} \cos(k-s)\varphi + A_{k-s} \sin(k-s)\varphi\right)$$
$$- \left(k + \tfrac{1}{2}s\right)\left(B_{k+s} \cos(k+s)\varphi + A_{k+s} \sin(k+s)\varphi\right)$$
$$= -\left(k - \tfrac{1}{2}s\right)\left(B_{k-s}(\cos k\varphi \cos s\varphi + \sin k\varphi \sin s\varphi)\right.$$
$$\left. + A_{k-s}(\sin k\varphi \cos s\varphi - \cos k\varphi \sin s\varphi)\right)$$
$$- \left(k + \tfrac{1}{2}s\right)\left(B_{k+s}(\cos k\varphi \cos s\varphi - \sin k\varphi \sin s\varphi)\right.$$
$$\left. + A_{k+s}(\sin k\varphi \cos s\varphi + \cos k\varphi \sin s\varphi)\right).$$

Identifying the coefficients of $\cos k\varphi \cos s\varphi$, $\sin k\varphi \sin s\varphi$, $\sin k\varphi \cos s\varphi$, and $\cos k\varphi \sin s\varphi$, we get the formulae for the Christoffel symbols in the case $0 < k < s$.

For $0 < k = s$, we have, for example,

$$\Gamma_{A_k A_k}^{B_l} = -([A_k, B_l]|A_k) = -\tfrac{1}{2}(k+l)\delta_{2k}^l = -\tfrac{3}{2}k\delta_{2k}^l.$$

The other expressions are proved in a similar way. $\qquad\square$

5. Stochastic parallel transport; symmetries of the noise

Consider for each $k \geq 0$ a $\mathbb{R}^2$-valued Brownian motion $\zeta_k(t) = (x_k(t), y_k(t))$; all these Brownian motions are taken to be independent. Choose a weight $\rho(k) \geq 0$ and consider the $\mathcal{G}$ valued process

$$p_t = \sum_{k>0} \rho(k)\left(x_k(t) \times A_k + y_k(t) \times B_k\right). \tag{5.1}$$

Consider the Stratonovitch SDE

$$d\psi_t = -\Gamma(dp_t) \circ \psi_t, \quad \psi_0 = \text{Identity}. \tag{5.2}$$

As the Γ are antisymmetric operators this equation takes formally its values in the unitary group of $\mathcal{G}$.

The geometric meaning of (5.2) is to describe in terms of the algebraic parallelism inherited from the group structure of G the Levi-Civita parallelism inherited from the Riemannian structure of G; for this reason we call (5.2) the *equation of stochastic parallel transport.*

Symmetries of the noise. The translation $\tau_\varphi : \theta \mapsto \theta + \varphi$ is a diffeomorphism

$$[(\tau_\varphi)_*(z)](\theta) = z(\theta - \varphi)$$

The collection $(\tau_\varphi)_*$, $\varphi \in S^1$, constitutes a unitary representation of S^1 on $\mathcal{G}$ which decomposes into irreducible components along the direct sum of two-dimensional subspaces

$$\bigoplus_{k>0} \mathcal{E}_k, \quad \mathcal{E}_k := (A_k, B_k), \quad \mathcal{E}_0 := A_0,$$

the action of $(\tau_\varphi)_*$ on $\mathcal{E}_k$ being the rotation

$$\mathcal{D}_k(\varphi) := \begin{pmatrix} \cos k\varphi & -\sin k\varphi \\ \sin k\varphi & \cos k\varphi \end{pmatrix}, \quad \mathcal{D}_0(\varphi) := \text{Identity}.$$

Furthermore τ_φ preserves the Lie algebra structure. The Christoffel symbols are derived from the Hilbertian structure and from the bracket structure of $\mathcal{G}$. Therefore they commute with τ_φ in the sense that

$$(\tau_\varphi)_*[\Gamma(\xi)(\eta)] = \Gamma((\tau_\varphi)_*\xi)[(\tau_\varphi)_*\eta)], \quad \xi, \eta \in \mathcal{G};$$

or, denoting $\Gamma(z)$ the antihermitian endomorphism of $\mathcal{G}$ defined by the Christoffel symbols, we have

$$\Gamma((\tau_\varphi)_*(z)) = (\tau_\varphi)_* \circ \Gamma(z) \circ (\tau_{-\varphi})_*.$$

Denote by $su(\mathcal{G})$ the vector space of antisymmetric operators on the Hilbert space $\mathcal{G}$.

PROPOSITION. *Let p_t the $\mathcal{G}$-valued process defined in (5.1) and set $(\tau_\varphi)_* p =: p_*^\varphi$; then p_*^φ and p have the same law.*

PROOF. The rotation $\mathcal{D}_k(\phi)$ preserves in law the Brownian motion on $\mathcal{E}_k$. □

COROLLARY. *The processes $(\tau_\varphi) \circ \psi_t \circ (\tau_{-\varphi})$ and ψ_t have the same law.*

PROOF. Denote by ψ_t^p the solution of (3.3) associated to the noise p_t. Then

$$(\tau_\varphi) \circ \psi_t \circ (\tau_{-\varphi}) = \psi_t^{p^\varphi} \qquad\qquad □$$

The Stratonovich SDE (5.2) corresponds to the Itô SDE

$$d\psi_t^p = \left(\boldsymbol{\Gamma}(dp) + \mathcal{B}\,dt\right)\psi_t,$$

$$\mathcal{B} = \sum_{k \geq 0} \frac{[\rho(k)]^2}{2}\left(\boldsymbol{\Gamma}(A_k) * \boldsymbol{\Gamma}(A_k) + \boldsymbol{\Gamma}(B_k) * \boldsymbol{\Gamma}(B_k)\right).$$

We get $\mathcal{B} = (\tau_\varphi)_* \circ \mathcal{B} \circ (\tau_{-\varphi})_*$, which implies that $\mathcal{B}$ diagonalizes in the basis $\bigoplus \mathcal{E}_k$. More precisely:

THEOREM. *The operator*

$$[\boldsymbol{\Gamma}(A_s)]^2 + [\boldsymbol{\Gamma}(B_s)]^2$$

is diagonal and on the mode k it has eigenvalue

$$\lambda_s(k) = -(4k^2 + s^2), \quad k > 2s.$$

PROOF. We have

$$[\boldsymbol{\Gamma}(A_s)]^2(A_k) = -\left(k - \tfrac{1}{2}s\right)\boldsymbol{\Gamma}(A_s)(B_{k-s}) - \left(k + \tfrac{1}{2}s\right)\boldsymbol{\Gamma}(A_s)(B_{k+s})$$
$$= -\left(k - \tfrac{1}{2}s\right)\left(\left(k - \tfrac{3}{2}s\right)A_{k-2s} + \left(k - \tfrac{1}{2}s\right)A_k\right)$$
$$\qquad -\left(k + \tfrac{1}{2}s\right)\left(\left(k + \tfrac{1}{2}s\right)A_k + \left(k + \tfrac{3}{2}s\right)A_{k+2s}\right),$$

$$[\boldsymbol{\Gamma}(B_s)]^2(A_k) = -\left(k - \tfrac{1}{2}s\right)\boldsymbol{\Gamma}(B_s)(A_{k-s}) + \left(k + \tfrac{1}{2}s\right)\boldsymbol{\Gamma}(B_s)(A_{k+s})$$
$$= -\left(k - \tfrac{1}{2}s\right)\left(-\left(k - \tfrac{3}{2}s\right)A_{k-2s} + \left(k - \tfrac{1}{2}sA_k\right)\right)$$
$$\qquad -\left(k + \tfrac{1}{2}s\right)\left(\left(k + \tfrac{1}{2}s\right)A_k - \left(k + \tfrac{3}{2}s\right)A_{k+2s}\right).$$

Hence

$$[\boldsymbol{\Gamma}(A_s)]^2(A_k) + [\boldsymbol{\Gamma}(B_s)]^2(A_k) = -2\left(k - \tfrac{1}{2}s\right)^2 - 2\left(k + \tfrac{1}{2}s\right)^2. \qquad \square$$

We want to take, as in [Cruzeiro et al. 2007], a finite-mode driven Brownian motion, which means that $\rho(k) = 0$ except for a finite number of values of k.

6. Control of ultraviolet divergence by the transfer energy matrix

THEOREM. *Let e be a trigonometric polynomial, and define*

$$\xi_k(t) = E\left([(\psi_t(e) \mid A_k)]^2 + [(\psi_t(e) \mid B_k)]^2\right).$$

Then $\xi(t)$ satisfies the ordinary differential equation

$$\frac{d\xi(t)}{dt} = \mathcal{A}(\xi(t)), \qquad (6.1)$$

where the matrix $\mathcal{A}$ has diagonal entries

$$\mathcal{A}_l^l = -4 \sum_k \rho(k)^2 \left(2l^2 + \tfrac{1}{2}k^2\right) - \tfrac{9}{8}l^2 \rho^2\left(\tfrac{1}{2}l\right)$$

and nondiagonal entries

$$A_s^l = 2 \sum_k \rho(k)^2 \big((l - \tfrac{1}{2}k)^2 \delta_s^{|k-l|} + 2(l + \tfrac{1}{2}k)^2 \delta_s^{k+l}\big) + \tfrac{9}{8} l^2 \rho^2(\tfrac{1}{2}l),$$

with $s, l > 0$. The sum of the coefficients in each column vanishes.

PROOF. We have, explicitly,

$$d\psi_t^{A_l} = -\sum_m (\Gamma_{A_k B_m}^{A_l} \psi_t^{B_m} \, odx_k(t) + \Gamma_{B_k A_m}^{A_l} \psi_t^{A_m} \, ody_k(t)).$$

By Itô calculus,

$$d(\psi_t^{A_l})^2 = 2\psi_t^{A_l} d\psi_t^{A_l} + d\psi_t^{A_l} . d\psi_t^{A_l},$$
$$d(\psi_t^{B_l})^2 = 2\psi_t^{B_l} d\psi_t^{A_l} + d\psi_t^{B_l} . d\psi_t^{B_l}.$$

Since we are interested in taking expectations we compute only the bounded variation part of this semimartingale. Considering the terms $0 < m \le k$,

$$\begin{aligned}
d\psi_t^{A_l} &= -\Gamma_{A_k B_{l-k}}^{A_l} \psi_t^{B_{l-k}} \, odx_k(t) - \Gamma_{B_k A_{l-k}}^{A_l} \psi_t^{A_{l-k}} \, ody_k(t) \\
&\quad - \Gamma_{A_k B_{l+k}}^{A_l} \psi_t^{B_{l+k}} \, odx_k(t) - \Gamma_{B_k A_{l+k}}^{A_l} \psi_t^{A_{l+k}} \, ody_k(t) \\
&= -(l - \tfrac{1}{2}k)\psi_t^{B_{l-k}} \, odx_k(t) - (l - \tfrac{1}{2}k)\psi_t^{A_{l-k}} \, ody_k(t) \\
&\quad - (l + \tfrac{1}{2}k)\psi_t^{B_{l+k}} \, odx_k(t) + (l + \tfrac{1}{2}k)\psi_t^{A_{l+k}} \, ody_k(t) \\
&\quad - \tfrac{3}{2} \sum_k \rho(k)k\psi_t^{B_k} \, odx_k(t) - \tfrac{3}{2} \sum_k \rho(k)k\psi_t^{A_k} \, ody_k(t).
\end{aligned}$$

Computing the Itô contractions, we obtain, for example, in the case of the first term,

$$\begin{aligned}
-(l - \tfrac{1}{2}k)\psi_t^{B_{l-k}} \, odx_k(t) &= -(l - \tfrac{1}{2}k)\psi_t^{B_{l-k}} \, dx_k(t) \\
&\quad - \tfrac{1}{2}\big((l - \tfrac{3}{2}k)\psi_t^{A_{l-2k}} - (l - \tfrac{1}{2}k)\psi_t^{A_l} + (\tfrac{3}{2}k - l)\psi_t^{A_{2k-l}}\big) \, dt.
\end{aligned}$$

We can check by explicit computation that all the nondiagonal contributions coming from these Itô contractions cancel in their contribution to the expectation of $\psi_t^{A_l} d\psi_t^{A_l} + \psi_t^{B_l} d\psi_t^{B_l}$. The diagonal ones, for the case $0 < k < m$, sum up to give

$$-2 \sum_k (2l^2 + \tfrac{1}{2}k^2)\psi_t^{A_l} \, dt.$$

The terms in $0 < k < m$ give the same expression. The contribution from $k = m$ gives

$$-\tfrac{3}{2}\rho(\tfrac{1}{2}l) \tfrac{1}{2} l \psi_t^{A_{1/2}l} \, dt.$$

Concerning the B_l component of ψ_t, namely

$$d\psi_t^{B_l} = -\sum_m \big(\Gamma_{A_k A_m}^{B_l}\, \psi_t^{A_m}\, odx_k(t) + \Gamma_{B_k B_m}^{B_l}\, \psi_t^{B_m}\, ody_k(t)\big),$$

analogous computations give rise to the expressions

$$-2\sum_k (2l^2 + \tfrac{1}{2}k^2)\psi_t^{B_l}\, dt$$

for $l < m$ and $m < l$, and

$$-\tfrac{3}{2}\rho(\tfrac{1}{2}l)\,\tfrac{1}{2}l\,\psi_t^{B_{1/2}l}\, dt$$

when $k = m$.

The nondiagonal terms of the transfer energy matrix come from computing the contractions $d\psi_t^{A_l}.\,d\psi_t^{A_l}$ and $d\psi_t^{B_l}.\,d\psi_t^{B_l}$. We have, when $0 < k \le l$,

$$d\psi_t^{A_l}.\,d\psi_t^{A_l} = \sum_k \rho(k)^2\big(\Gamma_{A_k B_{l-k}}^{A_l}\,\psi_t^{B_{l-k}}\big)^2 dt + \sum_k \rho(k)^2\big(\Gamma_{B_k A_{l-k}}^{A_l}\,\psi_t^{A_{l-k}}\big)^2 dt$$

$$+\sum_k \rho(k)^2\big(\Gamma_{A_k B_{l+k}}^{A_l}\,\psi_t^{B_{l+k}}\big)^2 dt + \sum_k \rho(k)^2\big(\Gamma_{B_k A_{l+k}}^{A_l}\,\psi_t^{A_{l+k}}\big)^2 dt$$

$$+\rho(\tfrac{1}{2}l)^2\,\Gamma_{A_{1/2}l B_{1/2}l}^{A_l}\big(\psi_t^{B_l/2}\big)^2 + \rho(\tfrac{1}{2}l)^2\,\Gamma_{B_{1/2}l A_{1/2}l}^{A_l}\big(\psi_t^{A_l/2}\big)^2$$

$$= \sum_k \rho(k)^2\big(l - \tfrac{1}{2}k\big)^2\big(\psi_t^{B_{l-k}}\big)^2 + \sum_k \rho(k)^2\big(l - \tfrac{1}{2}k\big)^2\big(\psi_t^{A_{l-k}}\big)^2$$

$$+\sum_k \rho(k)^2\big(l + \tfrac{1}{2}k\big)^2\big(\psi_t^{B_{l+k}}\big)^2 + \sum_k \rho(k)^2\big(l + \tfrac{1}{2}k\big)^2\big(\psi_t^{A_{l+k}}\big)^2$$

$$+\tfrac{9}{4}\rho(\tfrac{1}{2}l)^2\big(\tfrac{1}{2}l\big)^2\big(\psi_t^{B_l/2}\big)^2 + \tfrac{9}{4}\rho(\tfrac{1}{2}l)^2\big(\tfrac{1}{2}l\big)^2\big(\psi_t^{A_l/2}\big)^2.$$

Computing the corresponding terms for the indices $0 < l < k$ as well as the contractions $d\psi_t^{B_l}.\,d\psi_t^{B_l}$ gives the desired result. $\qquad\square$

7. Ultraviolet divergence and dissipativity of the associated jump process

The ordinary differential equation (6.1) can be integrated quite explicitly by the exponential $\exp(t\mathcal{A})$; nevertheless the effective computation of this exponential is not easy.

It was observed in [Airault and Malliavin 2006, Theorem (3.10)] that $\mathcal{A}$ can be also considered as the infinitesimal generator of a Dirichlet form; therefore its exponentiation is equivalent to construct the jump process associated to this Dirichlet form. Recall how this jump process was constructed in that theorem.

In order to shorten our discussion we shall sketch our proof in the special case where

$$\rho(1) = 1, \quad \rho(k) = 0, \quad k \ne 1.$$

Then the random walk $X(n)$ is a nearest neighbor random walk defined on $\mathbb{N}$, the set of positive integers, as follows:

If $X(n) = k$, $k > 2$ we have

$$\text{Prob}\{X(n+1) = k+1\} = p_k := \frac{1}{2}\left(1 + \frac{k}{4k^2+1}\right),$$

$$\text{Prob}\{X(n+1) = k-1\} = 1 - p_k.$$

The random walk is nonsymmetric, it has a drift $\simeq \frac{1}{k}$ pushing it to escape at infinity. This drift has a negligible effect in our discussion and we shall proceed as if the random walk was symmetric.

The jump process is defined as

$$\eta(t) := X(\varphi(t))$$

where the change of clock $\varphi(t)$ is the integer-valued function defined by

$$\sum_{n \leq \varphi(t)} \frac{1}{4[X(n)]^2+1} \times \Lambda_n \leq t < \sum_{n \leq \varphi(t)+1} \frac{1}{4[X(n)]^2+1} \times \Lambda_n,$$

where $\{\Lambda_k\}$ is a sequence of independent exponential times.

THEOREM. *The jump process is conservative. That is, $\varphi(t) < \infty$ almost surely; more precisely,*

$$E([X(\varphi(t))]^q) < \infty \quad \text{for all } q > 0. \tag{7.2}$$

PROOF. What follows is an improved methodology of proof compared to the one used in [Cruzeiro et al. 2007]. The proof of (7.2) will occupy us till the end of Section 7.

Let Ω_1 be the probability space of the random walk; then Ω_1 is a space generated by an infinite sequence of independent Bernoulli variables; let Ω_2 be the probability space generated by an infinite sequence of independent exponential variables. Then the probability space of the jump process is $\Omega_1 \times \Omega_2$. We denote by E^{ω_i} the expectation relatively to Ω_i, the other coordinate being fixed, and we write $\text{Prob}_i(A) := E^{\omega_i}(1_A)$.

We introduce a strictly increasing sequence of stopping times $T_1 < T_2 < \cdots < T_k < \cdots$ on the random walk by the following recursion: T_1 is the first time where the value starting from 1 it reaches 2; T_{k+1} is the first time after T_k where $X(T_{k+1})$ leaves the interval $\left(\frac{1}{2}X(T_k), 2X(T_k)\right)$; we have

$$X(T_k) = 2^{\xi_k}, \quad \xi_k \in \mathbb{N}.$$

Then ξ_k is an unsymmetric random walk on the set of positive integers. We construct on Ω_1 a new random walk $X^*(n)$ by taking

$$X^*(T_{k+1}) = 2X^*(T_k), \text{ then } X(n) \leq X^*(n); \quad \inf_{m>n} X^*(m) \geq \frac{1}{2}X^*(n).$$

Denote by $\varphi^*(t)$ the time change in the jump process associated to the random walk $X^*(*)$; we obtain a new jump process $\eta^*(t)$, defined on the same probability space as η, and we have

$$\eta(t) \leq 2\eta^*(t);$$

therefore it is sufficient to prove (7.2) for η^*. Introduce the functionals

$$\Phi(p) := \sum_{n \leq p} \frac{1}{4|X^*(n)|^2 + 1}, \quad \Psi(p) := \sum_{n \leq p} \frac{\Lambda_n}{4[X^*(n)]^2 + 1} \Lambda_n;$$

then $E^{\omega_2}(\Psi(p)) = \Phi(p)$.

We have

$$\Phi(T_{k+1}) - \Phi(T_k) \geq \frac{T_{k+1} - T_k}{2^{2(\xi_k + 2)} + 1} \tag{7.3}$$

THEOREM. $\text{Prob}\{\Phi(T_k) - \Phi(T_s) < t\} \leq \exp\left(\frac{-3(k-s)^{3/2}}{12\sqrt{12t}}\right)$, $k - s > 20(t+1)$.

PROOF. Denote by $\mathcal{S}$ the exit time of the random walk from the interval $I_k := (2^{\xi_k - 1}, 2^{\xi_k + 1})$ and for $0 < \lambda < 1$ being fixed, define on I_k the function

$$v(p) = E_p(\lambda^{\mathcal{S}});$$

then v takes the value 1 at the boundary of I_k; by the Bellman programming equation it satisfies

$$v(p) = \tfrac{1}{2}\lambda\big(v(p-1) + v(p+1)\big).$$

Define $\Delta f(n) := \tfrac{1}{2}\big(f(n+1) + f(n-1)\big) - f(n)$; then

$$\Delta v = (\lambda^{-1} - 1)v.$$

Define $f_a(n) := a^n$; then $\tfrac{1}{2}\big(f_a(n+1) + f_a(n-1)\big) - f_a(n) = cf_a(n)$, $c = \tfrac{1}{2}(a + a^{-1}) - 1$. We satisfy these two equations by imposing the condition

$$a^2 - 2\lambda^{-1}a + 1 = 0 \tag{7.4},$$

which has for roots η, η^{-1}, $\eta < 1$. We deduce that

$$v(n) = \alpha\eta^n + \beta\eta^{-n},$$

where α, β are chosen such that the boundary conditions for v are satisfied; we deduce that

$$E(\lambda^{T_{k+1} - T_k}) = v(2^{\xi_k}) < \frac{1}{\cosh(2^{\xi_k - 1}\log\eta)}$$

Writing this equality with $\lambda = 1 - r^{-1}2^{-2\xi_k}$ we get

$$\text{Prob}\{T_{k+1} - T_k \leq r2^{2\xi_k}\} \times (1 - r^{-1}2^{-2\xi_k})^{r2^{2\xi_k}} \leq \frac{1}{\cosh(2^{\xi_k - 1}\log\eta)}$$

where η is obtained from (7.4) and where $\lambda = 1 - r^{-1}2^{-2\xi_k}$, a relation which leads to the asymptotic formula

$$\eta \simeq 1 - \sqrt{2} \times r^{-1/2}2^{-\xi_k}.$$

Further,

$$\mathrm{Prob}\{T_{k+1} - T_k \leq r 2^{2\xi_k}\} \leq 2e \, \exp\left(-\frac{1}{\sqrt{2r}}\right).$$

Finally we have, using (7.3),

$$\mathrm{Prob}(\Phi(T_{k+1}) - \Phi(T_k)) \leq r) \leq 2e \exp\left(-\frac{1}{3\sqrt{r}}\right).$$

Denote by ν the law of $(\Phi(T_{k+1}) - \Phi(T_k))$. Then

$$E \exp(-c(\Phi(T_{k+1}) - \Phi(T_k)) = \int_0^\infty \exp(-\lambda y)\nu(dy);$$

integration by parts yields for this expression the bound

$$\int_0^\infty \lambda \exp(-\lambda c) \, \nu([0,c]) \, dc \leq 2e\lambda \int_0^\infty \exp\left(-\lambda c - \frac{1}{3\sqrt{c}}\right) dc$$

$$\leq 2e \exp\left(-\tfrac{1}{3}[\lambda]^{1/3}\right).$$

Since the $\Phi(T_{k+1}) - \Phi(T_k)$ are independent, we have

$$E(\exp(-\lambda(\Phi(T_k) - \Phi(T_s)))) \leq \exp\left(-\tfrac{1}{4}(k-s)[\lambda]^{1/3}\right), \quad \lambda > 16,$$

and

$$\mathrm{Prob}\{\Phi(T_k) - \Phi(T_s) < t\} \leq \inf_\lambda \exp\left(\lambda t - 14(k-s)[\lambda]^{1/3}\right)$$

$$\leq \exp\left(-\frac{3(k-s)^{3/2}}{12\sqrt{12t}}\right), \quad k-s > 20(t+1). \quad \square$$

LEMMA.

$$\mathrm{Prob}_2\left\{\frac{\Psi(T_{k+1}) - \Psi(T_k)}{\Phi(T_{k+1}) - \Phi(T_k)} \leq \frac{1}{2}\right\} \leq \exp\left(-\frac{T_{k+1} - T_k}{64}\right) \leq \exp\left(-\frac{2^k}{128}\right). \quad (7.5)$$

PROOF. Let $\xi > 0$ and let $S := \Psi(T_{k+1}) - \Psi(T_k)$. Then

$$\mathrm{Prob}\{S \leq a\} \times \exp(-a\xi) \leq E(\exp(-\xi S))$$

or

$$\mathrm{Prob}\{S \leq a\} \leq \inf_{\xi > 0} \exp(a\xi) \times E(\exp(-\xi S)).$$

We have

$$S = \sum_{T_k < n \leq T_{k+1}} \frac{1}{4[X^*(n)]^2 + 1} \times \Lambda_n$$

By the independence of the Λ_n we have

$$E^{\omega_2}(\exp(-\xi S)) = \exp\left(-\sum_{T_k < n \leq T_{k+1}} \log\left(1 + \frac{\xi}{4[X^*(n)]^2 + 1}\right)\right).$$

Now we use the inequality

$$\log(1 + u) \geq \tfrac{3}{4}u, \quad u \in \left[0, \tfrac{1}{4}\right],$$

obtaining

$$E^{\omega_2}(\exp(-\xi S)) \leq \exp(-\xi \tfrac{3}{4}(\Phi(T_{k+1}) - \Phi(T_k))) \quad \xi \in [0, \xi_0], \quad \xi_0 := 2^{2(k-1)}.$$

Taking

$$a = \tfrac{1}{2}(\Phi(T_{k+1}) - \Phi(T_k)), \quad \xi = \xi_0,$$

we get

$$\mathrm{Prob}\{S \leq a\} \leq \exp\left(-14\xi_0(\Phi(T_{k+1}) - \Phi(T_k))\right),$$

that is to say,

$$\tfrac{1}{4}\xi_0\left(\Phi(T_{k+1}) - \Phi(T_k)\right) > 2^{2(k-2)}2^{-2(k+1)}(T_{k+1} - T_k),$$

which concludes the proof of the lemma. $\qquad\square$

Now, starting from (7.5), Borel–Cantelli proves (7.2). $\qquad\square$

8. Towards stochastic fluid motion on the configuration space

The configuration space in Arnold's point of view is G, the diffeomorphism group of the circle. The last section has given rise to a solution of a stochastic Burgers equation on the moment space $\mathcal{G}$; in this section we shall start to integrate this solution from the moment space to the configuration space.

Covariance functionals. Baxendale and Harris [1986] have characterized classical stochastic flows in terms of their covariance. The construction we propose will depend upon the integration of a *delayed* SDE, in contrast to Baxendale and Harris, who develop their study in the framework of classical infinite-dimensional SDE. Nevertheless covariance estimates will be needed.

THEOREM. *Assume that the noise energy ρ has a finite support. Let $\psi_x(t)$ be the stochastic parallel transport defined in* (5.2).

(a) *The covariance is*

$$\mathcal{C}_{x,t}(\theta, \theta') =$$
$$\sum_k \left([\psi_x^*(t)(A_k)](\theta)\,[\psi_x^*(t)(A_k)](\theta') + [\psi_x^*(t)(B_k)](\theta)\,[\psi_x^*(t)(B_k)](\theta')\right)\rho(k).$$

(b) *Almost surely the map* $t \mapsto \mathcal{C}_{x,t}(*, *)$ *is a* $H^q(S^1 \times S^1)$ *continuous map.*

(c) $E(\mathcal{C}_{x,t}(\theta, \theta')) = \bar{\mathcal{C}}_t(\theta - \theta')$.

(d) $E\left(\sup_{\theta, \theta', t < T} \dfrac{\mathcal{C}_{x,t}(\theta, \theta) + \mathcal{C}_{x,t}(\theta', \theta') - 2\mathcal{C}_{x,t}(\theta, \theta')}{(\theta - \theta')^2} \right) < \infty.$

PROOF. Part (c) results from the corollary on page 174, and part (b) follows from (7.2) and the continuity property of Brownian martingales. Let

$$p(\theta, \theta') := \mathcal{C}_{x,t}(\theta, \theta) + \mathcal{C}_{x,t}(\theta', \theta') - 2\mathcal{C}_{x,t}(\theta, \theta'),$$

then $p(\theta, \theta) = 0$. Since $\left[(\partial p / \partial \theta)(\theta, \theta') \right]_{\theta = \theta'} = 0$, Taylor's formula gives

$$p(\theta, \theta') = (\theta - \theta')^2 \int_0^1 \frac{\partial^2 p}{\partial \theta^2}(\theta' + t(\theta - \theta'), \theta')(1 - t) \, dt. \qquad \square$$

The system of Itô flow equations is not closed. Denote by $\mathcal{G}^s$ the space of vector fields with values in the Sobolev space of vector fields in H^s. Then $t \mapsto y_t$ is an $\mathcal{G}^s$-valued semimartingale. We have to solve a Stratonovitch SDE

$$d_t g_{x,t}(\theta) = (od \, y_t)(g_{x,t}(\theta))$$

(see [Cruzeiro et al. 2007]); there appears then the Itô contraction

$$Y_t(g_{x,t}(\theta) \times \mathcal{C}_{x,t}(\theta, \theta) \, dt,$$

where

$$Y_t = \frac{\partial g_{x,t}}{\partial \theta}.$$

In order to write the Itô SDE driving the flow we must know the derivative of the flow itself, an so on: we have an unclosed system of Itô SDE.

A usual procedure of existence for SDE relies on the Itô formalism. We could try the following alternative approach: solutions of Stratonovitch SDE are limits of solutions of corresponding ordinary differential equations. Then it may be possible to implement this limiting procedure in the geometric context of the stochastic development.

References

[Airault and Malliavin 2006] H. Airault and P. Malliavin, "Quasi-invariance of Brownian measures on the group of circle homeomorphisms and infinite-dimensional Riemannian geometry", *J. Funct. Anal.* **241**:1 (2006), 99–142.

[Arnold 1966] V. Arnold, "Sur la géométrie différentielle des groupes de Lie de dimension infinie et ses applications à l'hydrodynamique des fluides parfaits", *Ann. Inst. Fourier (Grenoble)* **16**:1 (1966), 319–361.

[Baxendale and Harris 1986] P. Baxendale and T. E. Harris, "Isotropic stochastic flows", *Ann. Probab.* **14**:4 (1986), 1155–1179.

[Brenier 2003] Y. Brenier, "Topics on hydrodynamics and volume preserving maps", pp. 55–86 in *Handbook of mathematical fluid dynamics*, vol. II, North-Holland, Amsterdam, 2003.

[Constantin and Kolev 2002] A. Constantin and B. Kolev, "On the geometric approach to the motion of inertial mechanical systems", *J. Phys. A* **35**:32 (2002), R51–R79.

[Constantin and Kolev 2003] A. Constantin and B. Kolev, "Geodesic flow on the diffeomorphism group of the circle", *Comment. Math. Helv.* **78**:4 (2003), 787–804.

[Cruzeiro and Malliavin 1996] A.-B. Cruzeiro and P. Malliavin, "Renormalized differential geometry on path space: structural equation, curvature", *J. Funct. Anal.* **139**:1 (1996), 119–181.

[Cruzeiro et al. 2007] A.-B. Cruzeiro, F. Flandoli, and P. Malliavin, "Brownian motion on volume preserving diffeomorphisms group and existence of global solutions of 2D stochastic Euler equation", *J. Funct. Anal.* **242**:1 (2007), 304–326.

[Ebin and Marsden 1970] D. G. Ebin and J. Marsden, "Groups of diffeomorphisms and the motion of an incompressible fluid", *Ann. of Math.* (2) **92** (1970), 102–163.

[Malliavin 2007] P. Malliavin, "Itô atlas, its application to mathematical finance and to exponentiation of infinite dimensional Lie algebras", pp. 501–514 in *Stochastic analysis and applications: The Abel Symposium 2005* (Oslo, 2005), edited by F. E. Benth et al., Springer, Berlin, 2007.

ANA BELA CRUZEIRO
GRUPO DE FÍSICA MATEMÁTICA UL AND
DEPARTAMENTO DE MATEMÁTICA IST (TUL)
AV. ROVISCO PAIS
1049-001 LISBOA
PORTUGAL
abcruz@math.ist.utl.pt

PAUL MALLIAVIN
RUE SAINT LOUIS EN L'ÎLE
75004 PARIS
FRANCE
sli@ccr.jussieu.fr

Probability, Geometry and Integrable Systems
MSRI Publications
Volume **55**, 2008

A quick derivation of the loop equations for random matrices

N. M. ERCOLANI AND K. D. T-R MCLAUGHLIN

ABSTRACT. The *loop equations* of random matrix theory are a hierarchy of equations born of attempts to obtain explicit formulae for generating functions of map enumeration problems. These equations, originating in the physics of 2-dimensional quantum gravity, have lacked mathematical justification. The goal of this paper is to provide a complete and short proof, relying on a recently established complete asymptotic expansion for the random matrix theory partition function.

1. Background and preliminaries

The study of the unitary ensembles (UE) of random matrices [Mehta 1991], begins with a family of probability measures on the space of $N \times N$ Hermitian matrices. The measures are of the form

$$d\mu_t = \frac{1}{\widetilde{Z}_N} \exp\{-N \operatorname{Tr} [V_t(M)]\} \, dM,$$

where the function V_t is a scalar function, referred to as the potential of the external field, or simply the "external field" for short. Typically it is taken to be a polynomial, and written as follows:

$$V_t = V(\lambda; t_1, \ldots, t_\upsilon) = \frac{1}{2}\lambda^2 + \sum_{j=1}^{\upsilon} t_j \lambda^j$$

where the parameters $\{t_1, \ldots, t_\upsilon\}$ are assumed to be such that the integral converges. For example, one may suppose that υ is even, and $t_\upsilon > 0$. The *partition*

McLaughlin was supported in part by NSF grants DMS-0451495 and DMS-0200749, as well as a NATO Collaborative Linkage Grant "Orthogonal Polynomials: Theory, Applications, and Generalizations" no. PST.CLG.979738. Ercolani was supported in part by NSF grants DMS-0073087.

function $\tilde{Z}_N$, is the normalization factor which makes the UE measures be probability measures.

Expectations of conjugation invariant matrix random variables with respect to these measures can be reduced, via the Weyl integration formula, to an integration against a symmetric density over the eigenvalues which has a form proportional to (1-1), below:

$$\exp\left\{-N^2\left(\frac{1}{N}\sum_{j=1}^{N}V(\lambda_j; t_1,\ldots,t_\upsilon)-\frac{1}{N^2}\sum_{j\neq\ell}\log|\lambda_j-\lambda_\ell|\right)\right\}d^N\lambda. \quad (1\text{-}1)$$

These latter multiple integrals can be more compactly expressed in terms of kernels constructed from polynomials $\{p_\ell(\lambda)\}$ orthogonal with respect to the exponential weight $e^{-NV_t(\lambda)}$ [Mehta 1991]. For instance, the fundamental matrix moments $E\left(\mathrm{Tr}M^j\right)$, where E denotes expectation with respect to the measure $d\mu_t$, are expressed as

$$E\left(\mathrm{Tr}M^j\right) = N\int_{-\infty}^{\infty}\lambda^j\rho_N^{(1)}(\lambda)\,d\lambda, \quad (1\text{-}2)$$

where $\rho_N^{(1)}(\lambda)$ denotes the so-called *one-point function*

$$\rho_N^{(1)}(x) = \frac{d}{dx}\mathbb{E}\left(\frac{1}{N}\#\left\{j:\lambda_j\in(-\infty,x)\right\}\right) = \frac{1}{N}K_N(x,x), \quad (1\text{-}3)$$

with the kernel

$$K_N(x,y) = e^{-(N/2)\,(V_t(x)+V_t(y))}\sum_{\ell=0}^{N-1}p_\ell(x)p_\ell(y).$$

The symbol $\mathbb{E}$ denotes expectation with respect to the normalization of the measure (1-1) which is given by dividing this family of measures by the corresponding family of integrals:

$$Z_N(t_1,t_2,\ldots,t_\upsilon) = \quad\quad\quad\quad\quad\quad\quad\quad\quad\quad\quad\quad (1\text{-}4)$$

$$\int\cdots\int\exp\left\{-N^2\left(\frac{1}{N}\sum_{j=1}^{N}V(\lambda_j; t_1,\ldots,t_\upsilon)-\frac{1}{N^2}\sum_{j\neq\ell}\log|\lambda_j-\lambda_\ell|\right)\right\}d^N\lambda.$$

We will sometimes refer to the following set of $t = (t_1,\ldots,t_\upsilon)$ for which (1-4) converges. For any given $T > 0$ and $\gamma > 0$, define

$$\mathbb{T}(T,\gamma) = \left\{t\in\mathbb{R}^\upsilon : |t|\leq T,\ t_\upsilon > \gamma\sum_{j=1}^{\upsilon-1}|t_j|\right\}.$$

The leading order behavior of $Z_N(t_1, t_2, \ldots, t_\upsilon)$ is rather classical, and is known for a very wide class of external fields V (see [Johansson 1998], for example). We will require the following result.

THEOREM 1.1. *There is $T > 0$ and $\gamma > 0$ such that, for all $t \in \mathbb{T}(T, \gamma)$, the following statements hold:*

(i)
$$\lim_{N \to \infty} \frac{1}{N^2} \log\{Z_N(t_1, t_2, \ldots, t_\upsilon)\} = -I(t_1, \ldots, t_\upsilon), \qquad (1\text{-}5)$$

where $I(t_1, \ldots, t_\upsilon)$ is the infimum of

$$\int V(\lambda)\, d\mu(\lambda) - \iint \log|\lambda - \mu|\, d\mu(\lambda)\, d\mu(\eta) \qquad (1\text{-}6)$$

over all positive, normalized Borel measures μ.

(ii) *There is a unique measure μ_V that achieves this infimum. This measure is absolutely continuous with respect to Lebesgue measure, and*

$$d\mu_V = \psi\, d\lambda,$$
$$\psi(\lambda) = \frac{1}{2\pi} \chi_{(\alpha,\beta)}(\lambda) \sqrt{(\lambda - \alpha)(\beta - \lambda)}\, h(\lambda),$$

where $h(\lambda)$ is a polynomial of degree $\upsilon - 2$, which is strictly positive on the interval $[\alpha, \beta]$ (recall that the external field V is a polynomial of degree υ). The polynomial h is defined by

$$h(z) = \frac{1}{2\pi i} \oint \frac{V'(s)}{\sqrt{(s - \alpha)}\sqrt{(s - \beta)}} \frac{ds}{s - z}$$

where the integral is taken on a circle containing (α, β) and z in the interior, oriented counterclockwise.

(iii) *There exists a constant l, depending on V, such that the following variational equations are satisfied by μ_V:*

$$\int 2 \log|\lambda - \eta|^{-1} d\mu_V(\eta) + V(\lambda) \geq l \ \text{ for } \ \lambda \in \mathbb{R} \setminus \mathrm{supp}(\mu_V),$$
$$\int 2 \log|\lambda - \eta|^{-1} d\mu_V(\eta) + V(\lambda) = l \ \text{ for } \ \lambda \in \mathrm{supp}(\mu_V). \qquad (1\text{-}7)$$

(iv) *The endpoints α and β are determined by the equations*

$$\int_\alpha^\beta \frac{V'(s)}{\sqrt{(s - \alpha)(\beta - s)}}\, ds = 0, \qquad \int_\alpha^\beta \frac{s V'(s)}{\sqrt{(s - \alpha)(\beta - s)}}\, ds = 2\pi.$$

(v) *The endpoints $\alpha(t)$ and $\beta(t)$ are actually analytic functions of t, which possess smooth extensions to the closure of $\{t : t \in \mathbb{T}(T, \gamma)\}$. They also satisfy $-\alpha(0) = \beta(0) = 2$. In addition, the coefficients of the polynomial $h(\lambda)$ are also analytic functions of t, with smooth extensions to the closure of $\{t : t \in \mathbb{T}(T, \gamma)\}$, with*

$$h(\lambda, t = 0) = 1.$$

REMARKS. (1) The variational problem appearing in 1.1(i) is a fundamental component in the theory of random matrices, as well as integrable systems and approximation theory. It is well known, (see, for example, [Saff and Totik 1997]), that under general assumptions on V, the infimum is achieved at a unique measure μ_V, called the equilibrium measure. For external fields V that are analytic in a neighborhood of the real axis, and with sufficient growth at ∞, the equilibrium measure is supported on finitely many intervals, with density that is analytic on the interior of each interval, behaving at worst like a square root at each endpoint, (see [Deift et al. 1998] and [Deift et al. 1999]).

(2) The result in (1-5) is commonly known in the approximation theory litera-ture. For a proof, see [Johansson 1998].

(3) It will prove useful to adapt the following alternative presentation for the function ψ:

$$\psi(\lambda) = \frac{1}{2\pi i} R_+(\lambda) h(\lambda), \quad \lambda \in (\alpha, \beta), \tag{1-8}$$

where the function $R(\lambda)$ is defined via $R(\lambda)^2 = (\lambda - \alpha)(\lambda - \beta)$, with $R(\lambda)$ analytic in $\mathbb{C} \setminus [\alpha, \beta]$, and normalized so that $R(\lambda) \sim \lambda$ as $\lambda \to \infty$. The subscript $\pm$ in $R_\pm(\lambda)$ denotes the boundary value obtained from the upper (lower) half-plane.

The goal of this paper is to provide a rigorous justification for the loop equations, which originated in the physics of two-dimensional quantum gravity (see, for example, the survey [Di Francesco et al. 1995] and the references therein). More precisely, this entails

- proving that the quantity

$$\int_{-\infty}^{\infty} \frac{\rho_N^{(1)}(x)}{x - z} dx$$

  possesses a complete asymptotic expansion in even powers of N (see Theo-rem 2.3 and (3-11)), and
- establishing that the coefficients $P_g(z)$ in the asymptotic expansion (3-11) satisfy the hierarchy of nonlocal equations (3-13), which are the loop equa-tions.

Once the coefficients $P_g(z)$ are known to exist as analytic functions of z and the times t, they may be interpreted as generating functions for a collection of graphical enumeration problems for labelled maps, counted according to vertex valences and the genus of the underlying Riemann surface into which the maps are embedded. (See [Bessis et al. 1980], and also [Ercolani and McLaughlin 2003].) Because of the combinatorial connection and its use in 2-dimensional quantum gravity, obtaining explicit formulae has been a fundamental goal within the physics community of quantum gravity. The loop equations arose as a means to obtain explicit information (and possibly explicit formulae) for these coefficients, although without mathematical justification.

In Section 3 we will need to consider the Cauchy transform of the equilibrium measure

$$F(z) = \int_{-\infty}^{\infty} \frac{\psi(\lambda)}{z - \lambda} d\lambda, \quad z \in \mathbb{C}/\mathbb{R}.$$

It follows from differentiating the variational equations in Theorem 1.1(iii) that $F(z)$ solves the scalar Riemann–Hilbert problem

$$F_+(s) + F_-(s) = V'(s), \quad s \in [\alpha, \beta],$$
$$F_+(s) - F_-(s) = 0, \qquad s \in \mathbb{R}/[\alpha, \beta],$$

where the subscript $\pm$ in $F_\pm(\lambda)$ denotes the boundary value obtained from the upper (lower) half-plane, and $F(z) = 1/z + \mathcal{O}(z^{-2})$ as $z \to \infty$. From this it is straightforward to deduce that

$$2F(z) = V'(z) - R(z)h(z). \tag{1-9}$$

2. Large-N expansions

The fundamental theorem for establishing complete large N expansions of expectations of random variables related to eigenvalue statistics was developed in [Ercolani and McLaughlin 2003]. A concise statement of this result is:

THEOREM 2.1 [Ercolani and McLaughlin 2003]. *There exist $T > 0$ and $\gamma > 0$ such that for all $t \in \mathbb{T}(T, \gamma)$, the expansion*

$$\int_{-\infty}^{\infty} f(\lambda)\rho_N^{(1)}(\lambda) \, d\lambda = f_0 + N^{-2} f_1 + N^{-4} f_2 + \cdots, \tag{2-1}$$

holds, provided the function $f(\lambda)$ is C^∞ smooth and grows no faster than a polynomial for $\lambda \to \infty$. The coefficients f_j depend analytically on t for $t \in \mathbb{T}(T, \gamma)$, and the asymptotic expansion may be differentiated term by term with respect to t.

The complete details for the derivation of this result are presented in [Ercolani and McLaughlin 2003]; however, there are a few specifics presented there that we repeat here for use in subsequent sections and for general background information:

- The function $\rho_N^{(1)}$ has a full and uniform asymptotic expansion, which starts off as follows:

$$\rho_N^{(1)}(\lambda) = \psi(\lambda) + \mathcal{O}(N^{-1/2}). \tag{2-2}$$

- The specific form that this expansion takes depends very much on where one is looking; for example, for $\lambda \in (\alpha, \beta)$, the expansion takes the form:

$$\rho_N^{(1)}(\lambda) = \psi(\lambda) + \frac{1}{4\pi N}\left(\frac{1}{\lambda - \beta} - \frac{1}{\lambda - \alpha}\right)\cos\left(N\int_\lambda^\beta \psi(s)\,ds\right)$$
$$+ \frac{1}{N^2}\left[H(\lambda) + G(\lambda)\sin\left(N\int_\lambda^\beta \psi(s)\,ds\right)\right] + \cdots \tag{2-3}$$

in which $H(\lambda)$ and $G(\lambda)$ are locally analytic functions which are explicitly computable in terms of the original external field $V(\lambda)$.

In [Ercolani and McLaughlin 2003] the primary application of this theorem was to establish that a complete large N asymptotic expansion of 1-1 exists:

THEOREM 2.2. *There is $T > 0$ and $\gamma > 0$ so that for $t \in \mathbb{T}(T, \gamma)$, one has the $N \to \infty$ asymptotic expansion*

$$\log\left(\frac{Z_N(t)}{Z_N(0)}\right) = N^2 e_0(x, t) + e_1(x, t) + \frac{1}{N^2}e_2(x, t) + \cdots. \tag{2-4}$$

The meaning of this expansion is: if you keep terms up to order N^{-2h}, the error term is bounded by CN^{-2h-2}, where the constant C is independent of t for all $t \in \mathbb{T}(T, \gamma)$. For each j, the function $e_j(t)$ is an analytic function of the (complex) vector (t), in a neighborhood of (0). Moreover, the asymptotic expansion of derivatives of $\log(Z_N)$ may be calculated via term-by-term differentiation of the series above. □

REMARKS. (1) Bleher and Its [2005] recently carried out a similar asymptotic expansion of the partition function for a 1-parameter family of external fields. A very interesting aspect of their work is that they establish the nature of the asymptotic expansion of the partition through a critical phase transition.

(2) A subsequent application in [Ercolani, McLaughlin and Pierce 2006] is to develop a hierarchy of ordinary differential equations whose solutions determine recursively the coefficients e_g for potentials of the form $V(\lambda) = \lambda^2/2 + t\lambda^{2\nu}$.

(3) The asymptotic results in [Ercolani and McLaughlin 2003] were also used recently in [Gustavsson 2005] to establish that asymptotics of each individual eigenvalue have Gaussian fluctuations, regardless of whether one is in the bulk or near the edge of the spectrum (provided only that the eigenvalue number, when counted from the edge, grows to ∞).

In the present paper we will make use of a mild extension of Theorem 2.1, in which the function f is of the form $w(\lambda)/(\lambda - z)$, with z living outside the interval $[\alpha, \beta]$:

THEOREM 2.3. *For each $\delta > 0$, there exist $T > 0$ and $\gamma > 0$ such that for all $t \in \mathbb{T}(T, \gamma)$, the expansion*

$$\int_{-\infty}^{\infty} \left(\frac{w(\lambda)}{\lambda - z} \right) \rho_N^{(1)}(\lambda) \, d\lambda = w_0(z) + N^{-2} w_1(z) + N^{-4} w_2(z) + \cdots, \quad (2\text{-}5)$$

holds, provided $z \in \mathbb{C} \setminus [\alpha - \delta, \beta + \delta]$ and the function $w(\lambda)$ is analytic in a neighborhood of $\mathbb{R}$ and grows no faster than a polynomial as $\lambda \to \infty$. The coefficients w_j depend analytically on z and t for $t \in \mathbb{T}(T, \gamma)$, and possess convergent Laurent expansions for $z \to \infty$. Furthermore, the asymptotic expansion may be differentiated term by term.

For z bounded away from the real axis, this follows from Theorem 2.1. The mild extension to the case when z may be near the axis (but bounded away from the support $[\alpha, \beta]$) follows by exploiting analyticity to replace the integral along the real axis near z by a semicircular contour so that λ remains uniformly bounded away from z. Once the contour is such that λ is bounded away from z, the uniform asymptotic expansion for $\rho_N^{(1)}$ may be used. Since z is away from the support $[\alpha, \beta]$, the newly introduced semicircular contour is also bounded away from the support, and one may use arguments similar to those presented in [Ercolani and McLaughlin 2003, Observation 4.2] (where they were used for λ real and outside the interval $[\alpha, \beta]$) to show that $\rho_N^{(1)}(\lambda)$ is uniformly exponentially small on the semicircular contour, and also that the residue term obtained from deforming the contour is also uniformly exponentially small. We will leave these details for the interested reader.

3. Derivation of the loop equations

Denote the Green's function of a random matrix M by

$$G(z, M) = (z - M)^{-1}$$

and its trace by

$$g(z) = \mathrm{Tr}\, G(z).$$

We evaluate $\partial G_{kl}/\partial M_{ij}$ in two different ways to get a useful relation.

LEMMA 3.1. $E(G_{ki}G_{jl}) = N E(G_{kl}V'(M)_{ji})$.

PROOF. We begin by assuming $i \leq j$. Note that when $i < j$, $\partial/\partial M_{ij}$ denotes differentiation with respect to the *complex* variable M_{ij}. Since $G \cdot (z - M) = 1$, we have $\partial/\partial M_{ij} \ G \cdot (z - M) \equiv 0$; equivalently,

$$\partial G/\partial M_{ij} \cdot (z - M) - G \cdot E_{ij} = 0,$$

where E_{ij} is the elementary matrix with a 1 in the (i, j) entry and all other entries zero (note that $\partial M_{ji}/\partial M_{ij} = 0$ for $i < j$). It follows that

$$\partial G/\partial M_{ij} = G \cdot E_{ij} \cdot G;$$

in particular,

$$\partial G_{kl}/\partial M_{ij} = \left(G_{ki}G_{jl}\right),$$

and so

$$E\left(\partial G_{kl}/\partial M_{ij}\right) = E\left(G_{ki}G_{jl}\right).$$

On the other hand, integrating by parts yields

$$E\left(\partial G_{kl}/\partial M_{ij}\right) = \frac{1}{\widetilde{Z}_N} \int_{\mathscr{H}} \frac{\partial G_{kl}}{\partial M_{ij}} \exp\left(-N \ \mathrm{Tr}[V_t(M)]\right) dM$$

$$= -\frac{1}{\widetilde{Z}_N} \int_{\mathscr{H}} G_{kl} \frac{\partial}{\partial M_{ij}} \exp\left(-N\mathrm{Tr}[V_t(M)]\right) dM$$

$$= N\frac{1}{\widetilde{Z}_N} \int_{\mathscr{H}} G_{kl}\mathrm{Tr}\left(\nabla V(M) \cdot E_{ij}\right) \exp\left(-N\mathrm{Tr}[V_t(M)]\right) dM$$

$$= N\frac{1}{\widetilde{Z}_N} \int_{\mathscr{H}} G_{kl}V'(M)_{ji} \exp\left(-N\mathrm{Tr}[V_t(M)]\right) dM$$

$$= N E\left(G_{kl}V'(M)_{ji}\right).$$

Combining the two representations above for $E\left(\partial G_{kl}/\partial M_{ij}\right)$ gives the result. One proves the analogous relation when $i > j$ in a similar fashion.   □

PROPOSITION 3.2. $E((g(z))^2) = N E(\mathrm{Tr}(G \cdot V'(M)))$.   (3-1)

PROOF. This follows directly from the lemma by setting $i = k$ and $j = l$, summing over k and l and dividing by $1/N^2$, which yields

$$\sum_{k,l} E(G_{kk}G_{ll}) = N \sum_{k,l} E\left(G_{k,l}V'(M)_{lk}\right).   □$$

The relation (3-1) can be naturally regarded as a generating function for the second order matrix cumulants of M when written in the equivalent form

$$E((g(z))^2) - E((g(z)))^2 = N E\left(\text{Tr }(G \cdot V'(M))\right) - E((g(z)))^2. \quad (3\text{-}1)$$

To proceed further we will need to introduce some more notation. First, we will use the following general expression for the potential

$$V(z) = \sum_{j=0}^{\infty} t_j z^j$$

which is understood to have only finitely many nonzero t_j's. We also have the formal *vertex operator*

$$\frac{d}{dV} = -\sum_{j=0}^{\infty} \frac{1}{z^{j+1}} \frac{d}{dt_j}. \quad (3\text{-}2)$$

(A precise meaning for this formal relation will be given in the beginning of the next section.) This can be used to give a compact formal representation of a generating function for matrix moments in terms of the RM partition function (1-1):

$$\frac{d}{dV} \frac{1}{N^2} \log Z_N = \sum_{j=0}^{\infty} \frac{1}{z^{j+1}} E\left(\frac{1}{N}\text{Tr } M^j\right). \quad (3\text{-}3)$$

Asymptotic expansions. To make formal relations such as (3-3) meaningful we need to use some fundamental asymptotic facts. The trace of $G(z)$ has a standard integral representation in terms of the RM one-point function (1-3)

$$g(z) = N \int_{-\infty}^{\infty} \frac{\rho_N^{(1)}(\lambda)}{z - \lambda} d\lambda. \quad (3\text{-}4)$$

By boundedness and exponential decay of $\rho_N^{(1)}(\lambda)$, $g(z)$ has a valid asymptotic expansion in large z as

$$\int_{-\infty}^{\infty} \frac{\rho_N^{(1)}(\lambda)}{z - \lambda} d\lambda \sim \sum_{j=0}^{\infty} \frac{1}{z^{j+1}} E\left(\frac{1}{N}\text{Tr } M^j\right). \quad (3\text{-}5)$$

Thus (3-3) can be precisely understood as saying that for each m and large z,

$$\frac{d}{dV^{(m)}} \frac{1}{N^2} \log Z_N = \sum_{j=0}^{m-1} \frac{1}{z^{j+1}} \int_{-\infty}^{\infty} \frac{\lambda^j \rho_N^{(1)}(\lambda)}{z - \lambda} d\lambda$$

$$= \int_{-\infty}^{\infty} \frac{\rho_N^{(1)}(\lambda)}{z - \lambda} d\lambda + \mathbb{O}(z^{-(m+1)}), \quad (3\text{-}6)$$

where

$$\frac{d}{dV^{(m)}} = -\sum_{j=0}^{m-1} \frac{1}{z^{j+1}} \frac{d}{dt_j}.$$

In a similar sense we have for each m and large z the asymptotic equation

$E((g(z))^2) - E((g(z)))^2$

$$= \sum_{j,k=0}^{m} \frac{1}{z^{j+k+2}} \left(E\,(\mathrm{Tr}\,M^j \cdot \mathrm{Tr}\,M^k) - E\,(\mathrm{Tr}\,M^j) E\,(\mathrm{Tr}\,M^k) \right) + \mathbb{O}(z^{-(2m+3)})$$

$$= \frac{d}{dV^{(m)}} \frac{d}{dV^{(m)}} \frac{1}{N^2} \log Z_N + \mathbb{O}(z^{-(2m+3)})$$

$$= \frac{d}{dV^{(m)}} \int_{-\infty}^{\infty} \frac{\rho_N^{(1)}(\lambda)}{z-\lambda} d\lambda + \mathbb{O}(z^{-(2m+3)}). \tag{3-7}$$

In what follows, we will use d/dV instead of $d/dV^{(m)}$ but with the above asymptotic interpretation understood. In the rest of this section we need to establish that there are estimates controlling the errors in the asymptotic expansions (3-6) and (3-7) that remain valid uniformly as $N \to \infty$. To this end we first note that for (3-6) the error has the form

$$\frac{1}{z^{m+1}} \int_{-\infty}^{\infty} \frac{\lambda^m \rho_N^{(1)}(\lambda)}{z-\lambda} d\lambda$$

$$= \frac{1}{z^{m+1}} \left(f_0^{(m)}(z) + N^{-2} f_1^{(m)}(z) + N^{-4} f_2^{(m)}(z) + \cdots \right)$$

for $t \in \mathbb{T}(T, \gamma)$. The right-hand side is an asymptotic expansion valid uniformly (in $\mathbb{T}$), which follows from the fundamental Theorem 2.1. Similarly for (3-7) the error has the form

$$-\sum_{k=m}^{\infty} \frac{1}{z^{k+1}} \frac{d}{dt_k} \frac{1}{z^{m+1}} \left(f_0^{(m)}(z) + N^{-2} f_1^{(m)}(z) + N^{-4} f_2^{(m)}(z) + \ldots \right)$$

$$= \sum_{k=m}^{\upsilon} \frac{1}{z^{k+m+2}} \left(g_0^{(m)}(z) + N^{-2} g_1^{(m)}(z) + N^{-4} g_2^{(m)}(z) + \cdots \right),$$

in which the sum on the right is finite since V depends on only finitely many distinct t_k. We use here the fact stated in Theorem 2.1 that these asymptotic expansions in N can be differentiated term by term, preserving uniformity.

With the observations of this section we can express the relation (3-1) in terms of integral representations involving the one-point function:

$$\frac{d}{dV} \int_{-\infty}^{\infty} \frac{\rho_N^{(1)}(\lambda)}{z - \lambda} d\lambda$$

$$= N^2 \int_{-\infty}^{\infty} \frac{V'(\lambda)\rho_N^{(1)}(\lambda)}{z - \lambda} d\lambda - N^2 \left(\int_{-\infty}^{\infty} \frac{\rho_N^{(1)}(\lambda)}{z - \lambda} d\lambda \right)^2, \quad (3\text{-}8)$$

to be understood in the sense of an asymptotic expansion in large z whose co-efficients moreover have asymptotic expansions in *even* powers of N which are uniform in admissible t. We note that one consequence of this is that the two terms on the right-hand side of (3-8) cancel at leading order so that the difference has leading order $\mathbb{O}(1)$ for large N.

Loop equations. To prepare for the transformation to a recursive loop equation we parse the first integral on the right-hand side of (3-8) as

$$\int_{-\infty}^{\infty} \frac{V'(\lambda)\rho_N^{(1)}(\lambda)}{z - \lambda} d\lambda = \int_{\alpha-\delta}^{\beta+\delta} \frac{V'(\lambda)\rho_N^{(1)}(\lambda)}{z - \lambda} d\lambda + \mathbb{O}(e^{-cN}), \quad (3\text{-}9)$$

where $c > 0$ depends on the choice of the positive constant δ. The justification for the exponential error term is part of the proof of the fundamental asymptotic relation presented in [Ercolani and McLaughlin 2003].

From (3-9), we may further transform this term:

$$\int_{-\infty}^{\infty} \frac{V'(\lambda)\rho_N^{(1)}(\lambda)}{z - \lambda} d\lambda$$

$$= \frac{1}{2\pi i} \int_{\alpha-\delta}^{\beta+\delta} \left(\oint_{\mathscr{C}} \frac{V'(x)}{x - \lambda} dx \right) \frac{\rho_N^{(1)}(\lambda)}{z - \lambda} d\lambda + \mathbb{O}(e^{-cN}),$$

$$= \frac{1}{2\pi i} \oint_{\mathscr{C}} \int_{\alpha-\delta}^{\beta+\delta} \left(\frac{V'(x)}{x - \lambda} \right) \frac{\rho_N^{(1)}(\lambda)}{z - \lambda} d\lambda \, dx + \mathbb{O}(e^{-cN}),$$

$$= \frac{1}{2\pi i} \oint_{\mathscr{C}} \int_{\alpha-\delta}^{\beta+\delta} V'(x)\rho_N^{(1)}(\lambda) \left(\frac{1}{x - \lambda} - \frac{1}{z - \lambda} \right) \left(\frac{1}{z - x} \right) d\lambda \, dx + \mathbb{O}(e^{-cN})$$

$$= \frac{1}{2\pi i} \oint_{\mathscr{C}} \int_{\alpha-\delta}^{\beta+\delta} V'(x)\rho_N^{(1)}(\lambda) \left(\frac{1}{x - \lambda} \right) \left(\frac{1}{z - x} \right) d\lambda \, dx + \mathbb{O}(e^{-cN})$$

$$= \frac{1}{2\pi i} \oint_{\mathscr{C}} \frac{V'(x)}{z - x} \int_{\alpha-\delta}^{\beta+\delta} \frac{\rho_N^{(1)}(\lambda)}{x - \lambda} d\lambda \, dx + \mathbb{O}(e^{-cN}),$$

where on the first line we have expressed $V'(\lambda)$ as a loop integral à la Cauchy's Theorem, where the contour $\mathscr{C}$ encircles the interval $(\alpha-\delta, \beta+\delta)$, with z outside the contour of integration, and on the penultimate line one term has vanished by Cauchy's Theorem and analyticity.

Inserting the last expression above into (3-8), we have derived the final form of the loop equation generating function:

$$\frac{1}{2\pi i} \oint_{\mathscr{C}} \frac{V'(x)}{z-x} \int_{\alpha-\delta}^{\beta+\delta} \frac{\rho_N^{(1)}(\lambda)}{x-\lambda} d\lambda\, dx$$

$$= -N^{-2} \frac{d}{dV} \int_{\alpha-\delta}^{\beta+\delta} \frac{\rho_N^{(1)}(\lambda)}{z-\lambda} d\lambda - \left(\int_{\alpha-\delta}^{\beta+\delta} \frac{\rho_N^{(1)}(\lambda)}{z-\lambda} d\lambda\right)^2 + \mathbb{O}(e^{-cN}). \quad (3\text{-}10)$$

Using Theorem 2.3, the term in parentheses on the last line is easily seen to possess an asymptotic expansion in large N, each of whose coefficients possesses a Laurent expansion in large z:

$$\int_{\alpha-\delta}^{\beta+\delta} \frac{\rho_N^{(1)}(\lambda)}{z-\lambda} d\lambda \sim \sum_{g=0}^{\infty} N^{-2g}\, P_g(z). \quad (3\text{-}11)$$

We note that the terms $P_g(z)$ are independent of the parameter δ. Combining (3-6) and (2-4) we see that

$$P_g(z) = \frac{d}{dV} e_g(t) = \frac{d}{dV^{(\upsilon+1)}} e_g(t) = -\sum_{j=0}^{\upsilon} \frac{1}{z^{j+1}} \frac{de_g(t)}{dt_j}. \quad (3\text{-}12)$$

Inserting (3-11) into the loop equation generating function (3-10) yields the hierarchy of loop equations:

$$\frac{1}{2\pi i} \oint_{\mathscr{C}} \frac{V'(x)}{z-x} P_g(x)\, dx = -\frac{d}{dV} P_{g-1}(z) - \sum_{g'=0}^{g} P_{g'}(z) P_{g-g'}(z),$$

$$\frac{1}{2\pi i} \oint_{\mathscr{C}} \frac{(V'(x)-2P_0(x))}{z-x} P_g(x)\, dx = -\frac{d}{dV} P_{g-1}(z) - \sum_{g'=1}^{g-1} P_{g'}(z) P_{g-g'}(z),$$

$$\frac{1}{2\pi i} \oint_{\mathscr{C}} \frac{\psi(x)}{x-z} P_g(x)\, dx = \frac{d}{dV} P_{g-1}(z) + \sum_{g'=1}^{g-1} P_{g'}(z) P_{g-g'}(z),$$

$$(3\text{-}13)$$

where the transition to the final recursion formula is mediated by the identity (1-9):

$$2\pi i \psi(x) = V'(x) - 2 \int_{-\infty}^{\infty} \frac{\psi(\lambda)}{x-\lambda} d\lambda = V'(x) - 2P_0(x) \quad (3\text{-}14)$$

where $\psi(x)$ here is interpreted as the analytic extension of the density for the equilibrium measure off of the slit $[\alpha, \beta]$ as given by (1-8).

With this result in hand it is now possible to consider a recursive derivation of the terms P_g starting with P_0 as given by (3-14). These terms are related to the map enumeration generating functions, $e_g(t)$, through (3-12). In the physics literature there are instances in which loop equations are used to formally derive expressions for some of the e_g. In particular, we refer the reader to [Ambjørn et al. 1993].

A natural application of our derivation of (3-13) would be to the derivation of closed form expressions for $e_g(t)$ which extend our results in [Ercolani, McLaughlin and Pierce 2006] for potentials V depending only on a single nonzero time $t_{2\nu}$. We may also be able to use these equations to say something about the general qualitative and asymptotic behavior of the e_g. Finally, the $P_g(z, t)$ contain information about the large N asymptotic behavior of the matrix moments, such as (1-2), which could be used to explore whether or not the general unitary ensembles are asymptotically free.

References

[Ambjørn et al. 1993] J. Ambjørn, L. Chekhov, C. F. Kristjansen, and Y. Makeenko, "Matrix model calculations beyond the spherical limit", *Nuclear Phys. B* **404**:1-2 (1993), 127–172.

[Bessis et al. 1980] D. Bessis, C. Itzykson, and J. B. Zuber, "Quantum field theory techniques in graphical enumeration", *Adv. in Appl. Math.* **1**:2 (1980), 109–157.

[Bleher and Its 2005] P. M. Bleher and A. R. Its, "Asymptotics of the partition function of a random matrix model", *Ann. Inst. Fourier (Grenoble)* **55**:6 (2005), 1943–2000.

[Deift et al. 1998] P. Deift, T. Kriecherbauer, and K. T-R McLaughlin, "New results on the equilibrium measure for logarithmic potentials in the presence of an external field", *J. Approx. Theory* **95**:3 (1998), 388–475.

[Deift et al. 1999] P. Deift, T. Kriecherbauer, K. D. T-R McLaughlin, S. Venakides, and X. Zhou, "Uniform asymptotics for polynomials orthogonal with respect to varying exponential weights and applications to universality questions in random matrix theory", *Comm. Pure Appl. Math.* **52**:11 (1999), 1335–1425.

[Di Francesco et al. 1995] P. Di Francesco, P. Ginsparg, and J. Zinn-Justin, "2D gravity and random matrices", *Phys. Rep.* **254**:1-2 (1995), 1–133.

[Ercolani and McLaughlin 2003] N. M. Ercolani and K. D. T-R McLaughlin, "Asymptotics of the partition function for random matrices via Riemann-Hilbert techniques and applications to graphical enumeration", *Int. Math. Res. Not.* **2003**:14 (2003), 755–820.

[Ercolani, McLaughlin and Pierce 2006] N. M. Ercolani, K. D. T-R McLaughlin, and V. U. Pierce, "Random matrices, graphical enumeration and the continuum limit of Toda lattices", preprint, 2006. Available at arXiv:math-ph/0606010. To appear in *Comm. Math. Phys.*

[Gustavsson 2005] J. Gustavsson, "Gaussian fluctuations of eigenvalues in the GUE", *Ann. Inst. H. Poincaré Probab. Statist.* **41**:2 (2005), 151–178.

[Johansson 1998] K. Johansson, "On fluctuations of eigenvalues of random Hermitian matrices", *Duke Math. J.* **91**:1 (1998), 151–204.

[Mehta 1991] M. L. Mehta, *Random matrices*, 2nd ed., Academic Press, Boston, 1991.

[Saff and Totik 1997] E. B. Saff and V. Totik, *Logarithmic potentials with external fields*, Grundlehren der Math. Wissenschaften **316**, Springer, Berlin, 1997.

N. M. ERCOLANI
DEPARTMENT OF MATHEMATICS
UNIVERSITY OF ARIZONA
TUCSON, AZ 85721
UNITED STATES
  ercolani@math.arizona.edu

K. D. T-R MCLAUGHLIN
DEPARTMENT OF MATHEMATICS
UNIVERSITY OF ARIZONA
TUCSON, AZ 85721
UNITED STATES
  mcl@math.arizona.edu

Probability, Geometry and Integrable Systems
MSRI Publications
Volume 55, 2008

Singular solutions for geodesic flows
of Vlasov moments

JOHN GIBBONS, DARRYL D. HOLM, AND CESARE TRONCI

For Henry McKean, on the occasion of his 75th birthday

ABSTRACT. The Vlasov equation for the collisionless evolution of the single-particle probability distribution function (PDF) is a well-known example of coadjoint motion. Remarkably, the property of coadjoint motion survives the process of taking moments. That is, the evolution of the moments of the Vlasov PDF is also a form of coadjoint motion. We find that *geodesic* coadjoint motion of the Vlasov moments with respect to powers of the single-particle momentum admits singular (weak) solutions concentrated on embedded subspaces of physical space. The motion and interactions of these embedded subspaces are governed by canonical Hamiltonian equations for their geodesic evolution.

1. Introduction

The Vlasov equation. The evolution of N identical particles in phase space with coordinates (q_i, p_i) $i = 1, 2, \ldots, N$, may be described by an evolution equation for their joint probability distribution function. Integrating over all but one of the particle phase-space coordinates yields an evolution equation for the single-particle probability distribution function (PDF). This is the Vlasov equation.

The solutions of the Vlasov equation reflect its heritage in particle dynamics, which may be reclaimed by writing its many-particle PDF as a product of delta functions in phase space. Any number of these delta functions may be integrated out until all that remains is the dynamics of a single particle in the collective field of the others. In plasma physics, this collective field generates the total electromagnetic properties and the self-consistent equations obeyed by the single

particle PDF are the Vlasov–Maxwell equations. In the electrostatic approximation, these become the Vlasov–Poisson equations, which govern the statistical
distributions of particle systems ranging from integrated circuits (MOSFETS,
metal-oxide semiconductor field-effect transistors), to charged-particle beams,
to the distribution of galaxies in the Universe.

A class of singular solutions of the VP equations called the "cold plasma" solutions have a particularly beautiful experimental realization in the Malmberg–
Penning trap. In this experiment, the time average of the vertical motion closely
parallels the Euler fluid equations. In fact, the cold plasma singular Vlasov–
Poisson solution turns out to obey the equations of point-vortex dynamics in
an incompressible ideal flow. This coincidence allows the discrete arrays of
"vortex crystals" envisioned by J. J. Thomson for fluid vortices to be realized
experimentally as solutions of the Vlasov–Poisson equations. For a survey of
these experimental cold-plasma results see [Dubin and O'Neil 1990].

Vlasov moments. The Euler fluid equations arise by imposing a closure relation on the first three momentum moments, or p-moments of the Vlasov PDF
$f(p, q, t)$. The zeroth p-moment is the spatial density of particles. The first
p-moment is the mean momentum and its ratio with the zeroth p-moment is the
Eulerian fluid velocity. Introducing an expression for the fluid pressure in terms
of the density and velocity closes the system of p-moment equations, which
otherwise would possess a countably infinite number of dependent variables.

The operation of taking p-moments preserves the geometric nature of Vlasov's
equation. It's closure after the first p-moment results in Euler's useful and beautiful theory of ideal fluids. As its primary geometric characteristic, Euler's fluid
theory represents fluid flow as Hamiltonian geodesic motion on the space of
smooth invertible maps acting on the flow domain and possessing smooth inverses. These smooth maps (called diffeomorphisms) act on the fluid reference
configuration so as to move the fluid particles around in their container. And
their smooth inverses recall the initial reference configuration (or label) for the
fluid particle currently occupying any given position in space. Thus, the motion
of all the fluid particles in a container is represented as a time-dependent curve
in the infinite-dimensional group of diffeomorphisms. Moreover, this curve
describing the sequential actions of the diffeomorphisms on the fluid domain
is a special optimal curve that distills the fluid motion into a single statement.
Namely, "A fluid moves to get out of its own way as efficiently as possible."
Put more mathematically, fluid flow occurs along a curve in the diffeomorphism
group which is a geodesic with respect to the metric on its tangent space supplied
by its kinetic energy.

Given the beauty and utility of the solution behavior for Euler's equation
for the first p-moment, one is intrigued to know more about the dynamics of

the other moments of Vlasov's equation. Of course, the dynamics of the p-moments of the Vlasov–Poisson equation is one of the mainstream subjects of plasma physics and space physics.

Summary. This paper formulates the problem of Vlasov p-moments governed by *quadratic* Hamiltonians. This dynamics is a certain type of geodesic motion on the symplectomorphisms, rather than the diffeomorphisms. The symplectomorphisms are smooth invertible maps acting on the phase space and possessing smooth inverses. We shall consider the singular solutions of the geodesic dynamics of the Vlasov p-moments. Remarkably, these equations turn out to be related to integrable systems governing shallow water wave theory. In fact, when the Vlasov p-moment equations for geodesic motion on the symplectomorphisms are closed at the level of the first p-moment, their singular solutions are found to recover the peaked soliton of the integrable Camassa–Holm equation for shallow water waves [Camassa and Holm 1993].

Thus, geodesic symplectic dynamics of the Vlasov p-moments is found to possess singular solutions whose closure at the fluid level recovers the peakon solutions of shallow water theory. Being solitons, the peakons superpose and undergo elastic collisions in fully nonlinear interactions. The singular solutions for Vlasov p-moments presented here also superpose and interact nonlinearly as coherent structures.

The plan of the paper follows:

Section 2 defines the Vlasov p-moment equations and formulates them as Hamiltonian system using the Kupershmidt–Manin Lie–Poisson bracket. This formulation identifies the p-moment equations as coadjoint motion under the action of a Lie algebra $\mathfrak{g}$ on its dual Lie algebra $\mathfrak{g}^*$, in any number of spatial dimensions.

Section 3 derives variational formulations of the p-moment dynamics in both their Lagrangian and Hamiltonian forms.

Section 4 formulates the problem of geodesic motion on the symplectomorphisms in terms of the Vlasov p-moments and identifies the singular solutions of this problem, whose support is concentrated on delta functions in position space. In a special case, the truncation of geodesic symplectic motion to geodesic diffeomorphic motion for the first p-moment recovers the singular solutions of the Camassa–Holm equation.

Section 5 discusses how the singular p-moment solutions for geodesic symplectic motion are related to the cold plasma solutions. By symmetry under exchange of canonical momentum p and position q, the Vlasov q-moments are also found to admit singular (weak) solutions.

2. Vlasov moment dynamics

The Vlasov equation may be expressed as

$$\frac{\partial f}{\partial t} = \left[f, \frac{\delta h}{\delta f} \right] = \frac{\partial f}{\partial p} \frac{\partial}{\partial q} \frac{\delta h}{\delta f} - \frac{\partial f}{\partial q} \frac{\partial}{\partial p} \frac{\delta h}{\delta f} =: -\operatorname{ad}^*_{\delta h/\delta f} f. \qquad (2\text{-}1)$$

Here the canonical Poisson bracket $[\cdot, \cdot]$ is defined for smooth functions on phase space with coordinates (q, p) and $f(q, p, t)$ is the evolving Vlasov single-particle distribution function. The variational derivative $\delta h/\delta f$ is the single particle Hamiltonian.

A functional $g[f]$ of the Vlasov distribution f evolves according to

$$\frac{dg}{dt} = \iint \frac{\delta g}{\delta f} \frac{\partial f}{\partial t} \, dq \, dp = \iint \frac{\delta g}{\delta f} \left[f, \frac{\delta h}{\delta f} \right] dq \, dp$$

$$= -\iint f \left[\frac{\delta g}{\delta f}, \frac{\delta h}{\delta f} \right] dq \, dp =: -\left\langle\!\left\langle f, \left[\frac{\delta g}{\delta f}, \frac{\delta h}{\delta f} \right] \right\rangle\!\right\rangle =: \{ g, h \}$$

In this calculation boundary terms are neglected upon integrating by parts and the notation $\langle\!\langle \cdot, \cdot \rangle\!\rangle$ is introduced for the L^2 pairing in phase space. The quantity $\{ g, h \}$ defined in terms of this pairing is the Lie–Poisson Vlasov (LPV) bracket [Morrison 1980]. This Hamiltonian evolution equation may also be expressed as

$$\frac{dg}{dt} = \{ g, h \} = \left\langle\!\left\langle f, \operatorname{ad}_{\delta h/\delta f} \frac{\delta g}{\delta f} \right\rangle\!\right\rangle = -\left\langle\!\left\langle \operatorname{ad}^*_{\delta h/\delta f} f, \frac{\delta g}{\delta f} \right\rangle\!\right\rangle$$

which defines the Lie-algebraic operations ad and ad^* in this case in terms of the L^2 pairing on phase space $\langle\!\langle \cdot, \cdot \rangle\!\rangle \colon \mathfrak{s}^* \times \mathfrak{s} \mapsto \mathbb{R}$. Thus, the notation $\operatorname{ad}^*_{\delta h/\delta f} f$ in (2-1) expresses *coadjoint action* of $\delta h/\delta f \in \mathfrak{s}$ on $f \in \mathfrak{s}^*$, where $\mathfrak{s}$ is the Lie algebra of single particle Hamiltonian vector fields and $\mathfrak{s}^*$ is its dual under L^2 pairing in phase space. This is the sense in which the Vlasov equation represents coadjoint motion on the symplectomorphisms.

2.1. Dynamics of Vlasov q, p-moments. The phase space q, p-moments of the Vlasov distribution function are defined by

$$g_{\hat{m}m} = \iint f(q, p) \, q^{\hat{m}} p^m \, dq \, dp.$$

The q, p-moments $g_{\hat{m}m}$ are often used in treating the collisionless dynamics of plasmas and particle beams [Dragt et al. 1990]. This is usually done by considering low order truncations of the potentially infinite sum over phase space moments,

$$g = \sum_{\hat{m},m=0}^{\infty} a_{\hat{m}m} g_{\hat{m},m}, \qquad h = \sum_{\hat{n},n=0}^{\infty} b_{\hat{n}n} g_{\hat{n},n},$$

with constants $a_{\hat{m}m}$ and $b_{\hat{n}n}$, with $\hat{m}, m, \hat{n}, n = 0, 1, \ldots$. If h is the Hamiltonian, the sum over q, p-moments g evolves under the Vlasov dynamics according to the Poisson bracket relation

$$\frac{dg}{dt} = \{g, h\} = \sum_{\hat{m}, m, \hat{n}, n = 0}^{\infty} a_{\hat{m}m} b_{\hat{n}n} (\hat{m}m - \hat{n}n) g_{\hat{m}+\hat{n}-1, m+n-1}.$$

This Poisson bracket may be identified with the smooth Hamiltonian vector fields on p and q, by invoking the standard Lie-algebra antihomomorphism

$$X_H = \{\cdot, H\},$$

for any function $H(p, q)$, then noticing that the q, p-moments are linear functionals of the canonical variables. The symplectic invariants associated with Hamiltonian flows of the q, p-moments were discovered and classified in [Holm et al. 1990].

2.2. Dynamics of Vlasov p-moments.

The momentum moments, or "p-moments," of the Vlasov function are defined as

$$A_m(q, t) = \int p^m f(q, p, t) \, dp, \qquad m = 0, 1, \ldots.$$

That is, the p-moments are q-dependent integrals over p of the product of powers p^m, $m = 0, 1, \ldots$, times the Vlasov solution $f(q, p, t)$. We shall consider functionals of these p-moments defined by

$$g = \sum_{m=0}^{\infty} \iint \alpha_m(q) \, p^m f \, dq \, dp = \sum_{m=0}^{\infty} \int \alpha_m(q) \, A_m(q) \, dq =: \sum_{m=0}^{\infty} \langle A_m, \alpha_m \rangle,$$

$$h = \sum_{n=0}^{\infty} \iint \beta_n(q) \, p^n f \, dq \, dp = \sum_{n=0}^{\infty} \int \beta_n(q) \, A_n(q) \, dq =: \sum_{n=0}^{\infty} \langle A_n, \beta_n \rangle,$$

where $\langle \cdot, \cdot \rangle$ is the L^2 pairing on position space.

The functions α_m and β_n with $m, n = 0, 1, \ldots$ are assumed to be suitably smooth and integrable against the Vlasov p-moments. To assure these properties, one may relate the p-moments to the previous sums of Vlasov q, p-moments by choosing

$$\alpha_m(q) = \sum_{\hat{m}=0}^{\infty} a_{\hat{m}m} q^{\hat{m}}, \qquad \beta_n(q) = \sum_{\hat{n}=0}^{\infty} b_{\hat{n}n} q^{\hat{n}}.$$

For these choices of $\alpha_m(q)$ and $\beta_n(q)$, the sums of p-moments will recover the full set of Vlasov (q, p)-moments. Thus, as long as the q, p-moments of the distribution $f(q, p)$ continue to exist under the Vlasov evolution, one may assume that the dual variables $\alpha_m(q)$ and $\beta_n(q)$ are smooth functions whose

Taylor series expands the p-moments in the q, p-moments. These functions are dual to the p-moments $A_m(q)$ with $m = 0, 1, \ldots$ under the L^2 pairing $\langle \cdot, \cdot \rangle$ in the spatial variable q. In what follows we will assume *homogeneous* boundary conditions. This means, for example, that we will ignore boundary terms arising from integrations by parts.

2.3. Poisson bracket for Vlasov p-moments.

The Poisson bracket among the p-moments is obtained from the LPV bracket through explicit calculation:

$$
\begin{aligned}
\{ g, h \} &= - \sum_{m,n=0}^{\infty} \iint f \left[\alpha_m(q)\, p^m, \, \beta_n(q)\, p^n \right] dq\, dp \\
&= - \sum_{m,n=0}^{\infty} \iint \left[m\alpha_m \beta_n' - n\beta_n \alpha_m' \right] f\, p^{m+n-1}\, dq\, dp \\
&= - \sum_{m,n=0}^{\infty} \int A_{m+n-1}(q) \left[m\alpha_m \beta_n' - n\beta_n \alpha_m' \right] dq \\
&=: \sum_{m,n=0}^{\infty} \left\langle A_{m+n-1}, \, \mathrm{ad}_{\beta_n} \alpha_m \right\rangle \\
&= - \sum_{m,n=0}^{\infty} \int \left[n\beta_n A_{m+n-1}' + (m+n) A_{m+n-1} \beta_n' \right] \alpha_m\, dq \\
&=: - \sum_{m,n=0}^{\infty} \left\langle \mathrm{ad}_{\beta_n}^* A_{m+n-1}, \, \alpha_m \right\rangle,
\end{aligned}
$$

where we have integrated by parts and the symbols ad and ad* stand for the adjoint and coadjoint actions. This is done by again invoking the Lie-algebra antihomomorphism with the smooth Hamiltonian vector fields, since the smooth functions $\alpha_m(q)$ and $\beta_n(q)$ are assumed to possess convergent Taylor series.

Upon recalling the dual relations

$$
\alpha_m = \frac{\delta g}{\delta A_m} \quad \text{and} \quad \beta_n = \frac{\delta h}{\delta A_n}
$$

the LPV bracket in terms of the p-moments may be expressed as

$$
\begin{aligned}
\{ g, h \}(\{A\}) &= - \sum_{m,n=0}^{\infty} \int \frac{\delta g}{\delta A_m} \left[n \frac{\delta h}{\delta A_n} \frac{\partial}{\partial q} A_{m+n-1} + (m+n) A_{m+n-1} \frac{\partial}{\partial q} \frac{\delta h}{\delta A_n} \right] dq \\
&=: - \sum_{m,n=0}^{\infty} \left\langle A_{m+n-1}, \, \left[\!\left[\frac{\delta g}{\delta A_m}, \frac{\delta h}{\delta A_n} \right]\!\right] \right\rangle.
\end{aligned}
$$

This is the Kupershmidt–Manin Lie–Poisson (KMLP) bracket [Kupershmidt and Manin 1978], which is defined for functions on the dual of the Lie algebra with bracket

$$\llbracket \alpha_m, \beta_n \rrbracket = m\alpha_m \partial_q \beta_n - n\beta_n \partial_q \alpha_m.$$

This Lie algebra bracket inherits the Jacobi identity from its definition in terms of the canonical Hamiltonian vector fields. Thus, we have shown:

THEOREM 2.1 [Gibbons 1981]. *The operation of taking p-moments of Vlasov solutions is a Poisson map. It takes the LPV bracket describing the evolution of $f(q, p)$ into the KMLP bracket, describing the evolution of the p-moments $A_n(x)$.*

REMARK 2.2. A result related to theorem 2.1 for the Benney hierarchy [Benney 1966] was also noted by Lebedev and Manin [Lebedev and Manin 1979].

The evolution of a particular p-moment $A_m(q, t)$ is obtained from the KMLP bracket by

$$\frac{\partial A_m}{\partial t} = \{A_m, h\} = -\sum_{n=0}^{\infty} \left(n\frac{\delta h}{\delta A_n}\frac{\partial}{\partial q}A_{m+n-1} + (m+n)A_{m+n-1}\frac{\partial}{\partial q}\frac{\delta h}{\delta A_n} \right).$$

The KMLP bracket among the p-moments is given by

$$\{A_m, A_n\} = -n\frac{\partial}{\partial q}A_{m+n-1} - mA_{m+n-1}\frac{\partial}{\partial q},$$

expressed as a differential operator acting to the right. This operation is skew-symmetric under the L^2 pairing and the general KMLP bracket can then be written as (see [Gibbons 1981])

$$\{g, h\}(\{A\}) = \sum_{m,n=0}^{\infty} \int \frac{\delta g}{\delta A_m}\{A_m, A_n\}\frac{\delta h}{\delta A_n}\, dq,$$

so that

$$\frac{\partial A_m}{\partial t} = \sum_{n=0}^{\infty}\{A_m, A_n\}\frac{\delta h}{\delta A_n}.$$

2.4. Multidimensional treatment. We now show that the KMLP bracket and the equations of motion may be written in three dimensions in multi-index notation. By writing $p^{2n+1} = p^{2n}\, p$, and checking that

$$p^{2n} = \sum_{i+j+k=n} \frac{n!}{i!j!k!}\, p_1^{2i} p_2^{2j} p_3^{2k},$$

it is easy to see that the multidimensional treatment can be performed in terms of the quantities

$$p^\sigma =: p_1^{\sigma_1} p_2^{\sigma_2} p_3^{\sigma_3},$$

where $\sigma \in \mathbb{N}^3$. Let A_σ be defined as

$$A_\sigma (q, t) =: \int p^\sigma f (q, p, t) \, dp$$

and consider functionals of the form

$$g = \sum_\sigma \iint \alpha_\sigma (q) \, p^\sigma f (q, p, t) \, dq \, dp =: \sum_{\sigma \in \mathbb{N}^3} \langle A_\sigma, \alpha_\sigma \rangle,$$

$$h = \sum_\rho \iint \beta_\rho (q) \, p^\rho f (q, p, t) \, dq \, dp =: \sum_{\rho \in \mathbb{N}^3} \langle A_\rho, \beta_\rho \rangle.$$

With the notation

$$1_j := (0, ..., 1..., 0) \quad (1 \text{ in } j\text{-th position}),$$

so that $(1_j)_i = \delta_{ji}$. the ordinary LPV bracket leads to

$$\{g, h\} = -\sum_{\sigma, \rho} \iint f \left[\alpha_\sigma (q) \, p^\sigma, \beta_\rho (q) \, p^\rho \right] dq \, dp$$

$$= -\sum_{\sigma, \rho} \sum_j \iint f \left(\alpha_\sigma p^\rho \frac{\partial p^\sigma}{\partial p_j} \frac{\partial \beta_\rho}{\partial q_j} - \beta_\rho p^\sigma \frac{\partial p^\rho}{\partial p_j} \frac{\partial \alpha_\sigma}{\partial q_j} \right) dq \, dp$$

$$= -\sum_{\sigma, \rho} \sum_j \iint f \left(\sigma_j \alpha_\sigma p^\rho p^{\sigma - 1_j} \frac{\partial \beta_\rho}{\partial q_j} - \rho_j \beta_\rho p^\sigma p^{\rho - 1_j} \frac{\partial \alpha_\sigma}{\partial q_j} \right) dq \, dp$$

$$= -\sum_{\sigma, \rho} \sum_j \int A_{\sigma + \rho - 1_j} \left(\sigma_j \alpha_\sigma \frac{\partial \beta_\rho}{\partial q_j} - \rho_j \beta_\rho \frac{\partial \alpha_\sigma}{\partial q_j} \right) dq$$

$$=: \sum_{\sigma, \rho} \sum_j \langle A_{\sigma + \rho - 1_j}, (\text{ad}_{\beta_\rho})_j \, \alpha_\sigma \rangle$$

$$= -\sum_{\sigma, \rho} \sum_j \int \left(\rho_j \beta_\rho \frac{\partial}{\partial q_j} A_{\sigma + \rho - 1_j} + (\sigma_j + \rho_j) A_{\sigma + \rho - 1_j} \frac{\partial \beta_\rho}{\partial q_j} \right) \alpha_\sigma \, dq$$

$$=: -\sum_{\sigma, \rho} \sum_j \langle (\text{ad}_{\beta_\rho}^*)_j \, A_{\sigma + \rho - 1_j}, \alpha_\sigma \rangle,$$

where the sum extends to all $\sigma, \rho \in \mathbb{N}^3$.

The LPV bracket in terms of the p-moments may then be written as

$$\frac{\partial A_\sigma}{\partial t} = -\sum_{\rho \in \mathbb{N}^3} \sum_j \left(\mathrm{ad}^*_{\delta h/\delta A_\rho} \right)_j A_{\sigma+\rho+1_j}$$

where the Lie bracket is now expressed as

$$\left[\!\left[\frac{\delta g}{\delta A_\sigma}, \frac{\delta h}{\delta A_\rho} \right]\!\right]_j = \sigma_j \alpha_\sigma \frac{\partial}{\partial q_j} \frac{\delta h}{\delta A_\rho} - \rho_j \beta_\rho \frac{\partial}{\partial q_j} \frac{\delta g}{\delta A_\sigma}.$$

Moreover the evolution of a particular p-moment A_σ is obtained by

$$\frac{\partial A_\sigma}{\partial t} = \{A_\sigma, h\}$$
$$= -\sum_\rho \sum_j \left(\rho_j \frac{\delta h}{\delta A_\rho} \frac{\partial}{\partial q_j} A_{\sigma+\rho-1_j} + (\sigma_j + \rho_j) A_{\sigma+\rho-1_j} \frac{\partial}{\partial q_j} \frac{\delta h}{\delta A_\rho} \right)$$

and the KMLP bracket among the multidimensional p-moments is given in by

$$\{A_\sigma, A_\rho\} = -\sum_j \left(\sigma_j \frac{\partial}{\partial q_j} A_{\sigma+\rho-1_j} + \rho_j A_{\sigma+\rho-1_j} \frac{\partial}{\partial q_j} \right).$$

Inserting the previous operator in this multidimensional KMLP bracket yields

$$\{g, h\} (\{A\}) = \sum_{\sigma,\rho} \int \frac{\delta g}{\delta A_\sigma} \{A_\sigma, A_\rho\} \frac{\delta h}{\delta A_\rho} \, dq,$$

and the corresponding evolution equation becomes

$$\frac{\partial A_\sigma}{\partial t} = \sum_\rho \{A_\sigma, A_\rho\} \frac{\delta h}{\delta A_\rho}.$$

Thus, in multi-index notation, the form of the Hamiltonian evolution under the KMLP bracket is essentially unchanged in going to higher dimensions.

2.5. Applications of the KMLP bracket. The KMLP bracket was derived in the context of Benney long waves, whose Hamiltonian is

$$H_2 = \tfrac{1}{2}(A_2 + A_0^2).$$

This leads to the moment equations

$$\frac{\partial A_n}{\partial t} + \frac{\partial A_{n+1}}{\partial q} + n A_{n-1} \frac{\partial A_0}{\partial q} = 0$$

derived by Benney [1966] as a description of long waves on a shallow perfect fluid, with a free surface at $y = h(q, t)$. In his interpretation, the A_n were vertical moments of the horizontal component of the velocity $p(q, y, t)$:

$$A_n = \int_{y=0}^{h} p(q, y, t)^n \, dy.$$

The corresponding system of evolution equations for $p(q, y, t)$ and $h(q, t)$ is related by the hodograph transformation, $y = \int_{-\infty}^{p} f(q, p', t) \, dp'$, to the Vlasov equation

$$\frac{\partial f}{\partial t} + p \frac{\partial f}{\partial q} - \frac{\partial A_0}{\partial q} \frac{\partial f}{\partial p} = 0.$$

The most important fact about the Benney hierarchy is that it is completely integrable. This fact emerges from the following observation. Upon defining a function $\lambda(q, p, t)$ by the principal value integral,

$$\lambda(q, p, t) = p + P \int_{-\infty}^{\infty} \frac{f(q, p', t)}{p - p'} \, dp',$$

it is straightforward to verify [Lebedev and Manin 1979] that

$$\frac{\partial \lambda}{\partial t} + p \frac{\partial \lambda}{\partial q} - \frac{\partial A_0}{\partial q} \frac{\partial \lambda}{\partial p} = 0;$$

so that f and λ are advected along the same characteristics.

In higher dimensions, particularly $n = 3$, we may take the direct sum of the KMLP bracket, together with the Poisson bracket for an electromagnetic field (in the Coulomb gauge) where the electric field E and magnetic vector potential A are canonically conjugate; then the Hamiltonian

$$H_{MV} = \iint \left(\frac{1}{2m} |p - eA|^2 \right) f(p, q) \, d^n p \, d^n q$$
$$+ \int \left(\frac{1}{2} |E|^2 + \frac{1}{4} \sum_{i=1}^{n} \sum_{j=1}^{n} (A_{i,j} - A_{j,i})^2 \right) d^n q$$

yields the Maxwell–Vlasov (MV) equations for systems of interacting charged particles. For a discussion of the MV equations from a geometric viewpoint in the same spirit as the present approach, see [Cendra et al. 1998]. For discussions of the Lie-algebraic approach to the control and steering of charged particle beams, see [Dragt et al. 1990].

3. Variational principles and Hamilton–Poincaré formulation

In this section we show how the p-moment dynamics can be derived from Hamilton's principle both in the Hamilton–Poincaré and Euler–Poincaré forms. These variational principles are defined , respectively, on the dual Lie algebra $\mathfrak{g}^*$ containing the moments, and on the Lie algebra $\mathfrak{g}$ itself. For further details about these dual variational formulations, see [Cendra et al. 2003] and [Holm et al. 1998]. Summation over repeated indices is intended in this section.

3.1. Hamilton–Poincaré hierarchy.
We begin with the Hamilton–Poincaré principle for the p-moments written as

$$\delta \int_{t_i}^{t_j} dt \left(\langle A_n, \beta_n \rangle - H(\{A\}) \right) = 0$$

(where $\beta_n \in \mathfrak{g}$). We shall prove that this leads to the same dynamics as found in the context of the KMLP bracket. To this purpose, we must define the n-th p-moment in terms of the Vlasov distribution function. We check that

$$0 = \delta \int_{t_i}^{t_j} dt \left(\langle A_n, \beta_n \rangle - H(\{A\}) \right) = \int_{t_i}^{t_j} dt \left(\delta \{\!\{ f, p^n \beta_n \}\!\} - \left\langle\!\!\left\langle \delta f, \frac{\delta H}{\delta f} \right\rangle\!\!\right\rangle \right)$$

$$= \int_{t_i}^{t_j} dt \left(\left\langle\!\!\left\langle \delta f, \left(p^n \beta_n - \frac{\delta H}{\delta f} \right) \right\rangle\!\!\right\rangle + \{\!\{ f, \delta(p^n \beta_n) \}\!\} \right).$$

Now recall that any $g = \delta G / \delta f$ belonging to the Lie algebra $\mathfrak{s}$ of the symplectomorphisms (which also contains the distribution function itself) may be expressed as

$$g = \frac{\delta G}{\delta f} = p^m \frac{\delta G}{\delta A_m} = p^m \xi_m,$$

by the chain rule. Consequently, one finds the pairing relationship

$$\left\langle\!\!\left\langle \delta f, \left(p^n \beta_n - \frac{\delta H}{\delta f} \right) \right\rangle\!\!\right\rangle = \left\langle \delta A_n, \left(\beta_n - \frac{\delta H}{\delta A_n} \right) \right\rangle.$$

Next, recall from the general theory that variations on a Lie group induce variations on its Lie algebra of the form

$$\delta w = \dot{u} + [g, u]$$

where $u, w \in \mathfrak{s}$ and u vanishes at the endpoints. Writing $u = p^m \eta_m$ then yields

$$\int_{t_i}^{t_j} dt \, \{\!\{ f, \delta(p^n \beta_n) \}\!\} = \int_{t_i}^{t_j} dt \, \{\!\{ f, (\dot{u} + [p^n \beta_n, u]) \}\!\}$$

$$= - \int_{t_i}^{t_j} dt \left(\langle \dot{A}_m, \eta_m \rangle - \langle A_{n+m-1}, [\![\beta_n, \eta_m]\!] \rangle \right)$$

$$= - \int_{t_i}^{t_j} dt \left\langle (\dot{A}_m + \mathrm{ad}_{\beta_n}^* A_{m+n-1}), \eta_m \right\rangle.$$

Consequently, the Hamilton–Poincaré principle may be written entirely in terms of the moments as

$$\delta S = \int_{t_i}^{t_j} dt \left\{ \left\langle \delta A_n, \left(\beta_n - \frac{\delta H}{\delta A_n} \right) \right\rangle - \left\langle (\dot{A}_m + \mathrm{ad}^*_{\beta_n} A_{m+n-1}), \eta_m \right\rangle \right\} = 0.$$

This expression produces the inverse Legendre transform

$$\beta_n = \frac{\delta H}{\delta A_n}$$

(holding for hyperregular Hamiltonians). It also yields the equations of motion

$$\frac{\partial A_m}{\partial t} = -\mathrm{ad}^*_{\beta_n} A_{m+n-1},$$

which are valid for arbitrary variations δA_m and variations $\delta \beta_m$ of the form

$$\delta \beta_m = \dot{\eta}_m + \mathrm{ad}_{\beta_n} \eta_{m-n+1},$$

where the variations η_m satisfy vanishing endpoint conditions,

$$\eta_m|_{t=t_i} = \eta_m|_{t=t_j} = 0.$$

Thus, the Hamilton–Poincaré variational principle recovers the hierarchy of the evolution equations derived in the previous section using the KMLP bracket.

3.2. Euler–Poincaré hierarchy. The corresponding Lagrangian formulation of the Hamilton's principle now yields

$$\begin{aligned}
\delta \int_{t_i}^{t_j} L(\{\beta\}) \, dt &= \int_{t_i}^{t_j} \left\langle \frac{\delta L}{\delta \beta_m}, \delta \beta_m \right\rangle dt \\
&= \int_{t_i}^{t_j} \left\langle \frac{\delta L}{\delta \beta_m}, (\dot{\eta}_m + \mathrm{ad}_{\beta_n} \eta_{m-n+1}) \right\rangle dt \\
&= -\int_{t_i}^{t_j} \left(\left\langle \frac{\partial}{\partial t} \frac{\delta L}{\delta \beta_m}, \eta_m \right\rangle + \left\langle \mathrm{ad}^*_{\beta_n} \frac{\delta L}{\delta \beta_m}, \eta_{m-n+1} \right\rangle \right) dt \\
&= -\int_{t_i}^{t_j} \left(\left\langle \frac{\partial}{\partial t} \frac{\delta L}{\delta \beta_m}, \eta_m \right\rangle + \left\langle \mathrm{ad}^*_{\beta_n} \frac{\delta L}{\delta \beta_{m+n-1}}, \eta_m \right\rangle \right) dt \\
&= -\int_{t_i}^{t_j} \left\langle \left(\frac{\partial}{\partial t} \frac{\delta L}{\delta \beta_m} + \mathrm{ad}^*_{\beta_n} \frac{\delta L}{\delta \beta_{m+n-1}} \right), \eta_m \right\rangle dt,
\end{aligned}$$

upon using the expression previously found for the variations $\delta \beta_m$ and relabeling indices appropriately. The Euler–Poincaré equations may then be written as

$$\frac{\partial}{\partial t} \frac{\delta L}{\delta \beta_m} + \mathrm{ad}^*_{\beta_n} \frac{\delta L}{\delta \beta_{m+n-1}} = 0$$

with the same constraints on the variations as in the previous paragraph. Applying the Legendre transformation

$$A_m = \frac{\delta L}{\delta \alpha_m}$$

yields the Euler–Poincaré equations (for hyperregular Lagrangians). This again leads to the same hierarchy of equations derived earlier using the KMLP bracket.

To summarize, the calculations in this section have proved this result:

THEOREM 3.1. *With the above notation and hypotheses of hyperregularity the following statements are equivalent:*

(i) (*The Euler–Poincaré variational principle.*) *The curves* $\beta_n(t)$ *are critical points of the action*

$$\delta \int_{t_i}^{t_j} L(\{\beta\})\, dt = 0$$

for variations of the form

$$\delta \beta_m = \dot\eta_m + \mathrm{ad}_{\beta_n} \eta_{m-n+1},$$

in which η_m *vanishes at the endpoints*

$$\eta_m|_{t=t_i} = \eta_m|_{t=t_j} = 0$$

and the variations δA_n *are arbitrary.*

(ii) (*The Lie–Poisson variational principle.*) *The curves* $(\beta_n, A_n)(t)$ *are critical points of the action*

$$\delta \int_{t_i}^{t_j} \left(\langle A_n, \beta_n \rangle - H(\{A\}) \right) dt = 0$$

for variations of the form

$$\delta \beta_m = \dot\eta_m + \mathrm{ad}_{\beta_n} \eta_{m-n+1},$$

where η_m *satisfies endpoint conditions*

$$\eta_m|_{t=t_i} = \eta_m|_{t=t_j} = 0$$

and the variations δA_n *are arbitrary.*

(iii) *The* **Euler–Poincaré equations** *hold:*

$$\frac{\partial}{\partial t} \frac{\delta L}{\delta \beta_m} + \mathrm{ad}^*_{\beta_n} \frac{\delta L}{\delta \beta_{m+n-1}} = 0.$$

(iv) *The* **Lie–Poisson equations** *hold:*

$$\dot A_m = -\mathrm{ad}^*_{\delta H/\delta A_n} A_{m+n-1}.$$

For further details on the proof of this theorem we direct the reader to [Cendra et al. 2003]. An analogous result is also valid in the multidimensional case with slight modifications.

4. Quadratic Hamiltonians

4.1. Geodesic motion. We shall consider the problem of geodesic motion on the space of p-moments. For this, we define the Hamiltonian as the norm on the p-moment given by the following metric and inner product,

$$h = \tfrac{1}{2}\|A\|^2 = \tfrac{1}{2}\sum_{n,s=0}^{\infty}\iint A_n(q)G_{ns}(q,q')A_s(q')\,dq\,dq' \qquad (4\text{-}1)$$

The metric $G_{ns}(q,q')$ is chosen to be positive definite, so it defines a norm for $\{A\} \in \mathfrak{g}^*$. The corresponding geodesic equation with respect to this norm is found as in the previous section to be

$$\frac{\partial A_m}{\partial t} = \{A_m, h\} = -\sum_{n=0}^{\infty}\left(n\beta_n\frac{\partial}{\partial q}A_{m+n-1} + (m+n)A_{m+n-1}\frac{\partial}{\partial q}\beta_n\right), \quad (4\text{-}2)$$

with dual variables $\beta_n \in \mathfrak{g}$ defined by

$$\beta_n = \frac{\delta h}{\delta A_n} = \sum_{s=0}^{\infty}\int G_{ns}(q,q')A_s(q')\,dq' = \sum_{s=0}^{\infty}G_{ns}*A_s. \qquad (4\text{-}3)$$

Thus, evolution under (4-2) may be rewritten formally as (infinitesimal) coadjoint motion on $\mathfrak{g}^*$

$$\frac{\partial A_m}{\partial t} = \{A_m, h\} =: -\sum_{n=0}^{\infty}\mathrm{ad}^*_{\beta_n}A_{m+n-1}. \qquad (4\text{-}4)$$

The explicit identification of coAdjoint motion by the full group action on the dual Lie algebra is left for a future study. This system comprises an infinite system of nonlinear, nonlocal, coupled evolutionary equations for the p-moments. In this system, evolution of the m-th moment is governed by the potentially infinite sum of contributions of the velocities β_n associated with n-th moment sweeping the $(m+n-1)$-th moment by coadjoint action. Moreover, by equation (4-3), each of the β_n potentially depends nonlocally on all of the moments.

Equations (4-1) and (4-3) may be written in three dimensions in multi-index notation, as follows: the Hamiltonian is given by

$$h = \tfrac{1}{2}\|A\|^2 = \tfrac{1}{2}\sum_{\mu,\nu}\iint A_\mu(q,t)G_{\mu\nu}(q,q')A_\nu(q',t)\,dq\,dq'$$

so the dual variable is written as

$$\beta_\rho = \frac{\delta h}{\delta A_\rho} = \sum_\nu \iint G_{\rho\nu}(\boldsymbol{q},\boldsymbol{q}') A_\nu(\boldsymbol{q}',t) \, d\boldsymbol{q} \, d\boldsymbol{q}' = \sum_\nu G_{\rho\nu} * A_\nu.$$

4.2. Singular geodesic solutions. Remarkably, in any number of spatial dimensions, the geodesic equation (4-2) possesses exact solutions which are *singular*; that is, they are supported on delta functions in q-space.

THEOREM 4.1 (SINGULAR SOLUTION ANSATZ FOR GEODESIC FLOWS OF VLASOV p-MOMENTS). *Equation (4-2) admits singular solutions of the form*

$$A_\sigma(\boldsymbol{q},t) = \sum_{j=1}^{N} \int P_j^\sigma(\boldsymbol{q},t,a_j) \, \delta(\boldsymbol{q} - \boldsymbol{Q}_j(t,a_j)) \, da_j, \tag{4-5}$$

in which the integrals over coordinates a_j are performed over N embedded subspaces of the q-space and the parameters (Q_j, P_j) satisfy canonical Hamiltonian equations in which the Hamiltonian is the norm h in (4-1) evaluated on the singular solution Ansatz (4-5).

In one dimension, the coordinates a_j are absent and the singular solutions in (4-5) reduce to

$$A_s(q,t) = \sum_{j=1}^{N} P_j^s(q,t) \, \delta(q - Q_j(t)). \tag{4-6}$$

In order to show this is a solution in one dimension, one checks that these singular solutions satisfy a system of partial differential equations in Hamiltonian form, whose Hamiltonian couples all the moments

$$H_N = \frac{1}{2} \sum_{n,s=0}^{\infty} \sum_{j,k=1}^{N} P_j^s(Q_j(t),t) P_k^n(Q_k(t),t) G_{ns}(Q_j(t), Q_k(t)).$$

One forms the pairing of the coadjoint equation

$$\dot{A}_m = -\sum_{n,s} \mathrm{ad}^*_{G_{ns}*A_s} A_{m+n-1}$$

with a sequence of smooth functions $\{\varphi_m(q)\}$, so that

$$\langle \dot{A}_m, \varphi_m \rangle = \sum_{n,s} \langle A_{m+n-1}, \mathrm{ad}_{G_{ns}*A_s} \varphi_m \rangle$$

One expands each term and denotes $\tilde{P}_j(t) := P_j(Q_j, t)$:

$$\langle \dot{A}_m, \varphi_m \rangle = \sum_j \int dq\, \varphi_m(q) \frac{\partial}{\partial t} \left(P_j^m(q, t) \delta(q - Q_j) \right)$$

$$= \sum_j \int dq\, \varphi_m(q) \left(\delta(q - Q_j) \frac{\partial P_j^m}{\partial t} - P_j^m \dot{Q}_j \delta'(q - Q_j) \right)$$

$$= \sum_j \left(\frac{d \tilde{P}_j^m}{dt} \varphi_m(Q_j) + \tilde{P}_j^m \dot{Q}_j \varphi_m'(Q_j) \right)$$

Similarly, expanding

$$\langle A_{m+n-1}, \mathrm{ad}_{G_{ns} * A_s} \varphi_m \rangle$$

$$= \sum_{j,k} \int dq\, \tilde{P}_k^s\, P_j^{m+n-1} \delta(q - Q_j) \left(n \varphi_m' G_{ns}(q, Q_k) - m \varphi_m \frac{\partial G_{ns}(q, Q_k)}{\partial q} \right)$$

$$= \sum_{j,k} \tilde{P}_k^s \tilde{P}_j^{m+n-1} \left(n\, \varphi_m'(Q_j) G_{ns}(Q_j, Q_k) - m\, \varphi_m(Q_j) \frac{\partial G_{ns}(Q_j, Q_k)}{\partial Q_j} \right)$$

leads to

$$\tilde{P}_j^m \frac{dQ_j}{dt} = \sum_{n,s} \sum_k n\, \tilde{P}_k^s \tilde{P}_j^{m+n-1} G_{ns}(Q_j, Q_k),$$

$$\frac{d \tilde{P}_j^m}{dt} = -m \sum_{n,s} \sum_k \tilde{P}_k^s \tilde{P}_j^{m+n-1} \frac{\partial G_{ns}(Q_j, Q_k)}{\partial Q_j},$$

so we finally obtain equations for Q_j and $\tilde{P}_j$ in canonical form,

$$\frac{dQ_j}{dt} = \frac{\partial H_N}{\partial \tilde{P}_j}, \qquad \frac{d \tilde{P}_j}{dt} = -\frac{\partial H_N}{\partial Q_j}.$$

Remark about higher dimensions. The singular solutions (4-5) with the integrals over coordinates a_j exist in higher dimensions. The higher dimensional singular solutions satisfy a system of canonical Hamiltonian integral-partial differential equations, instead of ordinary differential equations.

5. Discussion

5.1. Remarks about EPSymp and connections with EPDiff. Importantly, geodesic motion for the p-moments is equivalent to geodesic motion for the Euler–Poincaré equations on the symplectomorphisms (EPSymp) given by the

Hamiltonian

$$H[f] = \frac{1}{2} \iint f(q, p, t) \mathcal{G}(q, p, q', p') f(q', p', t) \, dq \, dp \, dq' \, dp' \quad (5\text{-}1)$$

The equivalence with EPSymp emerges when the function $\mathcal{G}$ is written as

$$\mathcal{G}(q, q', p, p') = \sum_{n,m} p^n G_{nm}(q, q') p'^m.$$

Thus, whenever the metric $\mathcal{G}$ for EPSymp has a Taylor series, its solutions may be expressed in terms of the geodesic motion for the p-moments.

Moreover the distribution function corresponding to the singular solutions for the moments is a particular case of the *cold-plasma approximation*, given by

$$f(q, p, t) = \sum_j \rho_j(q, t) \delta(p - P_j(q, t)),$$

where in our case a summation is introduced and ρ is written as a Lagrangian particle-like density: $\rho_j(q, t) = \delta(q - Q_j(t))$.

To check this is a solution for the geodesic motion of the generating function, one repeats exactly the same procedure as for the moments, in order to find the Hamiltonian equations

$$\frac{dQ_j}{dt} = \frac{\partial}{\partial \tilde{P}_j} \frac{\delta H}{\delta f}(Q_j, \tilde{P}_j), \qquad \frac{d\tilde{P}_j}{dt} = \frac{\partial}{\partial Q_j} \frac{\delta H}{\delta f}(Q_j, \tilde{P}_j)$$

where $\tilde{P}_j = P_j \circ Q_j$ denotes the composition of the two functions P_j and Q_j. This recovers single particle motion for density ρ_j defined on a delta function.

As we shall show, these singular solutions of EPSymp are also solutions of the Euler–Poincaré equations on the diffeomorphisms (EPDiff), provided one truncates to consider only first order moments [Holm and Marsden 2005]. With this truncation, the singular solutions in the case of single-particle dynamics reduce in one dimension to the pulson solutions for EPDiff [Camassa and Holm 1993].

5.2. Exchanging variables in EPSymp.

One can show that exchanging the variables $q \leftrightarrow p$ in the single particle PDF leads to another nontrivial singular solution of EPSymp, which is different from those found previously. To see this, let f be given by

$$f(q, p, t) = \sum_j \delta(q - Q_j(p, t)) \delta(p - P_j(t)).$$

At this stage nothing has changed with respect to the previous solution since the generating function is symmetric with respect to q and p. However, inserting this expression in the definition of the m-th moment yields

$$A_m(q,t) = \sum_j P_j^m \, \delta(q - Q_j(P_j,t)),$$

which is quite different from the solutions found previously. One again obtains a canonical Hamiltonian structure for P_j and Q_j.

This second expression is an alternative parametrisation of the cold-plasma reduction above and it may be useful in situations where the composition $Q_j \circ P_j$ is more convenient than $P_j \circ Q_j$.

5.3. Remarks about truncations. The problem presented by the coadjoint motion equation (4-4) for geodesic evolution of p-moments under EPDiff needs further simplification. One simplification would be to modify the (doubly) infinite set of equations in (4-4) by truncating the Poisson bracket to a finite set. These moment dynamics may be truncated at any stage by modifying the Lie-algebra in the KMLP bracket to vanish for weights $m + n - 1$ greater than a chosen cut-off value.

5.4. Examples of simplifying truncations and specializations. For example, if we truncate the sums to $m, n = 0, 1, 2$ only, then equation (4-4) produces the coupled system of partial differential equations

$$\frac{\partial A_0}{\partial t} = -\mathrm{ad}^*_{\beta_1} A_0 - \mathrm{ad}^*_{\beta_2} A_1,$$

$$\frac{\partial A_1}{\partial t} = -\mathrm{ad}^*_{\beta_0} A_0 - \mathrm{ad}^*_{\beta_1} A_1 - \mathrm{ad}^*_{\beta_2} A_2,$$

$$\frac{\partial A_2}{\partial t} = -\mathrm{ad}^*_{\beta_0} A_1 - \mathrm{ad}^*_{\beta_1} A_2.$$

Expanding now the expression of the coadjoint operation

$$\mathrm{ad}^*_{\beta_h} A_{k+h-1} = (k+h) \, A_{k+h-1} \partial_q \beta_h + h \beta_h \partial_q A_{k+h-1}$$

and relabeling

$$\mathrm{ad}^*_{\beta_h} A_k = (k+1) \, A_k \partial_q \beta_h + h \beta_h \partial_q A_k$$

one calculates

$$\frac{\partial A_0}{\partial t} = -\partial_q (A_0 \beta_1) - 2 A_1 \partial_q \beta_2 - 2 \beta_2 \partial_q A_1,$$

$$\frac{\partial A_1}{\partial t} = -A_0 \partial_q \beta_0 - 2 A_1 \partial_q \beta_1 - \beta_1 \partial_q A_1 - 3 A_2 \partial_q \beta_2 - 2 \beta_2 \partial_q A_2,$$

$$\frac{\partial A_2}{\partial t} = -2 A_1 \partial_q \beta_0 - 3 A_2 \partial_q \beta_1 - \beta_1 \partial_q A_2.$$

We specialize to the case that each velocity depends only on its corresponding moment, so that $\beta_s = G * A_s$, $s = 0, 1, 2$. If we further specialize by setting A_0 and A_2 initially to zero, then these three equations reduce to the single equation

$$\frac{\partial A_1}{\partial t} = -\beta_1 \, \partial_q A_1 - 2A_1 \, \partial_q \beta_1.$$

Finally, if we assume that G in the convolution $\beta_1 = G * A_1$ is the Green's function for the operator relation

$$A_1 = (1 - \alpha^2 \partial_q^2)\beta_1$$

for a constant lengthscale α, then the evolution equation for A_1 reduces to the integrable Camassa–Holm (CH) equation [1993] in the absence of linear dispersion. This is the one-dimensional EPDiff equation, which has singular (peakon) solutions. Thus, after these various specializations of the EPDiff p-moment equations, one finds the integrable CH peakon equation as a specialization of the coadjoint moment dynamics of equation (4-4).

That such a strong restriction of the p-moment system leads to such an interesting special case bodes well for future investigations of the EPSymp p-moment equations. Further specializations and truncations of these equations will be explored elsewhere. Before closing, we mention one or two other open questions about the solution behavior of the p-moments of EPSymp.

6. Open questions for future work

Several open questions remain for future work. The first is whether the singular solutions found here will emerge spontaneously in EPSymp dynamics from a smooth initial Vlasov PDF. This spontaneous emergence of the singular solutions does occur for EPDiff. Namely, one sees the singular solutions of EPDiff emerging from *any* confined initial distribution of the dual variable. (The dual variable is fluid velocity in the case of EPDiff). In fact, integrability of EPDiff in one dimension by the inverse scattering transform shows that *only* the singular solutions (peakons) are allowed to emerge from any confined initial distribution in that case [Camassa and Holm 1993]. In higher dimensions, numerical simulations of EPDiff show that again only the singular solutions emerge from confined initial distributions. In contrast, the point vortex solutions of Euler's fluid equations (which are isomorphic to the cold plasma singular solutions of the Vlasov Poisson equation) while comprising an invariant manifold of singular solutions, do not spontaneously emerge from smooth initial conditions in Euler fluid dynamics. Nonetheless, something quite analogous to the singular solutions is seen experimentally for cold plasma in a Malmberg–Penning trap [Dubin and O'Neil 1990]. Therefore, one may ask which outcome will prevail for the singular solutions of EPSymp. Will they emerge from a confined smooth initial distribution, or will they only exist as an invariant manifold for special initial conditions? Of course, the interactions of these singular solutions for various metrics and the properties of their collective dynamics is a question for future work.

Geometric questions also remain to be addressed. In geometric fluid dynamics, Arnold and Khesin [1998] formulate the problem of symplectohydrodynamics, the symplectic counterpart of ordinary ideal hydrodynamics on the special diffeomorphisms SDiff. In this regard, the work of Eliashberg and Ratiu [1991] showed that dynamics on the symplectic group radically differs from ordinary hydrodynamics, mainly because the diameter of $\text{Symp}(M)$ is infinite, whenever M is a compact exact symplectic manifold with a boundary. Of course, the presence of boundaries is important in fluid dynamics. However, generalizing a result by Shnirelman [1985], Arnold and Khesin point out that the diameter of $\text{SDiff}(M)$ is finite for any compact simply connected Riemannian manifold M of dimension greater than two.

In the case under discussion here, the situation again differs from that envisioned by Eliashberg and Ratiu. The EPSymp Hamiltonian (5-1) determines geodesic motion on $\text{Symp}(T^*\mathbb{R}^3)$, which may be regarded as the restriction of the $\text{Diff}(T^*\mathbb{R}^3)$ group, so that the Liouville volume is preserved. The main difference in our case is that $M = T^*\mathbb{R}^3$ is not compact, so one of the conditions for the Eliashberg–Ratiu result does not hold. Thus, one may ask, what are the geometric properties of Symp acting on a symplectic manifold which is not compact? What remarkable differences if any remain to be found between Symp and SDiff in such a situation? Another intriguing possibility is that some relation of the work here may be found with the work of Bloch et al. on integrable geodesic flows on the symplectic group [Bloch et al. 2005]. A final question of interest is whether the present work might be linked with the Lie algebra structure of the BBGKY hierarchy [Marsden et al. 1984].

Yet another interesting case occurs when the particles undergoing Vlasov dynamics are confined in a certain region of position space. In this situation, again the phase space is not compact, since the momentum may be unlimited. The dynamics on a bounded spatial domain descends from that on the unbounded cotangent bundle upon taking the p-moments of the Hamiltonian vector field. Thus, in this topological sense *p-moments and q-moments are not equivalent*. In the present work, this distinction has been ignored by assuming either homogeneous or periodic boundary conditions.

Acknowledgements. This work was prepared for a meeting at UC Berkeley in honor of Henry McKean, to whom we are grateful for interesting and encouraging discussions over many years. We are grateful to J. E. Marsden, G. Pavliotis and A. Weinstein for correspondence and discussions in this matter. Tronci is also grateful to the TERA Foundation for Oncological Hadrontherapy and in particular to the working group at CERN (Geneva, Switzerland) for their lively interest. We would like to thank the European Science Foundation for partial support through the MSGAM program. Holm's work was partially supported by a Royal Society Wolfson award and by the US Department of Energy Office of Science ASCR program in Applied Mathematical Research.

References

[Arnold and Khesin 1998] V. I. Arnold and B. A. Khesin, *Topological methods in hydrodynamics*, Applied Mathematical Sciences **125**, Springer, New York, 1998.

[Benney 1966] D. J. Benney, "Long non-linear waves in fluid flows", *J. Math. and Phys.* **45** (1966), 52–63.

[Bloch et al. 2005] A. M. Bloch, A. Iserles, J. E. Marsden, and T. S. Ratiu, "A class of integrable geodesic flows on the symplectic group and the symmetric matrices", preprint, 2005. Available at arXiv:math-ph/0512093.

[Camassa and Holm 1993] R. Camassa and D. D. Holm, "An integrable shallow water equation with peaked solitons", *Phys. Rev. Lett.* **71**:11 (1993), 1661–1664.

[Cendra et al. 1998] H. Cendra, D. D. Holm, M. J. W. Hoyle, and J. E. Marsden, "The Maxwell-Vlasov equations in Euler-Poincaré form", *J. Math. Phys.* **39**:6 (1998), 3138–3157.

[Cendra et al. 2003] H. Cendra, J. E. Marsden, S. Pekarsky, and T. S. Ratiu, "Variational principles for Lie-Poisson and Hamilton-Poincaré equations", *Mosc. Math. J.* **3**:3 (2003), 833–867.

[Dragt et al. 1990] A. J. Dragt, F. Neri, G. Rangarajan, D. R. Douglas, L. M. Healy, and R. D. Ryne, "Lie algebraic treatment of linear and nonlinear beam dynamics", *Ann. Rev. Nucl. Part. Sci.* **38** (1990), 455–496.

[Dubin and O'Neil 1990] D. H. E. Dubin and T. M. O'Neil, "Trapped nonneutral plasmas, liquids, and crystals (the thermal equilibrium states)", *Rev. Mod. Phys.* **71**:1 (1990), 87–172.

[Eliashberg and Ratiu 1991] Y. Eliashberg and T. Ratiu, "The diameter of the symplectomorphism group is infinite", *Invent. Math.* **103**:2 (1991), 327–340.

[Gibbons 1981] J. Gibbons, "Collisionless Boltzmann equations and integrable moment equations", *Phys. D* **3**:3 (1981), 503–511.

[Holm and Marsden 2005] D. D. Holm and J. E. Marsden, "Momentum maps and measure-valued solutions (peakons, filaments, and sheets) for the EPDiff equation", pp. 203–235 in *The breadth of symplectic and Poisson geometry*, Progr. Math. **232**, Birkhäuser, Boston, 2005.

[Holm et al. 1990] D. D. Holm, W. P. Lysenko, and J. C. Scovel, "Moment invariants for the Vlasov equation", *J. Math. Phys.* **31**:7 (1990), 1610–1615.

[Holm et al. 1998] D. D. Holm, J. E. Marsden, and T. S. Ratiu, "The Euler-Poincaré equations and semidirect products with applications to continuum theories", *Adv. Math.* **137**:1 (1998), 1–81.

[Kupershmidt and Manin 1978] B. A. Kupershmidt and J. I. Manin, "Long wave equations with a free surface, II: The Hamiltonian structure and the higher equations", *Funktsional. Anal. i Prilozhen.* **12**:1 (1978), 25–37.

[Lebedev and Manin 1979] D. R. Lebedev and Y. I. Manin, "Conservation laws and Lax representation of Benney's long wave equations", *Phys. Lett. A* **74**:3-4 (1979), 154–156.

[Marsden et al. 1984] J. E. Marsden, P. J. Morrison, and A. Weinstein, "The Hamiltonian structure of the BBGKY hierarchy equations", pp. 115–124 in *Fluids and plasmas: geometry and dynamics* (Boulder, CO, 1983), edited by J. E. Marsden, Contemp. Math. **28**, Amer. Math. Soc., Providence, RI, 1984.

[Morrison 1980] P. J. Morrison, "The Maxwell–Vlasov equations as a continuous Hamiltonian system", *Phys. Lett. A* **80**:5-6 (1980), 383–386.

[Shnirelman 1985] A. I. Shnirelman, "The geometry of the group of diffeomorphisms and the dynamics of an ideal incompressible fluid", *Mat. Sb. (N.S.)* **128**:1 (1985), 82–109.

JOHN GIBBONS
DEPARTMENT OF MATHEMATICS
IMPERIAL COLLEGE LONDON
LONDON SW7 2AZ
UNITED KINGDOM
    j.gibbons@ic.ac.uk

DARRYL D. HOLM
DEPARTMENT OF MATHEMATICS
IMPERIAL COLLEGE LONDON
LONDON SW7 2AZ
UNITED KINGDOM
AND
COMPUTER AND COMPUTATIONAL SCIENCE DIVISION
LOS ALAMOS NATIONAL LABORATORY
LOS ALAMOS, NM, 87545
UNITED STATES
    d.holm@imperial.ac.uk, dholm@lanl.gov

CESARE TRONCI
DEPARTMENT OF MATHEMATICS
IMPERIAL COLLEGE LONDON
LONDON SW7 2AZ
UNITED KINGDOM
AND
TERA FOUNDATION FOR ONCOLOGICAL HADRONTHERAPY
11 VIA PUCCINI
NOVARA 28100
ITALY
    cesare.tronci@ic.ac.uk

Probability, Geometry and Integrable Systems
MSRI Publications
Volume **55**, 2008

Reality problems in the soliton theory

PETR G. GRINEVICH AND SERGEI P. NOVIKOV

Dedicated to Henry McKean

ABSTRACT. This is a survey article dedicated mostly to the theory of real regular finite-gap (algebro-geometrical) periodic and quasiperiodic sine-Gordon solutions. Long period this theory remained unfinished and ineffective, and by that reason practically had no applications. Even for such simple physical quantity as topological charge no formulas existed expressing it through inverse spectral data. A few years ago the present authors solved this problem and made this theory effective. This article contains description of the history and recent achievements. It describes also the reality problems for several other fundamental soliton systems.

1. Introduction

The most powerful method for constructing explicit periodic and quasiperiodic solutions of soliton equations is based on the finite-gap or algebro-geometric approach, developed by Novikov [1974], Dubrovin et al. [1976b], Its and Matveev [1975], Lax [1975], and McKean and van Moerbeke [1975] for $1 + 1$ systems, and extended by Krichever in 1976 for $2 + 1$ systems like KP. Already in 1976 new ideas were formulated on how to extend this approach to the $2 + 1$ systems associated with the spectral theory of the 2D Schrödinger operator restricted to one energy level; see [Manakov 1976; Dubrovin et al. 1976a]. These ideas were developed in 1980s by several people in Moscow's Novikov Seminar, as discussed see below. The "spectral data" characterizing the associated Lax-type operators consist of a Riemann surface (spectral curve)

Petr Grinevich was supported by the RFBR grant 05-01-01032-a and by the grant NSh-4182.2006.1 of the Presidential Council on Grants (Russia).

221

equipped with a selected set of points (divisor of poles, infinities). In the finite gap case this Riemann surface has finite genus, and the number of selected point is also finite. The algebro-geometric approach in particular allows one to write down explicit solutions in terms of the Riemann θ functions.

In modern literature very often the problem is assumed to be more or less completely solved if such formulas are derived. However, in some cases this belief is too naive and does not correspond to the needs of real life. It is often necessary to select physically or geometrically relevant classes of solutions corresponding to the source problem: for instance solutions satisfying certain reality conditions, or regular solutions, or bounded solutions. Is this easy or hard?

To reach this goal, the following problems must be solved.

- Problem 1. How to select solutions that are real for real (x, t).
- Problem 2. How to select real nonsingular solutions.
- Problem 3. How to select periodic solutions with a given period (or quasi-periodic solutions with a given group of quasiperiods).

REMARK. We call a solution *nonsingular* if it is nonsingular on the whole real Abel torus. It should remain nonsingular under action of all (real) higher flows from the corresponding integrable hierarchy. The generic x-direction is normally ergodic in the Abel torus, so this definition is equivalent to the standard one. However, for some specific values of constants of motion theoretically we may have solutions, which are regular in the standard sense, but blow-up under the action of the higher symmetries.

For some models like the Korteweg–de Vries equation (KdV), the defocusing noninear Schrödinger equation (NLS) and the Kadomtsev–Petviashvili 2 (KP2) equation, the selection of real and nonsingular solution is straightforward. But for many other models such as KP1, the focusing NLS, the sine-Gordon equation (SG), and the inverse scattering transformation for the Schrödinger operator based at one energy, the problem of selecting real solutions is difficult.

The theory of θ functions is complicated and ineffective. The complexity is hidden behind the simple notations in these formulas.

Our goal is to discuss in more detail the sine-Gordon equation

$$u_{tt} - u_{xx} + \sin u(x, t) = 0. \tag{1-1}$$

In the light-cone coordinates

$$x = 2(\xi + \eta), \quad t = 2(\xi - \eta), \tag{1-2}$$

it has the form

$$u_{\xi\eta} = 4 \sin u, \quad u = u(\xi, \eta). \tag{1-3}$$

According to our definition, the solution $u(x,t)$ is x-periodic with period T if $\exp\{iu\}$ is x-periodic with that period. For the function u we have

$$u(x+T,t) = u(x,t) + 2\pi n, \quad n \in \mathbb{Z}.$$

We call the quantity n the *topological charge* corresponding to the period T.

We call the ratio n/T the *density of topological charge*.

The density of topological charge can be naturally extended to all real generic regular finite-gap (quasiperiodic) solutions. It is the most basic conservation law.

PROBLEM. *Calculate the topological charge of real finite-gap solutions in terms of spectral data.*

We recall that the inverse scattering (spectral) data for the KdV and sine-Gordon systems consist of a Riemann surface (spectral curve) Γ of finite genus g and a collection of points (divisor) $D = \gamma_1 + \cdots + \gamma_g$. (For the NLS and some other systems number of poles may be different from genus.)

In the case of KdV or the finite-gap periodic Schrödinger operator $L = -\partial_x^2 + u(x)$, this surface Γ is hyperelliptic. In the case of the sine-Gordon equation the surface is also hyperelliptic, $\mu^2 = \lambda \prod_{i=1}^{2g}(\lambda - \lambda_i)$, with branching points $(0, \lambda_1, \ldots, \lambda_{2g}, \infty)$. However the classes of admissible Riemann surfaces and divisors for KdV and sine-Gordon are dramatically different, as we shall see below.

The θ-functional formulas for sine-Gordon were obtained in [Kozel and Kotlyarov 1976; Its and Kotlyarov 1976]. The reality problem remained unsolved. Indeed, the class of admissible Riemann surfaces was found in these works; see [Its and Kotlyarov 1976]. The nonzero finite branching points $(\lambda_1, \ldots, \lambda_{2g})$ can be either real negative $(\lambda_1, \ldots, \lambda_{2k}) \in \mathbb{R}$ or complex conjugate with nonzero imaginary part $\lambda_{2k+1} = \bar{\lambda}_{2k+2}, \ldots, \lambda_{2g-1} = \bar{\lambda}_{2g}$. However, no ideas were proposed where the poles are located on the Riemann surface.

In the early 1980s it was realized that this problem is nontrivial; see [McKean 1981; Dybrovin and Novikov 1982; 1982; Ercolani and Forest 1985; Ercolani et al. 1984]. For this reason, periodic finite-gap sine-Gordon theory lacked applications for a long time.

An important idea for how to describe position of poles for the real nonsingular solutions was in fact suggested by Cherednik [1980]. He was the first author who discovered (ineffectively) that for the given admissible real Riemann surface there can be many different real Abel tori generating real nonsingular quasiperiodic solutions. Their number is equal to 2^k where $2k$ is the number of negative real branching points. All real finite-gap solutions are nonsingular for sine-Gordon for the generic Riemann surface. His work was written in the abstract algebro-geometric form, and he never developed these ideas later.

Extending Cherednik's approach on the basis of "algebro-topological" ideas, Dubrovin and Novikov [1982] presented an interesting idea for how to calculate topological charge in terms of the "inverse spectral data". However, as pointed out in [Novikov 1984], there was a mistake in the argument; the proof of the formula proposed in [Dybrovin and Novikov 1982] worked only for a small neighborhood of some very special solutions. The problem remained open till 2001. The complete solution, confirming the Dubrovin–Novikov formula, was obtained in [Grinevich and Novikov 2001] as a development of the algebro-topological approach suggested in [Dybrovin and Novikov 1982]; see also [Grinevich and Novikov 2003a; 2003b]. In [Dubrovin and Natanzon 1982] and [Ercolani and Forest 1985], these components were described as the real subtori in the Jacobian variety $J(\Gamma)$. However this "θ-functional description", which does not involve a specific basis of cycles, did not lead to a formula for the topological charge. As we know now, a good formula for the topological charge can only be written in a very specific basis. We believe that using this basis of cycles one can deduce our formula from the θ-functional expression. It would be good to do that.

2. Physically relevant classes of solutions for the different soliton systems

The Korteweg–de Vries (KdV) equation

$$u_t + u_{xxx} - 6uu_x = 0, \quad u = u(x, t), \tag{2-1}$$

was originally derived in the theory of water waves. As discovered in early 1960s (see the introduction to [Novikov et al. 1984]), it naturally appears as a first nonvanishing correction for the dispersive nonlinear systems if dissipation can be neglected. In these models only real nonsingular solution are physically relevant.

Integration of the KdV equation is based on the "inverse scattering transform" for the one-dimensional Schrödinger operator

$$L = -\partial_x^2 + u(x, t). \tag{2-2}$$

The selection of real KdV solutions is straightforward.

(1) The spectral curve Γ defined by $\mu^2 = R_{2g+1}(\lambda)$ should be real. This means that $R_{2g+1}(\lambda) = \lambda^{2g+1} + \sum_{i=0}^{2g} p_i \lambda^i$ has real coefficients $p_i \in \mathbb{R}$; equivalently, all roots are either real or form complex conjugate pairs. Therefore we have a holomorphic involution $\tau : (\lambda, \mu) \rightarrow (\bar{\lambda}, -\bar{\mu})$ on Γ.

(2) The divisor D should be real with respect to τ: $\tau D = D$, or equivalently, the unordered set of points $\gamma_1, \ldots, \gamma_g$ is invariant with respect to τ. Of course, τ may interchange some of them.

Real nonsingular KdV solutions correspond to the following special spectral data:

(1) All branching points of λ_k of Γ are real and distinct. Assume that $\lambda_1 < \lambda_2 < \cdots < \lambda_{2g+1}$. Then τ has exactly $g + 1$ real ovals over the intervals $a_0 = (-\infty, \lambda_1]$, $a_1 = [\lambda_2, \lambda_3]$, $\ldots$, $a_g = [\lambda_{2g}, \lambda_{2g+1}]$.
(2) Each finite oval a_k, $1 \le k \le g$ contains exactly one divisor point $\gamma_k \in a_k$.

REMARK. A real curve of genus g may have at most $g + 1$ real oval. Curves with $g + 1$ real ovals (the greatest possible number) are called M-curves.

Generic finite-gap solutions are quasiperiodic with g incommensurable periods. How to select x-periodic solutions with prescribed period T? Avoiding any use of algebraic geometry and Riemann surfaces, a nice approach to the characterization of the strictly x-periodic solution in terms of the so-called quasimomentum map was developed by Marchenko and Ostrovskii [1975]. This map was studied in the quantum solid state physics literature in 1959 (see [Kohn 1959]). It is well-defined in the upper half-plane outside of some vertical edges. Its analytical properties were effectively used in [Marchenko and Ostrovskii 1975]. For example the approximation of x-periodic solution (potential) by the finite-gap ones periodic with the same period, was proved. Another approach, based on isoperiodic deformations of finite-gap potentials, was developed by Grinevich and Schmidt in 1995 [Grinevich and Schmidt 1995]. In the KdV case the isoperiodic deformations can be interpreted as the so-called Loewner equations for the corresponding conformal map. We point out that there exists a big literature, dedicated to the KdV solutions with real poles (rational solutions, singular trigonometric and elliptic solutions) — see [Airault et al. 1977], where these ideas were started. These solutions are very important from the mathematical point of view: for example, the dynamics of poles satisfies to the equations of the rational and elliptic Moser–Calogero models respectively. However, they are related neither to nonlinear wave problems nor to the spectral theory of the corresponding Schrödinger operators. So we do not discuss this literature in the present survey article.

The modified Korteweg-de Vries equation has the form:

$$v_t + v_{xxx} - 6v^2 v_x = 0, \quad v = v(x, t). \tag{2-3}$$

It is connected with KdV by the Miura transformation:

$$u(x, t) = v_x(x, t) + v^2(x, t). \tag{2-4}$$

The real nonsingular solutions are physically relevant.

The "complex" nonlinear Schrödinger equation (NLS) is a system of equations for the pair of independent complex functions $q = q(x, t)$, $r = r(x, t)$:

$$iq_t + q_{xx} + 2q^2 r = 0,$$
$$-ir_t + r_{xx} + 2r^2 q = 0. \tag{2-5}$$

This system has two natural real reductions: the defocusing NLS, with $r(x, y) = -\overline{q(x, y)}$, hence

$$iq_t + q_{xx} - 2|q|^2 q = 0, \tag{2-6}$$

and the self-focusing NLS, with $r(x, y) = \overline{q(x, y)}$, hence

$$iq_t + q_{xx} + 2|q|^2 q = 0. \tag{2-7}$$

These equation describes nonlinear media with dispersion relations depending on the square of the wave amplitude (see [Novikov et al. 1984]). Among the todays applications of NLS is the theory of light propagation in the fiber optics. The sign $+$ or $-$ is determined by the dispersion relation, and the qualitative behavior critically depends on it. From the mathematical point of view, the defocusing NLS system is much simpler because the linear Lax operator is self-adjoint. The focusing NLS is much more complicated. In both cases physical applications requires regular solutions.

The complex NLS spectral data are following: A hyperelliptic Riemann surface Γ with $2g+2$ finite branching points $\lambda_1, \ldots, \lambda_{2g+2}$ and $g+1$ divisor points $D = \gamma_1 + \cdots + \gamma_{g+1}$. In contrast with the KdV case, there is no branching at ∞.

Solutions of the defocusing NLS correspond to the following spectral data:

(1) Γ is real, i.e. the polynomial $R_{2g+2} = \prod_{k=1}^{2g+2}(\lambda - \lambda_k)$ has real coefficients. Γ is defined by $\mu^2 = R(\lambda)$. The antiholomorphic involution on Γ is defined by the map $\tau : \mathbb{C}^2 \to \mathbb{C}^2$ where $(\lambda, \mu) \to (\bar{\lambda}, -\bar{\mu})$.

(2) The divisor D is real with respect to τ: $\tau D = D$.

Selection of regular solutions is also very similar to the KdV case

(1) All branching points of Γ are real. Therefore Γ has $g + 1$ real ovals over the intervals $[\lambda_{2k-1}, \lambda_{2k}]$, $k = 1, \ldots, g+1$; that is, Γ is an M-curve.

(2) There is exactly one divisor point at each real oval.

The selection of x-periodic solutions is completely analogous to the KdV case.

We describe the data generating real solutions of the self-focusing NLS equations for regular spectral curves. By Cherednik's theorem [1980], these solutions are automatically nonsingular. The solutions corresponding to singular spectral curves can be obtained as proper degenerations. In contrast with the defocusing case, singular curves may generate regular x-quasiperiodic solutions.

(1) Γ is a real hyperelliptic surface of genus g with $2g+2$ $\lambda_1 < \lambda_2 < \cdots < \lambda_{2g+2}$ finite branching points. There are no branching points on the real line, so they form complex conjugate pairs. The antiholomorphic involution τ acts on the λ-plane as $\tau\lambda = \bar{\lambda}$. The points in Γ lying over the real line are invariant with respect to τ. Equivalently, $\tau : (\lambda, \mu) \to (\bar{\lambda}, \bar{\mu})$.

(2) There exists a meromorphic differential Ω satisfying these conditions:

- $\Omega = (1 + o(1))d\lambda$ at the infinite points of Γ.

- Ω is regular outside infinity. Therefore it has exactly $2g + 2$ zeroes.

- Let $D = \gamma_1 + \cdots + \gamma_{g+1}$. Then the divisor of zeroes of Ω is $D + \tau D$. Therefore, $D + \tau D = 2\infty_1 + 2\infty_2 - K$.

The sine-Gordon equation in the light-cone variables was derived in the end of the nineteenth century. It describes immersions of the negative curvature surfaces into $\mathbb{R}^3$. Assume that an asymptotic coordinate system is chosen (a coordinate system such that coordinate lines have zero normal curvature). The angle between the coordinate lines satisfy (1-3). This means that only real regular solutions such that $u(x, t) \neq 0$ mod π are relevant.

The sine-Gordon equation describes also dynamics of the Josephson junctions. In this model $u(x, t)$ is the phase difference between the contacts, therefore the real nonsingular solutions are relevant. However, according to the leading experts in the Superconductivity Theory, the problem always requires boundary problem, so we have to consider either the finite interval or the half-line.

The elliptic sinh-Gordon equation

$$u_{xx} + u_{yy} + 4H \sinh u = 0. \tag{2-8}$$

describes the constant mean curvature surfaces with genus equal to one, outside umbilic points (see the review in [Bobenko 1991]). The constant mean curvature tori have no umbilic points, therefore real nonsingular solutions should be selected. In contrast with soliton equations, all real smooth double-periodic solutions are automatically finite-gap here [Hitchin 1988; Pinkall and Sterling 1989]. This is a consequence of the following observations by Hitchin [1988]: all isospectral flows from the corresponding hierarchy are zero eigenfunctions of the linearized problem. But the linearized system is the two-dimensional (elliptic) Schrödinger operator, and it may have only finite-dimensional space of double-periodic zero eigenfunctions. This means that the hierarchy contains only finitely many linearly independent flows at this point. As a corollary the spectral curve has finite genus. A further development of this idea was used by Novikov and Veselov [1997], who showed that all periodic chains of Laplace transformations consisting of the two-dimensional double-periodic Schrödinger operators with regular coefficients are algebro-geometric (2D analogs of finite-gap operators).

The Boussinesq equation

$$u_t = \eta_x,$$
$$\eta_t = -\tfrac{1}{3}u_{xxx} + \tfrac{4}{3}uu_x \tag{2-9}$$

is used for describing the water waves. For physical applications it is necessary to select real nonsingular solutions. We point out that the problem of selecting such solutions in terms of the finite-gap data remains open.

The Kadomtsev–Petviashvili (KP) equation

$$(u_t + u_{xxx} - 6uu_x)_x + 3\alpha^2 u_{yy} = 0, \quad u = u(x, y, t), \quad \alpha^2 \in \mathbb{R}. \tag{2-10}$$

The auxiliary linear operator for KP has the form

$$L = \alpha \partial_y - \partial_x^2 + u(x, y, t). \tag{2-11}$$

If α is imaginary, we have the so-called KP1 equation, and L is the one-dimensional nonstationary Schrödinger operator. If α is real, L is the parabolic operator. In both cases the real nonsingular solutions are physically relevant only. The necessary and sufficient conditions for the finite-gap spectral data selecting the real nonsingular solutions were found by Dubrovin and Natanzon [1988].

Real nonsingular solutions of the KP-2 equation correspond to the following geometry:

(1) Γ is a algebraic surface of genus g with a marked point and an antiholomorphic involution τ such that the marked point is invariant under the action of τ. The marked point is the essential singularity of the wave function.

(2) τ has exactly $g + 1$ fixed oval, i.e. Γ in an M-curve with respect to τ. Denote the oval containing the essential singularity by a_0 and the other ovals by a_n, $n = 1, \ldots, g$.

(3) Each oval a_n, $n \neq 0$ contains exactly one divisor point.

In the case of the Kadomtsev–Petviashvili 1 equation the reality constraints on the spectral curve are exactly the same as in the KP 2 case, but the divisor D has a completely different description: There exists a meromorphic differential Ω with exactly one second-order pole located at the marked point such that the divisor of zeroes of Ω is exactly $D + \tau D$. Equivalently, $D + \tau D = 2\infty - K$, where ∞ denotes the marked point. Regular real solutions are generated by the data with the following extra constraint:

The pair (Γ, τ) is of separating type, i.e. after removing all real ovals Γ splits into 2 components.

An important example of "solvable" inverse spectral transform is the one-energy problem for the two-dimensional Schrödinger operator started in [Manakov 1976; Dubrovin et al. 1976a].

$$L = -\partial_x^2 - \partial_y^2 + u(x, y), \qquad (2\text{-}12)$$

It is well-known that the full set of scattering data for multidimensional Schrödinger operators $n > 1$ is overdetermined. A lot of people have studied this problem; we won't even quote this literature. However, the case $n = 2$ turned out to be very specific. Manakov, Dubrovin, Krichever and Novikov [Manakov 1976; Dubrovin et al. 1976a] started a completely new approach for this specific case, creating inverse scattering theory and the corresponding soliton theory associated with one selected energy level. A lot of work has been done since; see [Novikov and Veselov 1984; Veselov et al. 1985; Grinevich and Novikov 1988] and the review [Grinevich 2000] for additional references. In particular, in the first work [Dubrovin et al. 1976a] they defined the natural analogs of finite-gap potentials for the two-dimensional problem as the potentials, "finite-gap at one energy". Let $u(x, y)$ be double-periodic. Denote the dispersion relation by $\varepsilon_j(k_x, k_y)$. The Fermi-curve at the energy level E_0 is defined by:

$$\varepsilon_j(k_x, k_y) = E_0. \qquad (2\text{-}13)$$

Denote the complex continuation of the Fermi curve by Γ. The potential $u(x, y)$ is called *finite-gap at one energy* if Γ has finite genus.

For generic spectral data the operators constructed in [Dubrovin et al. 1976a] have generically a nonzero magnetic field, i.e. they have some extra first-order terms:

$$L = -\partial_x^2 - \partial_y^2 + A_1(x, y)\partial_x + A_2(x, y)\partial_y + u(x, y), \qquad (2\text{-}14)$$

It might happen that $H(x, y) \neq 0$, where $H(x, y) = \partial_x A_2(x, y) - \partial_y A_1(x, y)$. For physical applications it is important to select the case of "potential operators" $A_1(x, y) = A_2(x, y) = 0$ with real potential $u(x, y)$. Sufficient conditions on the spectral data leading to the potential operators were found by Novikov and Veselov [1984]. For double periodic potentials the existence of such forms is necessary; this follows from the direct spectral theory, developed by Krichever [1992]. For the generic regular quasiperiodic potentials "finite-gap for one energy level", this problem remains open. Selection of real potentials here is simple.

How to select the class of regular potentials in terms of algebro-geometrical spectral data? There is no complete solution to this problem. It was shown in [Novikov and Veselov 1984] that if the spectral curve is the so-called M-curve, then the potential $u(x, y)$ is regular, and the operator L is strictly positive (the selected energy level lies below the ground state). An alternative proof of the last statement was obtained by the authors in [Grinevich and Novikov 1988]. The complete characterization of the data generating strictly positive operators (with real regular potentials) was "more or less" clarified but some special features remain unproved rigorously.

If the selected energy level is located above the ground state, the topology of the spectral curve Γ become more complicated. Many classes of spectral data generating real nonsingular solutions were found by Natanzon (see the review in [Natanzon 1995]), but the classification is not complete till now.

3. The sine-Gordon equation

Connections between the sine-Gordon equation and the inverse scattering method were first established by G. Lamb [1971]. The modern approach was started by Ablowitz, Kaup, Newell and Segur [Ablowitz et al. 1973]. It is based on the following zero-curvature representation:[1]

$$\Psi_x = \tfrac{1}{4}(U + V)\Psi, \quad \Psi_t = \tfrac{1}{4}(U - V)\Psi, \tag{3-1}$$

where

$$U = U(\lambda, x, t) = \begin{bmatrix} i(u_x + u_t) & 1 \\ -\lambda & -i(u_x + u_t) \end{bmatrix}, \tag{3-2}$$

$$V = V(\lambda, x, t) = \begin{bmatrix} 0 & -\tfrac{1}{\lambda}e^{iu} \\ e^{-iu} & 0 \end{bmatrix}. \tag{3-3}$$

As we mentioned above, the finite-gap spectral data consist of

(1) a hyperelliptic Riemann surface Γ defined by $\mu^2 = \lambda \prod_{i=1}^{2g}(\lambda - \lambda_i)$, with $2g + 2$ branching points $(0, \lambda_1, \ldots \lambda_{2g}, \infty)$; and
(2) the divisor (a collection of points) $D = \gamma_1 + \cdots + \gamma_g$ in Γ.

In our text we always assume that the spectral curve Γ is *generic*, that is, all branching points are distinct.

The construction of complex sine-Gordon solutions is based on the following standard Lemma:

LEMMA 1. *For generic data Γ, D there exists a unique two-component vector-function $\Psi(\gamma, x, t)$ (the Baker–Akhiezer functions) such that:*

(1) *For fixed (x, t) the function $\Psi(\gamma, x, t)$ is meromorphic in the variable $\gamma \in \Gamma$ outside the points 0, ∞ and has at most 1-st order poles at the divisor points γ_k, $k = 1, \ldots, g$.*

---

[1]The zero-curvature representation for the sine-Gordon equation presented here was, in fact, first written by these authors in subsequent works.

(2) $\Psi(\gamma, x, t)$ has essential singularities at the points 0, ∞ with the following asymptotic:

$$\Psi(\gamma, x, t) = \begin{pmatrix} 1 + o(1) \\ i\sqrt{\lambda} + O(1) \end{pmatrix} e^{\frac{i\sqrt{\lambda}}{4}(x+t)} \quad as \ \lambda \to \infty, \tag{3-4}$$

$$\Psi(\gamma, x, t) = \begin{pmatrix} \phi_1(x, t) + o(1) \\ i\sqrt{\lambda}\phi_2(x, t) + O(\lambda) \end{pmatrix} e^{-\frac{i}{4\sqrt{\lambda}}(x-t)} \quad as \ \lambda \to 0, \tag{3-5}$$

with some $\phi_1(x, t)$, $\phi_2(x, t)$.

The sine-Gordon potential $u(x, t)$ is defined by

$$u(x, t) = i \ln \frac{\phi_2(x, t)}{\phi_1(x, t)}. \tag{3-6}$$

We denote by $\lambda_k(x, t)$ the projections of the zeroes of the first component of $\Psi(\gamma, x, t)$ to the λ-plane. Then

$$e^{iu(x,t)} = \prod_{j=0}^{g} (-\lambda_j(x, t)) \bigg/ \sqrt{\prod_{j=1}^{2g} E_j}. \tag{3-7}$$

REMARK. To be more precise, formulas (3-1)–(3-7) define simultaneously a pair of sine-Gordon solutions $u_1(x, t)$, $u_2(x, t)$, depending on the choice of the branch $1/\sqrt{(\lambda)}$ near the point $\lambda = 0$. They are connected by the following relation $u_2(x, t) = u_1(t, x) + \pi$. In the real case it is possible to fix a canonical branch by making the analytical continuation along the real line. This rule is unstable in the following sense: if we add a pair of complex conjugate branching points which are very close to the positive half-line (or, equivalently, open a resonant point), it is a small transformation in terms of the spectral data, but it exchanges u_1 with u_2.

The real sine-Gordon solutions (by Cherednik's lemma they are automatically regular [Čerednik 1980]) correspond to the following data:

(1) Γ is real, i.e. the branching points of Γ are either real, or form complex conjugate pairs. Therefore we have an antiholomorphic involutions $\tau : (\lambda, \mu) \to (\bar{\lambda}, \bar{\mu})$. Denote the number of real finite branching points by $2k + 1$.
(2) All real branching points lie in the negative half-line $\lambda \leq 0$. It is convenient to use following enumeration for the branching points different from 0 and ∞: $0 > \lambda_1 > \lambda_2 > \cdots > \lambda_{2k}, \lambda_{2k+1} = \bar{\lambda}_{2k+2}, \ldots, \lambda_{2g-1} = \bar{\lambda}_{2g}$.
(3) There exists a meromorphic differential Ω (Cherednik differential) with first order poles at 0, ∞, holomorphic on $\Gamma \backslash \{0, \infty\}$ with the zeroes at the points $\gamma_1, \ldots, \gamma_g, \tau\gamma_1, \ldots, \tau\gamma_g$ (or, equivalently the divisor D satisfy the relation $D + \tau D = 0 + \infty - K$).

As shown in [Čerednik 1980], the variety of all real potentials corresponding to the given spectral curve Γ consists of 2^k connected components. A characterization of these components in terms of the Abel tori was obtained in [Dubrovin and Natanzon 1982] but this technique did not led to the calculation of topological charge through the inverse spectral data.

Our calculation of the topological charge for the finite-gap sine-Gordon solutions is based on the following effective description of these components, for details of which see [Grinevich and Novikov 2001; 2003a; 2003b]:

Any meromorphic differential with first-order pole at ∞ can be written as

$$\Omega = c \left(1 - \frac{\lambda P_{g-1}(\lambda)}{R(\lambda)^{1/2}} \right) \frac{d\lambda}{2\lambda}, \tag{3-8}$$

where $P_{g-1}(\lambda)$ is a polynomial of degree at most $g - 1$. It is also natural to put $c = 1$. In case of the Cherednik differentials the set of zeroes is invariant with respect to τ. Therefore all coefficients of the polynomial $P_{g-1}(\lambda)$ are real.

Take an arbitrary real polynomial $P_{g-1}(\lambda)$. Is it possible to construct a real sine-Gordon solution corresponding to it? The necessary and sufficient condition is this: *the zeroes of Ω can be divided into two groups, $\{\gamma_1, \ldots, \gamma_g\}$ and $\{\gamma_{g+1}, \ldots, \gamma_{2g}\}$, such that $\tau \gamma_k = \gamma_{k+g}$, $k = 1, \ldots, g$.* Equivalently, a polynomial $P_{g-1}(\lambda)$ generates real SG solutions if and only if all real root of Ω have even multiplicity. In generic situation (all roots form distinct complex conjugate pairs) each polynomial $P_{g-1}(\lambda)$ generates 2^g different solutions. To choose one of them one has to say, which point to choose in each complex conjugate pair belonging to D (the second one belongs to τD). In degenerate cases (i.e. if there are real roots) the number of choices is smaller. Al these solutions associated with a given $P_{g-1}(\lambda)$ belong to the same real Abel torus.

DEFINITION. A polynomial $P_{g-1}(\lambda)$ (and the corresponding differential Ω) are called *admissible* if all real roots of Ω have even multiplicity.

Admissible polynomials $P_{g-1}(\lambda)$ can be characterized geometrically. We start by taking the graph of the functions

$$f_\pm(\lambda) = \pm \frac{\sqrt{R(\lambda)}}{\lambda}, \tag{3-9}$$

and coloring in black the domains

$$\lambda < 0, \ y^2 < \frac{R(\lambda)}{\lambda^2} \quad \text{and} \quad \lambda > 0, \ y^2 > \frac{R(\lambda)}{\lambda^2}, \tag{3-10}$$

as in Figure 1.

LEMMA 2. *The polynomial $P_{g-1}(\lambda)$ is admissible if and only if the graph of $P_{g-1}(\lambda)$ has no parts inside the black open domains.*

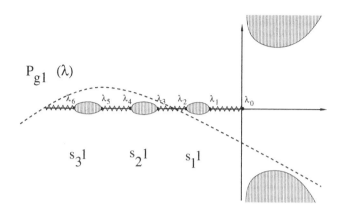

Figure 1.

If the graph does not touch these domains, we have no real divisor points. Real divisor points correspond to the case when the graph touches one of these domains but does not cross the boundary.

Each pair $\lambda_{2j-1}, \lambda_{2j}$ is connected by a black island. The graph of admissible $P_{g-1}(\lambda)$ should go above or below this island, so $P_{g-1}(\lambda) \neq 0$ on at all intervals $[\lambda_{2j}, \lambda_{2j-1}]$, $j \leq k$. We associate with an admissible polynomial $P_{g-1}(\lambda)$ a collection of numbers s_j, $j = 1, \ldots, k$ by the following rule: $s_j = 1$ if the graph of $P_{g-1}(\lambda)$ is positive in the interval $[\lambda_{2j}, \lambda_{2j-1}]$, and $s_j = -1$ otherwise. We call the set s_j the *topological type of the real solution*. There are exactly 2^k possible topological types. Elementary analytic estimates (see [Grinevich and Novikov 2003a]) show that all these components are nonempty. Each connected component is a real Abel torus, on which the x-dynamics defines a straight line. To calculate the density of the topological charge it is sufficient to know the direction of this line and the charges along the basic cycles. This follows from a simple analytic lemma:

LEMMA 3. *Let $u(\vec{X})$, $X \in \mathbb{R}^n$ be a smooth function in $\mathbb{R}^n$ such that $\exp(i u(\vec{X}))$ is single-valued on the torus $\mathbb{R}^n / \mathbb{Z}^n$. Equivalently, we have $\exp(i u(\vec{X} + \vec{N})) = \exp(i u(\vec{X}))$ for any integer vector $\vec{N}$, and*

$$u(X^1, X^2, \ldots, X^k + 1, \ldots, X^n) - u(X^1, X^2, \ldots, X^k, \ldots, X^n) = 2\pi n_k.$$

The numbers n_k are called the topological charges along the basic cycles $\mathfrak{A}_k$, $k = 1, \ldots, n$. Denote by $u(x)$ restriction of $u(\vec{X})$ to the strait line $\vec{X} = \vec{X}_0 + x \cdot \vec{v}$, $\vec{v} = (v^1, v^2, \ldots, v^n)$. Then the density of topological charge

$$\bar{n} := \lim_{T \to \infty} \frac{u(x + T) - u(x)}{2\pi T}$$

is well-defined; it does not depend on the point $\vec{X}_0$, and

$$\bar{n} = \sum_{k=1}^{n} n_k v^k.$$ (3-11)

The calculation of the direction vector for the x-dynamics is standard; see [Ercolani and Forest 1985], for example. Denote by ω^l the canonical basis of holomorphic differentials on Γ:

$$\omega^l = i \frac{\sum_{j=0}^{g-1} D_j^k \lambda^j}{\sqrt{R(\lambda)}} \, d\lambda, \quad D_j^k \in \mathbb{R}$$ (3-12)

Then for the components of the x-direction vector we have

$$U_k = \frac{1}{2}\left(D_{g-1}^k + D_0^k \Big/ \sqrt{\prod_{j=1}^{2g} E_j} \right).$$ (3-13)

To obtain a simple expression for the basic charges it is critical to use a proper basis of cycles in Γ. In [Grinevich and Novikov 2001; 2003a; 2003b] the authors used the following basis, first suggested in [Dybrovin and Novikov 1982]:

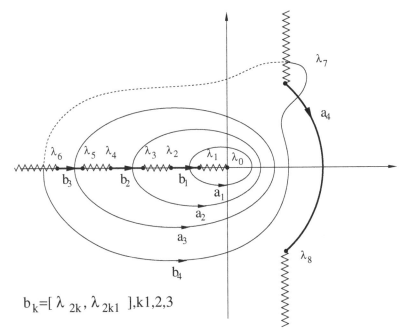

$$b_k = [\lambda_{2k}, \lambda_{2k1}], k1,2,3$$

Figure 2.

Here the cycles a_j, $j = 1, \ldots, k$ are ovals on the upper sheet of Γ, containing inside the points $\lambda_0 = 0, \lambda_1, \lambda_2, \ldots, \lambda_{2j-1}$. The cycle b_j, $1 \leq j \leq k$, lies over the interval $[\lambda_{2j}, \lambda_{2j-1}]$. The cycles a_j, $j = k+1, \ldots, g$, lie over paths connecting the pairs λ_{2j-1} and λ_{2j}. We assume that these cycles do not intersect each other, and the cycles a_j, $j = k+1, \ldots, g$, do not intersect the negative half-line. The cuts are shown by the zigzag lines. The upper sheet contains the half-line $\lambda > 0$, $\mu > 0$.

Consider a basic cycle $\mathfrak{A}_j$ on the real component of Jacoby torus, represented by the closed curve. The image of this cycle in Γ under the inverse Abel map is a closed oriented curve C_j, formed by the motion of the corresponding divisor points (it may have several connected components). The motion of an individual divisor point does not have to be periodic, after going along the cycle we may obtain a permutation of the divisor points. The curve C_j is homological to the cycle $a_j \in H_1(\Gamma, Z)$. It follows from (3-7) that the topological charge n_j along the cycle $\mathfrak{A}_k$ equals to the winding number of the curve C_j with respect to the point 0. Equivalently

$$n_j = \tilde{C}_j \circ \mathbb{R}_-, \qquad (3\text{-}14)$$

where $\circ$ denotes the intersection number, $\tilde{C}_j$ denotes the projection of C_j to the λ-plane, $\mathbb{R}_-$ is negative half-line with the standard orientation.

For each point of $\mathfrak{A}_j$ the corresponding divisor $\gamma_1, \ldots, \gamma_g$ is admissible. From the characterization of admissible divisors obtained above it is easy to show that the curve C_j does not touch the closed segments on the real line $[-\infty, \lambda_{2m}], \ldots,$ $[\lambda_3, \lambda_2], [\lambda_1, 0]$. Therefore any time the curve C_j crosses the negative half-line, it intersects one of the basic cycles b_j, $j = 1, \ldots, k$.

This information does not yet suffice to calculate the basic charge, because the orientation of the cycles b_j coincides with the orientation of the negative half-line at one sheet and they are opposite at the other one. For example, in Figure 3 we see two realizations of the cycle a_1, representing different topological types. a_1 is drawn at the upper sheet and a_1' is drawn at the lower one. We have $a_1 \circ b_1 = a_1' \circ b_1 = 1$, but $\tilde{a}_1 \circ \mathbb{R}_- = 1$, $\tilde{a}_1' \circ \mathbb{R}_- = -1$, therefore $n_1 = 1$ and $n_1 = -1$ for these cycles respectively.

Fortunately, the topological type contains information about the sheet where the intersection takes place:

LEMMA 4. *Assume that the cycles C_j intersects the negative half-line at the interval $(\lambda_{2l}, \lambda_{2l-1})$. Then orientations of b_l and $\mathbb{R}_-$ coincide in the intersection point if $(-1)^{l-1} s_l > 0$ and are opposite if $(-1)^{l-1} s_l < 0$.*

Combining all these results we obtain the final formula, expressed in the next theorem:

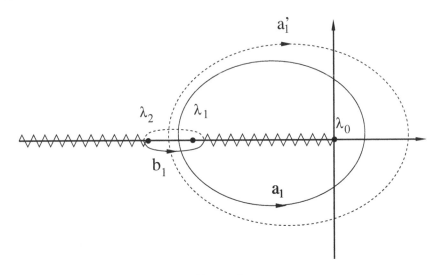

Figure 3.

THEOREM. *The density of topological charge for a real sine-Gordon solution is given by*

$$\bar{n} = \sum_{j=1}^{k} (-1)^{j-1} s_j U_j, \tag{3-15}$$

where the vector U_j is defined by (3-13).

References

[Ablowitz et al. 1973] M. J. Ablowitz, D. J. Kaup, A. C. Newell, and H. Segur, "Method for solving the sine-Gordon equation", *Phys. Rev. Lett.* **30** (1973), 1262–1264.

[Airault et al. 1977] H. Airault, H. P. McKean, and J. Moser, "Rational and elliptic solutions of the KdV equation and a related many-body problem", *Commun. Pure Appl. Math.* **30** (1977), 94–148.

[Bobenko 1991] A. I. Bobenko, "Surfaces of constant mean curvature and integrable equations", *Uspekhi Mat. Nauk* **46**:4 (1991), 3–42, 192.

[Čerednik 1980] I. V. Čerednik, "On the conditions of reality in "finite-gap integration"", *Dokl. Akad. Nauk SSSR* **252**:5 (1980), 1104–1108. In Russian; translated in *Sov. Phys. Dokl.* **25** (1980), 450–452.

[Dubrovin and Natanzon 1982] B. A. Dubrovin and S. M. Natanzon, "Real two-zone solutions of the sine-Gordon equation", *Funktsional. Anal. i Prilozhen.* **16**:1 (1982), 27–43. In Russian; translation in *Funct. Anal. Appl.* **16** (1982), 21–33.

[Dubrovin and Natanzon 1988] B. A. Dubrovin and S. M. Natanzon, "Real theta-function solutions of the Kadomtsev–Petviashvili equation", *Izv. Akad. Nauk SSSR Ser. Mat.* **52**:2 (1988), 267–286, 446.

[Dubrovin et al. 1976a] B. A. Dubrovin, I. M. Krichever, and N. S. P., "Schroedinger equation in magnetic field and Riemann surfaces", *Soviet Math. Dokl.* **17** (1976), 947–951.

[Dubrovin et al. 1976b] B. A. Dubrovin, V. B. Matveev, and S. P. Novikov, "Nonlinear equations of Korteweg–de Vries type, finite-band linear operators and Abelian varieties", *Uspehi Mat. Nauk* **31**:1 (1976), 55–136. In Russian; translated in *Russ. Math. Surveys* **31**:1 (1976), 59–146.

[Dybrovin and Novikov 1982] B. A. Dybrovin and S. P. Novikov, "Algebrogeometric Poisson brackets for real finite-range solutions of the sine-Gordon and nonlinear Schrödinger equations", *Dokl. Akad. Nauk SSSR* **267**:6 (1982), 1295–1300. In Russian; translated in *Sov. Math. Dokl.* **26**:3 (1982), 760–765.

[Ercolani and Forest 1985] N. M. Ercolani and M. G. Forest, "The geometry of real sine-Gordon wavetrains", *Comm. Math. Phys.* **99**:1 (1985), 1–49.

[Ercolani et al. 1984] N. Ercolani, M. G. Forest, and D. W. McLaughlin, "Modulational stability of two-phase sine-Gordon wavetrains", *Stud. Appl. Math.* **71**:2 (1984), 91–101.

[Grinevich 2000] P. G. Grinevich, "The scattering transform for the two-dimensional Schrödinger operator with a potential that decreases at infinity at fixed nonzero energy", *Uspekhi Mat. Nauk* **55**:6(336) (2000), 3–70. In Russian; translated in *Russian Math. Surveys*, **55**:6 (2000), 1015–1083.

[Grinevich and Novikov 1988] P. G. Grinevich and S. P. Novikov, "A two-dimensional "inverse scattering problem" for negative energies, and generalized-analytic functions, I: Energies lower than the ground state", *Funktsional. Anal. i Prilozhen.* **22**:1 (1988), 23–33. In Russian; translated in *Funct. Anal. Appl.*, **22** (1988), 19–27.

[Grinevich and Novikov 2001] P. G. Grinevich and S. P. Novikov, "Real finite-gap solutions of the sine-Gordon equation: a formula for the topological charge", *Uspekhi Mat. Nauk* **56**:5 (2001), 181–182. In Russian; translated in *Russ. Math. Surveys* **56**:5 (2001), 980–981.

[Grinevich and Novikov 2003a] P. G. Grinevich and S. P. Novikov, "Topological charge of the real periodic finite-gap sine-Gordon solutions", *Comm. Pure Appl. Math.* **56**:7 (2003), 956–978. Dedicated to the memory of Jürgen K. Moser.

[Grinevich and Novikov 2003b] P. G. Grinevich and S. P. Novikov, "Topological phenomena in the real periodic sine-Gordon theory", *J. Math. Phys.* **44**:8 (2003), 3174–3184. Integrability, topological solitons and beyond.

[Grinevich and Schmidt 1995] P. G. Grinevich and M. U. Schmidt, "Period preserving nonisospectral flows and the moduli space of periodic solutions of soliton equations", *Phys. D* **87**:1-4 (1995), 73–98.

[Hitchin 1988] N. Hitchin, "Harmonic maps from T^2 to S^3", pp. 103–112 in *Harmonic mappings, twistors, and σ-models* (Luminy, 1986), Adv. Ser. Math. Phys. **4**, World Sci. Publishing, Singapore, 1988.

[Its and Kotlyarov 1976] A. R. Its and V. P. Kotlyarov, "Explicit formulas for solutions of the Nonlinear Schrödinger equation", *Dokl. Akad. Nauk Ukrain. SSR Ser. A* no. 11 (1976), 965–968. In Russian.

[Its and Matveev 1975] A. R. Its and V. B. Matveev, "Hill operators with a finite number of lacunae", *Funkcional. Anal. i Priložen.* **9**:1 (1975), 69–70. In Russian; translated in *Funct. Anal. Appl.* **9** (1975), 65–66.

[Kohn 1959] W. Kohn, "Analytic properties of Bloch waves and Wannier functions", *Phys. Rev.* (2) **115** (1959), 809–821.

[Kozel and Kotlyarov 1976] V. A. Kozel and V. P. Kotlyarov, "Almost periodic solutions of the equation $u_{tt} - u_{xx} + \sin u = 0$", *Dokl. Akad. Nauk Ukrain. SSR Ser. A* no. 10 (1976), 878–881. In Russian.

[Krichever 1992] I. M. Krichever, *Perturbation theory in periodic problems for the two-dimensional integrable systems*, Soviet Sci. Rev. C: Math. Phys. Rev. **9**:2, Harwood, 1992.

[Lamb 1971] G. L. Lamb, Jr., "Analytical descriptions of ultrashort optical pulse propagation in a resonant medium", *Rev. Modern Phys.* **43** (1971), 99–124.

[Lax 1975] P. D. Lax, "Periodic solutions of the KdV equation", *Comm. Pure Appl. Math.* **28** (1975), 141–188.

[Manakov 1976] S. V. Manakov, "The method of the inverse scattering problem, and two-dimensional evolution equations", *Uspehi Mat. Nauk* **31**:5 (1976), 245–246. In Russian.

[Marchenko and Ostrovskii 1975] V. A. Marchenko and I. V. Ostrovskii, "A characterization of the spectrum of the Hill operator", *Mat. Sb. (N.S.)* **97**:4 (1975), 540–606. In Russian.

[McKean 1981] H. P. McKean, "The sine-Gordon and sinh-Gordon equations on the circle", *Comm. Pure Appl. Math.* **34**:2 (1981), 197–257.

[McKean and van Moerbeke 1975] H. P. McKean and P. van Moerbeke, "The spectrum of Hill's equation", *Invent. Math.* **30**:3 (1975), 217–274.

[Natanzon 1995] S. M. Natanzon, "Real nonsingular finite zone solutions of soliton equations", pp. 153–183 in *Topics in topology and mathematical physics*, Amer. Math. Soc. Transl. Ser. 2 **170**, Amer. Math. Soc., Providence, RI, 1995.

[Novikov 1974] S. P. Novikov, "A periodic problem for the Korteweg-de Vries equation. I", *Funkcional. Anal. i Priložen.* **8**:3 (1974), 54–66. In Russian; translated in *Funct. Anal. Appl.* **8** (1974), 236–246.

[Novikov 1984] S. P. Novikov, "Algebro-topological approach to reality problems. Real action variables in the theory of finite-gap solutions of the sine-Gordon equation", *Zap. Nauchn. Sem. Leningrad. Otdel. Mat. Inst. Steklov. (LOMI)* **133** (1984), 177–196. In Russian; translated in *J. Soviet Math.* **31**:6 (1985), 3373–3387.

[Novikov and Veselov 1984] S. P. Novikov and A. P. Veselov, "Finite-zone, two-dimensional Schrödinger operators", **30** (1984), 588–591, 705–708.

[Novikov and Veselov 1997] S. P. Novikov and A. P. Veselov, "Exactly solvable two-dimensional Schrödinger operators and Laplace transformations", pp. 109–132 in *Solitons, geometry, and topology: on the crossroad*, Amer. Math. Soc. Transl. Ser. 2 **179**, Amer. Math. Soc., Providence, RI, 1997.

[Novikov et al. 1984] S. Novikov, S. V. Manakov, L. P. Pitaevskiĭ, and V. E. Zakharov, *Theory of solitons: The inverse scattering method*, Consultants Bureau, New York, 1984.

[Pinkall and Sterling 1989] U. Pinkall and I. Sterling, "On the classification of constant mean curvature tori", *Ann. of Math.* (2) **130**:2 (1989), 407–451.

[Veselov et al. 1985] A. P. Veselov, I. M. Krichever, and S. P. Novikov, "Two-dimensional periodic Schrödinger operators and Prym's θ-functions", pp. 283–301 in *Geometry today* (Rome, 1984), Progr. Math. **60**, Birkhäuser, Boston, 1985.

PETR G. GRINEVICH
L. D. LANDAU INSTITUTE FOR THEORETICAL PHYSICS
pgg@landau.ac.ru

SERGEI P. NOVIKOV
INSTITUTE FOR PHYSICAL SCIENCE AND TECHNOLOGY
UNIVERSITY OF MARYLAND AT COLLEGE PARK AND
L. D. LANDAU INSTITUTE FOR THEORETICAL PHYSICS
novikov@ipst.umd.edu

Probability, Geometry and Integrable Systems
MSRI Publications
Volume **55**, 2008

Random walks and orthogonal polynomials: some challenges

F. ALBERTO GRÜNBAUM

To Henry, teacher and friend, with gratitude and admiration.

ABSTRACT. The study of several naturally arising "nearest neighbour" random walks benefits from the study of the associated orthogonal polynomials and their orthogonality measure. I consider extensions of this approach to a larger class of random walks. This raises a number of open problems.

1. Introduction

Consider a birth and death process, i.e., a discrete time Markov chain on the nonnegative integers, with a one step transition probability matrix $\mathbb{P}$. There is then a time-honored way of writing down the n-step transition probability matrix $\mathbb{P}^n$ in terms of the orthogonal polynomials associated to $\mathbb{P}$ and the spectral measure. This goes back to Karlin and McGregor [1957] and, as they observe, it is nothing but an application of the spectral theorem. One can find some precursors of these powerful ideas, see for instance [Harris 1952; Ledermann and Reuter 1954]. Inasmuch as this is such a deep and general result, it holds in many setups, such as a nearest neighbours random walk on the N th-roots of unity. In general this representation of $\mathbb{P}^n$ allows one to relate properties of the Markov chain, such as recurrence or other limiting behaviour, to properties of the orthogonality measure.

Mathematics Subject Classification: 33C45, 22E45.

Keywords: Matrix-valued orthogonal polynomials, Karlin–McGregor representation, Jacobi polynomials.

The author was supported in part by NSF Grant # DMS 0603901.

In the few cases when one can get one's hands on the orthogonality measure and the polynomials themselves this gives fairly explicit answers to various questions.

The two main drawbacks to the applicability of this representation (to be recalled below) are:

a) typically one cannot get either the polynomials or the measure explicitly.
b) the method is restricted to "nearest neighbour" transition probability chains that give rise to tridiagonal matrices and thus to orthogonal polynomials.

The challenge that we pose in this paper is very simple: to try to extend the class of Markov processes whose study can benefit from a similar association.

There is an important collection of papers that study in detail the cases where the entries in $\mathbb{P}$ depends linearly, quadratically or even rationally on the index n. We make no attempt to review these results, but we just mention that the linear case involves (associated) Laguerre and Meixner polynomials, and the other cases involve associated dual Hahn polynomials. For a very small sample of important sources dealing with this connection see [Chihara 1978; van Doorn 2003; Ismail et al. 1990].

The plan for this paper is as follows. In Section 2, we review briefly the approach of S. Karlin and J. McGregor. In Sections 3, 4, and 5, we consider a few examples of physically important Markov chains that happen to feature rather well known families of orthogonal polynomials. In Sections 6–10 we propose a way of extending this representation to the case of certain Markov chains where the one-step transition probability matrix is not necessarily tridiagonal. For concreteness we restrict ourselves to the case of pentadiagonal matrices or more generally block tridiagonal matrices. This is illustrated with some examples. A number of open problems are mentioned along the way; a few more are listed in Section 11. The material in Sections 2–5 is well known, while the proposal developed in Sections 3–10 appears to be new.

After this paper was completed we noticed that [Karlin and McGregor 1959] contains an explicit expression for the spectral matrix corresponding to the example that we treat in Section 10. The same example, as well as the connection with matrix valued orthogonal polynomials is discussed in [Dette et al. 2006]. See also [Grünbaum and de la Iglesia 2007] for a fruitful interaction with group representation theory.

2. The Karlin–McGregor representation

If we have

$$\mathbb{P}_{i,j} = \Pr\{X(n+1) = j \mid X(n) = i\}$$

for the 1-step transition probability of our Markov chain, and we put $p_i = \mathbb{P}_{i,i+1}$, $q_{i+1} = \mathbb{P}_{i+1,i}$, and $r_i = \mathbb{P}_{i,i}$ we get for the matrix $\mathbb{P}$, in the case of a birth and death process, the expression

$$\mathbb{P} = \begin{pmatrix} r_0 & p_0 & 0 & 0 & \\ q_1 & r_1 & p_1 & 0 & \\ 0 & q_2 & r_2 & p_2 & \\ & \ddots & \ddots & \ddots & \end{pmatrix}$$

We will assume that $p_j > 0$, $q_{j+1} > 0$ and $r_j \geq 0$ for $j \geq 0$. We also assume $p_j + r_j + q_j = 1$ for $j \geq 1$ and by putting $p_0 + r_0 \leq 1$ we allow for the state $j = 0$ to be an absorbing state (with probability $1 - p_0 - r_0$). Some of these conditions can be relaxed.

If we introduce the polynomials $Q_j(x)$ through the conditions $Q_{-1}(0) = 0$, $Q_0(x) = 1$ and, using the notation

$$Q(x) = \begin{pmatrix} Q_0(x) \\ Q_1(x) \\ \vdots \end{pmatrix},$$

we insist on the recursion relation

$$\mathbb{P}Q(x) = xQ(x),$$

we can prove the existence of a unique measure $\psi(dx)$ supported in $[-1, 1]$ such that

$$\pi_j \int_{-1}^{1} Q_i(x)Q_j(x)\psi(dx) = \delta_{ij},$$

and obtain the Karlin–McGregor representation formula

$$(\mathbb{P}^n)_{ij} = \pi_j \int_{-1}^{1} x^n Q_i(x)Q_j(x)\psi(dx).$$

Many general results can be obtained from this representation formula, some of which will be given for certain examples in the next three sections.

Here we just remark that the existence of

$$\lim_{n \to \infty} (\mathbb{P}^n)_{ij}$$

is equivalent to $\psi(dx)$ having no mass at $x = -1$. If this is the case this limit is positive exactly when $\psi(dx)$ has some mass at $x = 1$.

If one notices that $Q_n(x)$ is nothing but the determinant of the $(n+1) \times (n+1)$ upper-left corner of the matrix $xI - \mathbb{P}$, divided by the factor

$$p_0 p_1, \ldots, p_{n-1},$$

and one defines the polynomials $q_n(x)$ by solving the same three-term recursion relation satisfied by the polynomials $Q_n(x)$, but with the indices shifted by one, and the initial conditions $q_0(x) = 1$, $q_1(x) = (x - r_1)/p_1$, it becomes clear that the $(0, 0)$ entry of the matrix

$$(xI - \mathbb{P})^{-1}$$

should be given, except by the constant p_0, by the limit of the ratio

$$q_{n-1}(x)/Q_n(x).$$

On the other hand the same spectral theorem alluded to above establishes an intimate relation between

$$\lim_{n \to \infty} q_{n-1}(x)/Q_n(x)$$

and

$$\int_{-1}^{1} \frac{d\psi(\lambda)}{x - \lambda}.$$

We will see in some of the examples a probabilistic interpretation for the expression above in terms of generating functions.

The same connection with orthogonal polynomials holds in the case of a birth and death process with continuous time, and this has been extensively described in the literature. The discrete time situation discussed above is enough to illustrate the power of this method.

3. The Ehrenfest urn model

Consider the case of a Markov chain in the finite state space $0, 1, 2, \ldots, 2N$, where the matrix $\mathbb{P}$ given by

$$\begin{pmatrix} 0 & 1 & & & & & \\ \frac{1}{2N} & 0 & \frac{2N-1}{2N} & & & & \\ & \frac{2}{2N} & 0 & \frac{2N-2}{2N} & & & \\ & & \ddots & 0 & \ddots & & \\ & & & \ddots & \ddots & \ddots & \\ & & & & \ddots & 0 & \frac{1}{2N} \\ & & & & & \frac{2N}{2N} & 0 \end{pmatrix}.$$

This situation arises in a model introduced by P. and T. Ehrenfest [1907], in an effort to illustrate the issue that irreversibility and recurrence can coexist. The background here is, of course, the famous H-theorem of L. Boltzmann.

For a more detailed discussion of the model see [Feller 1967; Kac 1947]. This model has also been considered in dealing with a quantum mechanical version of a discrete harmonic oscillator by Schrödinger himself; see [Schrödinger and Kohlrausch 1926].

In this case the corresponding orthogonal polynomials (on a finite set) can be given explicitly. Consider the so called Krawtchouk polynomials, given by means of the (truncated) Gauss series

$$2\tilde{F}_1 \left(\begin{matrix} a,b \\ c \end{matrix} ; z \right) = \sum_0^{2N} \frac{(a)_n (b)_n}{n!(c)_n} z^n$$

with

$$(a)_n \equiv a(a+1)\ldots(a+n-1), \quad (a)_0 = 1.$$

The polynomials are given by

$$K_i(x) = {}_2\tilde{F}_1 \left(\begin{matrix} -i,-x \\ -2N \end{matrix} ; 2 \right)$$

$$x = 0, 1, \ldots, 2N; \quad i = 0, 1, \ldots, 2N$$

Observe that

$$K_0(x) \equiv 1, \, K_i(2N) = (-1)^i.$$

The orthogonality measure is read off from

$$\sum_{x=0}^{2N} K_i(x) K_j(x) \frac{\binom{2N}{x}}{2^{2N}} = \frac{(-1)^i i!}{(-2N)_i} \delta_{ij} \equiv \pi_i^{-1} \delta_{ij} \quad 0 \le i, j \le 2N.$$

These polynomials satisfy the second order difference equation

$$\tfrac{1}{2}(2N-i)K_{i+1}(x) - \tfrac{1}{2}2N K_i(x) + \tfrac{1}{2}i K_{i-1}(x) = -x K_i(x),$$

and this has the consequence that

$$\begin{pmatrix} 0 & 1 & & & & & \\ \frac{1}{2N} & 0 & \frac{2N-1}{2N} & & & & \\ & \frac{2}{2N} & 0 & \frac{2N-2}{2N} & & & \\ & & \ddots & 0 & \ddots & & \\ & & & \ddots & \ddots & \ddots & \\ & & & & \ddots & 0 & \frac{1}{2N} \\ & & & & & \frac{2N}{2N} & 0 \end{pmatrix} \begin{pmatrix} K_0(x) \\ K_1(x) \\ \vdots \\ K_{2N}(x) \end{pmatrix}$$

$$= \left(1 - \frac{x}{N}\right) \begin{pmatrix} K_0(x) \\ K_1(x) \\ \vdots \\ K_{2N}(x) \end{pmatrix}$$

any time that x is one of $0, 1, \ldots, 2N$. This means that the eigenvalues of the matrix $\mathbb{P}$ above are given by the values of $1 - (x/N)$ at these values of x, that is,

$$1, 1 - \frac{1}{N}, \ldots, -1,$$

and that the corresponding eigenvectors are the values of

$$[K_0(x), K_1(x), \ldots, K_{2N}(x)]^T$$

at these values of x.

Since the matrix $\mathbb{P}$ above is the one step transition probability matrix for our urn model we conclude that

$$(\mathbb{P}^n)_{ij} = \pi_j \sum_{x=0}^{2N} \left(1 - \frac{x}{N}\right)^n K_i(x) K_j(x) \frac{\binom{2N}{x}}{2^{2N}}.$$

We can use these expressions to rederive some results given in [Kac 1947].

We have

$$(\mathbb{P}^n)_{00} = \sum_{x=0}^{2N} \left(1 - \frac{x}{N}\right)^n \frac{\binom{2N}{x}}{2^{2N}}$$

and the "generating function" for these probabilities, defined by

$$U(z) \equiv \sum_{n=0}^{\infty} z^n (\mathbb{P}^n)_{00}$$

becomes

$$U(z) = \sum_{x=0}^{2N} \frac{N}{N(1-z) + xz} \frac{\binom{2N}{x}}{2^{2N}}.$$

In particular $U(1) = \infty$ and then the familiar "renewal equation" (see [Feller 1967]) given by

$$U(z) = F(z)U(z) + 1,$$

where $F(z)$ is the generating function for the probabilities f_n of returning from state 0 to state 0 for the first time in n steps

$$F(z) = \sum_{n=0}^{\infty} z^n f_n$$

gives

$$F(z) = 1 - \frac{1}{U(z)}$$

Therefore we have $F(1) = 1$, indicating that one returns to state 0 with probability one in finite time.

These results allow us to compute the expected time to return to state 0. This expected value is given by $F'(1)$, and we have

$$F'(z) = \frac{U'(z)}{U^2(z)}.$$

Since

$$U'(z) = \sum_{x=0}^{2N} \frac{N(N-x)}{(N(1-z)+xz)^2} \frac{\binom{2N}{x}}{2^{2N}}$$

we get $F'(1) = 2^{2N}$. The same method shows that any state $i = 0, \ldots, 2N$ is also recurrent and that the expected time to return to it is given by

$$\frac{2^{2N}}{\binom{2N}{i}}.$$

The moral of this story is clear: if $i = 0$ or $2N$, or if i is close to these values, meaning that we start from a state where most balls are in one urn, it will take on average a huge amount of time to get back to this state. On the other hand if $i = N$, that is, we are starting from a very balanced state, then we will (on average) return to this state fairly soon. Thus we see how the issues of irreversibility and recurrence are rather subtle.

In a very precise sense these polynomials are discrete analogs of those of Hermite in the case of the real line. For interesting material regarding this section the reader should consult [Askey 2005].

4. A Chebyshev-type example

The example below illustrates nicely how certain recurrence properties of the process are related to the presence of point masses in the orthogonality measure. This is seen by comparing the two integrals at the end of the section.

Consider the matrix

$$\mathbb{P} = \begin{pmatrix} 0 & 1 & 0 & \\ q & 0 & p & \\ 0 & q & 0 & p \\ & & \ddots & \ddots & \ddots \end{pmatrix}$$

with $0 \le p \le 1$ and $q = 1 - p$. We look for polynomials $Q_j(x)$ such that

$$Q_{-1}(x) = 0, \quad Q_0(x) = 1$$

and if $Q(x)$ denotes the vector

$$Q(x) \equiv \begin{pmatrix} Q_0(x) \\ Q_1(x) \\ Q_2(x) \\ \vdots \end{pmatrix}$$

we ask that we should have

$$\mathbb{P}Q(x) = xQ(x).$$

The matrix $\mathbb{P}$ can be conjugated into a symmetric one and in this fashion one can find the explicit expression for these polynomials.

We have

$$Q_j(x) = \left(\frac{q}{p}\right)^{j/2} \left((2-2p)T_j\left(\frac{x}{2\sqrt{pq}}\right) + (2p-1)U_j\left(\frac{x}{2\sqrt{pq}}\right)\right)$$

where T_j and U_j are the Chebyshev polynomials of the first and second kind.

If $p \geq 1/2$ we have

$$\left(\frac{p}{1-p}\right)^n \int_{-\sqrt{4pq}}^{\sqrt{4pq}} Q_n(x)Q_m(x)\frac{\sqrt{4pq-x^2}}{1-x^2}dx = \delta_{nm} \begin{cases} 2(1-p)\pi & \text{if } n = 0, \\ 2p(1-p)\pi & \text{if } n \geq 1, \end{cases}$$

while if $p \leq 1/2$ we get a new phenomenon, namely the presence of point masses in the spectral measure

$$\left(\frac{p}{1-p}\right)^n \left(\int_{-\sqrt{4pq}}^{\sqrt{4pq}} Q_n(x)Q_m(x)\frac{\sqrt{4pq-x^2}}{1-x^2}dx \right.$$
$$\left. +(2-4p)\pi[Q_n(1)Q_m(1) + Q_n(-1)Q_m(-1)]\right)$$
$$= \delta_{nm} \begin{cases} 2(1-p)\pi & \text{if } n = 0, \\ 2p(1-p)\pi & \text{if } n \geq 1. \end{cases}$$

From a probabilistic point of view these results are very natural.

5. The Hahn polynomials, Laplace and Bernoulli

As has been pointed out before, a limitation of this method is given by the sad fact that given the matrix $\mathbb{P}$ very seldom can one write down the corresponding polynomials and their orthogonality measure. In general there is no reason why physically interesting Markov chains will give rise to situations where these mathematical objects can be found explicitly.

The example below shows that one can get lucky: there is a very old model of the exchange of heat between two bodies going back to Laplace and Bernoulli, see [Feller 1967, p. 378]. It turns out that in this case the corresponding orthogonal polynomials can be determined explicitly.

The Bernoulli–Laplace model for the exchange of heat between two bodies consists of two urns, labeled 1 and 2. Initially there are W white balls in urn 1 and B black balls in urn 2. The transition mechanism is as follows: a ball is picked from each urn and these two balls are switched. It is natural to expect that eventually both urns will have a nice mixture of white and black balls.

The state of the system at any time is described by w, defined to be the number of white balls in urn 1. It is clear that we have, for $w = 0, 1, \ldots, W$

$$\mathbb{P}_{w,w+1} = \frac{W-w}{W}\frac{W-w}{B}, \quad \mathbb{P}_{w,w-1} = \frac{w}{W}\frac{B-W+w}{B},$$

$$\mathbb{P}_{w,w} = \frac{w}{W}\frac{W-w}{B} + \frac{W-w}{W}\frac{B-W+w}{B}.$$

Notice that

$$\mathbb{P}_{w,w-1} + \mathbb{P}_{w,w} + \mathbb{P}_{w,w+1} = 1.$$

Now introduce the dual Hahn polynomials by means of

$$R_n(\lambda(x)) = {}_3\tilde{F}_2\left(\begin{matrix} -n, -x, x-W-B-1 \\ -W, -W \end{matrix}\middle| 1\right)$$

$$n = 0, \ldots, W; \ x = 0, \ldots, W.$$

These polynomials depend in general on one more parameter.

Notice that these are polynomials of degree n in

$$\lambda(x) \equiv x(x-W-B-1).$$

One has

$$\mathbb{P}_{w,w-1}R_{w-1} + \mathbb{P}_{w,w}R_w + \mathbb{P}_{w,w+1}R_{w+1} = \left(1 - \frac{x(B+W-x+1)}{BW}\right)R_w.$$

This means that for each value of $x = 0, \ldots, W$ the vector

$$\begin{pmatrix} R_0(\lambda(x)) \\ R_1(\lambda(x)) \\ \vdots \\ R_W(\lambda(x)) \end{pmatrix}$$

is an eigenvector of the matrix $\mathbb{P}$ with eigenvalue $1 - \dfrac{x(B+W-x+1)}{BW}$. The relevant orthogonality relation is given by

$$\pi_j \sum_{x=0}^{W} R_i(\lambda(x)) R_j(\lambda(x)) \mu(x) = \delta_{ij}$$

with

$$\mu(x) = \frac{w!(-w)_x(-w)_x(2x - W - B - 1)}{(-1)^{x+1}x!(-B)_x(x - W - B - 1)_{w+1}}, \qquad \pi_j = \frac{(-w)_j}{j!}\frac{(-B)_{w-j}}{(w-j)!}.$$

The Karlin–McGregor representation gives

$$(\mathbb{P}^n)_{ij} = \pi_j \sum_{x=0}^{W} R_i(\lambda(x)) R_j(\lambda(x)) e^n(x) \mu(x)$$

with $e(x) = 1 - \dfrac{x(B + W - x + 1)}{BW}$.

These results can be used, once again, to get some quantitative results on this process.

Interestingly enough, these polynomials were considered in great detail by S. Karlin and J. McGregor [1961] and used by these authors in the context of a model in genetics describing fluctuations of gene frequency under the influence of mutation and selection. The reader will find useful remarks in [Diaconis and Shahshahani 1987].

6. The classical orthogonal polynomials and the bispectral problem

The examples discussed above illustrate the following point: quite often the orthogonal polynomials that are associated with important Markov chains belong to the small class of polynomials usually referred to as *classical*. By this one means that they satisfy not only three term recursion relations but that they are also the common eigenfunctions of some fixed (usually second order) differential operator. The search for polynomials of this kind goes back at least to [Bochner 1929]. In fact this issue is even older; see [Routh 1884] and also [Ismail 2005] for a more complete discussion.

In the context where both variables are continuous, this problem has been raised in [Duistermaat and Grünbaum 1986]. For a view of some related subjects see [Harnad and Kasman 1998]. The reader will find useful material in [Askey and Wilson 1985; Andrews et al. 1999; Ismail 2005].

7. Matrix-valued orthogonal polynomials

Here we recall a notion due to M. G. Krein [1949; 1971]. Given a self adjoint positive definite matrix-valued smooth weight function $W(x)$ with finite moments, we can consider the skew symmetric bilinear form defined for any pair of matrix-valued polynomial functions $P(x)$ and $Q(x)$ by the numerical matrix

$$(P, Q) = (P, Q)_W = \int_{\mathbb{R}} P(x) W(x) Q^*(x) dx,$$

where $Q^*(x)$ denotes the conjugate transpose of $Q(x)$. By the usual construction this leads to the existence of a sequence of matrix-valued orthogonal polynomials with nonsingular leading coefficient.

Given an orthogonal sequence $\{P_n(x)\}_{n\geq 0}$ one gets by the usual argument a three term recursion relation

$$xP_n(x) = A_n P_{n-1}(x) + B_n P_n(x) + C_n P_{n+1}(x), \tag{7-1}$$

where A_n, B_n and C_n are matrices and the last one is nonsingular.

We now turn our attention to an important class of orthogonal polynomials which we will call *classical matrix-valued orthogonal polynomials*. Very much as in [Duran 1997; Grünbaum et al. 2003; Grünbaum et al. 2005] we say that the weight function is *classical* if there exists a second order ordinary differential operator D with matrix-valued polynomial coefficients $A_j(x)$ of degree less or equal to j of the form

$$D = A_2(x)\frac{d^2}{dx^2} + A_1(x)\frac{d}{dx} + A_0(x), \tag{7-2}$$

such that for an orthogonal sequence $\{P_n\}$, we have

$$DP_n^* = P_n^* \Lambda_n, \tag{7-3}$$

where Λ_n is a real-valued matrix. This form of the eigenvalue equation (7-3) appears naturally in [Grünbaum et al. 2002] and differs only superficially with the form used in [Duran 1997], where one uses right handed differential operators.

During the last few years much activity has centered around an effort to produce families of matrix-valued orthogonal polynomials that would satisfy differential equations as those above. One of the examples that resulted from this search, see [Grünbaum 2003], will be particularly useful later on.

8. Pentadiagonal matrices and matrix-valued orthogonal polynomials

Given a pentadiagonal scalar matrix it is often useful to think of it either in its original unblocked form or as being made, let us say, of 2×2 blocks. These two ways of seeing a matrix, and the fact that matrix operations like multiplication can by performed "by blocks", has proved very important in the development of fast algorithms.

In the case of a birth and death process it is useful to think of a graph like

Suppose that we are dealing with a more complicated Markov chain in the same probability space, where the elementary transitions can go beyond "nearest neighbours". In such a case the graph may look as follows:

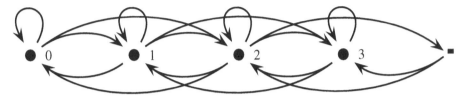

The matrix $\mathbb{P}$ going with the graph above is now pentadiagonal. By thinking of it in the manner mentioned above we get a block tridiagonal matrix. As an extra bonus, its off-diagonal blocks are triangular.

The graph

clearly corresponds to a general block tridiagonal matrix, with blocks of size 2×2.

If $\mathbb{P}_{i,j}$ denotes the (i,j)-block of $\mathbb{P}$ we can generate a sequence of 2×2 matrix-valued polynomials $Q_i(t)$ by imposing the three-term recursion of Section 8. Using the notation of Section 2, we would have

$$\mathbb{P}Q(x) = xQ(x),$$

where the entries of the column vector $Q(x)$ are now 2×2 matrices.

Proceeding as in the scalar case, this relation can be iterated to give

$$\mathbb{P}^n Q(x) = x^n Q(x),$$

and if we assume the existence of a weight matrix $W(x)$ as in Section 7, with the property

$$(Q_j, Q_j)\delta_{i,j} = \int_{\mathbb{R}} Q_i(x)W(x)Q_j^*(x)dx,$$

it is then clear that one can get an expression for the (i, j) entry of the block matrix $\mathbb{P}^n$ that would look exactly as in the scalar case, namely

$$(\mathbb{P}^n)_{ij}(Q_j, Q_j) = \int x^n Q_i(x) W(x) Q_j^*(x) dx.$$

Just as in the scalar case, this expression becomes useful when we can get our hands on the matrix-valued polynomials $Q_i(x)$ and the orthogonality measure $W(x)$. Notice that we have not discussed conditions on the matrix $\mathbb{P}$ to give rise to such a measure. For this issue the reader can consult [Durán and Polo 2002; Duran 1999] and the references in these papers.

The spectral theory of a scalar double-infinite tridiagonal matrix leads naturally to a 2×2 semi-infinite matrix. This has been looked at in terms of random walks in [Pruitt 1963]. In [Ismail et al. 1990] this work is elaborated further to get a formula that could be massaged to look like the right-hand side of the one above. See also the last section in [Karlin and McGregor 1959].

9. An explicit example

Consider the matrix-valued polynomials given by the three-term recursion relation

$$A_n \Phi_{n-1}(x) + B_n \Phi_n(x) + C_n \Phi_{n+1}(x) = t \Phi_n(x), \quad n \geq 0,$$

with

$$\Phi_{-1}(x) = 0, \quad \Phi_0(t) = I,$$

and where the entries in A_n, B_n, C_n are given by

$$A_n^{11} := \frac{n(\alpha+n)(\beta+2\alpha+2n+3)}{(\beta+\alpha+2n+1)(\beta+\alpha+2n+2)(\beta+2\alpha+2n+1)},$$

$$A_n^{12} := \frac{2n(\beta+1)}{(\beta+2n+1)(\beta+\alpha+2n+2)(\beta+2\alpha+2n+1)}, \quad A_n^{21} := 0,$$

$$A_n^{22} := \frac{n(\alpha+n+1)(\beta+2n+3)}{(\beta+2n+1)(\beta+\alpha+2n+2)(\beta+\alpha+2n+3)},$$

$$C_n^{11} := \frac{(\beta+n+2)(\beta+2n+1)(\beta+\alpha+n+2)}{(\beta+2n+3)(\beta+\alpha+2n+2)(\beta+\alpha+2n+3)},$$

$$C_n^{21} := \frac{2(\beta+1)(\beta+n+2)}{(\beta+2n+3)(\beta+\alpha+2n+3)(\beta+2\alpha+2n+5)}, \quad C_n^{12} := 0,$$

$$C_n^{22} := \frac{(\beta+n+2)(\beta+\alpha+n+3)(\beta+2\alpha+2n+3)}{(\beta+\alpha+2n+3)(\beta+\alpha+2n+4)(\beta+2\alpha+2n+5)},$$

$$B_n^{11} := 1 + \frac{n(\beta+n+1)(\beta+2n-1)}{(\beta+2n+1)(\beta+\alpha+2n+1)} - \frac{(n+1)(\beta+n+2)(\beta+2n+1)}{(\beta+2n+3)(\beta+\alpha+2n+3)}$$
$$- \frac{2(\beta+1)^2}{(\beta+2n+1)(\beta+2n+3)(\beta+2\alpha+2n+3)},$$

$$B_n^{12} := \frac{2(\beta+1)(\alpha+\beta+n+2)}{(\beta+2n+3)(\beta+\alpha+2n+2)(\beta+2\alpha+2n+3)},$$

$$B_n^{21} := \frac{2(\alpha+n+1)(\beta+1)}{(\beta+2n+1)(\beta+\alpha+2n+3)(\beta+2\alpha+2n+3)},$$

$$B_n^{22} := 1 + \frac{n(\beta+n+1)(\beta+2n+3)}{(\beta+2n+1)(\beta+\alpha+2n+2)} - \frac{(n+1)(\beta+n+2)(\beta+2n+5)}{(\beta+2n+3)(\beta+\alpha+2n+4)}$$
$$+ \frac{2(\beta+1)^2}{(\beta+2n+1)(\beta+2n+3)(\beta+2\alpha+2n+3)}.$$

Notice that the matrices A_n and C_n are upper and lower triangular respectively.
If the matrix $\Psi_0(x)$ is given by

$$\Psi_0(x) = \begin{pmatrix} 1 & 1 \\ 1 & \frac{(\beta+2\alpha+3)x}{\beta+1} - \frac{2(\alpha+1)}{\beta+1} \end{pmatrix}$$

one can see that the polynomials $\Phi_n(x)$ satisfy the orthogonality relation

$$\int_0^1 \Phi_i(x)W(x)\Phi_j^*(x)dx = 0 \quad \text{if } i \neq j,$$

where

$$W(x) = \Psi_0(x) \begin{pmatrix} (1-x)^\beta x^{\alpha+1} & 0 \\ 0 & (1-x)^\beta x^\alpha \end{pmatrix} \Psi_0^*(x).$$

The polynomials $\Phi_n(x)$ are classical in the sense that they are eigenfunctions
of a fixed second order differential operator. More precisely, we have

$$\mathscr{F}\Phi_n^* = \Phi_n^* \Lambda_n,$$

where $\Lambda_n = \text{diag}\left(-n^2-(\alpha+\beta+2)n+\alpha+1+\frac{1}{2}(\beta+1), -n^2-(\alpha+\beta+3)n\right)$ and

$$\mathscr{F} = x(1-x)\left(\frac{d}{dx}\right)^2$$
$$+ \begin{pmatrix} \frac{(\alpha+1)(\beta+2\alpha+5)}{\beta+2\alpha+3} - (\alpha+\beta+3)x & \frac{2\alpha+2}{2\alpha+\beta+3} + x \\ \frac{\beta+1}{\beta+2\alpha+3} & \frac{(\alpha+2)\beta+2\alpha^2+5\alpha+4}{\beta+2\alpha+3} - (\alpha+\beta+4)x \end{pmatrix}\frac{d}{dx}$$
$$+ \begin{pmatrix} \alpha+1+\frac{\beta+1}{2} & 0 \\ 0 & 0 \end{pmatrix} I.$$

As mentioned earlier this is the reason why this example has surfaced recently; see [Grünbaum 2003]. An explicit expression for the polynomials thenselves is given in [Tirao 2003, Corollary 3].

Now we observe that the entries of the corresponding pentadigonal matrix are all nonnegative, and that the sum of the entries on any given row are all equal to 1. This allows for an immediate probabilistic interpretation of the pentadiagonal matrix as the one step transition probability matrix for a Markov chain whose state space could be visualized in the graph

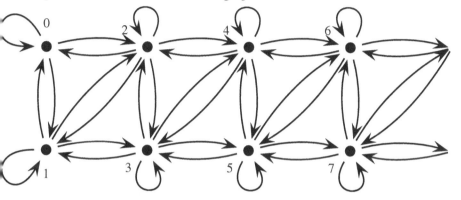

I find it remarkable that this example, which was produced for an entirely different purpose, should have this extra property. Finding an appropriate combinatorial mechanism, maybe in terms of urns, that goes along with this example remains an interesting challenge.

Two final observations dealing with these state spaces that can be analyzed using matrix-valued orthogonal polynomials. If we were using matrix-valued polynomials of size N we would have as state space a semiinfinite network consisting of N (instead of two) parallel collection of nonnegative integers with connections going from each of the N states on each vertical rung to every one in the same rung and the two neighbouring ones. The examples in [Grünbaum et al. 2002] give instances of this situation with a rather local connection pattern.

In the case of $N = 2$ one could be tempted to paraphrase a well known paper and say that "it has not escaped our notice that" some of these models could be used to study transport phenomena along a DNA segment.

10. Another example

Here we consider a different example of matrix-valued orthogonal polynomials whose block tridiagonal matrix can be seen as a scalar pentadiagonal matrix with nonnegative elements. In this case the sum of the elements in the rows of this scalar matrix is not identically one, but this poses no problem in terms of a Karlin–McGregor-type representation formula for the entries of the powers $\mathbb{P}^n$.

This example has the important property that the orthogonality weight matrix $W(x)$, as well as the polynomials themselves are explicitly known. This is again a classical situation; see [Castro and Grünbaum 2006].

Consider the block tridiagonal matrix

$$\begin{pmatrix} B_0 & I & & & \\ A_1 & B_1 & I & & \\ 0 & A_2 & B_2 & I & \\ & & \ddots & \ddots & \ddots \end{pmatrix}$$

with 2×2 blocks given by $B_0 = \frac{1}{2}I$, $B_n = 0$ if $n \geq 1$, and $A_n = \frac{1}{4}I$ if $n \geq 1$. In this case one can compute explicitly the matrix-valued polynomials P_n given by

$$A_n P_{n-1}(x) + B_n P_n(x) + P_{n+1}(x) = x P_n(x), \quad P_{-1}(x) = 0, \quad P_0(x) = I.$$

One gets

$$P_n(x) = \frac{1}{2^n} \begin{pmatrix} U_n(x) & -U_{n-1}(x) \\ -U_{n-1}(x) & U_n(x) \end{pmatrix}$$

where $U_n(x)$ is the n-th Chebyshev polynomial of the second kind.

The orthogonality measure is read off from the identity

$$\frac{4^i}{\pi} \int_{-1}^{1} P_i(x) \frac{1}{\sqrt{1-x^2}} \begin{pmatrix} 1 & x \\ x & 1 \end{pmatrix} P_j(x) dx = \delta_{ij} I.$$

We get, for $n = 0, 1, 2, \ldots$

$$\frac{4^i}{\pi} \int_{-1}^{1} x^n P_i(x) \frac{1}{\sqrt{1-x^2}} \begin{pmatrix} 1 & x \\ x & 1 \end{pmatrix} P_j(x) dx = (\mathbb{P}^n)_{ij},$$

where, as above, $(\mathbb{P}^n)_{ij}$ stands for the i, j block of the matrix $\mathbb{P}^n$.

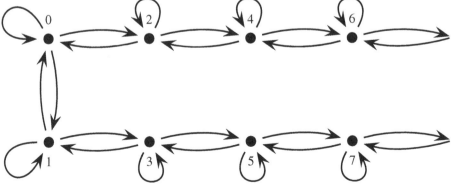

In this way one can compute the entries of the powers $\mathbb{P}^n$, with $\mathbb{P}^n$ thought of as a pentadiagonal matrix, namely

$$\mathbb{P} = \begin{pmatrix} 0 & \frac{1}{2} & 1 & 0 & 0 & 0 & \\ \frac{1}{2} & 0 & 0 & 1 & 0 & 0 & \ddots \\ \frac{1}{4} & 0 & 0 & 0 & 1 & 0 & \ddots \\ 0 & \frac{1}{4} & 0 & 0 & 0 & 1 & \ddots \\ & 0 & \frac{1}{4} & 0 & 0 & 0 & \ddots \\ & & 0 & \frac{1}{4} & 0 & 0 & \ddots \\ & & & \ddots & \ddots & \ddots & \ddots \end{pmatrix}.$$

This example goes along with the graph on the previous page.

11. A few more challenges

We have already pointed out a few challenges raised by our attempt to extend the Karlin–McGregor representation beyond its original setup. Here we list a few more open problems. The reader will undoubtedly come up with more.

Is there a natural version of the models introduced by Bernoulli–Laplace and by P. and T. Ehrenfest whose solution features matrix-valued polynomials?

Is it possible to modify the simplest Chebyshev-type examples in [Duran 1999] to accommodate cases where some of the blocks in the tridiagonal matrix give either absorption or reflection boundary conditions?

One could consider the emerging class of polynomials of several variables and find here interesting instances where the state space is higher-dimensional. For a systematic study of polynomials in several variables one should consult [Dunkl and Xu 2001] as well as the monograph [Macdonald 2003], on Macdonald polynomials of various kinds. A look at the pioneering work of Tom Koornwinder (see [Koornwinder 1975], for instance) is always a very good idea.

∗   ∗   ∗   ∗

After this paper was finished I came up with two independent sources of multivariable polynomials of the type alluded to in the previous paragraph. One is the series of papers by Hoare and Rahman [1979; 1984; 1988; ≥ 2007]. The other deals with the papers [Milch 1968; Iliev and Xu 2007; Geronimo and Iliev 2006].

In queueing theory one finds the notion of quasi-birth-and-death processes; see [Latouche and Ramaswami 1999; Neuts 1989]. Within those that are non-homogeneous one could find examples where the general approach advocated here might be useful.

References

[Andrews et al. 1999] G. E. Andrews, R. Askey, and R. Roy, *Special functions*, Encyclopedia of Mathematics and its Applications **71**, Cambridge University Press, Cambridge, 1999.

[Askey 2005] R. Askey, "Evaluation of Sylvester type determinants using orthogonal polynomials", pp. 1–16 in *Advances in analysis*, World Scientific, Singapore, 2005.

[Askey and Wilson 1985] R. Askey and J. Wilson, "Some basic hypergeometric orthogonal polynomials that generalize Jacobi polynomials", *Mem. Amer. Math. Soc.* **54**:319 (1985), iv+55.

[Bochner 1929] S. Bochner, "Über Sturm-Liouvillesche Polynomsysteme", *Math. Z.* **29**:1 (1929), 730–736.

[Castro and Grünbaum 2006] M. Castro and F. A. Grünbaum, "The algebra of matrix valued differential operators associated to a given family of matrix valued orthogonal polynomials: five instructive examples", *Internat. Math. Res. Notices* **2006**:7 (2006), 1–33.

[Chihara 1978] T. S. Chihara, *An introduction to orthogonal polynomials*, Gordon and Breach, New York, 1978.

[Dette et al. 2006] H. Dette, B. Reuther, W. J. Studden, and M. Zygmunt, "Matrix measures and random walks with a block tridiagonal transition matrix", *SIAM J. Matrix Anal. Appl.* **29**:1 (2006), 117–142.

[Diaconis and Shahshahani 1987] P. Diaconis and M. Shahshahani, "Time to reach stationarity in the Bernoulli-Laplace diffusion model", *SIAM J. Math. Anal.* **18**:1 (1987), 208–218.

[van Doorn 2003] E. A. van Doorn, "On associated polynomials and decay rates for birth-death processes", *J. Math. Anal. Appl.* **278**:2 (2003), 500–511.

[Duistermaat and Grünbaum 1986] J. J. Duistermaat and F. A. Grünbaum, "Differential equations in the spectral parameter", *Comm. Math. Phys.* **103**:2 (1986), 177–240.

[Dunkl and Xu 2001] C. F. Dunkl and Y. Xu, *Orthogonal polynomials of several variables*, Encyclopedia of Mathematics and its Applications **81**, Cambridge University Press, Cambridge, 2001.

[Duran 1997] A. J. Duran, "Matrix inner product having a matrix symmetric second order differential operator", *Rocky Mountain J. Math.* **27**:2 (1997), 585–600.

[Duran 1999] A. J. Duran, "Ratio asymptotics for orthogonal matrix polynomials", *J. Approx. Theory* **100**:2 (1999), 304–344.

[Durán and Polo 2002] A. J. Durán and B. Polo, "Gaussian quadrature formulae for matrix weights", *Linear Algebra Appl.* **355** (2002), 119–146.

[Ehrenfest 1907] P. Ehrenfest and T. Eherenfest, "Über zwei bekannte Einwände gegen das Boltzmannsche H-Theorem", *Phys. Z.* **8** (1907), 311–314.

[Feller 1967] V. Feller, *An introduction to probability theory and its applications*, vol. 1, 3rd ed., Wiley, New York, 1967.

[Geronimo and Iliev 2006] J. Geronimo and P. Iliev, "Two variable deformations of the Chebyshev measure", preprint, 2006. Available at arXiv.org/abs/math/0612489.

[Grünbaum 2003] F. A. Grünbaum, "Matrix valued Jacobi polynomials", *Bull. Sci. Math.* **127**:3 (2003), 207–214.

[Grünbaum and de la Iglesia 2007] F. A. Grünbaum and M. de la Iglesia, "Matrix valued orthogonal polynomials arising from group representation theory and a family of quasi-birth-and-death processes", preprint, 2007. Submitted.

[Grünbaum et al. 2002] F. A. Grünbaum, I. Pacharoni, and J. Tirao, "Matrix valued spherical functions associated to the complex projective plane", *J. Funct. Anal.* **188**:2 (2002), 350–441.

[Grünbaum et al. 2003] F. A. Grünbaum, I. Pacharoni, and J. Tirao, "Matrix valued orthogonal polynomials of the Jacobi type", *Indag. Math. (N.S.)* **14**:3-4 (2003), 353–366.

[Grünbaum et al. 2005] F. A. Grünbaum, I. Pacharoni, and J. Tirao, "Matrix valued orthogonal polynomials of Jacobi type: the role of group representation theory", *Ann. Inst. Fourier (Grenoble)* **55**:6 (2005), 2051–2068.

[Harnad and Kasman 1998] J. Harnad and A. Kasman (editors), *The bispectral problem* (Montreal, 1997), CRM Proceedings & Lecture Notes **14**, American Mathematical Society, Providence, RI, 1998.

[Harris 1952] T. E. Harris, "First passage and recurrence distributions", *Trans. Amer. Math. Soc.* **73** (1952), 471–486.

[Hoare and Rahman 1979] M. R. Hoare and M. Rahman, "Distributive processes in discrete systems", *Phys. A* **97**:1 (1979), 1–41.

[Hoare and Rahman 1984] M. R. Hoare and M. Rahman, "Cumulative Bernoulli trials and Krawtchouk processes", *Stochastic Process. Appl.* **16**:2 (1984), 113–139.

[Hoare and Rahman 1988] M. R. Hoare and M. Rahman, "Cumulative hypergeometric processes: a statistical role for the $_nF_{n-1}$ functions", *J. Math. Anal. Appl.* **135**:2 (1988), 615–626.

[Hoare and Rahman ≥ 2008] M. Hoare and M. Rahman, "A probabilistic origin for a new class of bivariate polynomials", Technical report. To appear.

[Iliev and Xu 2007] P. Iliev and Y. Xu, "Discrete orthogonal polynomials and difference equations of several variables", *Adv. Math.* **212**:1 (2007), 1–36.

[Ismail 2005] M. E. H. Ismail, *Classical and quantum orthogonal polynomials in one variable*, Encyclopedia of Mathematics and its Applications **98**, Cambridge University Press, Cambridge, 2005.

[Ismail et al. 1990] M. E. H. Ismail, D. R. Masson, J. Letessier, and G. Valent, "Birth and death processes and orthogonal polynomials", pp. 229–255 in *Orthogonal polynomials* (Columbus, OH, 1989), edited by P. Nevai, NATO Adv. Sci. Inst. Ser. C Math. Phys. Sci. **294**, Kluwer Acad. Publ., Dordrecht, 1990.

[Kac 1947] M. Kac, "Random walk and the theory of Brownian motion", *Amer. Math. Monthly* **54** (1947), 369–391.

[Karlin and McGregor 1957] S. Karlin and J. McGregor, "The classification of birth and death processes", *Trans. Amer. Math. Soc.* **86** (1957), 366–400.

[Karlin and McGregor 1959] S. Karlin and J. McGregor, "Random walks", *Illinois J. Math.* **3** (1959), 66–81.

[Karlin and McGregor 1961] S. Karlin and J. L. McGregor, "The Hahn polynomials, formulas and an application", *Scripta Math.* **26** (1961), 33–46.

[Koornwinder 1975] T. Koornwinder, "Two-variable analogues of the classical orthogonal polynomials", pp. 435–495 in *Theory and application of special functions*, edited by R. Askey, Math. Res. Center, Univ. Wisconsin, Publ. **35**, Academic Press, New York, 1975.

[Krein 1949] M. Krein, "Infinite J-matrices and a matrix-moment problem", *Doklady Akad. Nauk SSSR (N.S.)* **69** (1949), 125–128.

[Krein 1971] M. G. Krein, "Fundamental aspects of the representation theory of hermitian operators with deficiency index (m, m)", *Amer. Math. Soc. Transl.* (2) **97** (1971).

[Latouche and Ramaswami 1999] G. Latouche and V. Ramaswami, *Introduction to matrix analytic methods in stochastic modeling*, ASA-SIAM Series on Statistics and Applied Probability, Soc. Ind. Appl. Math., Philadelphia, 1999.

[Ledermann and Reuter 1954] W. Ledermann and G. E. H. Reuter, "Spectral theory for the differential equations of simple birth and death processes", *Philos. Trans. Roy. Soc. London. Ser. A.* **246** (1954), 321–369.

[Macdonald 2003] I. G. Macdonald, *Affine Hecke algebras and orthogonal polynomials*, Cambridge Tracts in Mathematics **157**, Cambridge University Press, Cambridge, 2003.

[Milch 1968] P. R. Milch, "A multi-dimensional linear growth birth and death process", *Ann. Math. Statist.* **39** (1968), 727–754.

[Neuts 1989] M. F. Neuts, *Structured stochastic matrices of $M/G/1$ type and their applications*, vol. 5, Dekker, New York, 1989.

[Pruitt 1963] W. E. Pruitt, "Bilateral birth and death processes", *Trans. Amer. Math. Soc.* **107** (1963), 508–525.

[Routh 1884] E. Routh, "On some properties of certain solutions of a differential equation of the second order", *Proc. London Math. Soc.* **16** (1884), 245–261.

[Schrödinger and Kohlrausch 1926] E. Schrödinger and F. Kohlrausch, "Das Ehrenfestche Model der H-Kurve", *Phys. Z.* **27** (1926), 306–313.

[Tirao 2003] J. A. Tirao, "The matrix-valued hypergeometric equation", *Proc. Natl. Acad. Sci. USA* **100**:14 (2003), 8138–8141.

F. ALBERTO GRÜNBAUM
DEPARTMENT OF MATHEMATICS
UNIVERSITY OF CALIFORNIA
BERKELEY, CA 94720
UNITED STATES
grunbaum@math.berkeley.edu

Probability, Geometry and Integrable Systems
MSRI Publications
Volume **55**, 2008

Integration of pair flows of the Camassa–Holm hierarchy

ENRIQUE LOUBET

To Henry McKean with my admiration and respect on the occasion
of his seventy-fifth birthday

ABSTRACT. We present the integration of the "pair" flows associated to the
Camassa–Holm (CH) hierarchy i.e., an explicit exact formula for the update
of the initial velocity profile in terms of initial data when run by the flow
associated to a Hamiltonian which (up to a constant factor) is given by the
sum of the reciprocals of the squares of any two eigenvalues of the underlying
spectral problem. The method stems from the integration of "individual" flows
of the CH hierarchy described in [Loubet 2006; McKean 2003], and is seen
to be more general in scope in that it may be applied when considering more
complex flows (e.g., when the Hamiltonian involves an arbitrary number of
eigenvalues of the associated spectral problem) up to when envisaging the full
CH flow itself which is nothing but a superposition of commuting individual
actions. Indeed, by incorporating piece by piece into the Hamiltonian the
distinct eigenvalues describing the spectrum associated to the initial profile,
we may recover McKean's Fredholm determinant formulas [McKean 2003]
expressing the evolution of initial data when acted upon by the full CH flow.
We also give account of the large-time (and limiting remote past and future)
asymptotics and obtain (partial) confirmation of the thesis about soliton genesis
and soliton interaction raised in [Loubet 2006].

Keywords: integrable systems, soliton traveling waves, spectral theory, Darboux transformations, asymptotic analysis.

Research partially supported by the Swiss National Science Foundation. This is gratefully acknowledged.

261

1. Introduction

The equation of Camassa and Holm (CH) [1993; 1994] is an approximate one-dimensional description of unidirectional propagation of long waves in shallow water. In dimensionless space-time variables it reads

$$\frac{\partial m}{\partial t} + (mD + Dm)(v) = 0 \tag{1}$$

in which $D = \partial/\partial x = (\cdot)'$, $m = (1 - D^2)v$ and at any given time t in $\mathbb{R}$, the real valued function $v = v(t, \cdot)$ represents the fluid velocity (or equivalently the height of the water's free surface above flat bottom). It is an infinite dimensional integrable bi-Hamiltonian system i.e., (1) is equivalent to

$$\frac{\partial m}{\partial t} = \mathcal{J}(m)\left(\frac{\partial H_{\mathrm{CH}}}{\partial m}\right) = \{m, H_{\mathrm{CH}}\}_{\mathcal{J}} = \{m, H_{\mathrm{CH}}^+\}_{\mathcal{K}} = \mathcal{K}\left(\frac{\partial H_{\mathrm{CH}}^+}{\partial m}\right)$$

with Hamiltonians

$$-H_{\mathrm{CH}} := \tfrac{1}{2} \int_{-\infty}^{+\infty} [v^2 + (v')^2] \quad \text{and} \quad -H_{\mathrm{CH}}^+ := \tfrac{1}{2} \int v[v^2 + (v')^2]$$

linked, via their corresponding functional gradients, by the Lenard raising/lowering rule [McKean 1993] as in $\mathcal{J}(m)(\partial H_{\mathrm{CH}}/\partial m) = \mathcal{K}(\partial H_{\mathrm{CH}}^+/\partial m)$. The pair

$$\mathcal{J}(m) := mD + Dm \qquad \text{and} \qquad \mathcal{K} := D(1 - D^2)$$

of *skew* operators (with respect to the H^0-inner product) being employed to define (via the H^0-inner product) a pair of *compatible* Poisson brackets so that, for a suitable class of functionals defined on phase space, a Lie algebra is specified (see [Loubet 2006] for more details.) Moreover, just like most integrable nonlinear evolution equations, CH equation (1) is also equivalent to the compatibility condition of an *overdetermined linear system* comprising the so called Lax pair; an evolution problem

$$\frac{\partial f}{\partial t} = \frac{1}{2}\frac{\partial v}{\partial x} f - \left(v + \frac{1}{2\lambda}\right)\frac{\partial f}{\partial x} \tag{2}$$

and a spectral problem, the acoustic equation with "potential" or "mass" m,

$$(1/4 - D^2)f = \lambda m f, \tag{3}$$

where λ and f denote, respectively, the eigenvalue and its associated eigenfunction. Here, compatibility means enforcing the matching of mixed derivatives i.e., $(f^{\bullet})'' = (f'')^{\bullet}$ where $\partial/\partial t = (\cdot)^{\bullet}$. It follows that (1) preserves the spectral characteristics of (3) i.e., CH flow is *isospectral*.

For real summable m the spectrum of (3) is purely discrete and simple i.e., $\mathrm{spec}(m) = \{\lambda_j(m) \in \mathbb{R}, \ j = \ldots, -1, 0, 1, \ldots\}$ where λ_j is a real value for which there exists a unique normalized solution f_j of (3) in H^1:

$$|f_j|_1^2 := \int \lambda_j m f_j^2 = \int [f_j'^2 + \tfrac{1}{4} f_j^2] = 1.$$

Most significant is that, within the class of summable m, CH flow is nothing but a *superposition of commuting* individual actions. Indeed, this opens the possibility to analyze CH flow via the accumulated effects that each of its constituents entail, the latter being presumed to be simpler to describe. And, indeed, the flows based upon a Hamiltonian of the form $H = 1/\lambda$ where λ is any eigenvalue of (3) turned out to be manageable [Loubet 2006]. More specifically, our goal was to elucidate as many qualitative properties of the full CH flow as possible from a direct and detailed analysis of the changes that each of its constitutive components produce when acting on generic initial data. We paid particular attention to how much could be said about the emergence of solitons for the CH flow by tracking down the effects of its individual actions. This investigation was possible from a careful analysis of explicit exact formulas for the updates of a generic initial profile run by any such elementary flow when expressed in terms of its private "Lagrangian" scale. Denoting by $\mathbb{X}_H$ the Hamiltonian vector field associated to the Hamiltonian H, $\phi_{\mathbb{X}_H}^t$ the corresponding flow map describing the updates $m := (\phi_{\mathbb{X}_H}^t m_0)$ at time t of the elements m_0 in phase space — here the class of real valued summable functions — and $\phi_{\mathbb{X}_H *}^t$ the flow that it induces on functionals of m_0, a "Lagrangian" scale is specified by

$$\frac{\partial \mathscr{L}_H^t}{\partial t} = -\left(\phi_{\mathbb{X}_H *}^t \frac{\partial H}{\partial m_0} \right) \circ \mathscr{L}_H^t, \qquad \mathscr{L}_H^0 = \mathrm{id}$$

in which $\partial H / \partial m_0$ denotes the functional gradient. We have proved:

THEOREM 1. *The Hamiltonian flow of the CH hierarchy arising from* $H = 1/\lambda$ *(where* λ *is an arbitrary eigenvalue of the acoustic equation* $f_0'' = (1/4 - \lambda m_0) f_0$ *which is associated to the summable initial data* $m_0 = v_0 - v_0''$) *is integrated explicitly in terms of the latter with the help of its private "Lagrangian" scale specified by*

$$\partial \mathscr{L}_H^t / \partial t = -(\phi_{\mathbb{X}_H *}^t \partial H / \partial m_0) \circ \mathscr{L}_H^t, \qquad \mathscr{L}_H^0 = \mathrm{id},$$

and three "theta" functions (each of which depends on t *and a spatial variable denoted by* $\cdot$ *if unspecified, e.g.,* $\vartheta = \vartheta(t, \cdot; \lambda m_0)$ *and so on), namely*

$$
\begin{matrix} \vartheta_- \\ \vartheta \\ \vartheta_+ \end{matrix} = 1 + (e^t - 1) \int_{-\infty}^{\cdot} \begin{cases} (f_0' - \tfrac{1}{2} f_0)^2 \\ \lambda m_0 f_0^2 \\ (f_0' + \tfrac{1}{2} f_0)^2. \end{cases} \tag{4}
$$

To wit,

$$
(\phi_{\mathbb{X}_H}^t v_0) \circ \mathscr{L}_H^t = \frac{\vartheta_+}{\vartheta_-}\left(\frac{v_0 - v_0'}{2}\right) + \frac{\vartheta_-}{\vartheta_+}\left(\frac{v_0 + v_0'}{2}\right)
$$
$$
+ \frac{\sqrt{\vartheta_-' \vartheta_+'}}{\lambda \vartheta_- \vartheta_+}(\vartheta_+ - \vartheta_-) + \frac{\vartheta}{2\lambda \vartheta_- \vartheta_+}(\vartheta_+ - \vartheta_-)',
$$

$$
\frac{\partial(\phi_{\mathbb{X}_H}^t v_0) \circ \mathscr{L}_H^t}{\partial \mathscr{L}_H^t} = -\frac{\vartheta_+}{\vartheta_-}\left(\frac{v_0 - v_0'}{2}\right) + \frac{\vartheta_-}{\vartheta_+}\left(\frac{v_0 + v_0'}{2}\right)
$$
$$
+ \frac{\sqrt{\vartheta_-' \vartheta_+'}}{\lambda \vartheta_- \vartheta_+}(\vartheta_+ - \vartheta_-)' + \frac{\vartheta}{2\lambda \vartheta_- \vartheta_+}(\vartheta_+ + \vartheta_-)'
$$
$$
+ \frac{\vartheta_-' \vartheta_+'}{\lambda \vartheta_- \vartheta_+ \vartheta}(\vartheta_+ - \vartheta_-)
$$

or, equivalently,

$$
(\phi_{\mathbb{X}_H}^t m_0) \circ \mathscr{L}_H^t = \left(\frac{\vartheta_- \vartheta_+}{\vartheta^2}\right)^2 m_0.
$$

It is remarkable that *all* updated expressions arising in the study of individual flows are given in terms of the theta functions (4). In fact, McKean had previously integrated the CH equation on the line by means of a triple of "theta-like" Fredholm determinants [McKean 2003]. The nomenclature is prompted from the fact that these determinants as well as their individual theta functions counterparts (4) satisfy a number of properties which are reminiscent of those met by Riemann's theta function together with its translates. Notably, there is only one theta-like (determinant) function, the others being produced from it by infinitesimal addition [McKean 2001]. Moreover, these (determinants) functions satisfy curious algebraic identities among themselves [McKean 2003]; the most significant one being

$$
\vartheta^2 = \vartheta_- \vartheta_+ + \vartheta_-' \vartheta_+ - \vartheta_- \vartheta_+'.
$$

In this paper, we will show that these underlying algebraic structures prevail when considering composite flows of a particular but sufficiently general class (see Theorem 2 below). Our aim is to offer a detailed account of the integration of the aforementioned pair flows associated with the CH hierarchy and discuss their large-time asymptotics. It will become clear to the reader that exactly the same method can be applied to integrate composite flows arising from Hamiltonians involving an arbitrary number of eigenvalues up to the full CH Hamiltonian which, when m is summable, satisfies

$$-H_{CH} = \tfrac{1}{2} \int mG * m = \frac{1}{4} \sum \frac{1}{\lambda_n^2}$$

where $G = e^{-|\cdot|}/2$ is the Green's function $(1 - D^2)G = \delta$, i.e., $G * m = v$; the sum accounting for all spectrum. Henceforth, we will focus in leading the reader along the hints and observations embodying the concatenation of lucky occurrences that culminate in the final expressions that substantiate the following main result.

THEOREM 2. *The Hamiltonian flow of the CH hierarchy arising from* $H = (1/\lambda_+^2 + 1/\lambda_-^2)/4$ *(in which* $\lambda_\pm$ *denote an arbitrary pair of eigenvalues of the acoustic equation* $(f_\pm^0)'' = (1/4 - \lambda_\pm m_0) f_\pm^0$ *which is associated to the summable initial data* $m_0 = v_0 - v_0''$*) is integrated explicitly in terms of the latter with the help of its private "Lagrangian" scale, specified by*

$$\partial \mathscr{L}_H^t / \partial t = -(\phi_{\chi_H}^t * \partial H / \partial m_0) \circ \mathscr{L}_H^t, \qquad \mathscr{L}_H^0 = \mathrm{id}$$

and three "theta" determinants (each of which depends on t and a spatial variable denoted by $\cdot$ if unspecified, e.g., $\Theta = \Theta(t, \cdot; \lambda m_0)$ and so on), namely

$$\begin{matrix} \Theta_- \\ \Theta \\ \Theta_+ \end{matrix} := \det \left[\mathrm{Id} + \mathscr{E}(t, \Lambda) \left\{ \begin{matrix} \int_{-\infty}^{\cdot} \Upsilon_{-,0} \otimes \Upsilon_{-,0} \\ \Lambda \Phi_0 \\ \int_{-\infty}^{\cdot} \Upsilon_{+,0} \otimes \Upsilon_{+,0} \end{matrix} \right. \right]. \tag{5}$$

where Id= *the* 2×2 *identity matrix,*

$$\Lambda := \begin{pmatrix} \lambda_- & 0 \\ 0 & \lambda_+ \end{pmatrix}, \qquad \mathscr{E}(t, \Lambda) := (e^{t/(2\Lambda)} - \mathrm{Id}),$$

$$\Upsilon_{-,0} := \begin{pmatrix} (f_-^0)' - \tfrac{1}{2} f_-^0 \\ (f_+^0)' - \tfrac{1}{2} f_+^0 \end{pmatrix}, \qquad \Upsilon_{+,0} := \begin{pmatrix} (f_-^0)' + \tfrac{1}{2} f_-^0 \\ (f_+^0)' + \tfrac{1}{2} f_+^0 \end{pmatrix},$$

and

$$\Phi_0 := \int_{-\infty}^{\cdot} m_0 f_0 \otimes f_0 = \begin{pmatrix} \varphi_-^0/\lambda_- & \varphi_0 \\ \varphi_0 & \varphi_+^0/\lambda_+ \end{pmatrix}$$

where $f_0 := (f_-^0, f_+^0)^\dagger$ *and*

$$\varphi_0 := \int_{-\infty}^{\cdot} m_0 f_-^0 f_+^0; \qquad \varphi_\pm^0 := \int_{-\infty}^{\cdot} \lambda_\pm m_0 (f_\pm^0)^2.$$

To wit,

$$(\phi^t_{\mathbb{X}_H} v_0) \circ \mathscr{L}^t_H = \frac{\Theta_-}{\Theta_+} \left(\frac{v_0 + v'_0}{2} + \tfrac{1}{2} \Upsilon^\dagger_{+,0} (\mathscr{M}\Lambda)^{-1} \mathscr{E} \Upsilon_{+,0} \right)$$
$$+ \frac{\Theta_+}{\Theta_-} \left(\frac{v_0 - v'_0}{2} - \tfrac{1}{2} \Upsilon^\dagger_{-,0} (\mathscr{M}\Lambda)^{-1} \mathscr{E} \Upsilon_{-,0} \right)$$

$$\frac{\partial(\phi^t_{\mathbb{X}_H} v_0) \circ \mathscr{L}^t_H}{\partial \mathscr{L}^t_H} = \frac{\Theta_-}{\Theta_+} \left(\frac{v_0 + v'_0}{2} + \tfrac{1}{2} \Upsilon^\dagger_{+,0} (\mathscr{M}\Lambda)^{-1} \mathscr{E} \Upsilon_{+,0} \right)$$
$$- \frac{\Theta_+}{\Theta_-} \left(\frac{v_0 - v'_0}{2} - \tfrac{1}{2} \Upsilon^\dagger_{-,0} (\mathscr{M}\Lambda)^{-1} \mathscr{E} \Upsilon_{-,0} \right)$$

where $\mathscr{M}(t, \cdot) := \mathrm{Id} + \mathscr{E}(t, \Lambda) \Lambda \Phi_0$ *(so that* $\Theta = \det \mathscr{M}$*) or, equivalently,*

$$(\phi^t_{\mathbb{X}_H} m_0) \circ \mathscr{L}^t_H = \left(\frac{\Theta_- \Theta_+}{\Theta^2} \right)^2 m_0.$$

The algebraic similitude of the formulas in Theorem 1 and Theorem 2 might, in part, be at the core of why most interesting features pertaining to the full flow are already reflected at the level of its components. Furthermore, we interpret this fact as stronger evidence substantiating the nature, interplay, and relevance of individual flows to the understanding of the underlying mechanisms that are involved in soliton formation and soliton interactions.

Indeed, the explicit formulas of Theorem 1 were shown to be valuable while conducting the large-time asymptotic analysis in that they afforded a mathematical treatment to establish the eventual emergence (provided we waited long enough) of a soliton escaping to infinity at a speed commensurable to the eigenvalue characterizing the individual flow at play. See [Loubet 2006].

THEOREM 3. *Assume* $\lambda > 0$, *and let the real summable initial data* $m_0 = v_0 - v''_0$ *be such that* $m_0 = o(1)$ *for* $x \ll 0$ *and disposed as in* $\mathrm{sign}\{\lambda m_0(x)\} = \mathrm{sign}\{x\}$. *Then, provided that we wait long enough, we see that, to leading order, the velocity profile (run by the individual flow arising from the Hamiltonian* $H = 1/\lambda$*) shapes itself like the escaping soliton. In symbols:*

$$\mathscr{L}^t_H(x) \in [\mathscr{L}^t_H(\mathrm{R}_-(t, \kappa)), \mathscr{L}^t_H(\mathrm{R}_+(t, \kappa))] \quad \textit{for all } x,$$
$$\left| [(\phi^t_{\mathbb{X}_H} v_0) \circ \mathscr{L}^t_H](x) - \frac{1}{2\lambda} e^{-|\mathscr{L}^t_H(x) - \mathscr{L}^t_H(\mathrm{R}_0(t))|} \right| = o(1) \quad \textit{as } t \uparrow +\infty,$$

where

$$\mathcal{L}_H^t\left(R_-(t,\kappa)\right) + t = -\log\left(\frac{1+\kappa}{\kappa}\right) + o(1)$$

$$\mathcal{L}_H^t\left(R_0(t)\right) + t = o(1) \qquad\qquad as \quad t \uparrow +\infty,$$

$$\mathcal{L}_H^t\left(R_+(t,\kappa)\right) + t = +\log\left(\frac{1+\kappa}{\kappa}\right) + o(1)$$

and R_0, $R_\pm$ are, for sufficiently large times, defined respectively by

$$\vartheta(t, R_0(t)) := 0, \quad and \quad \vartheta_\pm(t, R_\mp(t,\kappa)) := 1 + \kappa^{\pm 1}$$

κ being a nonnegative parameter.

Note that the signature disposition $\mathrm{sign}\{\lambda m_0(x)\} = \mathrm{sign}\{x\}$ on initial data m_0 guarantees breakdown (i.e., $v' \downarrow -\infty$ for some $0 < t < +\infty$; see [McKean 1998; McKean 2004]) or, what is the same thing, the vanishing of $\vartheta(t, \cdot)$ for a sufficiently large time

$$t > T := \log\left[1 + \left(\int_{-\infty}^0 -\lambda m_0 f_0^2\right)^{-1}\right]$$

at a unique site $R_0(t) < 0$. Actually, the fact that the soliton of Theorem 3 (escaping to $-\infty$) has its peak (for $t > T$) precisely at the root $R_0(t)$ of $\vartheta(t, \cdot) = 0$ is merely accidental. Indeed, even in the case where no breakdown occurs (i.e., where ϑ no longer vanishes), one can adapt the analysis as in [Loubet 2006] to conclude about the genesis a soliton moving at the right of the origin (see concluding remarks in that reference). As the direction and speed of soliton propagation are given, respectively, by the signature and magnitude of the underlying eigenvalue, we see that similar large-time asymptotic behavior (as that following the results of Theorem 3 describing events way ahead into the future) would take place when going far back into the past.

What is more, the algebraic robustness of our formulas as $t \to \pm\infty$ offered further quantitative confirmation of the qualitative description of soliton genesis [Loubet $\geq$ 2007a; $\geq$ 2007b]. Indeed, as $t \to \pm\infty$, the soliton that emerged escapes to infinity leaving behind a stationary profile $\lim_{t\to\pm\infty}[(\phi_{\mathcal{X}_H}^t v_0) \circ \mathcal{L}_H^t]$. We have corroborated these facts from the energetic and spectral standpoints with the help of the exact limiting formulas describing the latter, and we have shown that the energy of the residual profile is less than the energy of the initial profile by an amount that corresponds exactly to the energy that is embodied (at any time $|t| < +\infty$) by the escaping soliton. On the other hand, the isospectrality of individual flows, $\mathrm{spec}\left[(\phi_{\mathcal{X}_H}^t m_0) \circ \mathcal{L}_H^t\right] = \mathrm{spec}\, m_0$, though true for any given $|t| < +\infty$, ceases to hold in the limits $t \to \pm\infty$. Indeed, the maps

$$m_0 \mapsto \lim_{t\to\pm\infty} \left[(\phi_{\mathcal{X}_H}^t m_0) \circ \mathcal{L}_H^t\right]$$

from initial to residual profiles have a Darboux-type character in that precisely the eigenvalue λ of (3) associated to m_0 from which the underlying individual flow was based upon is excised, i.e., it is no longer part of

$$\mathrm{spec}\Big(\lim_{t\to\pm\infty}[(\phi^t_{\mathbb{X}_H} m_0)\circ\mathscr{L}^t_H]\Big),$$

the discrete spectrum associated to the stationary profiles. In short, the discrete train of solitons generated by a suitable superposition of finitely many individual actions should be regarded as a caricature of the infinite soliton train describing the large-time asymptotics of the full CH model. Indeed, under current evidence [Loubet 2006], it is hard to disbelieve that the aforementioned pair flows (with $\lambda_- < 0 < \lambda_+$) will eventually give rise to symmetrically disposed pairs of solitons escaping from the origin, each with a fixed speed (which must be) regulated by their corresponding eigenvalue. We also elaborate on some of these themes in the present paper. Nonetheless, a rigorous mathematical verification of this intuitive picture may not be, a priori, as simple to establish as in the case of individual flows. On the one hand, the formulas pertaining to the pair flows involve theta-like determinants which in principle are harder to manipulate. More significantly, the waiting times before solitons occur need now to be distinguished quantitatively and not merely qualitatively ("long enough") as before. Indeed, the asymptotic analysis that is related to a Hamiltonian depending on a couple of distinct spectral values would require a precise estimate of the patience one must bear — which should depend somehow on the ratio of the intervening eigenvalues — before it is possible to detect a slower developing soliton trailing behind a faster sibling. In any case, even if such an attempt is proved to be successful — which in our opinion would constitute an instructive exercise — once we move up to the next stage in complexity, say, when considering "quadruple" flows and beyond, there is little hope that we would be able to discern the large-time asymptotics directly from the corresponding theta-like determinants of matrices of higher rank with components depending (rationally) on the eigenvalues. In short, we believe that a new approach is required to reach a concise mathematical understanding of soliton train formation associated to the CH equation. Be that as it may, our formulas might spur useful potential numerical experiments that might shed light into aspects of the genesis of solitons and their interactions (prior to their escape at infinity) that may encourage new promising strategies. Indeed, the algebraic similarity of the recipes that arise in each of the cases that we have considered so far, with the Fredholm analogues which McKean [2003] employed to give account of CH on the line, cannot simply be accidental.

2. Preparations

2.1. Identification of the pair flow. Let $H = (1/\lambda_+^2 + 1/\lambda_-^2)/4$ be the Hamiltonian corresponding (up to the constant factor $1/4$) to the reciprocal of the squares of *any* pair of eigenvalues $\lambda_\pm$ of the spectral problem (3 with real summable m_0), its associated H^1 eigenfunctions $f_\pm^0$ normalized as in

$$|f_\pm^0|_1^2 = \int \lambda_\pm m_0 (f_\pm^0)^2 = \int [((f_\pm^0)')^2 + \tfrac{1}{4}(f_\pm^0)^2] = 1. \tag{6}$$

(Here we have used the notation established on page 272.) A routine computation establishes that the H^0-functional gradient of the reciprocal of any eigenvalue (spectral invariant) is given by the square of the associated (normalized) eigenfunction, so that

$$\frac{\partial H}{\partial m_0} = \frac{1}{2\lambda_-}(f_-^0)^2 + \frac{1}{2\lambda_+}(f_+^0)^2.$$

Hence, the Hamiltonian pair flow is regulated by

$$m^{\bullet} = (mD + Dm)\left(\frac{1}{2\lambda_-}f_-^2 + \frac{1}{2\lambda_+}f_+^2\right), \quad m(0, \cdot) = m_0 \tag{7}$$

where $f_\pm := (\phi_{\mathbb{X}_H}^t * f_\pm^0)$ denote the normalized time t updates of $f_\pm^0$.

2.2. Induced flow on eigenfunctions. For summable m_0, the spectrum of (3) is discrete and simple [Loubet 2006]. Hence, as eigenfunctions are well-defined functionals of m_0, their variation is to be inferred from that of the potential, e.g., the evolution of the updates $f_\pm$ of the normalized eigenfunctions $f_\pm^0$ is dictated from that of m_0 and the normalization constraint. More precisely, it is prescribed by the solution of the inhomogeneous acoustic problem — which arises after taking the Lie derivative of the original acoustic problem along the vector field $\mathbb{X}_H$ i.e., after differentiating with respect to t the (time t) updated acoustic problem associated to $f_\pm$ — which conforms to the preservation of normalization. Let

$$\mathcal{A}_\pm := [1/4 - D^2 - \lambda_\pm m].$$

Then, by (7), the motion $\mathbb{X}_H[m](f_\pm)$ of the updates $f_\pm$ satisfies

$$\mathcal{A}_\pm(\mathbb{X}_H[m](f_\pm)) = \lambda_\pm f_\pm(\mathbb{X}_H[m]) = \lambda_\pm f_\pm \mathcal{I}\left(\frac{1}{2\lambda_-}f_-^2 + \frac{1}{2\lambda_+}f_+^2\right),$$

where $\mathcal{J} \equiv \mathcal{J}(m)$. Now, from the results in [Loubet 2006] pertaining to individual flows, we know that

$$\Omega_\pm := \tfrac{1}{2} f_\pm - f_\pm \int_{-\infty}^{\cdot} \lambda_\pm m f_\pm^2 = f_\pm(\tfrac{1}{2} - \varphi_\pm),$$

$$\Pi_\pm := -f_\mp \int_{-\infty}^{\cdot} \lambda_\pm m f_- f_+ = -\lambda_\pm f_\mp \varphi$$

satisfy $\mathcal{A}_\pm(\Omega_\pm) = \lambda_\pm f_\pm \mathcal{J}(f_\pm^2)$ and $\mathcal{A}_\pm(\Pi_\pm) = \lambda_\pm f_\pm \mathcal{J}(f_\mp^2)$. Then, as $\mathcal{A}_\pm$ and $\mathcal{J}$ are linear, we have

$$\mathcal{A}_\pm\left(\frac{1}{2\lambda_\pm}\Omega_\pm + \frac{1}{2\lambda_\mp}\Pi_\pm\right) = \lambda_\pm f_\pm \mathcal{J}\left(\frac{1}{2\lambda_-}f_-^2 + \frac{1}{2\lambda_+}f_+^2\right).$$

In these expressions

$$\varphi_\pm := (\phi_{\mathbb{X}_H}^t * \varphi_\pm^0) \qquad \text{and} \qquad \varphi := (\phi_{\mathbb{X}_H}^t * \varphi_0)$$

denote the time t updates of, respectively,

$$\varphi_\pm^0 := \int_{-\infty}^{\cdot} \lambda_\pm m_0 (f_\pm^0)^2 \quad \text{and} \quad \varphi_0 := \int_{-\infty}^{\cdot} m_0 f_-^0 f_+^0, \tag{8}$$

as the action of the (induced) flow commutes with integration,

$$(\phi_{\mathbb{X}_H}^t * \lambda_\pm m_0 (f_\pm^0)^2) = \lambda_\pm (\phi_{\mathbb{X}_H}^t m_0)(\phi_{\mathbb{X}_H}^t * f_\pm^0)^2 \equiv \lambda_\pm m f_\pm^2$$

and so on. Hence, we are tempted to declare that

$$\overset{\bullet}{f_\pm} = \mathbb{X}_H[m](f_\pm) := \frac{1}{2\lambda_\pm}\Omega_\pm + \frac{1}{2\lambda_\mp}\Pi_\pm = \frac{f_\pm}{2\lambda_\pm}(\tfrac{1}{2} - \varphi_\pm) - \frac{\lambda_\pm f_\mp}{2\lambda_\mp}\varphi. \tag{9}$$

To convince ourselves that this is the correct recipe, we need to check whether or not, under such evolution, the norm is preserved. The verification is simple: Let $\mathcal{N}_\pm(t) := \int \lambda_\pm m f_\pm^2$. Then, according to the "tentative" prescription (9),

$$\overset{\bullet}{\mathcal{N}_\pm} = \int\left[\lambda_\pm f_\pm^2 \mathcal{J}\left(\frac{f_-^2}{2\lambda_-} + \frac{f_+^2}{2\lambda_+}\right) + 2\lambda_\pm m f_\pm\left(\frac{f_\pm}{2\lambda_\pm}(\tfrac{1}{2} - \varphi_\pm) - \frac{\lambda_\pm f_\mp}{2\lambda_\mp}\varphi\right)\right].$$

As $\mathcal{J}$ is skew-symmetric and $f_\pm$ vanish at infinity, the integral of $f_\pm^2 \mathcal{J}(f_\mp^2)$ vanishes. On the other hand, $f_\pm^2 \mathcal{J}(f_\mp^2) = (m f_-^2 f_+^2 + (\lambda_\pm - \lambda_\mp)\varphi^2)'$, and since eigenfunctions associated to different eigenvalues are orthogonal ($\int m f_- f_+ = 0$) the last display reduces to

$$\overset{\bullet}{\mathcal{N}_\pm} = \frac{1}{2\lambda_\pm}\mathcal{N}_\pm(1 - \mathcal{N}_\pm)$$

from where it is plain that $\mathcal{N}_\pm(t) \equiv 1$ for every $|t| < +\infty$ since $\mathcal{N}_\pm(0) = 1$. In other words, the right-hand side of (9) dictates the evolution of $f_\pm$ that conforms to normalization.

2.3. Constants of motion and private Lagrangian scale. In addition to the infinite number of independent functionals in involution which are preserved by any flow of the CH hierarchy as follows from Magri's observation [1978] that *compatibility of brackets is equivalent to saying that the class of raisable functions is one and the same as the class of lowerable ones* and the Lenard scheme (starting from the lower/upper pair H_{CH} and H_{CH}^+) alluded to in the introduction, the CH equation has another infinite collection of fundamental invariants. They are defined as follows. Every nice functional H defined on phase space gives rise to a flow $\phi_{\mathbb{X}_H}^t$ i.e., a one-parameter group of diffeomorphisms of a domain of phase space into itself characterized by the solution curves $m := (\phi_{\mathbb{X}_H}^t m_0)$ starting at m_0 of the dynamical system — *in an underlying "original" spatial scale* x — associated to the (locally Lipschitz) Hamiltonian vector field: $m^{\bullet} := \mathbb{X}_H[m] = \mathcal{F}(\partial H/\partial m)$. It also gives rise to a *new* "Lagrangian" scale $\mathcal{L}_H^t(x) =: \bar{x}(t, x)$ characterized by $\bar{x}^{\bullet} = -(\partial H/\partial m) \circ \bar{x}$, $\bar{x}(0, x) = x$. That is, at any given time t up to breakdown [McKean 1998; 2004], the map $\bar{x}(t, \cdot) = \mathcal{L}_H^t$ is a diffeomorphism of the real line issuing from the identity. The upshot being that, at any time t before (possible) breakdown and for every x in $\mathbb{R}$, $[(\phi_{\mathbb{X}_H}^t m_0) \circ \mathcal{L}_H^t](x) \times [\partial \mathcal{L}_H^t(x)/\partial x]^2$ is a constant of motion $\equiv m_0(x)$. The verification is straightforward (see Remark 4 below). In particular, for the pair flow in question the associated Lagrangian scale obeys

$$\bar{x}^{\bullet} = -\left(\frac{1}{2\lambda_-} f_-^2 + \frac{1}{2\lambda_+} f_+^2 \right) \circ \bar{x}, \quad \bar{x}(0, \cdot) = \mathrm{id}. \tag{10}$$

REMARK 4. The fundamental invariant

$$[(\mathcal{L}_H^t)']^2 (\phi_{\mathbb{X}_H}^t m_0) \circ \mathcal{L}_H^t = m_0 \tag{11}$$

—or $(\bar{x}')^2 m \circ \bar{x} = m_0$ for short—in combination with the explicit form of the Green's function $G = e^{-|\cdot|}/2$ through which $v = G * m$ is connected to m show that the integration of *any* flow of the CH hierarchy boils down to the determination of the associated Lagrangian scale. Indeed, the formulas of Theorem 2 are obtained essentially following the explicit characterization of $\mathcal{L}_H^t$. The invariants (11) originate from intrinsic symmetries [Khesin and Misiołek 2003]. Indeed, CH (1) satisfies the least-action principle as it is a reexpression of geodesic flow on the group of smooth orientation preserving compressible diffeomorphisms on the line with respect to the right-invariant H^1-metric assimilated as the energy [Misiołek 1998]. On the other hand, Noether's theorem guarantees the existence of a first integral from each one-parameter subgroup that leaves the energy functional unchanged. By right-invariance, the elements of every orbit emanating from the identity constitute such a subgroup, and since these are plenty (one such for each initial direction in the tangent space at the identity alias the Lie algebra associated to the group); the corresponding infinite collection of associated invariants turns out to be embodied in a (one-parameter) identity (11). In other words, the CH equation (1) is nothing but a reexpression of the (time) invariance of the right-hand side of (11). More precisely, the Euler-Lagrange equation describing the critical points of the right-invariant H^1-energy functional reads

$$[(\mathscr{L}^t_{H_{\mathrm{CH}}})']^2 \cdot (1) \circ \mathscr{L}^t_{H_{\mathrm{CH}}} = (11)^{\bullet}_{H=H_{\mathrm{CH}}} = 0.$$

in which $\mathscr{L}^t_{H_{\mathrm{CH}}}$ is the "true" Lagrangian scale, that is,

$$(\mathscr{L}^t_{H_{\mathrm{CH}}})^{\bullet} = v \circ \mathscr{L}^t_{H_{\mathrm{CH}}} \quad \text{and} \quad \mathscr{L}^0_{H_{\mathrm{CH}}} = \mathrm{id}.$$

McKean [2003] (see also [Loubet 2006]) made the crucial observation that the first integral (11) remains in force for *all* other flows (i.e., flows originating from *any* Hamiltonian H) of the CH hierarchy provided that in each case a suitable "Lagrangian" scale is employed.

NOTATION. From now on, $m(t, x)$ will be short for the more cumbersome expression $(\phi^t_{\mathbb{X}_H} m_0)(x)$ describing the evaluation at x of the time t update of the solution curve starting at m_0, of the dynamical system (in phase space) defined by the Hamiltonian vector field $\mathbb{X}_H$, as explained in Section 2.3. Similarly, we will denote by

$$f_{\pm}(t, x) := (\phi^t_{\mathbb{X}_H *} f^0_{\pm})(x)$$

the evaluations at x of the (normalized time t) updates $(\phi^t_{\mathbb{X}_H *} f^0_{\pm})$ of the normalized eigenfunctions $f^0_{\pm}$ associated, respectively, to the spectral parameters $\lambda_{\pm}$, and so on. Moreover, whenever not confusing, we will occasionally omit the explicit dependence and write m and $f_{\pm}$ plain for brevity. In other words, all expressions with an upper/lower index "0" refer to (and hence involve purely) initial data whereas their counterparts where the label "0" is dropped account implicitly for their (time t) updates when acted upon by the (induced pair) flow. We emphasize that all initial and updated expressions are *functions* of an underlying independent spatial variable (denoted by $\cdot$ if left out), unspecified unless explicitly stated otherwise. In addition, in an effort to avoid unnecessary details which should be clear from the context, we omit writing explicitly the dummy variables and the differentials intervening in the integrands of integral expressions (for example, $\int_{-\infty}^{\cdot} e^y m_0$ is short for $\int_{-\infty}^{\cdot} e^y m_0(y) dy$ and so on.) Also, sometimes the same expressions will be used to denote both the functions and their evaluations at x. For example, depending on the context, when we write $\int^{\bar{x}}$ we might mean the function $\int^{\bar{x}(t,\cdot)}$, as in (13) below, or its evaluation at x, namely, $\int^{\bar{x}(t,x)}$, as in (19a). Finally, identities employing subscript $\pm$ are short for *two* such expressions, one with subscript $+$ and one with $-$.

3. The road to integration: lucky facts

3.1. Building up integrable expressions. As pointed out in Remark 4, the integration of the pair flows succumbs to the computation of their associated Lagrangian scale. This suggests that the key to integration is to play around with "sensible" objects involving the latter. The idea is to look for expressions incorporating the Lagrangian scale and functions of interest, whose evolution (under the pair flow) leads to integrable formulas from which to infer subsequently the integration of the items we actually care about. As in the case pertaining to individual flows [Loubet 2006] we start by analyzing whether

$$(\phi^t_{\chi_H} * \varphi^0_\pm) \circ \mathscr{L}^t_H := \left(\phi^t_{\chi_H} * \int_{-\infty}^{\cdot} \lambda_\pm m_0 (f^0_\pm)^2 \right) \circ \mathscr{L}^t_H$$

i.e., the composition of the time t update $\varphi_\pm$ of $\varphi^0_\pm$ with the diffeomorphism on the line given by the Lagrangian scale at t (i.e., $\mathscr{L}^t_H$) can be expressed in an alternative closed form. To this end, we compute the time derivative of $\varphi_\pm \circ \bar{x} = \int_{-\infty}^{\bar{x}} \lambda_\pm m f^2_\pm$ and explore to what extent the resulting equation is integrable. Direct computation using (7) and (9) yields

$$[\varphi_\pm \circ \bar{x}]^\bullet = \lambda_\pm (m f^2_\pm) \circ \bar{x} \cdot \bar{x}^\bullet$$
$$+ \int_{-\infty}^{\bar{x}} \left[\lambda_\pm f^2_\pm \mathscr{I} \left(\frac{1}{\lambda_-} f^2_- + \frac{1}{\lambda_+} f^2_+ \right) + 2\lambda_\pm m f_\pm \left(\frac{f_\pm}{2\lambda_\pm} (\tfrac{1}{2} - \varphi_\pm) - \frac{\lambda_\pm f_\mp}{2\lambda_\mp} \varphi \right) \right].$$

As $f^2_\pm \mathscr{I}(f^2_\pm) = (m f^4_\pm)'$ and $f^2_\pm \mathscr{I}(f^2_\mp) = (m f^2_- f^2_+ + (\lambda_\pm - \lambda_\mp)\varphi^2)'$ it follows by (10) after cancellations and appropriate identifications that

$$[\varphi_\pm \circ \bar{x}]^\bullet = \left(\frac{1}{2\lambda_\pm} \varphi_\pm (1 - \varphi_\pm) - \frac{\lambda_\pm}{2} \varphi^2 \right) \circ \bar{x}. \tag{12a}$$

In sharp contrast with the analogue equation arising in the study of individual flows [Loubet 2006], equation (12a) *is not* an ODE, as the presence of the term $\lambda_\pm \varphi^2/2$ shows; see (8). It accounts for the mutual interaction of the underlying individual flows comprising the pair flow. Hence, as it stands, equation (12a) can only be useful if we manage somehow to determine, a priori, $\varphi \circ \bar{x}$. Now, analogous manipulations to the ones leading to (12a) show that

$$[\varphi \circ \bar{x}]^\bullet = \left(\left[\frac{1}{2\lambda_-} (\tfrac{1}{2} - \varphi_-) + \frac{1}{2\lambda_+} (\tfrac{1}{2} - \varphi_+) \right] \varphi \right) \circ \bar{x}. \tag{12b}$$

Neither of the *coupled* equations (12a) and (12b) pertaining to the evolution under the pair flow of the Lagrangian-scaled valued updates of $\varphi^0_\pm$, respectively

φ_0, are seen to be integrable when looked at individually, but if we combine these expressions suitably in a (symmetric) matrix as in

$$\Phi_0 := \begin{pmatrix} \frac{1}{\lambda_-}\varphi_-^0 & \varphi_0 \\ \varphi_0 & \frac{1}{\lambda_+}\varphi_+^0 \end{pmatrix} = \int_{-\infty}^{\cdot} m_0 f_0 \otimes f_0, \qquad (12c)$$

where $f_0 := (f_-^0, f_+^0)^{\dagger}$, we learn that (12a) and (12b) translate into

$$[\Phi \circ \bar{x}]^{\bullet} = \left(-\tfrac{1}{2}\Phi^2 + \tfrac{1}{4}[\Lambda^{-1}\Phi + \Phi\Lambda^{-1}]\right) \circ \bar{x}, \qquad (12d)$$

where $\Phi := (\phi_{\mathbb{X}_H *}^t \Phi_0)$ and $\Lambda := \mathrm{diag}(\lambda_-, \lambda_+)$. This is the kind of luck that we were after: all we have to do now is solve an ODE! The last term on the right-hand side of (12d) suggests an ansatz of the form $\Phi \circ \bar{x} := e^{t/(4\Lambda)}\mathcal{G}(t, \cdot)e^{t/(4\Lambda)}$. By direct computation, the latter is seen to satisfy

$$[\Phi \circ \bar{x}]^{\bullet} = \frac{1}{4}[\Lambda^{-1}\Phi + \Phi\Lambda^{-1}] \circ \bar{x} + e^{t/(4\Lambda)} \overset{\bullet}{\mathcal{G}}(t, \cdot)e^{t/(4\Lambda)}.$$

We would be done if we could find $\mathcal{G}(t, x)$ such that $\mathcal{G}(0, x) = \Phi_0(x)$ and $e^{t/(4\Lambda)}\mathcal{G}^{\bullet}(t, \cdot)e^{t/(4\Lambda)} = -\tfrac{1}{2}[\Phi \circ \bar{x}]^2$. Now, the derivative of the inverse of a matrix is *quadratic*, i.e., $(\mathbb{O}^{-1})^{\bullet} = -\mathbb{O}^{-1}\mathbb{O}^{\bullet}\mathbb{O}^{-1}$, so that if $\mathcal{G}(t, x) := \mathcal{P}(x)\mathbb{O}^{-1}(t, x)$ then $e^{t/(4\Lambda)}\mathcal{G}^{\bullet}(t, \cdot)e^{t/(4\Lambda)} = -[\Phi \circ \bar{x}]e^{-t/(4\Lambda)}\mathbb{O}^{\bullet}\mathbb{O}^{-1}e^{t/(4\Lambda)}$, and thus, it would suffice to find $\mathbb{O}$ such that $e^{-t/(4\Lambda)}\mathbb{O}^{\bullet}\mathbb{O}^{-1}e^{t/(4\Lambda)} = \tfrac{1}{2}\Phi \circ \bar{x}$ or, what is the same thing, $\mathbb{O}^{\bullet} = \tfrac{1}{2}e^{t/(2\Lambda)}\mathcal{P} = (\Lambda e^{t/(2\Lambda)}\mathcal{P})^{\bullet}$. This implies that $\mathbb{O}(t, x) = \Lambda e^{t/(2\Lambda)}\mathcal{P}(x) + \mathcal{D}(x, \Lambda)$ for some 2×2 matrix $\mathcal{D}$. Finally, we observe that the initial constraint, $\Phi_0 = \mathcal{P}(\Lambda\mathcal{P} + \mathcal{D})^{-1}$, can be met by setting $\mathcal{D}(\cdot, \Lambda) := \mathrm{Id} - \Lambda\mathcal{P}$ and $\mathcal{P} := \Phi_0$. In short, the solution of (12d) is given by

$$\Phi \circ \bar{x} = e^{t/(4\Lambda)}\Phi_0[\mathrm{Id} + (e^{t/(2\Lambda)} - \mathrm{Id})\Lambda\Phi_0]^{-1}e^{t/(4\Lambda)}. \qquad (13)$$

It will be helpful to introduce a shorthand and record the latter (or its evaluation at x: see Notation on page 272) as

$$[\Phi \circ \bar{x}](x) = \mathcal{T}(t)\Phi_0(x)[\mathcal{M}(t, x)]^{-1}\mathcal{T}(t),$$

where

$$\mathcal{M}(t, x) := \mathrm{Id} + \mathcal{E}(t, \Lambda)\Lambda\Phi_0(x) = \mathrm{Id} + \mathcal{C}(t, \Lambda)\Phi_0(x) \qquad (14a)$$

in which

$$\mathcal{C}(t, \Lambda) := \mathcal{E}(t, \Lambda)\Lambda, \quad \mathcal{E}(t, \Lambda) := (\mathcal{T}^2(t) - \mathrm{Id}), \quad \mathcal{T}(t) := e^{t/(4\Lambda)}. \quad (14b)$$

For reasons that will be clear in a moment, we also need to investigate whether the evolution of other (Lagrangian-scaled valued updates of) integral expressions involving the eigenfunctions $f_{\pm}^0$, which are associated to the pair of eigenvalues $\lambda_{\pm}$ that define the flow and the so called "improper" eigenfunctions $e^{\pm \cdot/2}$ of the acoustic equation ($\lambda = 0$ is not in the spectrum of (3 with summable m_0)) admit alternative spellings. In other words, we play the same game as before with the exception that this time we look at the evolution of the truncated integrals

$$\mathcal{F}^{\downarrow}_{0,\pm} := \int_{-\infty}^{\cdot} \lambda_{\pm} m_0 f^0_{\pm} e^{y/2} = -e^{\cdot/2}((f^0_{\pm})' - \tfrac{1}{2} f^0_{\pm}) = -[f^0_{\pm}, e^{\cdot/2}],$$

$$\mathcal{F}^{\uparrow}_{0,\pm} := \int_{\cdot}^{+\infty} \lambda_{\pm} m_0 f^0_{\pm} e^{-y/2} = e^{-\cdot/2}((f^0_{\pm})' + \tfrac{1}{2} f^0_{\pm}) = [f^0_{\pm}, e^{-\cdot/2}],$$

(15a)

where the bracket $[f, g]$ is short for the Wronskian $f'g - fg'$. These expressions can also be written in integrated form in terms of Wronskians. It develops after some work that $\mathcal{F}^{\downarrow}_{\pm} := (\phi^t_{X_H} * \mathcal{F}^{\downarrow}_{0,\pm})$ and $\mathcal{F}^{\uparrow}_{\pm} := (\phi^t_{X_H} * \mathcal{F}^{\uparrow}_{0,\pm})$ satisfy

$$[\mathcal{F}^{\downarrow}_{\pm} \circ \bar{x}]^{\bullet} = \left(\frac{1}{2\lambda_{\pm}} (\tfrac{1}{2} - \varphi_{\pm}) \mathcal{F}^{\downarrow}_{\pm} - \frac{\lambda_{\pm}}{2\lambda_{\mp}} \varphi \mathcal{F}^{\downarrow}_{\mp} \right) \circ \bar{x},$$

$$[\mathcal{F}^{\uparrow}_{\pm} \circ \bar{x}]^{\bullet} = \left(\frac{1}{2\lambda_{\pm}} (\tfrac{1}{2} - \varphi_{\pm}) \mathcal{F}^{\uparrow}_{\pm} - \frac{\lambda_{\pm}}{2\lambda_{\mp}} \varphi \mathcal{F}^{\uparrow}_{\mp} \right) \circ \bar{x}.$$

To compute the solutions of these coupled systems of equations (and hence produce the desired tentative new spellings), we are led to pack the expressions (15a) into the vectors

$$\mathcal{F}^{\downarrow}_0 := \begin{pmatrix} \mathcal{F}^{\downarrow}_{0,-}/\lambda_- \\ \mathcal{F}^{\downarrow}_{0,+}/\lambda_+ \end{pmatrix} = -\Lambda^{-1} \begin{pmatrix} [f^0_-, e^{\cdot/2}] \\ [f^0_+, e^{\cdot/2}] \end{pmatrix} = -e^{\cdot/2} \Lambda^{-1} \Upsilon_{-,0},$$

$$\mathcal{F}^{\uparrow}_0 := \begin{pmatrix} \mathcal{F}^{\uparrow}_{0,-}/\lambda_- \\ \mathcal{F}^{\uparrow}_{0,+}/\lambda_+ \end{pmatrix} = \Lambda^{-1} \begin{pmatrix} [f^0_-, e^{-\cdot/2}] \\ [f^0_+, e^{-\cdot/2}] \end{pmatrix} = e^{-\cdot/2} \Lambda^{-1} \Upsilon_{+,0},$$

(15b)

where $\Upsilon_{\pm,0} := \pm(\tfrac{1}{2} \pm D) f_0$, i.e.,

$$\Upsilon_{-,0} := \begin{pmatrix} (f^0_-)' - \tfrac{1}{2} f^0_- \\ (f^0_+)' - \tfrac{1}{2} f^0_+ \end{pmatrix} \quad \text{and} \quad \Upsilon_{+,0} := \begin{pmatrix} (f^0_-)' + \tfrac{1}{2} f^0_- \\ (f^0_+)' + \tfrac{1}{2} f^0_+ \end{pmatrix},$$

(15c)

so that $\mp(\tfrac{1}{2} \pm D)\Upsilon_{\mp,0} = \Lambda m_0 f_0$. Indeed, the evolution of their respective components (as displayed lines above) is gathered nicely in the system of *uncoupled* ODE's

$$[\mathcal{F}^{\downarrow} \circ \bar{x}]^{\bullet} = \{ \tfrac{1}{2} (\tfrac{1}{2} \Lambda^{-1} - \Phi) \mathcal{F}^{\downarrow} \} \circ \bar{x},$$

$$[\mathcal{F}^{\uparrow} \circ \bar{x}]^{\bullet} = \{ \tfrac{1}{2} (\tfrac{1}{2} \Lambda^{-1} - \Phi) \mathcal{F}^{\uparrow} \} \circ \bar{x},$$

$\Phi \circ \bar{x}$ being already known from (13). Since $\mathcal{F}^{\downarrow} := (\phi^t_{X_H} * \mathcal{F}^{\downarrow}_0)$ and $\mathcal{F}^{\uparrow} := (\phi^t_{X_H} * \mathcal{F}^{\uparrow}_0)$ differ only in their initial values ($\mathcal{F}^{\downarrow}_0$ and $\mathcal{F}^{\uparrow}_0$), we only need to deal further with either of them, say $\mathcal{F}^{\downarrow}$. Direct computation shows that the educated guesses

$$\mathscr{F}^{\downarrow} \circ \bar{x} = e^{t/(4\Lambda)}[(\mathcal{M}(t, \cdot))^{\dagger}]^{-1}\mathscr{F}_0^{\downarrow},$$
$$\mathscr{F}^{\uparrow} \circ \bar{x} = e^{t/(4\Lambda)}[(\mathcal{M}(t, \cdot))^{\dagger}]^{-1}\mathscr{F}_0^{\uparrow}, \qquad (16)$$

with $\mathcal{M}$ as in (14a), provide the answer. Indeed, the time derivative of the right-hand side of the second line yields

$$\mathscr{T}^{\bullet}(\mathcal{M}^{\dagger})^{-1}\mathscr{F}_0^{\downarrow} - \mathscr{T}(\mathcal{M}^{\dagger})^{-1}(\mathcal{M}^{\dagger})^{\bullet}(\mathcal{M}^{\dagger})^{-1}\mathscr{F}_0^{\downarrow}.$$

As $\mathscr{T}^{\bullet} = \frac{1}{4}\Lambda^{-1}\mathscr{T}$ and $(\mathcal{M}^{\dagger})^{\bullet} = 2\Phi_0\Lambda\mathscr{T}\mathscr{T}^{\bullet} = \frac{1}{2}\Phi_0\mathscr{T}^2$ — see (14a) and (14b) — the latter reduces to

$$\left(\tfrac{1}{4}\Lambda^{-1} - \tfrac{1}{2}\mathscr{T}(\mathcal{M}^{\dagger})^{-1}\Phi_0\mathscr{T}\right)[\mathscr{F}^{\downarrow} \circ \bar{x}].$$

The verification is completed by appealing to the identity

$$(\mathcal{M}^{\dagger})^{-1}\Phi_0 = \Phi_0\mathcal{M}^{-1} \qquad (17)$$

and the preliminary integration result (13) so that

$$\mathscr{T}(\mathcal{M}^{\dagger})^{-1}\Phi_0\mathscr{T} = \mathscr{T}\Phi_0\mathcal{M}^{-1}\mathscr{T} \equiv \Phi \circ \bar{x}.$$

4. Determination of the Lagrangian scale

The trick to get an explicit formula for the Lagrangian scale $\bar{x} = \bar{x}(t, x)$ of section 2.3 in terms of time t, the original spatial scale x, and initial data, is to "peel off," in an orderly fashion, the integrated expressions of section 3. More precisely, we start by differentiating with respect to the underlying variable x both sides of the identity (see (13), (12c), (8), and the notation clarifications on page 272)

$$\int_{-\infty}^{\bar{x}} m\boldsymbol{f} \otimes \boldsymbol{f} = [\Phi \circ \bar{x}](x) = e^{t/(4\Lambda)}\Phi_0(x)[\mathcal{M}(t, x)]^{-1}e^{t/(4\Lambda)}$$

where $\boldsymbol{f} := (f_-, f_+)^{\dagger} = (\phi_{\mathbb{X}_H}^t * \boldsymbol{f}_0)$ is the update of $\boldsymbol{f}_0 := (f_-^0, f_+^0)^{\dagger}$. Writing $\mathcal{M} \equiv \mathrm{Id} + \mathscr{C}\Phi_0$ for simplicity as in (14a), we get

$$(m\boldsymbol{f} \otimes \boldsymbol{f}) \circ \bar{x} \cdot \bar{x}' = e^{t/(4\Lambda)}(\Phi_0\mathcal{M}^{-1})'e^{t/(4\Lambda)}.$$

From (17) and the equality $\mathcal{M}^{\dagger} = \mathrm{Id} + \Phi_0\mathscr{C}$ ($\mathscr{C}$ and Φ_0 being symmetric) we see that $(\Phi_0\mathcal{M}^{-1})'$ is equal to

$$(\mathrm{Id} - \Phi_0\mathcal{M}^{-1}\mathscr{C})\Phi_0'\mathcal{M}^{-1} = (\mathrm{Id} - (\mathcal{M}^{\dagger})^{-1}\Phi_0\mathscr{C})\Phi_0'\mathcal{M}^{-1} = (\mathcal{M}^{\dagger})^{-1}\Phi_0'\mathcal{M}^{-1}.$$

As $\Phi_0' = m_0\boldsymbol{f}_0 \otimes \boldsymbol{f}_0$, the latter in combination with the fundamental invariant (11), reduce (by associativity and linearity) the previous to last display to $m_0/\bar{x}'$ times

$$(\boldsymbol{f} \otimes \boldsymbol{f}) \circ \bar{x} = \bar{x}'(e^{t/(4\Lambda)}(\mathcal{M}^{\dagger})^{-1}\boldsymbol{f}_0) \otimes (e^{t/(4\Lambda)}(\mathcal{M}^{\dagger})^{-1}\boldsymbol{f}_0).$$

As m_0 is not identically zero and since $\bar{x}' > 0$ (at least for small times; see (10)), we use linearity once more to recognize that

$$f \circ \bar{x} = \sqrt{\bar{x}'} e^{t/(4\Lambda)} [(\mathcal{M}(t, \cdot))^\dagger]^{-1} f_0. \tag{18}$$

It is not surprising that the determination of the Lagrangian-scaled update of f_0 given by the left-hand side of (18) would follow once we have an explicit formula for $\bar{x}$. In fact (18) should be interpreted the other way around, namely, as a step towards the determination of $\bar{x}$: by equation (10), the trace identity

$$\mathrm{Tr}\big(\tfrac{1}{2}\Lambda^{-1}(f \otimes f) \circ \bar{x}\big) = \Big(\frac{1}{2\lambda_-} f_-^2 + \frac{1}{2\lambda_+} f_+^2\Big) \circ \bar{x} \equiv -\bar{x}^\bullet,$$

in combination with the *partial* result (18) leads to

$$\bar{x}^\bullet + \Gamma(t, x)\bar{x}' = 0,$$

in which $\Gamma(t, x)$ is short for the trace of

$$\tfrac{1}{2}\Lambda^{-1}(e^{t/(4\Lambda)}[(\mathcal{M}(t, x))^\dagger]^{-1} f_0(x)) \otimes (e^{t/(4\Lambda)}[(\mathcal{M}(t, x))^\dagger]^{-1} f_0(x)).$$

This is a first order linear evolution equation from which, in principle, the Lagrangian scale can be computed, $\bar{x}(0, x) = x$ being known. But to find an explicit expression for the solution of this seemingly trivial equation (say, by the method of characteristics) is not an easy matter. We actually take a different route that bumps into yet another piece of grace. The "problem" with (18) is that we do not have sufficient information about the shape of $f \circ \bar{x}$ to infer that of $\bar{x}$. To fix this, we apply the previous method to identities that incorporate the improper eigenfunctions $e^{\pm \cdot /2}$ ($\lambda = 0$ is not in $\mathrm{spec}(m_0)$ of (3) for summable m_0) whose shape is explicit and, more importantly, *fixed* for all times, improper eigenfunctions being insensitive to the potential. As a matter of fact, the computation of the scale $\bar{x}$ is not particularly sensitive to the actual shape of improper eigenfunctions but rather, and this is the key, to the fact that they are inverses of one another as we now show. Focus on the (evaluations at x) of the Lagrangian-scaled updates of (15a) for which we have found explicit alternative expressions (16), namely,

$$\int_{-\infty}^{\bar{x}} e^{y/2} m f = [\mathscr{F}^\downarrow \circ \bar{x}](x) = e^{t/(4\Lambda)}[(\mathcal{M}(t, x))^\dagger]^{-1} \mathscr{F}_0^\downarrow(x), \tag{19a}$$

$$\int_{\bar{x}}^{+\infty} e^{-y/2} m f = [\mathscr{F}^\uparrow \circ \bar{x}](x) = e^{t/(4\Lambda)}[(\mathcal{M}(t, x))^\dagger]^{-1} \mathscr{F}_0^\uparrow(x). \tag{19b}$$

Now, keeping in mind identity (11) i.e., $(\bar{x})'^2 m \circ \bar{x} = m_0$ for short, definitions (14a) and (12c), the fact that $f_0 \otimes f_0 = f_0 f_0^\dagger$ (for column vectors f_0) and the partial result (18), the x-derivative of both ends of (19a) yields (omitting here and there the explicit x-dependence)

$$\frac{m_0}{\bar{x}'}[f \circ \bar{x}]e^{\bar{x}/2} = \frac{m_0}{\bar{x}'}[f \circ \bar{x}]\sqrt{\bar{x}'}e^{x/2}\big(1 - e^{-x/2}(\mathscr{C}f_0)^\dagger((\mathcal{M}^\dagger)^{-1}\mathscr{F}_0^\downarrow)\big).$$

A similar expression follows from differentiating (19a) instead (the next display embodies both of them). Hence, dropping the common factor $m_0[f \circ \bar{x}]/\bar{x}'$, which is assumed to be different from zero (as is the case at least for small times), by linearity and appealing to the characterizations (15b) and (15c) involving the initial data, we have

$$e^{\pm \bar{x}/2} = \sqrt{\bar{x}'}e^{\pm x/2}\big(1 + (\mathscr{C}f_0)^\dagger((\Lambda\mathcal{M}^\dagger)^{-1}\Upsilon_{\mp,0})\big).$$

It is clear that the desired expressions for $\bar{x}'$ and in fact $\bar{x}$ can be obtained simply by *multiplying*, respectively, *dividing* the above identities. But before we actually do that, we wish to find more palatable expressions for the intervening factors. The coefficients on the right-hand side of the last display are of the form $1 + a^\dagger b$ (for column vectors a and b), so we can invoke the identity

$$1 + a^\dagger b \equiv 1 + a \cdot b = \det[\mathrm{Id} + a \otimes b]$$

to express them in terms of *determinants*. To wit,

$$1 + (\mathscr{C}f_0)^\dagger((\Lambda\mathcal{M}^\dagger)^{-1}\Upsilon_{\mp,0}) = \det[\mathrm{Id} + \mathscr{E}(\Phi_0\Lambda + f_0 \otimes \Upsilon_{\mp,0})]/\det\mathcal{M}, \quad (20)$$

where in the last step we used the formula for the determinant of a product of matrices and the fact that $\mathrm{Id} + a \otimes (\Lambda\mathcal{M}^\dagger)^{-1}b$ equals

$$\mathrm{Id} + a[(\Lambda\mathcal{M}^\dagger)^{-1}b]^\dagger = \mathrm{Id} + a \otimes b(\mathcal{M}\Lambda)^{-1} = [\mathcal{M}\Lambda + a \otimes b]\Lambda^{-1}\mathcal{M}^{-1},$$

so that by (14a)

$$\mathrm{Id} + (\mathscr{C}f_0) \otimes ((\Lambda\mathcal{M}^\dagger)^{-1}\Upsilon_{\mp,0}) = \Lambda(\mathrm{Id} + \mathscr{E}(\Phi_0\Lambda + f_0 \otimes \Upsilon_{\mp,0}))\Lambda^{-1}\mathcal{M}^{-1}.$$

Moreover, recalling (12c) we have by linearity, the acoustic equation (3), and definitions (15c) that $\Phi_0\Lambda$ is equal to

$$\int_{-\infty}^{\cdot} f_0 \otimes (m_0\Lambda f_0) = \int_{-\infty}^{\cdot} f_0 \otimes (1/4 - D^2)f_0 = \mp \int_{-\infty}^{\cdot} f_0 \otimes (\tfrac{1}{2} \pm D)\Upsilon_{\mp,0}.$$

Hence, upon integrating by parts, we see that

$$\Phi_0\Lambda + f_0 \otimes \Upsilon_{\mp,0} = \int_{-\infty}^{\cdot} \Upsilon_{\mp,0} \otimes \Upsilon_{\mp,0}.$$

Now, substituting this expression into (20) and recalling the definitions of the (evaluations at x of the) theta determinants (5) (bear in mind that all expressions following (19a) are assumed to be implicitly evaluated at x), we learn that

$$e^{\pm \bar{x}/2} = \sqrt{\bar{x}'} e^{\pm x/2} \frac{\Theta_{\mp}}{\Theta}.$$

Finally, as advertised, looking at the ratio and the product of the latter identities we get

$$e^{\bar{x}} = e^x \frac{\Theta_-}{\Theta_+}, \quad \text{respectively} \quad \bar{x}' = \frac{\Theta^2}{\Theta_- \Theta_+}. \tag{21}$$

REMARK 5. Actually, these identities are equivalent. Indeed, $\bar{x}(t, \pm\infty) = \pm\infty$ since $\Theta_\pm(t, -\infty) = 1$ and $\Theta_\pm(t, +\infty) = e^{t(\lambda_-^{-1} + \lambda_+^{-1})/2}$ by inspection of (5) and the normalizations of $f_\pm^0$; see (6). Also, note that $\Theta_\pm$ vanish nowhere, being the determinants of matrices

$$\mathcal{M}_\pm(t, x) := \mathrm{Id} + \mathcal{E}(t, \Lambda) \int_{-\infty}^x \Upsilon_{\pm,0} \otimes \Upsilon_{\pm,0} \tag{22}$$

(see (14b) and (15c)) with positive definite associated quadratic forms.

REMARK 6. Differentiating with respect to x the first expression in (21) and using the substitution afforded by the second we get the curious quadratic identity

$$\Theta^2 = \Theta_- \Theta_+ + \Theta'_- \Theta_+ - \Theta_- \Theta'_+$$

relating the theta determinants; compare with [McKean 2003; Loubet 2006].

REMARK 7. The integration of the pair flow is now more or less completed. Indeed, substituting the second identity of (21) into identity (11) yields

$$(\phi_{\mathcal{X}_H}^t m_0) \circ \mathcal{L}_H^t = \left(\frac{\Theta_- \Theta_+}{\Theta^2} \right)^2 m_0.$$

Hence, all that remains is to compute $(\phi_{\mathcal{X}_H}^t v_0) \circ \mathcal{L}_H^t = (\mathrm{G} * (\phi_{\mathcal{X}_H}^t m_0)) \circ \mathcal{L}_H^t$ as is done in Appendix B of [Loubet 2006]. Nonetheless, in the next section we will present a more direct route to the formulas in Theorem 2.

REMARK 8. The logarithmic (time) derivative of the first identity in (21) shows that $\bar{x}^{\bullet} = (\log \det(\mathcal{M}_- \mathcal{M}_+^{-1}))^{\bullet}$. Hence, upon invoking the identity $(\log \det \mathcal{Q})^{\bullet} = \mathrm{Tr}(\mathcal{Q}^{\bullet} \mathcal{Q}^{-1})$ valid for any differentiable square matrix $\mathcal{Q}$, we obtain an alternative description of the right-hand side of (10). To wit,

$$\mathrm{Tr}(\tfrac{1}{2} \Lambda^{-1} (f \otimes f)) \circ \bar{x} = \left(\frac{1}{2\lambda_-} f_-^2 + \frac{1}{2\lambda_+} f_+^2 \right) \circ \bar{x} = \mathrm{Tr}(\mathcal{M}_+^{\bullet} \mathcal{M}_+^{-1} - \mathcal{M}_-^{\bullet} \mathcal{M}_-^{-1}).$$

We invite the reader to check this, starting directly from (22) with the help of (21) and the identity preceding (18). More significantly, upon substituting (21) into (18), we obtain an exact formula for the Lagrangian-valued time t update of f_0. To wit,

$$f \circ \bar{x} = \frac{\Theta}{\sqrt{\Theta_- \Theta_+}} e^{t/(4\Lambda)} [(\mathcal{M}(t, \cdot))^\dagger]^{-1} f_0, \tag{23a}$$

or, componentwise,

$$f_\pm \circ \bar{x} = \frac{1}{\sqrt{\Theta_- \Theta_+}} e^{t/(4\lambda_\pm)} \big(f_\pm^0 + (e^{t/(2\lambda_\mp)} - 1)[f_\pm^0 \varphi_\mp^0 - \lambda_\mp f_\mp^0 \varphi_0] \big). \tag{23b}$$

Similarly, one can compute the Lagrangian-valued (time t) updates of eigenfunctions of (3) other than $f_\pm$ but that is not our purpose here.

5. Integration of the pair flow

In this section we finally show how to exploit the integrated expressions of section 3 and section 4 — notably the characterization of the Lagrangian scale (21) — to obtain the explicit formulas of Theorem 2. The trick is to stick to the successful algorithm that was used systematically in the bulk of the previous sections and apply it to

$$\mathcal{V}_0^\downarrow := \int_{-\infty}^{\cdot} e^y m_0 = e^{\cdot}(v_0 - v_0'), \quad \mathcal{V}_0^\uparrow := \int_{\cdot}^{+\infty} e^{-y} m_0 = e^{-\cdot}(v_0 + v_0').$$

It develops (using (15a) and (16)) that

$$[\mathcal{V}^\downarrow \circ \bar{x}]^\bullet = -\tfrac{1}{2}[(\mathcal{F}_-^\downarrow/\lambda_-)^2 + (\mathcal{F}_+^\downarrow/\lambda_+)^2] \circ \bar{x} \equiv -\tfrac{1}{2}|\mathcal{F}^\downarrow \circ \bar{x}|^2,$$

$$[\mathcal{V}^\uparrow \circ \bar{x}]^\bullet = \tfrac{1}{2}[(\mathcal{F}_-^\uparrow/\lambda_-)^2 + (\mathcal{F}_+^\uparrow/\lambda_+)^2] \circ \bar{x} \equiv \tfrac{1}{2}|\mathcal{F}^\uparrow \circ \bar{x}|^2,$$

where $\mathcal{V}^\downarrow := (\phi_{\mathbb{X}_H}^t * \mathcal{V}_0^\downarrow)$, and $\mathcal{V}^\uparrow := (\phi_{\mathbb{X}_H}^t * \mathcal{V}_0^\uparrow)$. Closer inspection of the squares of the Euclidean norms in the right-hand side of the equations above via the preliminary integrations (i.e., alternative spellings) of $\mathcal{F}^\downarrow \circ \bar{x}$ and $\mathcal{F}^\uparrow \circ \bar{x}$ reveals the last piece of the puzzle. Here is how. By identities (16), keeping in mind the definitions (14a), and omitting writing explicitly the dependence on independent variables, we have

$$|\mathcal{F}^\downarrow \circ \bar{x}|^2 = (\mathcal{T}(\mathcal{M}^\dagger)^{-1}\mathcal{F}_0^\downarrow)^\dagger(\mathcal{T}(\mathcal{M}^\dagger)^{-1}\mathcal{F}_0^\downarrow) = (\mathcal{F}_0^\downarrow)^\dagger \mathcal{M}^{-1}\mathcal{T}^2(\mathcal{M}^\dagger)^{-1}\mathcal{F}_0^\downarrow$$

$$|\mathcal{F}^\uparrow \circ \bar{x}|^2 = (\mathcal{T}(\mathcal{M}^\dagger)^{-1}\mathcal{F}_0^\uparrow)^\dagger(\mathcal{T}(\mathcal{M}^\dagger)^{-1}\mathcal{F}_0^\uparrow) = (\mathcal{F}_0^\uparrow)^\dagger \mathcal{M}^{-1}\mathcal{T}^2(\mathcal{M}^\dagger)^{-1}\mathcal{F}_0^\uparrow.$$

Now, since $\tfrac{1}{2}\mathcal{T}^2\Phi_0 = \mathcal{M}^\bullet$ by (14a), and because $(\mathcal{M}^{-1})^\bullet = -\mathcal{M}^{-1}\mathcal{M}^\bullet\mathcal{M}^{-1}$,

$$-\tfrac{1}{2}\mathcal{M}^{-1}\mathcal{T}^2(\mathcal{M}^\dagger)^{-1} = -\mathcal{M}^{-1}\mathcal{M}^\bullet\Phi_0^{-1}(\mathcal{M}^\dagger)^{-1} = (\mathcal{M}^{-1})^\bullet\mathcal{M}\Phi_0^{-1}(\mathcal{M}^\dagger)^{-1}.$$

Moreover, by identity (17)

$$\mathcal{M}\Phi_0^{-1}(\mathcal{M}^\dagger)^{-1} = \mathcal{M}(\mathcal{M}^\dagger\Phi_0)^{-1} = \mathcal{M}(\Phi_0\mathcal{M})^{-1} = \Phi_0^{-1}.$$

Altogether, from the last two displays and the fact that the functions Φ_0, $\mathscr{F}_0^\downarrow$ and $\mathscr{F}_0^\uparrow$ are independent of t, we learn that $-\frac{1}{2}|\mathscr{F}^\downarrow \circ \bar{x}|^2$ and $\frac{1}{2}|\mathscr{F}^\uparrow \circ \bar{x}|^2$ can be written as *time derivatives* i.e.,

$$[\mathcal{V}^\downarrow \circ \bar{x}]^\bullet = [(\mathscr{F}_0^\downarrow)^\dagger (\Phi_0 \mathcal{M})^{-1} \mathscr{F}_0^\downarrow]^\bullet,$$
$$[\mathcal{V}^\uparrow \circ \bar{x}]^\bullet = [-(\mathscr{F}_0^\uparrow)^\dagger (\Phi_0 \mathcal{M})^{-1} \mathscr{F}_0^\uparrow]^\bullet.$$

Integrating the latter with respect to the time variable from 0 to t yields

$$e^{\bar{x}}(v - v') \circ \bar{x} = \mathcal{V}^\downarrow \circ \bar{x} = e^{\cdot}(v_0 - v_0') + (\mathscr{F}_0^\downarrow)^\dagger \{(\Phi_0 \mathcal{M})^{-1} - \Phi_0^{-1}\} \mathscr{F}_0^\downarrow,$$
$$e^{-\bar{x}}(v + v') \circ \bar{x} = \mathcal{V}^\uparrow \circ \bar{x} = e^{-\cdot}(v_0 + v_0') - (\mathscr{F}_0^\uparrow)^\dagger \{(\Phi_0 \mathcal{M})^{-1} - \Phi_0^{-1}\} \mathscr{F}_0^\uparrow,$$

where, as before, the underlying spatial variable (denoted by $\cdot$) is left unspecified. By associativity, definitions (14a) and linearity, it is immediate to see that

$$(\Phi_0 \mathcal{M})^{-1} - \Phi_0^{-1} = \mathcal{M}^{-1}(\mathrm{Id} - \mathcal{M})\Phi_0^{-1} = \mathcal{M}^{-1}(-\mathscr{C}\Phi_0)\Phi_0^{-1} = -\mathcal{M}^{-1}\mathscr{C}.$$

Now we use in the last-but-one pair of identities the connection between the scales (21) (with the understanding that in the latter the scale x is now to be left unspecified i.e., $e^{\bar{x}} = e^{\cdot} \Theta_-/\Theta_+$). Together with linearity and the identifications $e^{-\cdot/2}\mathscr{F}_0^\downarrow = -\Lambda^{-1}\Upsilon_{-,0}$ and $e^{\cdot/2}\mathscr{F}_0^\uparrow = \Lambda^{-1}\Upsilon_{+,0}$ of (15b), in combination with the preceding identity and the equality $\mathscr{C}\Lambda^{-1} = \mathscr{E}$, from (14b), this yields

$$(v - v') \circ \bar{x} = \frac{\Theta_+}{\Theta_-}\left(v_0 - v_0' - \Upsilon_{-,0}^\dagger (\mathcal{M}\Lambda)^{-1}\mathscr{E}\Upsilon_{-,0}\right),$$
$$(v + v') \circ \bar{x} = \frac{\Theta_-}{\Theta_+}\left(v_0 + v_0' + \Upsilon_{+,0}^\dagger (\mathcal{M}\Lambda)^{-1}\mathscr{E}\Upsilon_{+,0}\right).$$

Finally, taking half of the sum, respectively, of the difference of these identities reproduce the punch line formulas of Theorem 2, and we are done.

6. Large-time asymptotics and limiting behavior

If $\lambda_- < 0 < \lambda_+$, the Lagrangian-scaled updates of eigenfunctions associated to the eigenvalues $\lambda_\pm$ (cf. (23a) and (23b)) vanish as $t \to \pm\infty$. This prompts that $\lambda_\pm$ are *excised* from the spectrum associated with the *residual* profiles. Indeed, $e^{t/(2\lambda_\mp)} = o(1)$ as $t \to \pm\infty$, and thus by (22) and (15c),

$$\Theta_\pm := \det \mathcal{M}_\pm(t, \cdot) = \begin{cases} e^{t/(2\lambda_+)}(\theta_\pm^{+\infty} + o(1)) & \text{as} \quad t \uparrow +\infty, \\ e^{t/(2\lambda_-)}(\theta_\pm^{-\infty} + o(1)) & \text{as} \quad t \downarrow -\infty, \end{cases}$$

where

$$\begin{matrix} \theta_\pm^{+\infty} \\ \theta_\pm^{-\infty} \end{matrix} := \Phi_\pm - \begin{cases} [a_\pm(\varphi_-^0 - 1) + b_\pm \varphi_+^0] \\ [a_\pm \varphi_-^0 + b_\pm(\varphi_+^0 - 1)] \end{cases} \tag{24}$$

are the corresponding limiting/stationary theta counterparts, $\varphi_\pm^0$ being defined in (8), and

$$\Phi_\pm := \lambda_+ f_+^0 \varphi_+^0 ((f_-^0)' \pm \tfrac{1}{2} f_-^0) + \lambda_- f_-^0 \varphi_-^0 ((f_+^0)' \pm \tfrac{1}{2} f_+^0) + (1 + \lambda_- \lambda_+) \varphi_-^0 \varphi_+^0,$$

$$a_\pm := \varphi_+^0 + f_+^0 ((f_+^0)' \pm \tfrac{1}{2} f_+^0) = \int_{-\infty}^{\cdot} ((f_+^0)' \pm \tfrac{1}{2} f_+^0)^2,$$

$$b_\pm := \varphi_-^0 + f_-^0 ((f_-^0)' \pm \tfrac{1}{2} f_-^0) = \int_{-\infty}^{\cdot} ((f_-^0)' \pm \tfrac{1}{2} f_-^0)^2.$$

Hence, by (23b) we have

$$\lim_{t \to \pm\infty} [(\phi_{\mathcal{X}_H}^t * f_+^0) \circ \mathcal{L}_H^t] = 0, \quad \text{and} \quad \lim_{t \to \pm\infty} [(\phi_{\mathcal{X}_H}^t * f_-^0) \circ \mathcal{L}_H^t] = 0.$$

On the other hand, using (14a) and (12c) we see that

$$\Theta := \det \mathcal{M}(t, \cdot) = e^{t/(2\lambda_\pm)} (\theta^{\pm\infty} + o(1)) \quad \text{as } t \to \pm\infty, \tag{25}$$

where

$$\theta^{\pm\infty} := \varphi_\pm^0 (1 - \varphi_\mp^0) + \lambda_- \lambda_+ \varphi_0^2. \tag{26}$$

In other words, either limits (in the remote past or future) give rise to stationary Lagrangian scales (cf. (21))

$$(\mathcal{L}_H^{\pm\infty})' := \lim_{t \to \pm\infty} \frac{\Theta^2}{\Theta_- \Theta_+} = \frac{(\theta^{\pm\infty})^2}{\theta_-^{\pm\infty} \theta_+^{\pm\infty}},$$

or, what is the same,

$$e^{\mathcal{L}_H^{\pm\infty}} = e \cdot \frac{\theta_-^{\pm\infty}}{\theta_+^{\pm\infty}}$$

and residual potentials

$$\lim_{t \to \pm\infty} [(\phi_{\mathcal{X}_H}^t m_0) \circ \mathcal{L}_H^t] = (\phi_{\mathcal{X}_H}^{\pm\infty} m_0) \circ \mathcal{L}_H^{\pm\infty} = \left(\frac{\theta_-^{\pm\infty} \theta_+^{\pm\infty}}{(\theta^{\pm\infty})^2} \right)^2 m_0.$$

NOTE. It is amusing to check directly from the definitions (24), (26), and (8) that (dropping the upper indexes $\pm\infty$)

$$\theta^2 = \theta_- \theta_+ + \theta'_- \theta_+ - \theta_- \theta'_+.$$

i.e., that the algebraic structure of the identity in Remark 6 relating the theta-determinants remains valid (in the limits $t \to \pm\infty$) for either of their stationary analogues.

Either of the residual profiles can be computed from the corresponding residual potentials via the Green's function as in (cf. Remark 7),

$$\lim_{t\to\pm\infty} [(\phi^t_{\mathcal{X}_H} v_0) \circ \mathscr{L}^t_H] = (\phi^{\pm\infty}_{\mathcal{X}_H} v_0) \circ \mathscr{L}^{\pm\infty}_H = (G * \phi^{\pm\infty}_{\mathcal{X}_H} m_0) \circ \mathscr{L}^{\pm\infty}_H,$$

but it is more efficient to infer them directly by taking (respectively) the limits as $t \to \pm\infty$ of the formulas of Theorem 2. Indeed, for $\lambda_- < 0 < \lambda_+$, we have[1]

$$\lim_{t\to\pm\infty} \frac{\Theta_\pm}{\Theta_\mp} = \frac{\theta^{\pm\infty}_\pm}{\theta^{\pm\infty}_\mp}$$

where $\theta^{\pm\infty}_\pm$ are given by (24) (see also (8)). On the other hand, we can check that, in the notation of (14a), (12c) and (8),

$$\Psi_\pm(\cdot, \Lambda) := \lim_{t\to\pm\infty} [(\mathcal{M}(t,\cdot)\Lambda)^{-1}\mathcal{E}(t,\Lambda)] = \begin{cases} \dfrac{1}{\theta+\infty} \begin{pmatrix} -\varphi^0_+/\lambda_- & \varphi_0 \\ \varphi_0 & (1-\varphi^0_-)/\lambda_+ \end{pmatrix} \\[2em] \dfrac{1}{\theta-\infty} \begin{pmatrix} (1-\varphi^0_+)/\lambda_- & \varphi_0 \\ \varphi_0 & -\varphi^0_-/\lambda_+ \end{pmatrix} \end{cases}$$

where $\theta^{\pm\infty}$ are given by (26) (see also (8)). Altogether, it follows directly from the formulas of Theorem 2 that

$$\lim_{t\to\pm\infty} [(\phi^t_{\mathcal{X}_H} v_0) \circ \mathscr{L}^t_H] = \frac{\theta^{\pm\infty}_-}{\theta^{\pm\infty}_+} \left(\frac{v_0 + v'_0}{2} + \tfrac{1}{2}\Upsilon^\dagger_{+,0}\Psi_\pm\Upsilon_{+,0} \right) \tag{27}$$
$$+ \frac{\theta^{\pm\infty}_+}{\theta^{\pm\infty}_-} \left(\frac{v_0 - v'_0}{2} - \tfrac{1}{2}\Upsilon^\dagger_{-,0}\Psi_\pm\Upsilon_{-,0} \right),$$

$$\lim_{t\to\pm\infty} \left[\frac{\partial(\phi^t_{\mathcal{X}_H} v_0) \circ \mathscr{L}^t_H}{\partial \mathscr{L}^t_H} \right] = \frac{\theta^{\pm\infty}_-}{\theta^{\pm\infty}_+} \left(\frac{v_0 + v'_0}{2} + \tfrac{1}{2}\Upsilon^\dagger_{+,0}\Psi_\pm\Upsilon_{+,0} \right)$$
$$- \frac{\theta^{\pm\infty}_+}{\theta^{\pm\infty}_-} \left(\frac{v_0 - v'_0}{2} - \tfrac{1}{2}\Upsilon^\dagger_{-,0}\Psi_\pm\Upsilon_{-,0} \right).$$

The reader is invited to check from these equations that the H^1-energy associated to either of the stationary profiles (27) falls short of the one associated to the initial profile v_0 by *exactly* an amount that is equal to the sum of the energies that, at any given time $|t| < +\infty$, each of the solitons escaping (respectively) at speeds $1/(2\lambda_\pm)$ embody. But there is more, one can verify, adapting the general method of sections 3 and 4, that the limits (as $t \to \pm\infty$) of the Lagrangian-scaled updates of the remaining eigenfunctions that we refer to at the end of Remark 8 do not vanish. In fact, one can check that the latter constitute a basis in H^1. In

---

[1] In $\theta^{\pm\infty}_\pm$, the signature of the upper index $\pm\infty$ indicates which of the limits $t \to \pm\infty$ is meant, while the lower indexes merely distinguish which of the theta functions, θ_- or θ_+, is being referred to; cf. Notation on page 272.

short, as in the large-time asymptotics pertaining to the individual flows where it is shown that the eigenvalue defining the flow at play is excised [Loubet $\geq$ 2007a; $\geq$ 2007b], in the case of pair flows, we have evidence that the maps from initial to residual profiles, as described in the introduction, are also of Darboux-type with the difference that two eigenvalues (the ones involved in the Hamiltonian defining the flow) are excised instead of just one.

7. Conclusion

Closer inspection to the bulk of sections 3 and 4 shows that the method therein employed, can be adapted to produce analogous explicit exact formulas for the updates of profiles when run by flows of the CH hierarchy associated to Hamiltonians of the form $\sum_{|j| \leq N} 1/(4\lambda_j^2)$ with arbitrary N in $\mathbb{Z}^+$, all the way up — with due technical precautions in order to guarantee convergence, etc. — to (the limiting case where $N \uparrow +\infty$ corresponding to) the full CH flow [McKean 2003]. Moreover, the present analysis suggests that all these expressions will be sufficiently robust to afford (at least) a quantitative description of soliton train development. Nonetheless, it remains to explore in more detail how manageable all these expressions really are in helping reveal any more qualitative and quantitative phenomena pertaining to soliton emergence and soliton interaction.

References

[Camassa and Holm 1993] R. Camassa and D. D. Holm, "An integrable shallow water equation with peaked solitons", *Phys. Rev. Lett.* **71**:11 (1993), 1661–1664.

[Camassa et al. 1994] R. Camassa, D. D. Holm, and M. Hyman, "A new integrable shallow water equation", *Adv. Appl. Math.* **31** (1994), 1–33.

[Khesin and Misiołek 2003] B. Khesin and G. Misiołek, "Euler equations on homogeneous spaces and Virasoro orbits", *Adv. Math.* **176**:1 (2003), 116–144.

[Loubet 2006] E. Loubet, "Genesis of solitons arising from individual flows of the Camassa-Holm hierarchy", *Comm. Pure Appl. Math.* **59**:3 (2006), 408–465.

[Loubet $\geq$ 2008a] E. Loubet, Asymptotic limits of individual flows of the KdV hierarchy, back to Darboux via addition. In preparation.

[Loubet $\geq$ 2008b] E. Loubet, Extinction of solitons arising from individual flows of the Camassa-Holm hierarchy: the rise of a novel Darboux like transform. In preparation.

[Magri 1978] F. Magri, "A simple model of the integrable Hamiltonian equation", *J. Math. Phys.* **19**:5 (1978), 1156–1162.

[McKean 1993] H. P. McKean, "Compatible brackets in Hamiltonian dynamics", in *Important developments in soliton theory*, edited by A. S. Fokas and V. E. Zakharov, Springer, Berlin, 1993.

[McKean 1998] H. P. McKean, "Breakdown of a shallow water equation", *Asian J. Math.* **2**:4 (1998), 867–874. Mikio Sato: a great Japanese mathematician of the twentieth century.

[McKean 2001] H. P. McKean, "Addition for the acoustic equation", *Comm. Pure Appl. Math.* **54**:10 (2001), 1271–1288.

[McKean 2003] H. P. McKean, "Fredholm determinants and the Camassa-Holm hierarchy", *Comm. Pure Appl. Math.* **56**:5 (2003), 638–680.

[McKean 2004] H. P. McKean, "Breakdown of the Camassa-Holm equation", *Comm. Pure Appl. Math.* **57**:3 (2004), 416–418.

[Misiołek 1998] G. Misiołek, "A shallow water equation as a geodesic flow on the Bott-Virasoro group", *J. Geom. Phys.* **24**:3 (1998), 203–208.

ENRIQUE LOUBET
INSTITUT FÜR MATHEMATIK
UNIVERSITÄT ZÜRICH
WINTERTHURERSTRASSE 190
CH-8057 ZÜRICH
SWITZERLAND
eloubet@math.unizh.ch

Probability, Geometry and Integrable Systems
MSRI Publications
Volume **55**, 2008

Landen Survey

DANTE V. MANNA AND VICTOR H. MOLL

To Henry, who provides inspiration, taste and friendship

ABSTRACT. Landen transformations are maps on the coefficients of an integral that preserve its value. We present a brief survey of their appearance in the literature.

1. In the beginning there was Gauss

In the year 1985, one of us had the luxury of attending a graduate course on *Elliptic Functions* given by Henry McKean at the Courant Institute. Among the many beautiful results he described in his unique style, there was a calculation of Gauss: take two positive real numbers a and b, with $a > b$, and form a new pair by replacing a with the arithmetic mean $(a+b)/2$ and b with the geometric mean $\sqrt{ab}$. Then iterate:

$$a_{n+1} = \frac{a_n + b_n}{2}, \quad b_{n+1} = \sqrt{a_n b_n} \tag{1-1}$$

starting with $a_0 = a$ and $b_0 = b$. Gauss [1799] was interested in the initial conditions $a = 1$ and $b = \sqrt{2}$. The iteration generates a sequence of algebraic numbers which rapidly become impossible to describe explicitly; for instance,

$$a_3 = \frac{1}{2^3} \left((1 + \sqrt[4]{2})^2 + 2\sqrt{2}\sqrt[8]{2}\sqrt{1 + \sqrt{2}} \right) \tag{1-2}$$

is a root of the polynomial

$$G(a) = 16777216a^8 - 16777216a^7 + 5242880a^6 - 10747904a^5$$
$$+ 942080a^4 - 1896448a^3 + 4436a^2 - 59840a + 1.$$

Keywords: Integrals, arithmetic-geometric mean, elliptic integrals.

The numerical behavior is surprising; a_6 and b_6 agree to 87 digits. It is simple to check that

$$\lim_{n \to \infty} a_n = \lim_{n \to \infty} b_n. \tag{1-3}$$

See (6-1) for details. This common limit is called the *arithmetic-geometric mean* and is denoted by $\mathrm{AGM}(a, b)$. It is the explicit dependence on the initial condition that is hard to discover.

Gauss computed some numerical values and observed that

$$a_{11} \sim b_{11} \sim 1.198140235, \tag{1-4}$$

and then he *recognized* the reciprocal of this number as a numerical approximation to the elliptic integral

$$I = \frac{2}{\pi} \int_0^1 \frac{dt}{\sqrt{1 - t^4}}. \tag{1-5}$$

It is unclear to the authors how Gauss recognized this number: he simply knew it. (Stirling's tables may have been a help; [Borwein and Bailey 2003] contains a reproduction of the original notes and comments.) He was particularly interested in the evaluation of this definite integral as it provides the length of a lemniscate. In his diary Gauss remarked, *'This will surely open up a whole new field of analysis'* [Cox 1984; Borwein and Borwein 1987].

Gauss' procedure to find an analytic expression for $\mathrm{AGM}(a, b)$ began with the elementary observation

$$\mathrm{AGM}(a, b) = \mathrm{AGM}\left(\frac{a + b}{2}, \sqrt{ab}\right) \tag{1-6}$$

and the homogeneity condition

$$\mathrm{AGM}(\lambda a, \lambda b) = \lambda \mathrm{AGM}(a, b) . \tag{1-7}$$

He used (1-6) with $a = (1 + \sqrt{k})^2$ and $b = (1 - \sqrt{k})^2$, with $0 < k < 1$, to produce

$$\mathrm{AGM}(1 + k + 2\sqrt{k}, 1 + k - 2\sqrt{k}) = \mathrm{AGM}(1 + k, 1 - k). \tag{1-8}$$

He then used the homogeneity of AGM to write

$$\mathrm{AGM}(1 + k + 2\sqrt{k}, 1 + k - 2\sqrt{k}) = \mathrm{AGM}((1 + k)(1 + k^*), (1 + k)(1 - k^*))$$
$$= (1 + k)\mathrm{AGM}(1 + k^*, 1 - k^*),$$

with

$$k^* = \frac{2\sqrt{k}}{1 + k}. \tag{1-9}$$

This resulted in the functional equation

$$\text{AGM}(1+k, 1-k) = (1+k)\,\text{AGM}(1+k^*, 1-k^*). \qquad (1\text{-}10)$$

In his analysis of (1-10), Gauss substituted the power series

$$\frac{1}{\text{AGM}(1+k, 1-k)} = \sum_{n=0}^{\infty} a_n k^{2n} \qquad (1\text{-}11)$$

into (1-10) and solved an infinite system of nonlinear equations to produce

$$a_n = 2^{-2n} \binom{2n}{n}^2. \qquad (1\text{-}12)$$

Then he recognized the series as that of an elliptic integral to obtain

$$\frac{1}{\text{AGM}(1+k, 1-k)} = \frac{2}{\pi} \int_0^{\pi/2} \frac{dx}{\sqrt{1-k^2 \sin^2 x}}. \qquad (1\text{-}13)$$

This is a remarkable tour de force.

The function

$$K(k) = \int_0^{\pi/2} \frac{dx}{\sqrt{1-k^2 \sin^2 x}} \qquad (1\text{-}14)$$

is the *elliptic integral of the first kind*. It can also be written in the algebraic form

$$K(k) = \int_0^1 \frac{dt}{\sqrt{(1-t^2)(1-k^2 t^2)}}. \qquad (1\text{-}15)$$

In this notation, (1-10) becomes

$$K(k^*) = (1+k)K(k). \qquad (1\text{-}16)$$

This is the *Landen transformation* for the complete elliptic integral. John Landen [1775], the namesake of the transformation, studied related integrals: for example,

$$\kappa := \int_0^1 \frac{dx}{\sqrt{x(1-x^2)}}. \qquad (1\text{-}17)$$

He derived identities such as

$$\kappa = \varepsilon + \sqrt{\varepsilon^2 - \pi}, \quad \text{where } \varepsilon := \int_0^{\pi/2} \sqrt{2 - \sin^2 \theta}\, d\theta, \qquad (1\text{-}18)$$

proven mainly by suitable changes of variables in the integral for ε. In [Watson 1933] the reader will find a historical account of Landen's work, including the above identities.

The reader will find in [Borwein and Borwein 1987] and [McKean and Moll 1997] proofs in a variety of styles. In trigonometric form, the Landen transformation states that

$$G(a,b) = \int_0^{\pi/2} \frac{d\theta}{\sqrt{a^2 \cos^2 \theta + b^2 \sin^2 \theta}} \tag{1-19}$$

is invariant under the change of parameters

$$(a,b) \mapsto \left(\frac{a+b}{2}, \sqrt{ab}\right).$$

D. J. Newman [1985] presents a very clever proof: the change of variables $x = b \tan \theta$ yields

$$G(a,b) = \frac{1}{2} \int_{-\infty}^{\infty} \frac{dx}{\sqrt{(a^2 + x^2)(b^2 + x^2)}}. \tag{1-20}$$

Now let $x \mapsto x + \sqrt{x^2 + ab}$ to complete the proof. Many of the above identities can now be searched for and proven on a computer [Borwein and Bailey 2003].

2. An interlude: the quartic integral

The evaluation of definite integrals of rational functions is one of the standard topics in Integral Calculus. Motivated by the lack of success of symbolic languages, we began a systematic study of these integrals. *A posteriori*, one learns that even rational functions are easier to deal with. Thus we start with one having *a power of a quartic* in its denominator. The evaluation of the identity

$$\int_0^{\infty} \frac{dx}{(x^4 + 2ax^2 + 1)^{m+1}} = \frac{\pi}{2^{m+3/2} (a+1)^{m+1/2}} P_m(a), \tag{2-1}$$

where

$$P_m(a) = \sum_{l=0}^{m} d_l(m) a^l \tag{2-2}$$

with

$$d_l(m) = 2^{-2m} \sum_{k=l}^{m} 2^k \binom{2m - 2k}{m - k} \binom{m + k}{m} \binom{k}{l}, \tag{2-3}$$

was first established in [Boros and Moll 1999b].

A standard hypergeometric argument yields

$$P_m(a) = P_m^{(\alpha,\beta)}(a), \tag{2-4}$$

where

$$P_m^{(\alpha,\beta)}(a) = \sum_{k=0}^{m}(-1)^{m-k}\binom{m+\beta}{m-k}\binom{m+k+\alpha+\beta}{k}2^{-k}(a+1)^k \quad (2\text{-}5)$$

is the classical Jacobi polynomial; the parameters α and β are given by $\alpha = m+\frac{1}{2}$ and $\beta = -m-\frac{1}{2}$. A general description of these functions and their properties are given in [Abramowitz and Stegun 1972]. The twist here is that they depend on m, which means most of the properties of P_m had to be proven from scratch. For instance, P_m satisfies the recurrence

$$P_m(a) = \frac{(2m-3)(4m-3)a}{4m(m-1)(a-1)}P_{m-2}(a) - \frac{(4m-3)a(a+1)}{2m(m-1)(a-1)}P'_{m-2}(a)$$
$$+ \frac{4m(a^2-1)+1-2a^2}{2m(a-1)}P_m(a).$$

This *cannot* be obtained by replacing $\alpha = m+\frac{1}{2}$ and $\beta = -m-\frac{1}{2}$ in the standard recurrence for the Jacobi polynomials. The reader will find in [Amdeberhan and Moll 2007] several different proofs of (2-1).

The polynomials $P_m(a)$ makes a surprising appearance in the expansion

$$\sqrt{a+\sqrt{1+c}} = \sqrt{a+1}\left(1 - \sum_{k=1}^{\infty}\frac{(-1)^k}{k}\frac{P_{k-1}(a)\,c^k}{2^{k+1}(a+1)^k}\right) \quad (2\text{-}6)$$

as described in [Boros and Moll 2001a]. The special case $a = 1$ appears in [Bromwich 1926], page 191, exercise 21. Ramanujan had a more general expression, but only for the case $c = a^2$:

$$(a+\sqrt{1+a^2})^n = 1 + na + \sum_{k=2}^{\infty}\frac{b_k(n)a^k}{k!}, \quad (2\text{-}7)$$

where, for $k \geq 2$,

$$b_k(n) = \begin{cases} n^2(n^2-2^2)(n^2-4^2)\cdots(n^2-(k-2)^2) & \text{if } k \text{ is even,} \\ n(n^2-1^2)(n^2-3^2)\cdots(n^2-(k-2)^2) & \text{if } k \text{ is odd.} \end{cases} \quad (2\text{-}8)$$

This result appears in [Berndt and Bowman 2000] as Corollary 2 to Entry 14 and is machine-checkable, as are many of the identities in this section.

The coefficients $d_l(m)$ in (2-3) have many interesting properties:

• They form a *unimodal sequence*: there exists an index $0 \leq m^* \leq m$ such that $d_j(m)$ increases up to $j = m^*$ and decreases from then on. See [Boros and Moll 1999a] for a proof of the more general statement: *If $P(x)$ is a polynomial*

with nondecreasing, nonnegative coefficients, then the coefficient sequence of
$P(x + 1)$ *is unimodal.*

- They form a *log-concave sequence*: define the operator

$$\mathcal{L}(\{a_k\}) : = \{a_k^2 - a_{k-1}a_{k+1}\}$$

acting on sequences of positive real numbers. A sequence $\{a_k\}$ is called log-concave if its image under $\mathcal{L}$ is again a sequence of positive numbers; i.e. $a_k^2 - a_{k-1}a_{k+1} \geq 0$. Note that this condition is satisfied if and only if the sequence $\{b_k : = \log(a_k)\}$ is concave, hence the name. We refer the reader to [Wilf 1990] for a detailed introduction. The log-concavity of $d_l(m)$ was established in [Kauers and Paule 2007] using computer algebra techniques: in particular, cylindrical algebraic decompositions as developed in [Caviness and Johnson 1998] and [Collins 1975].

- They produce interesting polynomials: in [Boros et al. 2001] one finds the representation

$$d_l(m) = \frac{A_{l,m}}{l! \, m! \, 2^{m+l}}, \tag{2-9}$$

with

$$A_{l,m} = \alpha_l(m) \prod_{k=1}^m (4k - 1) - \beta_l(m) \prod_{k=1}^m (4k + 1). \tag{2-10}$$

Here α_l and β_l are polynomials in m of degrees l and $l - 1$, respectively. For example, $\alpha_1(m) = 2m + 1$ and $\beta_1(m) = 1$, so that the coefficient of the linear term of $P_m(a)$ is

$$d_1(m) = \frac{1}{m! \, 2^{m+1}} \left((2m + 1) \prod_{k=1}^m (4k - 1) - \prod_{k=1}^m (4k + 1) \right). \tag{2-11}$$

J. Little [2005] established the remarkable fact that the polynomials $\alpha_l(m)$ and $\beta_l(m)$ have all their roots on the vertical line $\operatorname{Re} m = -\frac{1}{2}$.

When we showed this to Henry, he simply remarked: *the only thing you have to do now is to let $l \to \infty$ and get the Riemann hypothesis.* The proof in [Little 2005] consists in a study of the recurrence

$$y_{l+1}(s) = 2sy_l(s) - \left(s^2 - (2l - 1)^2\right)y_{l-1}(s), \tag{2-12}$$

satisfied by $\alpha_l((s-1)/2)$ and $\beta_l((s-1)/2)$. There is no Number Theory in the proof, so it is not likely to connect to the Riemann zeta function $\zeta(s)$, but one never knows.

The arithmetical properties of $A_{l,m}$ are beginning to be elucidated. We have shown that their 2-adic valuation satisfies

$$v_2(A_{l,m}) = v_2((m + 1 - l)_{2l}) + l, \tag{2-13}$$

where $(a)_k = a(a+1)(a+2)\cdots(a+k-1)$ is the Pochhammer symbol. This expression allows for a combinatorial interpretation of the block structure of these valuations. See [Amdeberhan et al. 2007] for details.

3. The incipient rational Landen transformation

The clean analytic expression in (2-1) is not expected to extend to rational functions of higher order. In our analysis we distinguish according to the domain of integration: the finite interval case, mapped by a bilinear transformation to $[0, \infty)$, and the whole line. In this section we consider the definite integral,

$$U_6(a,b;c,d,e) = \int_0^\infty \frac{cx^4 + dx^2 + e}{x^6 + ax^4 + bx^2 + 1} \, dx, \qquad (3\text{-}1)$$

as the simplest case on $[0, \infty)$. The case of the real line is considered below. The integrand is chosen to be even by necessity: *none of the techniques in this section work for the odd case*. We normalize two of the coefficients in the denominator in order to reduce the number of parameters. The standard approach for the evaluation of (3-1) is to introduce the change of variables $x = \tan \theta$. This leads to an intractable trigonometric integral.

A different result is obtained if one first symmetrizes the denominator: we say that a polynomial of degree d is *reciprocal* if $Q_d(1/x) = x^{-d}Q_d(x)$, that is, the sequence of its coefficients is a palindrome. Observe that if Q_d is any polynomial of degree d, then

$$T_{2d}(x) = x^d Q_d(x) Q_d(1/x) \qquad (3\text{-}2)$$

is a reciprocal polynomial of degree $2d$. For example, if

$$Q_6(x) = x^6 + ax^4 + bx^2 + 1. \qquad (3\text{-}3)$$

then

$$T_{12}(x) = x^{12} + (a+b)x^{10} + (a+b+ab)x^8$$
$$+ (2 + a^2 + b^2)x^6 + (a+b+ab)x^4 + (a+b)x^2 + 1.$$

The numerator and denominator in the integrand of (3-1) are now scaled by $x^6 Q_6(1/x)$ to produce a new integrand with reciprocal denominator:

$$U_6 = \int_0^\infty \frac{S_{10}(x)}{T_{12}(x)} \, dx, \qquad (3\text{-}4)$$

where we write

$$S_{10}(x) = \sum_{j=0}^{5} s_j x^{2j} \text{ and } T_{12}(x) = \sum_{j=0}^{6} t_j x^{2j}. \qquad (3\text{-}5)$$

The change of variables $x = \tan\theta$ now yields

$$U_6 = \int_0^{\pi/2} \frac{S_{10}(\tan\theta)\,\cos^{10}(\theta)}{T_{12}(\tan\theta)\,\cos^{12}(\theta)}\,d\theta. \tag{3-6}$$

Now let $w = \cos 2\theta$ and use $\sin^2\theta = \frac{1}{2}(1-w)$ and $\cos^2\theta = \frac{1}{2}(1+w)$ to check that the numerator and denominator of the new integrand,

$$S_{10}(\tan\theta)\,\cos^{10}\theta = \sum_{j=0}^{5} s_j \sin^{2j}\theta\,\cos^{10-2j}\theta \tag{3-7}$$

and

$$T_{12}(\tan\theta)\,\cos^{12}\theta = \sum_{j=0}^{6} t_j \sin^{2j}\theta\,\cos^{12-2j}\theta = 2^{-6}\sum_{j=0}^{6} t_j (1-w)^j (1+w)^{6-j},$$

are both polynomials in w. The mirror symmetry of T_{12}, reflected in $t_j = t_{6-j}$, shows that the new denominator is an *even* polynomial in w. The symmetry of cosine about $\pi/2$ shows that the terms with odd power of w have a vanishing integral. Thus, with $\psi = 2\theta$, and using the symmetry of the integrand to reduce the integral from $[0, \pi]$ to $[0, \pi/2]$, we obtain

$$U_6 = \int_0^{\pi/2} \frac{r_4 \cos^4\psi + r_2\cos^2\psi + r_0}{q_6\cos^6\psi + q_4\cos^4\psi + q_2\cos^2\psi + q_0}\,d\psi. \tag{3-8}$$

The parameters r_j, q_j have explicit formulas in terms of the original parameters of U_6. This even rational function of $\cos\psi$ can now be expressed in terms of $\cos 2\psi$ to produce (letting $\theta \leftarrow 2\psi$)

$$U_6 = \int_0^{\pi} \frac{\alpha_2\cos^2\theta + \alpha_1\cos\theta + \alpha_0}{\beta_3\cos^3\theta + \beta_2\cos^2\theta + \beta_1\cos\theta + \beta_0}\,d\theta. \tag{3-9}$$

The final change of variables $y = \tan\frac{\theta}{2}$ yields a new rational form of the integrand:

$$U_6 = \int_0^{\infty} \frac{c_1 y^4 + d_1 y^2 + e_1}{y^6 + a_1 y^4 + b_1 y^2 + 1}\,dy. \tag{3-10}$$

Keeping track of the parameters, we have established:

THEOREM 3.1. *The integral*

$$U_6 = \int_0^{\infty} \frac{cx^4 + dx^2 + e}{x^6 + ax^4 + bx^2 + 1}\,dx \tag{3-11}$$

is invariant under the change of parameters

$$a_1 \leftarrow \frac{ab + 5a + 5b + 9}{(a+b+2)^{4/3}}, \qquad b_1 \leftarrow \frac{a+b+6}{(a+b+2)^{2/3}},$$

for the denominator parameters and

$$c_1 \leftarrow \frac{c+d+e}{(a+b+2)^{2/3}}, \quad d_1 \leftarrow \frac{(b+3)c+2d+(a+3)e}{a+b+2}, \quad e_1 \leftarrow \frac{c+e}{(a+b+2)^{1/3}}$$

for those of the numerator.

Theorem 3.1 is the precise analogue of the elliptic Landen transformation (1-1) for the case of a rational integrand. We call (3-12) a *rational Landen transformation*. This construction was first presented in [Boros and Moll 2000].

3.1. Even rational Landen transformations. More generally, there is a similar transformation of coefficients for *any even rational integrand*; details appear in [Boros and Moll 2001b]. We call these *even rational Landen Transformations*. The obstruction in the general case comes from (3-7); one does not get a polynomial in $w = \cos 2\theta$.

The method of proof for even rational integrals can be summarized as follows.

1) Start with an even rational integral:

$$U_{2p} = \int_0^\infty \frac{\text{even polynomial in } x}{\text{even polynomial in } x} \, dx. \tag{3-12}$$

2) Symmetrize the denominator to produce

$$U_{2p} = \int_0^\infty \frac{\text{even polynomial in } x}{\text{even reciprocal polynomial in } x} \, dx. \tag{3-13}$$

The degree of the denominator is doubled.

3) Let $x = \tan \theta$. Then

$$U_{2p} = \int_0^{\pi/2} \frac{\text{polynomial in } \cos 2\theta}{\text{even polynomial in} \cos 2\theta} \, d\theta. \tag{3-14}$$

4) Symmetry produced the vanishing of the integrands with an odd power of $\cos \theta$ in the numerator. We obtain

$$U_{2p} = \int_0^{\pi/2} \frac{\text{even polynomial in } \cos 2\theta}{\text{even polynomial in } \cos 2\theta} \, d\theta. \tag{3-15}$$

5) Let $\psi = 2\theta$ to produce

$$U_{2p} = \int_0^\pi \frac{\text{even polynomial in } \cos \psi}{\text{even polynomial in } \cos \psi} \, d\psi. \tag{3-16}$$

Using symmetry this becomes an integral over $[0, \pi/2]$.

6) Let $y = \tan \psi$ and use $\cos \psi = 1/\sqrt{1+y^2}$ to obtain

$$U_{2p} = \int_0^\infty \frac{\text{even polynomial in } y}{\text{even polynomial in } y} \, dy. \qquad (3\text{-}17)$$

The degree of the denominator is half of what it was in Step 5.

Keeping track of the degrees one checks that the degree of the new rational function is the same as the original one, with new coefficients that appear as functions of the old ones.

4. A geometric interpretation

We now present a geometric foundation of the general even rational Landen transformation (3-12) using the theory of Riemann surfaces. The text [Springer 2002] provides an introduction to this theory, including definitions of objects we will refer to here. The sequence of transformations in Section 3 can be achieved in one step by relating $\tan 2\theta$ to $\tan \theta$. For historical reasons (this is what we did first) we present the details with *cotangent* instead of tangent.

Consider the even rational integral

$$I = \int_0^\infty R(x) \, dx = \frac{1}{2} \int_{-\infty}^\infty R(x) \, dx. \qquad (4\text{-}1)$$

Introduce the new variable

$$y = R_2(x) = \frac{x^2 - 1}{2x}, \qquad (4\text{-}2)$$

motivated by the identity $\cot 2\theta = R_2(\cot \theta)$. The function $R_2 : \mathbb{R} \to \mathbb{R}$ is a two-to-one map. The sections of the inverse are

$$x = \sigma_\pm(y) = y \pm \sqrt{y^2 + 1}. \qquad (4\text{-}3)$$

Splitting the original integral as

$$I = \int_{-\infty}^0 R(x) \, dx + \int_0^\infty R(x) \, dx \qquad (4\text{-}4)$$

and introducing $x = \sigma_+(y)$ in the first and $x = \sigma_-(y)$ in the second integral, yields

$$I = \int_{-\infty}^\infty (R_+(y) + R_-(y)) \, dy \qquad (4\text{-}5)$$

where

$$R_+(y) = R(\sigma_+(y)) + R(\sigma_-(y)),$$
$$R_-(y) = \frac{y}{\sqrt{y^2 + 1}} (R(\sigma_+(y)) - R(\sigma_-(y))). \qquad (4\text{-}6)$$

A direct calculation shows that R_+ and R_- are rational functions of degree at most that of R.

The change of variables $y = R_2(x)$ converts the meromorphic differential $\varphi = R(x)\,dx$ into

$$R(\sigma_+(y))\frac{d\sigma_+}{dy} + R(\sigma_-(y))\frac{d\sigma_-}{dy}$$

$$= \left((R(\sigma_+) + R(\sigma_-)) + \frac{y(R(\sigma_+) - R(\sigma_-))}{\sqrt{y^2+1}} \right) dy$$

$$= (R_+(y) + R_-(y))\,dy.$$

The general situation is this: start with a finite ramified cover $\pi : X \to Y$ of Riemann surfaces and a meromorphic differential φ on X. Let $U \subset Y$ be a simply connected domain that contains no critical values of π, and let $\sigma_1, \ldots, \sigma_k : U \to X$ be the distinct sections of π. Define

$$\pi_*\varphi\big|_U = \sum_{j=1}^{k} \sigma_j^*\varphi. \tag{4-7}$$

In [Hubbard and Moll 2003] we show that this construction preserves analytic 1-forms, that is, if φ is an analytic 1-form in X then $\pi_*\varphi$ is an analytic 1-form in Y. Furthermore, for any rectifiable curve γ on Y, we have

$$\int_\gamma \pi_*\varphi = \int_{\pi^{-1}\gamma} \varphi. \tag{4-8}$$

In the case of projective space, this leads to:

LEMMA 4.1. *If* $\pi : \mathbb{P}^1 \to \mathbb{P}^1$ *is analytic, and* $\varphi = R(z)\,dz$ *with* R *a rational function, then* $\pi_*\varphi$ *can be written as* $R_1(z)\,dz$ *with* R_1 *a rational function of degree at most the degree of* R.

This is the generalization of the fact that the integrals in (4-1) and (4-5) are the same.

5. A further generalization

The procedure described in Section 3 can be extended with the rational map R_m, defined by the identity

$$\cot m\theta = R_m(\cot \theta). \tag{5-1}$$

Here $m \in \mathbb{N}$ is arbitrary greater or equal than 2. We present some elementary properties of the rational function R_m.

PROPOSITION 5.1. *The rational function* R_m *satisfies:*

1) *For $m \in \mathbb{N}$ define*

$$P_m(x) := \sum_{j=0}^{\lfloor m/2 \rfloor} (-1)^j \binom{m}{2j} x^{m-2j},$$

$$Q_m(x) := \sum_{j=0}^{\lfloor (m-1)/2 \rfloor} (-1)^j \binom{m}{2j+1} x^{m-(2j+1)}.$$

Then $R_m := P_m/Q_m$.

2) *The function R_m is conjugate to $f_m(x) := x^m$ via $M(x) := \dfrac{x+i}{x-i}$; that is, $R_m = M^{-1} \circ f_m \circ M$.*

3) *The polynomials P_m and Q_m have simple real zeros given by*

$$p_k := \cot \frac{(2k+1)\pi}{2m} \quad \text{for } 0 \le k \le m-1,$$

$$q_k := \cot \frac{k\pi}{m} \quad\quad \text{for } 1 \le k \le m-1.$$

If we change the domain to the entire real line, we can, using the rational substitutions $R_m(x) \mapsto x$, produce a rational Landen transformation for an arbitrary integrable rational function $R(x) = B(x)/A(x)$ for each integer value of m. The result is a new list of coefficients, from which one produces a second rational function $R^{(1)}(x) = J(x)/H(x)$ with

$$\int_{-\infty}^{\infty} \frac{B(x)}{A(x)} \, dx = \int_{-\infty}^{\infty} \frac{J(x)}{H(x)} \, dx. \tag{5-2}$$

Iteration of this procedure yields a sequence x_n, that has a limit x_∞ with convergence of order m, that is,

$$\|x_{n+1} - x_\infty\| \le C \|x_n - x_\infty\|^m. \tag{5-3}$$

We describe this procedure here in the form of an algorithm; proofs appear in [Manna and Moll 2007a].

Lemma 4.1 applied to the map $\pi(x) = R_m(x)$, viewed as ramified cover of $\mathbb{P}^1$, guarantees the existence of a such new rational function $R^{(1)}$. The question of effective computation of the coefficients of J and H is discussed below. In particular, we show that all these calculations can be done symbolically.

• **Algorithm for deriving rational Landen transformations**

Step 1. The initial data is a rational function $R(x) := B(x)/A(x)$. We assume that A and B are polynomials with real coefficients and A has no real zeros and write

$$A(x) : = \sum_{k=0}^{p} a_k x^{p-k} \text{ and } B(x) : = \sum_{k=0}^{p-2} b_k x^{p-2-k}. \tag{5-4}$$

Step 2. Choose a positive integer $m \geq 2$.

Step 3. Introduce the polynomial

$$H(x) : = \text{Res}_z(A(z), P_m(z) - x Q_m(z)) \tag{5-5}$$

and write it as

$$H(x) : = \sum_{l=0}^{p} e_l x^{p-l}. \tag{5-6}$$

The polynomial H is thus defined as the determinant of the Sylvester matrix which is formed of the polynomial coefficients. As such, the coefficients e_l of $H(x)$ themselves are integer polynomials in the a_i. Explicitly,

$$e_l = (-1)^l a_0^m \prod_{j=1}^{p} Q_m(x_j) \times \sigma_l^{(p)}(R_m(x_1), R_m(x_2), \ldots, R_m(x_p)), \tag{5-7}$$

where $\{x_1, x_2, \ldots, x_p\}$ are the roots of A, each written according to multiplicity. The functions $\sigma_l^{(p)}$ are the elementary symmetric functions in p variables defined by

$$\prod_{l=1}^{p}(y - y_l) = \sum_{l=0}^{p}(-1)^l \sigma_l^{(p)}(y_1, \ldots, y_p) y^{p-l}. \tag{5-8}$$

It is possible to compute the coefficients e_l symbolically from the coefficients of A, without the knowledge of the roots of A.

Also define

$$E(x) : = H(R_m(x)) \times Q_m(x)^p. \tag{5-9}$$

Step 4. The polynomial A divides E and we denote the quotient by Z. The coefficients of Z are integer polynomials in the a_i.

Step 5. Define the polynomial $C(x) : = B(x)Z(x)$.

Step 6. There exists a polynomial $J(x)$, whose coefficients have an explicit formula in terms of the coefficients c_j of $C(x)$, such that

$$\int_{-\infty}^{\infty} \frac{B(x)}{A(x)} dx = \int_{-\infty}^{\infty} \frac{J(x)}{H(x)} dx. \tag{5-10}$$

This new integrand is the rational function whose existence is guaranteed by Lemma 4.1. The explicit computation of the coefficients of J can be found in [Manna and Moll 2007a]. This is the *rational Landen transformation* of order m.

EXAMPLE 5.1. Completing the algorithm with $m = 3$ and the rational function

$$R(x) = \frac{1}{ax^2 + bx + c}, \tag{5-11}$$

produces the result stated below. Notice that the values of the iterates are ratios of integer polynomials of degree 3, as was stated above. The details of this example appear in [Manna and Moll 2007b].

THEOREM 5.2. *The integral*

$$I = \int_{-\infty}^{\infty} \frac{dx}{ax^2 + bx + c} \tag{5-12}$$

is invariant under the transformation

$$a \mapsto \frac{a}{\Delta}\big((a+3c)^2 - 3b^2\big), \quad b \mapsto \frac{b}{\Delta}\big(3(a-c)^2 - b^2\big), \quad c \mapsto \frac{c}{\Delta}\big((3a+c)^2 - 3b^2\big), \tag{5-13}$$

where $\Delta : = (3a + c)(a + 3c) - b^2$. *The condition* $b^2 - 4ac < 0$, *imposed to ensure convergence of the integral, is preserved by the iteration.*

EXAMPLE 5.2. In this example we follow the steps described above in order to produce a rational Landen transformation of order 2 for the integral

$$I = \int_{-\infty}^{\infty} \frac{b_0 x^4 + b_1 x^3 + b_2 x^2 + b_3 x + b_4}{a_0 x^6 + a_1 x^5 + a_2 x^4 + a_3 x^3 + a_4 x^2 + a_5 x + a_6} \, dx. \tag{5-14}$$

Recall that the algorithm starts with a rational function $R(x)$ and produces a new function $\mathfrak{L}_2(R(x))$ satisfying

$$\int_{-\infty}^{\infty} R(x) \, dx = \int_{-\infty}^{\infty} \mathfrak{L}_2(R(x)) \, dx. \tag{5-15}$$

Step 1. The initial data is $R(x) = B(x)/A(x)$ with

$$A(x) = a_0 x^6 + a_1 x^5 + a_2 x^4 + a_3 x^3 + a_4 x^2 + a_5 x + a_6, \tag{5-16}$$

and

$$B(x) = b_0 x^4 + b_1 x^3 + b_2 x^2 + b_3 x + b_4. \tag{5-17}$$

The parameter p is the degree of A, so $p = 6$.

Step 2. We choose $m = 2$ to produce a method of order 2. The algorithm employs the polynomials $P_2(z) = z^2 - 1$ and $Q_2(z) = 2z$.

Step 3. The polynomial

$$H(x): \ = \text{Res}_z(A(z), z^2 - 1 - 2xz) \tag{5-18}$$

is computed with the Mathematica command `Resultant` to obtain

$$H(x) = e_0 x^6 + e_1 x^5 + e_2 x^4 + e_3 x^3 + e_4 x^2 + e_5 x + e_6, \tag{5-19}$$

where

$$
\begin{aligned}
e_0 &= 64 a_0 a_6, \\
e_1 &= -32(a_0 a_5 - a_1 a_6), \\
e_2 &= 16(a_0 a_4 - a_1 a_5 + 6 a_0 a_6 + a_2 a_6), \\
e_3 &= -8(a_0 a_3 - a_1 a_4 + 5 a_0 a_5 + a_2 a_5 - 5 a_1 a_6 - a_3 a_6), \\
e_4 &= 4(a_0 a_2 - a_1 a_3 + 4 a_0 a_4 + a_2 a_4 - 4 a_1 a_5 - a_3 a_5 + 9 a_0 a_6 + 4 a_2 a_6 + a_4 a_6), \\
e_5 &= -2(a_0 a_1 - a_1 a_2 + 3 a_0 a_3 + a_2 a_3 - 3 a_1 a_4 - a_3 a_4 + 5 a_0 a_5 \\
&\qquad + 3 a_2 a_5 + a_4 a_5 - 5 a_1 a_6 - 3 a_3 a_6 - a_5 a_6), \\
e_6 &= (a_0 - a_1 + a_2 - a_3 + a_4 - a_5 + a_6)(a_0 + a_1 + a_2 + a_3 + a_4 + a_5 + a_6).
\end{aligned}
\tag{5-20}
$$

The polynomial $H(x)$ is the denominator of the integrand $\mathfrak{L}_2(R(x))$ in (5-15).

In Step 3 we also define

$$E(x) = H(R_2(x))Q_2^6(x) = H\left(\frac{x^2 - 1}{2x}\right) \cdot (2x)^6. \tag{5-21}$$

The function $E(x)$ is a polynomial of degree 12, written as

$$E(x) = \sum_{k=0}^{12} \alpha_k x^{12-k}. \tag{5-22}$$

Using the expressions for e_j in (5-20) in (5-21) yields

$$
\begin{aligned}
\alpha_0 &= \ \ \alpha_{12} = 64 a_0 a_6, \\
\alpha_1 &= -\alpha_{11} = -64(a_0 a_5 - a_1 a_6), \\
\alpha_2 &= \ \ \alpha_{10} = 64(a_0 a_4 - a_1 a_5 + a_2 a_6), \\
\alpha_3 &= -\alpha_9 = -64(a_0 a_3 - a_1 a_4 + a_2 a_5 - a_3 a_6), \\
\alpha_4 &= \ \ \alpha_8 = 64(a_0 a_2 - a_1 a_3 + a_2 a_4 - a_3 a_5 + a_4 a_6), \\
\alpha_5 &= -\alpha_7 = -64(a_0 a_1 - a_1 a_2 + a_2 a_3 - a_3 a_4 + a_4 a_5 - a_5 a_6), \\
\alpha_6 &= \ \ \ \ \ 64(a_0^2 - a_1^2 + a_2^2 - a_3^2 + a_4^2 - a_5^2 + a_6^2).
\end{aligned}
\tag{5-23}
$$

Step 4. The polynomial $A(x)$ always divides $E(x)$. The quotient is denoted by $Z(x)$. The values of α_j given in (5-23) produce

$$Z(x) = 64(a_0 - a_1 x + a_2 x^2 - a_3 x^3 + a_4 x^4 - a_5 x^5 + a_6 x^6). \qquad (5\text{-}24)$$

Step 5. Define the polynomial $C(x) := B(x)Z(x)$. In this case, C is of degree 10, written as

$$C(x) = \sum_{k=0}^{10} c_k x^{10-k}, \qquad (5\text{-}25)$$

and the coefficients c_k are given by

$$
\begin{aligned}
c_0 &= 64 a_6 b_0, \\
c_1 &= -64(a_5 b_0 - a_6 b_1), \\
c_2 &= 64(a_4 b_0 - a_5 b_1 + a_6 b_2), \\
c_3 &= -64(a_3 b_0 - a_4 b_1 + a_5 b_2 - a_6 b_3), \\
c_4 &= 64(a_2 b_0 - a_3 b_1 + a_4 b_2 - a_5 b_3 + a_6 b_4), \\
c_5 &= -64(a_1 b_0 - a_2 b_1 + a_3 b_2 - a_4 b_3 + a_5 b_4), \qquad (5\text{-}26) \\
c_6 &= 64(a_0 b_0 - a_1 b_1 + a_2 b_2 - a_3 b_3 + a_4 b_4), \\
c_7 &= 64(a_0 b_1 - a_1 b_2 + a_2 b_3 - a_3 b_4), \\
c_8 &= 64(a_0 b_2 - a_1 b_3 + a_2 b_4), \\
c_9 &= 64(a_0 b_3 - a_1 b_4), \\
c_{10} &= 64 a_0 b_4.
\end{aligned}
$$

Step 6 produces the numerator $J(x)$ of the new integrand $\mathfrak{L}_2(R(x))$ from the coefficients c_j given in (5-26). The function $J(x)$ is a polynomial of degree 4, written as

$$J(x) = \sum_{k=0}^{4} j_k x^{4-k}. \qquad (5\text{-}27)$$

Using the values of (5-26) we obtain

$$
\begin{aligned}
j_0 &= 32(a_6 b_0 + a_0 b_4), \qquad\qquad\qquad\qquad\qquad\qquad\qquad (5\text{-}28) \\
j_1 &= -16(a_5 b_0 - a_6 b_1 + a_0 b_3 - a_1 b_4), \\
j_2 &= 8(a_4 b_0 + 3a_6 b_0 - a_5 b_1 + a_0 b_2 + a_6 b_2 - a_1 b_3 + 3a_0 b_4 + a_2 b_4), \\
j_3 &= -4(a_3 b_0 + 2a_5 b_0 + a_0 b_1 - a_4 b_1 - 2a_6 b_1 - a_1 b_2 + a_5 b_2 \\
&\qquad\qquad\qquad\qquad + 2a_0 b_3 + a_3 b_3 - a_6 b_3 - 2a_1 b_4 - a_3 b_4), \\
j_4 &= 2(a_0 b_0 + a_2 b_0 + a_4 b_0 + a_6 b_0 - a_1 b_1 - a_3 b_1 - a_5 b_1 \\
&\qquad\qquad + a_0 b_2 + a_2 b_2 + a_4 b_2 + a_6 b_2 - a_1 b_3 \\
&\qquad\qquad\qquad - a_3 b_3 - a_5 b_3 + a_0 b_4 + a_2 b_4 + a_4 b_4 + a_6 b_4).
\end{aligned}
$$

The explicit formula used to compute the coefficients of J can be found in [Manna and Moll 2007a].

The new rational function is

$$\mathcal{L}_2(R(x)): = \frac{J(x)}{H(x)}, \tag{5-29}$$

with $J(x)$ given in (5-27) and $H(x)$ in (5-19). The transformation is

$$\frac{b_0x^4 + b_1x^3 + b_2x^2 + b_3x + b_4}{a_0x^6 + a_1x^5 + a_2x^4 + a_3x^3 + a_4x^2 + a_5x + a_6}$$

$$\mapsto \frac{j_0x^4 + j_1x^3 + j_2x^2 + j_3x + j_4}{e_0x^6 + e_1x^5 + e_2x^4 + e_3x^3 + e_4x^2 + e_5x + e_6}.$$

The numerator coefficients are given in (5-20) and the denominator ones in (5-28), explicitly as polynomials in the coefficients of the original rational function. The generation of these polynomials is a completely symbolic procedure.

The first two steps of this algorithm, applied to the definite integral

$$\int_{-\infty}^{\infty} \frac{dx}{x^6 + x^3 + 1} = \frac{\pi}{9}(2\sqrt{3}\cos(\pi/9) + \sqrt{3}\cos(2\pi/9) + 3\sin(2\pi/9)), \tag{5-30}$$

produces the identities

$$\int_{-\infty}^{\infty} \frac{dx}{x^6 + x^3 + 1} = \int_{-\infty}^{\infty} \frac{2(16x^4 + 12x^2 + 2x + 2)}{64x^6 + 96x^4 + 36x^2 + 3}\,dx$$

$$= \int_{-\infty}^{\infty} \frac{4(2816x^4 - 1024x^3 + 8400x^2 - 884x + 5970)}{12288x^6 + 59904x^4 + 87216x^2 + 39601}\,dx.$$

6. The issue of convergence

The convergence of the double sequence (a_n, b_n) appearing in the elliptic Landen transformation (1-1) is easily established. Assume $0 < b_0 \leq a_0$, then the arithmetic-geometric inequality yields $b_n \leq b_{n+1} \leq a_{n+1} \leq a_n$. Also

$$0 \leq a_{n+1} - b_{n+1} = \frac{1}{2} \frac{(a_n - b_n)^2}{(\sqrt{a_n} + \sqrt{b_n})^2}. \tag{6-1}$$

This shows a_n and b_n have a common limit: $M = \mathrm{AGM}(a, b)$, the arithmetic-geometric of a and b. The convergence is quadratic:

$$|a_{n+1} - M| \leq C|a_n - M|^2, \tag{6-2}$$

for some constant $C > 0$ independent of n. Details can be found in [Borwein and Borwein 1987].

The Landen transformations produce maps on the space of coefficients of the integrand. In this section, we discuss the convergence of the rational Landen transformations. This discussion is divided in two cases:

Case 1: the half-line. Let $R(x)$ be an even rational function, written as $R(x) = P(x)/Q(x)$, with

$$P(x) = \sum_{k=0}^{p-1} b_k x^{2(p-1-k)}, \quad Q(x) = \sum_{k=0}^{p} a_k x^{2(p-k)}, \tag{6-3}$$

and $a_0 = a_p = 1$. The *parameter space* is

$$\mathfrak{P}_{2p}^+ = \{(a_1, \ldots, a_{p-1}; b_0, \ldots, b_{p-1})\} \subset \mathbb{R}^{p-1} \times \mathbb{R}^p. \tag{6-4}$$

We write

$$\boldsymbol{a} : = (a_1, \ldots, a_{p-1}), \quad \boldsymbol{b} : = (b_0, \ldots, b_p). \tag{6-5}$$

Define

$$\Lambda_{2p} = \left\{ (a_1, \ldots, a_{p-1}) \in \mathbb{R}^{p-1} : \int_0^\infty R(x)\, dx \text{ is finite} \right\}. \tag{6-6}$$

Observe that the convergence of the integral depends only on the parameters in the denominator.

The Landen transformations provide a map

$$\Phi_{2p} : \mathfrak{P}_{2p}^+ \to \mathfrak{P}_{2p}^+ \tag{6-7}$$

that preserves the integral. Introduce the notation

$$\boldsymbol{a}_n = (a_1^{(n)}, \ldots, a_{p-1}^{(n)}) \text{ and } \boldsymbol{b}_n = (b_0^{(n)}, \ldots, b_p^{(n)}), \tag{6-8}$$

where

$$(\boldsymbol{a}_n, \boldsymbol{b}_n) = \Phi_{2p}(\boldsymbol{a}_{n-1}, \boldsymbol{b}_{n-1}) \tag{6-9}$$

are the iterates of the map Φ_{2p}.

The result that one expects is this:

THEOREM 6.1. *The region Λ_{2p} is invariant under the map Φ_{2p}. Moreover*

$$\boldsymbol{a}_n \to \left(\binom{p}{1}, \binom{p}{2}, \ldots, \binom{p}{p-1} \right), \tag{6-10}$$

and there exists a number L, that depends on the initial conditions, such that

$$\boldsymbol{b}_n \to \left(\binom{p-1}{0} L, \binom{p-1}{1} L, \ldots, \binom{p-1}{p-1} L \right). \tag{6-11}$$

This is equivalent to saying that the sequence of rational functions formed by the Landen transformations, converge to $L/(x^2 + 1)$.

This was established in [Hubbard and Moll 2003] using the geometric language of Landen transformations which, while unexpected, is satisfactory.

THEOREM 6.2. *Let φ be a 1-form, holomorphic in a neighborhood of $\mathbb{R} \subset \mathbb{P}^1$. Then*

$$\lim_{n \to \infty} (\pi_*)^n \varphi = \frac{1}{\pi} \left(\int_{-\infty}^{\infty} \varphi \right) \frac{dz}{1 + z^2}, \qquad (6\text{-}12)$$

where the convergence is uniform on compact subsets of U, the neighborhood in the definition of π_*.

The proof is detailed for the map $\pi(z) = \dfrac{z^2 - 1}{2z} = R_2(z)$, but it extends without difficulty to the generalization R_m.

Theorem 6.2 can be equivalently reformulated as:

THEOREM 6.3. *The iterates of the Landen transformation starting at $(a_0, b_0) \in \mathfrak{P}_{2p}^+$ converge (to the limit stated in Theorem 6.1) if and only if the integral formed by the initial data is finite.*

It would be desirable to establish this result by purely dynamical techniques. This has been established only for the case $p = 3$. In that case the Landen transformation for

$$U_6 := \int_0^\infty \frac{cx^4 + dx^2 + e}{x^6 + ax^4 + bx^2 + 1} \, dx \qquad (6\text{-}13)$$

is

$$a_1 \leftarrow \frac{ab + 5a + 5b + 9}{(a + b + 2)^{4/3}}, \quad b_1 \leftarrow \frac{a + b + 6}{(a + b + 2)^{2/3}}, \qquad (6\text{-}14)$$

coupled with

$$c_1 \leftarrow \frac{c + d + e}{(a+b+2)^{2/3}}, \quad d_1 \leftarrow \frac{(b+3)c + 2d + (a+3)e}{a + b + 2}, \quad e_1 \leftarrow \frac{c + e}{(a+b+2)^{1/3}}.$$

The region

$$\Lambda_6 = \{(a, b) \in \mathbb{R}^2 : U_6 < \infty\} \qquad (6\text{-}15)$$

is described by the discriminant curve $\mathfrak{R}$, the zero set of the polynomial

$$R(a, b) = 4a^3 + 4b^3 - 18ab - a^2b^2 + 27. \qquad (6\text{-}16)$$

This zero set, shown in Figure 1, has two connected components: the first one $\mathfrak{R}_+$ contains $(3, 3)$ as a cusp and the second one $\mathfrak{R}_-$, given by $R_-(a, b) = 0$, is disjoint from the first quadrant. The branch $\mathfrak{R}_-$ is the boundary of the set Λ_6.

The identity

$$R(a_1, b_1) = \frac{(a - b)^2 \, R(a, b)}{(a + b + 2)^4}, \qquad (6\text{-}17)$$

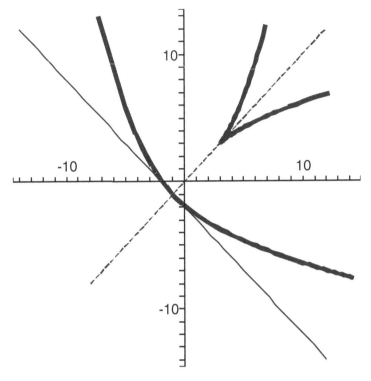

Figure 1. The zero locus of $R(a, b)$.

shows that $\partial\mathfrak{R}$ is invariant under Φ_6. By examining the effect of this map along lines of slope -1, we obtain a direct parametrization of the flow *on* the discriminant curve. Indeed, this curve is parametrized by

$$a(s) = \frac{s^3 + 4}{s^2} \text{ and } b(s) = \frac{s^3 + 16}{4s}. \tag{6-18}$$

Then

$$\varphi(s) = \left(\frac{4(s^2 + 4)^2}{s(s + 2)^2}\right)^{1/3} \tag{6-19}$$

gives the image of the Landen transformation Φ_6; that is,

$$\Phi_6(a(s), b(s)) = (a(\varphi(s)), b(\varphi(s))). \tag{6-20}$$

The map Φ_6 has three fixed points: $(3, 3)$, that is superattracting, a saddle point P_2 on the lower branch $\mathfrak{R}_-$ of the discriminant curve, and a third unstable spiral below this lower branch. In [Chamberland and Moll 2006] we prove:

THEOREM 6.4. *The lower branch of the discriminant curve is the curve Λ_6. This curve is also the global unstable manifold of the saddle point P_2. Therefore the iterations of Φ_6 starting at (a, b) converge if and only if the integral*

U_6, formed with the parameters (a, b), is finite. Moreover, $(a_n, b_n) \to (3, 3)$ quadratically and there exists a number L such that $(c_n, d_n, e_n) \to (1, 2, 1)L$.

The next result provides an analogue of the AGM (1-13) for the rational case. The main differences here are that our iterates converge to an *algebraic* number and we achieve *order-m* convergence.

Case 2: The whole line. This works for any choice of positive integer m. Let $R(x)$ be a rational function, written as $R(x) = B(x)/A(x)$. Assume that the co-efficients of A and B are real, that A has no real zeros and that $\deg B \leq \deg A - 2$. These conditions are imposed to guarantee the existence of

$$I = \int_{-\infty}^{\infty} R(x)\, dx. \qquad (6\text{-}21)$$

In particular A must be of even degree, and we write

$$A(x) = \sum_{k=0}^{p} a_k x^{p-k} \text{ and } B(x) = \sum_{k=0}^{p-2} b_k x^{p-2-k}. \qquad (6\text{-}22)$$

We can also require that $\deg(\gcd(A, B)) = 0$.

The class of such rational functions will be denoted by $\mathfrak{R}_p$.

The algorithm presented in Section 5 provides a transformation on the parameters

$$\mathfrak{P}_p := \{a_0, a_1, \ldots, a_p; b_0, b_1, \ldots, b_{p-2}\} = \mathbb{R}^{p+1} \times \mathbb{R}^{p-1} \qquad (6\text{-}23)$$

of $R \in \mathfrak{R}_p$ that preserves the integral I. In fact, we produce a family of maps, indexed by $m \in \mathbb{N}$,

$$\mathfrak{L}_{m,p} : \mathfrak{R}_p \to \mathfrak{R}_p,$$

such that

$$\int_{-\infty}^{\infty} R(x)\, dx = \int_{-\infty}^{\infty} \mathfrak{L}_{m,p}(R(x))\, dx. \qquad (6\text{-}24)$$

The maps $\mathfrak{L}_{m,p}$ induce a *rational Landen transformation*

$$\Phi_{m,p} : \mathfrak{P}_p \to \mathfrak{P}_p \qquad (6\text{-}25)$$

on the parameter space: we simply list the coefficients of $\mathfrak{L}_{m,p}(R(x))$.

The original integral is written in the form

$$I = \frac{b_0}{a_0} \int_{-\infty}^{\infty} \frac{x^{p-2} + b_0^{-1} b_1 x^{p-3} + b_0^{-1} b_2 x^{p-4} + \cdots + b_0^{-1} b_{p-2}}{x^p + a_0^{-1} a_1 x^{p-1} + a_0^{-1} a_2 x^{p-2} + \cdots + a_0^{-1} a_p}\, dx. \qquad (6\text{-}26)$$

The Landen transformation generates a sequence of coefficients,

$$\mathfrak{P}_{p,n} := \{a_0^{(n)}, a_1^{(n)}, \ldots, a_p^{(n)}; b_0^{(n)}, b_1^{(n)}, \ldots, b_{p-2}^{(n)}\}, \qquad (6\text{-}27)$$

with $\mathfrak{P}_{p,0} = \mathfrak{P}_p$ as in (6-23). We expect that, as $n \to \infty$,

$$x_n := \left(\frac{a_1^{(n)}}{a_0^{(n)}}, \frac{a_2^{(n)}}{a_0^{(n)}}, \ldots, \frac{a_p^{(n)}}{a_0^{(n)}}, \frac{b_1^{(n)}}{b_0^{(n)}}, \frac{b_2^{(n)}}{b_0^{(n)}}, \ldots, \frac{b_{p-2}^{(n)}}{b_0^{(n)}} \right) \tag{6-28}$$

converges to

$$x_\infty := \left(0, \binom{q}{1}, 0, \binom{q}{2}, \ldots, \binom{q}{q}; 0, \binom{q-1}{1}, 0, \binom{q-1}{2}, \ldots, \binom{q-1}{q-1} \right), \tag{6-29}$$

where $q = p/2$. Moreover, we should have

$$\|x_{n+1} - x_\infty\| \le C \|x_n - x_\infty\|^m. \tag{6-30}$$

The invariance of the integral then shows that

$$\frac{b_0^{(n)}}{a_0^{(n)}} \to \frac{1}{\pi} I. \tag{6-31}$$

This produces an iterative method to evaluate the integral of a rational function. The method's convergence is of order m.

The convergence of these iterations, and in particular the bound (6-30), can be established by the argument presented in Section 4. Thus, the transformation $\mathfrak{L}_{m,p}$ leads to a sequence that has order-m convergence. We expect to develop these ideas into an efficient numerical method for integration.

We choose to measure the convergence of a sequence of vectors to 0 in the L_2-norm,

$$\|v\|_2 = \frac{1}{\sqrt{2p-2}} \left(\sum_{k=1}^{2p-2} \|v_k\|^2 \right)^{1/2}, \tag{6-32}$$

and also the L_∞-norm,

$$\|v\|_\infty = \text{Max} \{ \|v_k\| : 1 \le k \le 2p-2 \}. \tag{6-33}$$

The rational functions appearing as integrands have rational coefficients, so, as a measure of their complexity, we take the largest number of digits of these coefficients. This appears in the column marked *size*.

The tables on the next page illustrate the iterates of rational Landen transformations of order 2, 3 and 4, applied to the example

$$I = \int_{-\infty}^{\infty} \frac{3x+5}{x^4 + 14x^3 + 74x^2 + 184x + 208} \, dx = -\frac{7\pi}{12}.$$

The first column gives the L_2-norm of $u_n - u_\infty$, the second its L_∞-norm, the third presents the relative error in (6-31), and in the last column we give the size

n	L_2-norm	L_∞-norm	Error	Size
1	58.7171	69.1000	1.02060	5
2	7.444927	9.64324	1.04473	10
3	4.04691	5.36256	0.945481	18
4	1.81592	2.41858	1.15092	41
5	0.360422	0.411437	0.262511	82
6	0.0298892	0.0249128	0.0189903	164
7	0.000256824	0.000299728	0.0000362352	327
8	1.92454×10^{-8}	2.24568×10^{-8}	1.47053×10^{-8}	659
9	1.0823×10^{-16}	1.2609×10^{-16}	8.2207×10^{-17}	1318

Table 1. Method of order 2.

n	L_2-norm	L_∞-norm	Error	Size
1	15.2207	20.2945	1.03511	8
2	1.97988	1.83067	0.859941	23
3	0.41100	0.338358	0.197044	69
4	0.00842346	0.00815475	0.00597363	208
5	5.05016×10^{-8}	5.75969×10^{-8}	1.64059×10^{-9}	626
6	1.09651×10^{-23}	1.02510×10^{-23}	3.86286×10^{-24}	1878
7	1.12238×10^{-70}	1.22843×10^{-70}	8.59237×10^{-71}	5634

Table 2. Method of order 3.

n	L_2-norm	L_∞-norm	Error	Size
1	7.44927	9.64324	1.04473	10
2	1.81592	2.41858	1.15092	41
3	0.0298892	0.0249128	0.0189903	164
4	1.92454×10^{-8}	2.249128×10^{-8}	1.47053×10^{-8}	659
5	3.40769×10^{-33}	3.96407×10^{-33}	2.56817×10^{-33}	2637

Table 3. Method of order 4.

of the rational integrand. At each step, we verify that the new rational function integrates to $-7\pi/12$.

As expected, for the method of order 2, we observe quadratic convergence in the L_2-norm and also in the L_∞-norm. The size of the coefficients of the integrand is approximately doubled at each iteration.

EXAMPLE 6.1. A method of order 3 for the evaluation of the quadratic integral

$$I = \int_{-\infty}^{\infty} \frac{dx}{ax^2 + bx + c}, \tag{6-34}$$

has been analyzed in [Manna and Moll 2007b]. We refer to Example 5.1 for the explicit formulas of this Landen transformation, and define the iterates accordingly. From there, we prove that the error term,

$$e_n : = \left(a_n - \tfrac{1}{2}\sqrt{4ac - b^2}, \ b_n, \ c_n - \tfrac{1}{2}\sqrt{4ac - b^2}\right) \tag{6-35}$$

satisfies $e_n \to 0$ as $n \to \infty$, with cubic rate:

$$\|e_{n+1}\| \le C\|e_n\|^3. \tag{6-36}$$

The proof of convergence is elementary. Therefore

$$(a_n, b_n, c_n) \to \left(\sqrt{ac - b^2/4}, 0, \sqrt{4ac - b^2/4}\right), \tag{6-37}$$

which, in conjunction with (6-34), implies that

$$I = \frac{2}{\sqrt{4ac - b^2}} \int_{-\infty}^{\infty} \frac{dx}{x^2 + 1}, \tag{6-38}$$

exactly as one would have concluded by completing the square. Unlike completing the square, our method extends to a general rational integral over the real line.

7. The appearance of the AGM in diverse contexts

The (elliptic) Landen transformation

$$a_1 \leftarrow \tfrac{1}{2}(a + b), \quad b_1 \leftarrow \sqrt{ab} \tag{7-1}$$

leaving invariant the elliptic integral

$$G(a, b) = \int_0^{\pi/2} \frac{d\varphi}{\sqrt{a^2 \cos^2 \varphi + b^2 \sin^2 \varphi}} \tag{7-2}$$

appears in many different forms. In this last section we present a partial list of them.

7.1. The elliptic Landen transformation. For the lattice $\mathbb{L} = \mathbb{Z} \oplus \omega \mathbb{Z}$, introduce the *theta-functions*

$$\vartheta_3(x,\omega) := \sum_{n=-\infty}^{\infty} z^{2n} q^{n^2}, \quad \vartheta_4(x,\omega) := \sum_{n=-\infty}^{\infty} (-1)^n z^{2n} q^{n^2}, \quad (7\text{-}3)$$

where $z = e^{\pi i x}$ and $q = e^{\pi i \omega}$. The condition $\text{Im}\,\omega > 0$ is imposed to ensure convergence of the series. These functions admit a variety of remarkable identities. In particular, the *null-values* (those with $x = 0$) satisfy

$$\vartheta_4^2(0, 2\omega) = \vartheta_3(0, \omega)\vartheta_4(0, \omega), \quad \vartheta_3^2(0, 2\omega) = \tfrac{1}{2}\big(\vartheta_3^2(0, \omega) + \vartheta_4^2(0, \omega)\big),$$

and completely characterize values of the AGM, leading to the earlier result [Borwein and Borwein 1987]. Grayson [1989] has used the doubling of the period ω to derive the arithmetic-geometric mean from the cubic equations describing the corresponding elliptic curves. See Chapter 3 in [McKean and Moll 1997] for more information. P. Sole et al. [1995; 1998] have proved generalizations of these identities using lattice enumeration methods related to binary and ternary codes.

7.2. A time-one map. We now present a deeper and more modern version of a result known to Gauss: given a sequence of points $\{x_n\}$ on a manifold X, decide whether there is a differential equation

$$\frac{dx}{dt} = V(x), \quad (7\text{-}4)$$

starting at x_0 such that $x_n = x(n, x_0)$. Here $x(t, x_0)$ is the unique solution to (7-4) satisfying $x(0, x_0) = x_0$. Denote by

$$\phi_{\text{ellip}}(a, b) = \big(\tfrac{1}{2}(a + b), \sqrt{ab}\big) \quad (7\text{-}5)$$

the familiar elliptic Landen transformation. Now take $a, b \in \mathbb{R}$ with $a > b > 0$. Use the null-values of the theta functions to find unique values (τ, ρ) such that

$$a = \rho\vartheta_3^2(0, \tau), \quad b = \rho\vartheta_4^2(0, \tau). \quad (7\text{-}6)$$

Finally define

$$x_{\text{ellip}}(t) = (a(t), b(t)) = \rho\big(\vartheta_3^2(0, 2^t\tau), \vartheta_4^2(0, 2^t\tau)\big), \quad (7\text{-}7)$$

with $x_{\text{ellip}}(0) = (a, b)$. The remarkable result is [Deift 1992]:

THEOREM 7.1 (DEIFT, LI, PREVIATO, TOMEI). *The map $t \to x_{\text{ellip}}(t)$ is an integrable Hamiltonian flow on X equipped with an appropriate symplectic structure. The Hamiltonian is the complete elliptic integral $G(a, b)$ and the* *angle is (essentially the logarithm of) the second period of the elliptic curve associated with τ. Moreover*

$$x_{\text{ellip}}(k) = \phi_{\text{ellip}}^k(a, b).\tag{7-8}$$

Thus the arithmetic-geometric algorithm is the time-one map of a completely integrable Hamiltonian flow.

Notice that this theorem shows that the result in question respects some additional structures whose invention postdates Gauss.

A natural question is whether the map (3-12) appears as a time-one map of an interesting flow.

7.3. A quadruple sequence. Several variations of the elliptic Landen appear in the literature. Borchardt [1876] considers the four-term quadratically convergent iteration

$$a_{n+1} = \frac{a_n + b_n + c_n + d_n}{4}, \quad b_{n+1} = \frac{\sqrt{a_n b_n} + \sqrt{c_n d_n}}{2},$$
$$c_{n+1} = \frac{\sqrt{a_n c_n} + \sqrt{b_n d_n}}{2}, \quad d_{n+1} = \frac{\sqrt{a_n d_n} + \sqrt{b_n c_n}}{2},\tag{7-9}$$

starting with $a_0 = a$, $b_0 = b$, $c_0 = c$ and $d_0 = d$. The common limit, denoted by $G(a, b, c, d)$, is given by

$$\frac{1}{G(a, b, c, d)} = \frac{1}{\pi^2} \int_0^{\alpha_3} \int_{\alpha_1}^{\alpha_2} \frac{(x - y)\, dx\, dy}{\sqrt{R(x)R(y)}},\tag{7-10}$$

where $R(x) = x(x - \alpha_0)(x - \alpha_1)(x - \alpha_2)(x - \alpha_3)$ and the numbers α_j are given by explicit formulas in terms of the parameters a, b, c, d. Details are given in [Mestre 1991].

The initial conditions $(a, b, c, d) \in \mathbb{R}^4$ for which the iteration converges has some interesting invariant subsets. When $a = b$ and $c = d$, we recover the AGM iteration (1-1). In the case that $b = c = d$, we have another invariant subset, linking to an iterative mean described below.

7.4. Variations of AGM with hypergeometric limit. Let $N \in \mathbb{N}$. The analysis of

$$a_{n+1} = \frac{a_n + (N-1)b_n}{N} \quad \text{and} \quad c_{n+1} = \frac{a_n - b_n}{N},\tag{7-11}$$

with $b_n = (a_n^N - c_n^N)^{1/N}$, is presented in [Borwein and Borwein 1991]. All the common ingredients appear there: a common limit, fast convergence, theta functions and sophisticated iterations for the evaluation of π. The common limit is denoted by $\mathrm{AG}_N(a, b)$. The convergence is of order N and the limit is identified for small N: we have, for $0 < k < 1$,

$$\frac{1}{\mathrm{AG}_2(1,k)} = {}_2F_1(1/2, 1/2; 1; 1-k^2), \quad \frac{1}{\mathrm{AG}_3(1,k)} = {}_2F_1(1/3, 2/3; 1; 1-k^2),$$

where

$$_2F_1(a, b; c; x) = \sum_{k=0}^{\infty} \frac{(a)_k\, (b)_k}{(c)_k\, k!} x^k \tag{7-12}$$

is the classical hypergeometric function. There are integral representations of these as well which parallel (1-13); see [Borwein et al. 2004a], Section 6.1 for details.

Other hypergeometric values appear from similar iterations. For example,

$$a_{n+1} = \frac{a_n + 3b_n}{4} \quad \text{and} \quad b_{n+1} = \sqrt{b_n(a_n + b_n)/2}, \tag{7-13}$$

have a common limit, denoted by $A_4(a, b)$. It is given by

$$\frac{1}{A_4(1,k)} = {}_2F_1^2(1/4, 3/4; 1; 1-k^2). \tag{7-14}$$

To compute π quartically, start at $a_0 = 1$, $b_0 = (12\sqrt{2}-16)^{1/4}$. Now compute a_n from two steps of AG_2:

$$a_{n+1} = \frac{a_n + b_n}{2}, \quad \text{and} \quad b_{n+1} = \left(\frac{a_n b_n^3 + b_n a_n^3}{2}\right)^{1/4}. \tag{7-15}$$

Then

$$\pi = \lim_{n\to\infty} 3a_{n+1}^4 \left(1 - \sum_{j=0}^{n} 2^{j+1}(a_j^4 - a_{j+1}^4)\right)^{-1} \tag{7-16}$$

with $|a_{n+1} - \pi| \le C|a_n - \pi|^4$, for some constant $C > 0$. This is much better than the partial sums of

$$\pi = 4 \sum_{k=0}^{\infty} \frac{(-1)^k}{2k + 1}. \tag{7-17}$$

The sequences (a_n), (b_n) defined by the iteration

$$a_{n+1} = \frac{a_n + 2b_n}{3}, \quad b_{n+1} = \left(\frac{b_n(a_n^2 + a_n b_n + b_n^2)}{3}\right)^{1/3}, \tag{7-18}$$

starting at $a_0 = 1$, $b_0 = x$ are analyzed in [Borwein and Borwein 1990]. They have a common limit $F(x)$ given by

$$\frac{1}{F(x)} = {}_2F_1(1/3, 2/3; 1; 1 - x^3). \tag{7-19}$$

7.5. Iterations where the limit is harder to find. J. Borwein and P. Borwein [1989] studied the iteration of

$$(a,b) \rightarrow \left(\frac{a+3b}{4}, \frac{\sqrt{ab}+b}{2} \right), \tag{7-20}$$

and showed the existence of a common limit $B(a_0, b_0)$. Define $B(x) = B(1, x)$. The study of the iteration (7-20) is based on the functional equation

$$B(x) = \frac{1+3x}{4} B\left(\frac{2(\sqrt{x}+x)}{1+3x} \right). \tag{7-21}$$

and a parametrization of the iterates by theta functions [Borwein and Borwein 1989]. The complete analysis of (7-20) starts with the purely computational observation that

$$B(x) \sim \frac{\pi^2}{3} \log^{-2}(x/4) \quad \text{as } x \to 0. \tag{7-22}$$

H. H. Chan, K. Chua and P. Sole [Heng Huat Chan and Sole 2002] identified the limiting function as

$$B(x) = \left({}_2F_1\left(\frac{1}{3}, \frac{1}{6}; 1; 27\frac{x(1-x)^2}{(1+3x)^3} \right) \right)^{-2}, \tag{7-23}$$

valid for $x \in \left(\frac{2}{3}, 1 \right)$. A similar hypergeometric expression gives $B(x)$ for $x \in \left(0, \frac{2}{3} \right)$.

7.6. Fast computation of elementary functions. The fast convergence of the elliptic Landen recurrence (1-1) to the arithmetic-geometric mean provides a method for numerical evaluation of the elliptic integral $G(a, b)$. The same idea provides for the fast computation of elementary functions. For example, in [Borwein and Borwein 1984] we find the estimate

$$\left| \log x - \left(G(1, 10^{-n}) - G(1, 10^{-n}x) \right) \right| < n10^{-2(n-1)}, \tag{7-24}$$

for $0 < x < 1$ and $n \geq 3$.

7.7. A continued fraction. The continued fraction

$$R_\eta(a, b) = \cfrac{a}{\eta + \cfrac{b^2}{\eta + \cfrac{4a^2}{\eta + \cfrac{9b^2}{\eta + \cdots}}}}, \tag{7-25}$$

has an interesting connection to the AGM. In their study of the convergence of $R_\eta(a, b)$, J. Borwein, R. Crandall and G. Fee [Borwein et al. 2004b] established the identity

$$R_\eta \left(\frac{a+b}{2}, \sqrt{ab} \right) = \frac{1}{2} \left(R_\eta(a,b) + R_\eta(b,a) \right). \qquad (7\text{-}26)$$

This identity originates with Ramanujan; the similarity with AGM is now direct.

The continued fraction converges for positive real parameters, but for a, $b \in \mathbb{C}$ the convergence question is quite delicate. For example, the even/odd parts of $R_1(1, i)$ converge to distinct limits. See [Borwein et al. 2004b; 2004c] for more details.

7.8. Elliptic Landen with complex initial conditions. The iteration of (1-1) with a_0, $b_0 \in \mathbb{C}$ requires a choice of square root at each step. Let a, $b \in \mathbb{C}$ be nonzero and assume $a \neq \pm b$. A square root c of ab is called the *right choice* if

$$\left| \frac{a+b}{2} - c \right| \leq \left| \frac{a+b}{2} + c \right|. \qquad (7\text{-}27)$$

It turns out that in order to have a limit for (1-1) one has to make the right choice for all but finitely many indices $n \geq 1$. This is described in detail in [Cox 1984].

7.9. Elliptic Landen with p-adic initial conditions. Let p be a prime and a, b be nonzero p-adic numbers. To guarantee that the p-adic series

$$c = a \sum_{i=0}^{\infty} \binom{\frac{1}{2}}{i} \left(\frac{b}{a} - 1 \right)^i \qquad (7\text{-}28)$$

converges, and thus defines a p-adic square root of ab, one must assume

$$b/a \equiv 1 \bmod p^\alpha, \qquad (7\text{-}29)$$

where $\alpha = 3$ for $p = 2$ and 1 otherwise. The corresponding sequence defined by (1-1) converges for $p \neq 2$ to a common limit: the p-adic AGM. In the case $p = 2$ one must assume that the initial conditions satisfy $b/a \equiv 1 \bmod 16$. In the case $b/a \equiv 1 \bmod 8$ but not 1 modulo 16, the corresponding sequence (a_n, b_n) does not converge, but the sequence of so-called *absolute invariants*

$$j_n = \frac{2^8 (a_n^4 - a_n^2 b_n^2 + b_n^4)^3}{a_n^4 b_n^4 (a_n^2 - b_n^2)^2} \qquad (7\text{-}30)$$

converges to a 2-adic integer. Information about these issues can be found in [Henniart and Mestre 1989]. D. Kohel [2003] has proposed a generalization of the AGM for elliptic curves over a field of characteristic $p \in \{2, 3, 5, 7, 13\}$. Mestre [2000] has developed an AGM theory for ordinary hyperelliptic curves over a field of characteristic 2. This has been extended to nonhyperelliptic curves of genus 3 curves by Lehavi and Ritzenhaler [2007]. An algorithm for counting points for ordinary elliptic curves over finite fields of characteristic $p > 2$ based on the AGM is presented in [Carls 2004].

7.10. Higher genus AGM. An algorithm analogue to the AGM for abelian integrals of genus 2 was discussed by Richelot [1836; 1837] and Humberdt [1901]. Some details are discussed in [Bost and Mestre 1988]. The case of abelian integrals of genus 3 can be found in [Lehavi and Ritzenhaler 2007].

Gauss was correct: his numerical calculation (1-4) *has grown in many unexpected directions.*

Acknowledgements

The second author acknowledges the partial support of NSF-DMS 0409968. The authors thank Jon Borwein for many comments that led to an improvement of the manuscript.

References

[Abramowitz and Stegun 1972] M. Abramowitz and I. Stegun, *Handbook of mathematical functions with formulas, graphs and mathematical tables*, Dover, New York, 1972.

[Amdeberhan and Moll 2007] T. Amdeberhan and V. Moll, "A formula for a quartic integral: a survey of old proofs and some new ones", *The Ramanujan Journal* (2007).

[Amdeberhan et al. 2007] T. Amdeberhan, D. Manna, and V. Moll, "The 2-adic valuation of a sequence arising from a rational integral", *Preprint* (2007).

[Berndt and Bowman 2000] B. Berndt and D. Bowman, "Ramanujan's short unpublished manuscript on integrals and series related to Euler's constant", *Canadian Math. Soc. Conf. Proc.* **27** (2000), 19–27.

[Borchardt 1876] C. W. Borchardt, "Uber das arithmetisch-geometrische Mittel aus vier Elementen", *Berl. Monatsber.* **53** (1876), 611–621.

[Boros and Moll 1999a] G. Boros and V. Moll, "A criterion for unimodality", *Elec. Jour. Comb.* **6** (1999), 1–6.

[Boros and Moll 1999b] G. Boros and V. Moll, "An integral hidden in Gradshteyn and Ryzhik", *Jour. Comp. Applied Math.* **106** (1999), 361–368.

[Boros and Moll 2000] G. Boros and V. Moll, "A rational Landen transformation. The case of degree 6", pp. 83–89 in *Analysis, geometry, number theory: the mathematics of Leon Ehrenpreis*, edited by E. L. Grinberg et al., Contemporary Mathematics **251**, American Mathematical Society, 2000.

[Boros and Moll 2001a] G. Boros and V. Moll, "The double square root, Jacobi polynomials and Ramanujan's master theorem", *Jour. Comp. Applied Math.* **130** (2001), 337–344.

[Boros and Moll 2001b] G. Boros and V. Moll, "Landen transformation and the integration of rational functions", *Math. Comp.* **71** (2001), 649–668.

[Boros et al. 2001] G. Boros, V. Moll, and J. Shallit, "The 2-adic valuation of the coefficients of a polynomial", *Scientia* **7** (2001), 37–50.

[Borwein and Bailey 2003] J. M. Borwein and D. H. Bailey, *Mathematics by experiment: plausible reasoning in the 21st century*, A K Peters, Natick, MA, 2003.

[Borwein and Borwein 1984] J. M. Borwein and P. B. Borwein, "The arithmetic-geometric mean and fast computation of elementary functions", *SIAM Review* **26** (1984), 351–366.

[Borwein and Borwein 1987] J. M. Borwein and P. B. Borwein, *Pi and the AGM: A study in analytic number theory and computational complexity*, Wiley, New York, 1987.

[Borwein and Borwein 1989] J. M. Borwein and P. B. Borwein, "On the mean iteration $(a, b) \leftarrow ((a + 3b)/4, (\sqrt{ab} + b)/2)$", *Math. Comp.* **53** (1989), 311–326.

[Borwein and Borwein 1990] J. M. Borwein and P. B. Borwein, "A remarkable cubic mean iteration", pp. 27–31 in *Computational methods and function theory*, edited by S. Ruscheweyh et al., Lectures Notes in Mathematics **1435**, Springer, New York, 1990.

[Borwein and Borwein 1991] J. M. Borwein and P. B. Borwein, "A cubic counterpart of Jacobi's identity and the AGM", *Trans. Amer. Math. Soc.* **323** (1991), 691–701.

[Borwein et al. 2004a] J. M. Borwein, D. H. Bailey, and R. Girgensohn, *Experimentation in mathematics: computational paths to discovery*, A K Peters, Natick, MA, 2004.

[Borwein et al. 2004b] J. M. Borwein, R. Crandall, and G. Fee, "On the Ramanujan AGM fraction, I: the real parameter case", *Experimental Math.* **13** (2004), 275–285.

[Borwein et al. 2004c] J. M. Borwein, R. Crandall, and G. Fee, "On the Ramanujan AGM fraction, II: the complex-parameter case", *Experimental Math.* **13** (2004), 287–295.

[Bost and Mestre 1988] J. B. Bost and J. F. Mestre, "Moyenne arithmetico-géometrique et périodes des courbes de genre 1 et 2", **38** (1988), 36–64.

[Bromwich 1926] T. J. Bromwich, *An introduction to the theory of infinite series*, 2nd ed., MacMillan, London, 1926.

[Carls 2004] R. Carls, *A generalized arithmetic geometric mean*, Ph.D. thesis, Gronigen, 2004.

[Caviness and Johnson 1998] B. Caviness and J. R. Johnson, *Quantifier elimination and cylindrical algebraic decomposition*, Texts and Monographs in Symbolic Computation, Springer, New York, 1998.

[Chamberland and Moll 2006] M. Chamberland and V. Moll, "Dynamics of the degree six Landen transformation", *Discrete and Dynamical Systems* **15** (2006), 905–919.

[Collins 1975] G. E. Collins, "Quantifier elimination for the elementary theory of real closed fields by cylindrical algebraic decomposition", *Lecture Notes in Computer Science* **33** (1975), 134–183.

[Cox 1984] D. Cox, "The arithmetic-geometric mean of Gauss", *L'Enseigement Mathematique* **30** (1984), 275–330.

[Deift 1992] P. Deift, "Continuous versions of some discrete maps or what goes on when the lights go out", *Jour. d'Analyse Math.* **58** (1992), 121–133.

[Gauss 1799] K. F. Gauss, "Arithmetisch-geometrisches Mittel", *Werke* **3** (1799), 361–432.

[Grayson 1989] D. Grayson, "The arithogeometric mean", *Arch. Math* **52** (1989), 507–512.

[Heng Huat Chan and Sole 2002] K. S. C. Heng Huat Chan and P. Sole, "Quadratic iterations to π associated with elliptic functions to the cubic and septic base", *Trans. Amer. Math. Soc.* **355** (2002), 1505–1520.

[Henniart and Mestre 1989] G. Henniart and J. F. Mestre, "Moyenne arithmetico-géométrique p-adique", *C. R. Acad. Sci. Paris* **308** (1989), 391–395.

[Hubbard and Moll 2003] J. Hubbard and V. Moll, "A geometric view of rational Landen transformation", *Bull. London Math. Soc.* **35** (2003), 293–301.

[Humbert 1901] G. Humbert, "Sur la transformation ordinaire des fonctions abéliennes", *J. Math.* **7** (1901).

[Kauers and Paule 2007] M. Kauers and P. Paule, "A computer proof of Moll's log-concavity conjecture", *Proc. Amer. Math. Soc.* **135** (2007), 3847–3856.

[Kohel 2003] D. Kohel, "The AGM-$X_0(N)$ Heegner point lifting algorithm and elliptic curve point counting", pp. 124–136 in *Proceedings of ASI-ACRYPT'03*, Lecture Notes in Computer Science **2894**, Springer, 2003.

[Landen 1775] J. Landen, "An investigation of a general theorem for finding the length of any arc of any conic hyperbola, by means of two elliptic arcs, with some other new and useful theorems deduced therefrom", *Philos. Trans. Royal Soc. London* **65** (1775), 283–289.

[Lehavi and Ritzenhaler 2007] D. Lehavi and C. Ritzenhaler, An explicit formula for the arithmetic geometric mean in genus 3, 2007.

[Little 2005] J. Little, "On the zeroes of two families of polynomials arising from certain rational integrals", *Rocky Mountain Journal* **35** (2005), 1205–1216.

[Manna and Moll 2007a] D. Manna and V. Moll, "Rational Landen transformations on $\mathbb{R}$", *Math. Comp.* **76** (2007), 2023–2043.

[Manna and Moll 2007b] D. Manna and V. Moll, "A simple example of a new class of Landen transformations", *Amer. Math. Monthly* **114** (2007), 232–241.

[McKean and Moll 1997] H. McKean and V. Moll, *Elliptic curves: function theory, geometry, arithmetic*, Cambridge University Press, New York, 1997.

[Mestre 1991] J. F. Mestre, "Moyenne de Borchardt et intégrales elliptiques", *C. R. Acad. Sci. Paris* **313** (1991), 273–276.

[Mestre 2000] J. F. Mestre, "Lettre addressée à Gaudry et Harley", (2000). Available at http://www.math.jussieu.fr/~mestre.

[Newman 1985] D. J. Newman, "A simplified version of the fast algorithm of Brent and Salamin", *Math. Comp.* **44** (1985), 207–210.

[Richelot 1836] F. Richelot, "Essai sur une méthode generale pour determiner la valeur des intégrales ultra-elliptiques, fondée sur des transformations remarquables de ces transcendents", *C. R. Acad. Sci. Paris* **2** (1836), 622–627.

[Richelot 1837] F. Richelot, "De transformatione integralium abelianorum primi ordinis commentatio", *J. reine angew. Math.* **16** (1837), 221–341.

[Sole 1995] P. Sole, "D_4, E_6, E_8 and the AGM", pp. 448–455 in *Applied algebra, algebraic algorithms and error-correcting codes : 11th international symposium, AAECC-11* (Paris, 1995), edited by G. Cohen et al., Lecture Notes in Computer Science **948**, Springer, New York, 1995.

[Sole and Loyer 1998] P. Sole and P. Loyer, "U_n lattices, construction B, and the AGM iterations", *Europ. J. Combinatorics* **19** (1998), 227–236.

[Springer 2002] G. Springer, *Introduction to Riemann surfaces*, 2nd ed., American Mathematical Society, 2002.

[Watson 1933] G. N. Watson, "The marquis and the land-agent", *Math. Gazette* **17** (1933), 5–17.

[Wilf 1990] H. S. Wilf, *generatingfunctionology*, Academic Press, 1990.

DANTE V. MANNA
DEPARTMENT OF MATHEMATICS AND STATISTICS
DALHOUSIE UNIVERSITY
HALIFAX, NOVA SCOTIA
CANADA, B3H 3J5
dantemanna@gmail.com

VICTOR H. MOLL
DEPARTMENT OF MATHEMATICS
TULANE UNIVERSITY
NEW ORLEANS, LA 70118
UNITED STATES
vhm@math.tulane.edu

Probability, Geometry and Integrable Systems
MSRI Publications
Volume **55**, 2008

Lines on abelian varieties

EMMA PREVIATO

ABSTRACT. We study the function field of a principally polarized abelian variety from the point of view of differential algebra. We implement in a concrete case the following result of I. Barsotti, which he derived from what he called the prostapheresis formula and showed to characterize theta functions: the logarithmic derivatives of the theta function along *one* line generate the function field. We outline three interpretations of the differential algebra of theta functions in the study of commutative rings of partial differential operators.

Henry McKean was one of the earliest contributors to the field of "integrable PDEs", whose origin for simplicity we shall place in the late 1960s. One way in which Henry conveyed the stunning and powerful discovery of a linearizing change of variables was by choosing Isaiah 40:3-4 as an epigram for [McKean 1979]: *The voice of him that crieth in the wilderness, Prepare ye the way of the Lord, make straight in the desert a highway for our God. Every valley shall be exalted and every mountain and hill shall be made low: and the crooked shall be made straight and the rough places plain.* Thus, on this contribution to a volume intended to celebrate Henry's many fundamental achievements on the occasion of his birthday, my title. I use the word line in the extended sense of "linear flow", of course, since no projective line can be contained in an abelian variety — the actual line resides in the universal cover.

This article is concerned primarily with classical theta functions, with an appendix to report on a daring extension of the concept to infinite-dimensional tori, also initiated by Henry. Thirty years (or forty, if you regard the earliest experiments by E. Fermi, J. Pasta and S. Ulam, then M. D. Kruskal and N. J. Zabusky, as more than an inspiration in the discovery of solitons; see [Previato 2008] for references) after the ground was broken in this new field, in my view one of the main remaining questions in the area of theta functions as related to PDEs, is still that of straight lines, both on abelian varieties and on Grassmann manifolds (the two objects of greatest interest to geometers in the nineteenth century!). On a Jacobian $\mathrm{Pic}^0(X)$, where X is a Riemann surface of genus g (which we also call a "curve", for brevity), there is a line which is better than any other. That is, after choosing a point on the curve. Whatever point is chosen, the sequence of hyperosculating vectors to the Abel image of the curve in the Jacobian at that point can be taken as the flows of the KP hierarchy, according the Krichever's inverse spectral theory. As a side remark, also related to KP, on a curve not all points are created equal. For a Weierstrass point, there are more independent functions in the linear systems nP for small n than there are for generic curves, which translates into early vanishing of (combinations of) KP flows, giving rise to n-th KdV-reduction hierarchies of a sort; other special differential-algebraic properties would obtain if $(2g-2)P$ is a canonical divisor K_X [Matsutani and Previato 2008]. However, on a general (principally polarized) abelian variety, "there should be complete democracy".[1] My central question is: What line, or lines, are important to the study of differential equations satisfied by the theta function?

In this paper I put together a number of different proposed constructions and ground them in a common project: use the differential equations for the theta function along a generic line in an abelian variety, to characterize abelian varieties, give in particular generalized KP equations, and interpret these PDEs as geometric constraints that define the image of infinite-dimensional flag manifolds in $\mathbb{P}B$, where B is a bosonic space. These topics are developed section-by-section as follows: Firstly, Barsotti proved (in an essentially algebraic way) that on any abelian variety[2] there exists a direction such that the set of derivatives of sufficiently high order of the logarithm of the theta function along that direction generates the function field of the abelian variety. Moreover, he characterized theta functions by a system of ordinary differential equations, polynomial in that direction. These facts have been found hard to believe by sufficiently many experts to whom I quoted them, that it may be of some value (if only entertainment value), to give a brutally "honest", boring and painstaking proof in this paper, for small dimension. This gives me the excuse for advertising a different line of work on differential equations for theta functions (Section 1). Then, I propose to link this problem of lines and the other outstanding problem of algebro-geometric PDEs, which was the theme of my talk at the workshop reported in this volume: commuting partial differential operators (PDOs). There is a classification of (maximal-)commutative rings of ordinary differential operators, and their isospectral deformations are in fact the KP flows. In more than one variable

---

[1] I quote this nice catchphrase without attribution, this being the reaction to the assertions of Section 1 evinced by an expert whom I hadn't warned he would be "on record".

[2] Assume for simplicity that it is irreducible; let me also beg forgiveness if in this introduction I do not specify all possible degenerate cases which Barsotti must except in his statements, namely extensions of abelian varieties by a number of multiplicative or additive 1-dimensional groups.

very little is known, though several remarkable examples have appeared. The two theories that I will mention here were proposed by Sato (and implemented by Nakayashiki) and Parshin. Nakayashiki's work produced commuting matrix partial differential operators, but has the advantage of giving differential equations for theta functions. Since Barsotti's equations characterize theta functions, I believe that it would be profitable to identify Nakayashiki's equations, which were never worked out explicitly, among Barsotti's (Section 2). Parshin's construction produces (in principle, though recent work by his students shows that essential constraints must be introduced) deformations of scalar PDOs; in his setting, it is possible to generalize the Krichever map. It is a generalization of the Krichever map which constitutes the last link I would like to propose. Parshin sends a surface and a line bundle on it to a flag manifold; Arbarello and De Concini generalize the Krichever map and embed the general abelian variety and a line bundle on it into a projective space where Sato's Grassmannian is a submanifold, the image of Jacobians. My proposal is to characterize the image of the abelian varieties, in both Parshin's and Arbarello–De Concini's maps, by Barsotti's equations (Section 3). In conclusion, some concrete constructions are touched upon (Section 4). In a much too short Appendix, I reference Henry McKean's contribution on infinite-genus Riemann surfaces.

1. Incomplete democracy

Lines in Jacobians. Jacobians are special among principally polarized abelian varieties (ppav's), in that they contain a curve that generates the torus as a subgroup. For any choice of point on the curve, there is a specific line on the torus, which one expects to have special properties: indeed, the hyperosculating tangents to the embedding of the curve in the Jacobian given by that chosen point, give a sequence of flows satisfying the KP hierarchy. The KP equations provide an analytic proof that the tangent line (more precisely, its projection modulo the period lattice) cannot be contained in the theta divisor (no geometric proof has been given to date), while the order of vanishing of the theta function at the point (first given in connection to the KP equation as a sum of codimensions of a stratification of Sato's Grassmannian) was recently interpreted geometrically [Birkenhake and Vanhaecke 2003].

More geometrically yet, the Riemann approach links linear series on the curve to differential equations on the Jacobian, and again these lines play a very special role. I give two examples only. I choose these because both authors pose specific open problems (concerning indeed the special role of Jacobians among ppav's, known as "Schottky problem"), through the theory of special linear series. The subvarieties of such special linear series are acquiring increasing importance in providing exact solutions to Hamiltonian systems; see [Eilbeck et al. 2007] and references therein.

EXAMPLE. In one among his many contributions to these problems, Gunning [1986] produced in several, essentially different ways, differential equations satisfied by level-two theta functions. These are mainly limits, after J. Fay, of addition formulas, and this depends crucially on the tangent direction to the curve (at any variable point), *the* line. Gunning's focus is the study of the "Wirtinger varieties", roughly speaking, the images under the Kummer map of the W_k ($1 \le k \le g$), which in turn are images in the Jacobian of the k-fold symmetric products of the curve, via differential equations and thetanulls. For example, he proves the following (his notation for level-2 theta functions is ϑ_2):

If S is the subspace of dimension dim $S = \binom{g+1}{2} + 1$ spanned by the vectors $\vartheta_2(0)$ and $\partial_{jk}\vartheta_2(0)$ for all (j, k), then the projectivization of this subspace contains the Kummer image of the surface $W_1 - W_1$, so it has intersection with the Kummer variety of dimension higher than expected, as soon as $g \ge 4$.

So little is known about these important subvarieties, that Welters [1986] states the following as an open problem: *Does there exist a relationship between* $\{a \in \mathrm{Pic}^0 X \mid a + W_d^r \subset W_d^{r-k}\}$ *and* $W_k^0 - W_k^0$ $(0 \le k \le r, 0 \le d \le g-1)$? He had previously shown that

$$W_1^0 - W_1^0 = \{a \in \mathrm{Pic}^0 X \mid a + W_{g-1}^1 \subset W_{g-1}^0\},$$

where the notation W_d^r is the classical one for linear series of degree d and (projective) dimension at least r; g_d^r denotes a linear series of degree d and projective dimension r.

EXAMPLE. It is intriguing that Mumford, in his book devoted to applications of theta functions to integrable systems, states as an open problem [1984, Chapter IIIb, §3]: If V is the vector space spanned by

$$\left\{ \vartheta^2(z), \vartheta(z) \cdot \frac{\partial^2 \vartheta}{\partial z_i \partial z_j} - \frac{\partial \vartheta}{\partial z_i} \cdot \frac{\partial \vartheta}{\partial z_j} \right\}$$

and B is the set of "decomposition functions" $\vartheta(z - a) \cdot \vartheta(z + a)$, does the intersection of V and B equal the set $\{\vartheta(z - \int_p^q) \cdot \vartheta(z + \int_p^q)\}$, where p, q are any two points of the curve? As Mumford notes, this is equivalent to asking: If $a \in \mathrm{Jac}\, X$ is such that for all $w \in W_{g-1}^1$, either $w + a$ or $w - a$ is in W_{g-1}^0, does a belong to $W_1 - W_1$? The latter is settled by Welters (*loc. cit.*), showing that indeed, for $g \ge 4$ (for smaller genus the statement should be modified and still holds when it makes sense),

$$X - X = \cap_{\xi \in W_{g-1}^1} \left((W_{g-1}^0)_{-\xi} + (W_{g-1}^0)_{\xi - K_X} \right)$$

(as customary, subscript denotes inverse image under translation in the Picard group and K_X the canonical divisor), unless X is trigonal, for which it was known:

$$\bigcap_{\xi \in W_{g-1}^1} (W_{g-1}^0)_{-\xi} + (W_{g-1}^0)_{\xi-K_X} = (W_3^0 - g_3^1) \cup (g_3^1 - W_3^0).$$

On the enumerative side, Beauville [1982] shows that the sum of all the divisor classes in W_d^r is a multiple of the canonical divisor, provided r and d satisfy $g = (r+1)(g+r-d)$. The proof uses nontrivial properties of the Chow ring of the Jacobian, and it would be nice to find an interpretation in terms of theta functions.

A line of attack to these problems is suggested in [Jorgenson 1992a; 1992b], where theta functions defined on the W_k's are related to algebraic functions, generalizing the way that the Weierstrass points are defined in terms of ranks of matrices of holomorphic differentials. In a related way, techniques of expansion of the sigma function (associated to theta) along the curve, yield differential equations; see [Eilbeck et al. 2007].

Barsotti lines. However, on a general abelian variety, there should be "complete democracy", the catchphrase, in reaction to my report on Barsotti's result, that I am appropriating. Barsotti showed — in a way which is exquisitely algebraic (and almost, though not quite, valid for any characteristic of the field of coefficients), based on his theory of "hyperfields" for describing abelian varieties (developed in the fifties and only partly translated by his school into standard language), and independent of the periods — that *one* line suffices, to produce the differential field of the abelian variety. Barsotti's approach was aimed at a characterization of functions which he called "theta type", and this means generalized theta, pertaining to a product of tori as well as group extensions by a number of copies of the additive and multiplicative group of the field.

I will phrase this important result, along with a sketch of the proof, reintroducing the period lattice, though aware that Barsotti would disapprove of this naive approach, and I will give an "honest" proof in the (trigonal) case of genus 3, the last case when all (indecomposable) ppav's are Jacobians, yet the first case in which several experts reacted to Barsotti's result with "complete disbelief" (not in the sense of deeming Barsotti wrong, but rather, in intrigued astonishment that the democracy of lines should allow for such a property).

Barsotti is concerned with *abelian group varieties*, our abelian varieties, which he studies locally by rings of formal power series $k\{u_1, u_2, \ldots, u_n\} = k\{u\}$, which we will take to be the convergent power series in n indeterminates, $\mathbb{C}[\![u]\!]$, as usual abbreviating by u the n-tuple of variables. The context below will accommodate both cases, that u signify an n-tuple or a single variable. We follow

Barsotti's notation for derivatives: $d_i = \partial/\partial u_i$, and in case $r = (r_1, \ldots, r_n)$ is a multi-index, $d_r = (r!)^{-1} d^r = (r_1!)^{-1} \cdots (r_n!)^{-1} d_1^{r_1} \cdots d_n^{r_n}$; also, $|r| := \sum_{i=1}^{n} r_i$, n-tuples of indices are ordered componentwise, and if different sets of indeterminates appear, d_{ur} will denote derivatives with respect to the u-variables. The symbol $\mathcal{Q}(-)$ generally associates to an integral domain its field of fractions. The notation is abbreviated: $k\{u\} := \mathcal{Q}(k[u])$.

THEOREM 1 [Barsotti 1983, Theorem 3.7]. *A function $\vartheta(u) \in k\{u\}$ is such that*

$$\vartheta(u+v)\vartheta(u-v) \in k\{u\} \otimes k\{v\} \tag{1}$$

if and only if it has the property

$$F(u, v, w) := \frac{\vartheta(u+v+w)\vartheta(u)\vartheta(v)\vartheta(w)}{\vartheta(u+v)\vartheta(u+w)\vartheta(v+w)} \in \mathcal{Q}(k\{u\} \otimes k\{v\} \otimes k\{w\}). \tag{2}$$

Barsotti regarded this as the main result of [Barsotti 1983]. He had called (1) the prostapheresis formula[3] and (2) the condition for being "theta-type". His ultimate goal was to produce a theory of theta functions that could work over any field, and in doing so, he analyzed the fundamental role of the addition formulas; indeed, H. E. Rauch, in his review of [Barsotti 1970] (MR0302655 – Mathematical Reviews **46** #1799) exclaims, of the fact that (2) characterizes classical theta functions for $k = \mathbb{C}$, "This ... result is, to this reviewer, new and beautiful and crowns a conceptually and technically elegant paper". In order to appreciate the scope of (1) and (2), we have to put them to the use of computing dimensions of vector spaces spanned by their derivatives. To me (I may be missing something more profound, of course) the *segue* from properties of type (1) or (2) into dimensions of spaces of derivatives is this: u (the n-tuple) gives us local coordinates on the abelian variety; we understand analytic functions by computing coefficients of their Taylor expansions (derivatives) and the finite dimensionality corresponds to the fact that, while *a priori* the LHS belongs to $k\{u, v] := k\{u\}\overline{\otimes}k\{v\}$, which denotes the completion of the tensor product $k\{u\} \otimes k\{v\}$, only finitely many tensors suffice. The precise statement is this:

LEMMA [Barsotti 1983, 2.1]. *A function $\varphi(u, v)$ in $k\{u, v]$ belongs to*

$$k\{u\} \otimes k\{v\}$$

if and only if the vector space U spanned over k by the derivatives $d_{vr}\varphi(u, 0)$ has finite dimension. If such is the case, the vector space V spanned over k by the derivatives $d_{ur}\varphi(0, v)$ has the same dimension, and $\varphi(u, v) \in U \otimes V$.

---

[3]"We are indebted to the Arab mathematician Ibn Jounis for having proposed, in the XIth century a method, called prostapheresis, to replace the multiplication of two sines by a sum of the same functions", according to Papers on History of Science, by Xavier Lefort, Les Instituts de Recherche sur l'Enseignement des Mathématiques, Nantes.

To understand the theta-type functions as analytic functions, we also need to introduce certain numerical invariants.

DEFINITION. We denote by C_ϑ the smallest subfield of $k\{u\}$ containing k and such that $F(u, v, w) \in C_\vartheta\{v, w\}$. Note that C_ϑ is generated over k by the $d_r \log \vartheta$ for $|r| \geq 2$. This fact has already nontrivial content, in the classical case; the function field of an abelian variety is generated by the second and higher logarithmic derivatives of the Riemann theta function. The transcendence degree transcϑ is transc(C_ϑ/k) and the dimension dim ϑ is the dimension (in the sense of algebraic varieties) of the smallest local subring of $k\{u\}$ whose quotient field contains a theta-type function associated to (namely, as usual, differing from by a quadratic exponential) ϑ. I am giving a slightly inaccurate definition of dimension, for in his algebraic theory Barsotti had introduced more sophisticated objects than subrings; but I will limit myself, for the purposes of the results of this paper, to the case of "nondegenerate" thetas, which Barsotti defines as satisfying dim $\vartheta = n$. The inequality transc $\vartheta \geq$ dim ϑ always holds and Barsotti calls ϑ a "theta function" when equality holds.

The next result is the root of all mystery. Here Barsotti demonstrates that in fact, the function field of the abelian variety could be generated by the derivatives of a theta function along fewer than m directions, m being the dimension of the abelian variety.

THEOREM 2 [Barsotti 1983, 2.4]. *For a nondegenerate theta-type*

$$\vartheta(u) \in k\{u_1, \cdots, u_n\},$$

there exist a nondegenerate theta $\theta(v) \in k\{v_1, \cdots, v_m\}$, $m \geq n$, and $c_{ij} \in k$, $1 \leq i \leq n$; $1 \leq j \leq m$, such that the matrix $[c_{ij}]$ has rank n, and $\vartheta(u) = \theta(x_1, \cdots, x_m)$ where $x_i = \sum_j c_{ij} u_j$. The induced homomorphism of $k\{v\}$ onto $k\{u\}$ induces an isomorphism between C_θ and C_ϑ. Conversely, given a compact abelian variety A of dimension m, for any $0 < n < m$ there is a holomorphic theta-type $\vartheta(u_1, \cdots u_n)$ such that C_ϑ is the function field of A, and is generated over k by a finite number of $d_r \log \vartheta$ with $|r| \geq 2$.

The example. Several experts have suggested (without producing details, as far as I know) that the statement may be believable in the case of a hyperelliptic Jacobian, but is already startling in the $g = 3$, nonhyperelliptic case, and this is the example I report. This is current work which I happen to be involved in for totally unrelated reasons; to summarize the motivation and goals in much too brief a manner, it is work concerned with addition formulae for a function associated to theta over a stratification of the theta divisor related to the abel image of the symmetric powers of the curve. Repeating the preliminaries would be quite lengthy and, more importantly, detract from the focus of this paper, so aside from indispensable notation I take the liberty of referring to [Eilbeck et al. 2007].

The key idea goes back to Klein and was developed by H. F. Baker over a long period (see especially [Baker 1907], where he collected and systematized this work). To generalize the theory of elliptic functions to higher-genus curves, these authors started with curves of special (planar) type, for which they expressed algebraically as many of the abelian objects as possible, differentials of first and second kind, Jacobi inversion formula, and ultimately, equations for the Kummer variety (in terms of theta-nulls) and linear flows on the Jacobian. In the process, they obtained or introduced important PDEs to characterize the abelian functions in question, and anecdotally, even produced, in the late 1800s, exact solutions to the KdV and KP hierarchy, without of course calling them by these names. I just need to quote certain PDEs satisfied by these "generalized abelian functions", but I will mention the methods by which these can be obtained. Firstly, the simplest function to work with, for reasons of local expansion at the origin, is called "sigma", it is associated to Riemann's theta function, and its normalized (almost-)period matrix satisfies generalized Legendre relations, being the matrix of periods of suitable bases of differentials of first and second kind. The definition of sigma is not explicit and considerable computer algebra is involved, genus-by-genus. The $g = 3$ case I need here is explicitly reported in [Eilbeck et al. 2007], but had been obtained earlier (by Ônishi, for instance).

In the suitable normalization, the "last" holomorphic differential ω_g always gives rise to the KP flow, namely the abelian vector $(0,\ldots,0,1)$ in the coordinates $(u_1,\cdots,u_g) = \int_{g\infty}^{\sum_{i=1}^{g}(x_i,y_i)} \omega$, $\omega = (\omega_1,\cdots,\omega_g)$, simply because of the given orders of zero of the basis of differentials at the point ∞ of the curve, in the affine (x, y) plane, which is also chosen as the point of tangency of the KP flow to the abel image of the curve (indeed, in [Eilbeck et al. 2007] the Boussinesq equation is derived, as expected for the cyclic trigonal case). It is for this reason that I choose this direction for the Barsotti variable u.

Now the role of Barsotti's theta is played by $\sigma(u_1, u_2, u_3)$ — associated to a Riemann theta function with half-integer characteristics, and explicitly given in [Eilbeck et al. 2007, (3.8)] — and the role of the Weierstrass $\wp$-function, by the abelian functions $\wp_{ij}(u) = -\frac{\partial^2}{\partial u_i \partial u_j} \log \sigma(u)$; we label the higher derivatives the same way,

$$\wp_{ijk}(u) = \frac{\partial}{\partial u_k} \wp_{ij}(u), \quad \wp_{ijk\ell}(u) = \frac{\partial}{\partial u_\ell} \wp_{ijk}(u),$$

(et cetera, but I only need the first four in my proof).

Barsotti's statement now amounts to this: the function $\wp_{33}(0, 0, u_3)$ together with all its derivatives in the u_3 variable, generate the function field of the Jacobian. Here's the boring proof! First, the work in [Eilbeck et al. 2007] (and a series of papers that preceded it): It is straightforward to expand σ in terms of a local parameter on the curve, for example,

$$u_1 = \frac{1}{5}u_3{}^5 + \cdots, \quad u_2 = \frac{1}{2}u_3{}^2 + \cdots$$

and

$$x(u_1, u_2, u_3) = \frac{1}{u_3{}^3} + \cdots, \quad y(u_1, u_2, u_3) = \frac{1}{u_3{}^4} + \cdots.$$

where $P \mapsto \int_\infty^P \omega := u(P)$, so $x(P)$ and $y(P)$ are viewed as functions of $u(P) = (u_1, u_2, u_3)$; the image of the curve implicitly defines any of the three coordinates as functions of one only. Next one expands σ as a function of (u_1, u_2, u_3), and with the aid of computer algebra, obtains PDEs for the abelian functions. For example, the identity

$$\wp_{3333} = 6\wp_{33}^2 - 3\wp_{22}$$

implies the Boussinesq equation for the function $\wp_{33}$, as expected. It is by using these differential equations, worked out in [Eilbeck et al. 2007] up to four indices (Appendix B), that I prove Barsotti's result. As a shorcut, I record a basis of the space 3Θ where Θ (this notation slightly differs from the one chosen in that reference) is the divisor of the σ function. If we can get this basis of abelian functions, we are sure to generate the function field of the Jacobian, since by the classical Lefschetz theorem the 3Θ-divisor map is an embedding. Lemma 8.1 in [Eilbeck et al. 2007] provides the following basis of 27 elements: $\{1, \wp_{11}, \wp_{12}, \wp_{13}, \wp_{22}, \wp_{23}, \wp_{33}, Q_{1333}, \wp_{111}, \wp_{112}, \wp_{113}, \wp_{122}, \wp_{123}, \wp_{133}, \wp_{222}, \wp_{223}, \wp_{233}, \wp_{333}, \partial_1 Q_{1333}, \partial_2 Q_{1333}, \partial_3 Q_{1333}, \wp^{[11]}, \wp^{[12]}, \wp^{[13]}, \wp^{[22]}, \wp^{[23]}, \wp^{[33]}\}$, where

$$Q_{ijk\ell}(u) = \wp_{ijk\ell}(u) - 2(\wp_{ij}\wp_{k\ell} + \wp_{ik}\wp_{j\ell} + \wp_{i\ell}\wp_{jk})(u)$$

and $\wp^{[ij]}$ is the determinant of the complementary (i, j)-minor of $[\wp_{ij}]_{3\times3}$. It is easy, by substituting in the equations given in [Eilbeck et al. 2007], to see that if we can obtain all the 2-index $\wp$ functions, then we can write the necessary 3, 4, and 5-index functions in the given basis. By definition of the Barsotti line, we have $\wp_{33}$, which gives us $\wp_{22}$ by the Boussinesq relation given above (we are allowed to take derivatives with respect to u_3). The one that seemed most difficult to obtain was $\wp_{23}$, and I argued as follows: Denote by F the differential field in the variable u_3 generated over $\mathbb{C}$ by $\wp_{33}$; as we saw it contains $\wp_{333}$ and $\wp_{22}$. Now F_{ij} denotes the field generated over F by adding $\wp_{ij}$. Eliminating $\wp_{13}$ from the two equations

$$\wp_{333}^2 = \wp_{23}^2 + 4\wp_{13} - 4\wp_{33}\wp_{22} + 4\wp_{33}^3$$

and

$$Q_{2233} = 4\wp_{13} + 3\lambda_3\wp_{23} + 2\lambda_2,$$

we see that F_{23} is an extension of degree at most 2 of F; then, either equation says that $\wp_{13}$ belongs to F_{23}. Now, we would like to say that F_{23} is also at most a cubic extension of F, and for that, use the equation

$$\wp_{223}\wp_{233} = 2\wp_{23}^3 + 2\wp_{22}\wp_{23}\wp_{33}$$
$$+ 2\lambda_1 + 4\wp_{23}\wp_{13} + 2\wp_{23}\lambda_2 + 2\lambda_3\wp_{13} + 2\lambda_3\wp_{23}^2 + \lambda_3\wp_{22}\wp_{33}.$$

However, we can't quite control $\wp_{233}$, so we also bring in the equations

$$\wp_{333}\wp_{223} = 2\wp_{33}\wp_{23}^2 + \wp_{33}\lambda_3\wp_{23} - 2\wp_{22}^2 + \tfrac{2}{3}\wp_{1333} + 2\wp_{33}^2\wp_{22}$$

and

$$\wp_{233}^2 = 4\wp_{33}\wp_{23}^2 + 4\wp_{33}\lambda_3\wp_{23} + \wp_{22}^2 - \tfrac{4}{3}\wp_{1333} + 4\wp_{33}\lambda_2 + 8\wp_{33}\wp_{13}.$$

The first says that $\wp_{1333}$ is at most degree two (over F) in $\wp_{23}$; now using the cubic (and substituting for $\wp_{1333}$ in it), we see that $\wp_{23}$ satisfies an equation of degree 3 over $F[\wp_{233}]$, but from the second equation, $\wp_{233}$ is in an extension of degree at most 2 of F_{23}, so if F_{23} and $F[\wp_{233}]$ were disjoint, their join would have degree 4 and $\wp_{23}$ could not have degree 3 over $F[\wp_{233}]$. This shows that $\wp_{233}$ is in F_{23}, and now the cubic together with the quadratic equation yield $\wp_{23} \in F$. The proof that all other $\wp_{ij}$ are also in F is now much easier, again using several of the equations given in [Eilbeck et al. 2007]. If there is an easier proof, it beats me, for now at least.

REMARK. In a letter of reply to my querie (February 6, 1987), which I would translate, were it not for fear of misrepresenting as a conjecture what he only intended to offer as a possibility for my pursuing, Barsotti wrote that it might be that for generic $(c_1, \cdots, c_m)$, suitably high derivatives of $\log \vartheta(c_1 u, \cdots, c_m u)$ generate the function field of the abelian variety. In my view, this would not only restore democracy, but give a beautiful technique for stratifying the moduli space of abelian varieties according to the "special" parameters $c_1, \cdots, c_m$ whose line fails to generate, and which might correspond to tangent vectors to an abelian variety of smaller dimension (I claim all blame for this additional thought, but see Section 4 below). In the case of the "purely trigonal" curve above, we know that "elliptic solitons" can occur [Eilbeck et al. 2001], so does my proof say that even though σ is an elliptic function in the u_3 direction, still u_3 is a Barsotti direction? I don't think so; my proof requires obtaining $\wp_{23}$ from algebraic equations with coefficients in F, for example, but those coefficients depend on the λ_i's and there is no reason why for special values of λ_i's the equations shouldn't become trivial identities (while they patently can be solved for $\wp_{23}$ when the λ_i's are generic).

I can state with certainty at least that at the time of his tragic demise in 1987 Barsotti was very keen on pursuing these ideas [Scorza Dragoni 1988].

2. Barsotti's and Nakayashiki's equations

Barsotti equations include KP. Barsotti [1983; 1985] then proceeded to characterize abelian varieties. Again, I give a sketchy rendition of his results, which glosses over the technical issues of decomposable or degenerate abelian varieties. These are both important and subtle (for instance, the results have to be modified if you take for ϑ a polynomial!) but since this paper does not make substantial use of those exceptional cases, my goal is to give a geometric understanding of the generic situation. Calling "holomorphic" a theta-type function whose divisor $\mathrm{div}\vartheta$ is effective, Barsotti obtains:

THEOREM 3 [Barsotti 1983, 4.1]. *A nonzero function $\vartheta(u) \in \mathbb{C}[\![u_1, \ldots, u_n]\!]$ is holomorphic theta-type if and only if all differential polynomials*

$$H_{r,s}(\vartheta(u)) = \sum_{\substack{p+q=s \\ i+j=r}} (-1)^{i+p} d_i(\vartheta)d_p(\vartheta)d_j(\vartheta)d_q(\vartheta)$$

span a finite dimensional $\mathbb{C}$-vector space. In this case, if $\{U_0, \ldots, U_h\}$ is a basis, the field $\mathbb{C}(\ldots, \vartheta^{-3}U_i, \ldots)$ is the same as the field of the abelian variety associated to ϑ, $H_{r,s}$ in turn are holomorphic theta-type and their divisors are linearly equivalent to $3\mathrm{div}\vartheta$.

Finally, by Taylor-series expansion, Barsotti writes a set of universal PDEs that characterize abelian varieties, and because of the "incomplete democracy" result, such PDEs can be produced for any positive number of variables less than or equal to the dimension of the abelian variety, in particular, one!

THEOREM 4 [Barsotti 1983, 5.5; Barsotti 1985, 12.2]. *For the universal differential polynomials with rational coefficients $P_{2k}(y_2, y_4, \ldots, y_{2k})$ defined by*

$$\vartheta(u+v)\vartheta(u-v) = 2\vartheta^2(u) \sum_{r=0}^{\infty} P_{2r}(\vartheta(u))v^{2r}$$

the same criterion as the above for $H_{r,s}$ holds. In particular, for the case of one variable ($n = 1$), the $P_0 = \frac{1}{2}, \ldots, P_{2k}(y_2, y_4, \ldots, y_{2k})$ are given by

$$P_{2r}(\vartheta) = \sum_j 2^{|j|-1}(j!)^{-1}\vartheta_2^{j_1}\vartheta_4^{j_2} \ldots \vartheta_{2r}^{j_r},$$

where the sum is over the multi-indices $j \geq 0$ such that $j_1 + 2j_2 + \cdots + r\,j_r = r$.

Barsotti [1985] wrote examples of a PDE version of his result, suggesting that it would be interesting to determine explicitly these PDEs from the ones in one variable and the vector field $\partial/\partial u$, and in [1989] he conjectures that the KP equation in his notation become

$$12 P_{400}(\ldots, \vartheta_i, \ldots) - 3 P_{020}(\ldots, \vartheta_i, \ldots) - 2 P_{101}(\ldots, \vartheta_i, \ldots) = 0.$$

Nakayashiki's generalized KP. A generalization of the KP equation as deformation of commutative rings of PDOs was long sought-after, and Nakayashiki [1991] did in fact produce such rings, in g variables for generic (thus, not Jacobians if $g \geq 4$) abelian varieties A of dimension g (as well as more general cases), as $(g! \times g!)$ matrix operators. He constructed modules over such rings that deform according to a generalized KP hierarchy, though he did not pursue explicit equations for bases of such modules, which have the form

$$N_{ct}(n) = \sum_{s \in \mathbb{Z}^g / n\mathbb{Z}^g} \mathbb{C}[t] \frac{\vartheta \begin{bmatrix} s/n \\ 0 \end{bmatrix} (nz + c - (x' \cdot d - x_1, x'))}{\vartheta^n(z)}$$

$$\times \exp\left(-\sum_{i=1}^{g} \sum_{m \geq \delta_{i1}} t_{m,(i)} \frac{(-1)^m}{m!} \left(u_{m,(i)} + d_i(1 - \delta_{i1})u_{m+1,(1)}\right)\right),$$

where we have set $x_1 = t_{1,(1)}$, $x_i = t_{0,(i)}$ for $2 \leq i \leq g$, and $d = (d_2, \ldots, d_g) \in \mathbb{C}^{g-1}$ is such that at the point of the theta divisor we are considering (the elements of the module N_{ct} are in the stalk of a sheaf, defined via a cocycle by the vector $c \in \mathbb{C}^g$, the "initial condition" for the hierarchy) the g-tuple $(\zeta_1^{-1}, \zeta_1^{-1}\zeta_i + d_i)_{i=2,\ldots,g})$ gives local coordinates; moreover x' denotes the vector $(x_2, \ldots, x_g)$ if $x = (x_1, \ldots, x_g)$, while (i) denotes the $(g-1)$-tuple $(0, \ldots, 0, 1, 0, \ldots)$ with a 1 in the $(i-1)$-st position, and $(1) = (0, \ldots, 0)$; finally, $u_{i_1, \ldots, i_g}$ denotes $\partial_{z_1}^{i_1} \ldots \partial_{z_g}^{i_g} \log \vartheta(z)$.

The differential equations are obtained as follows. Firstly, we denote by $\mathcal{P}$ the ring of microdifferential operators, defined by Sato [1989] via the choice of a codirection dx_1, which can be taken to correspond to an equation $x_1 = 0$ for the theta divisor

$$\mathcal{D} = \mathbb{C}[[t_1, \ldots, t_g]][\partial_1, \ldots, \partial_g] \subset \mathcal{P} = \mathbb{C}[t][[\partial_1^{-1}, \partial_1^{-1}\partial_2 \ldots, \partial_1^{-1}\partial_g]][\partial_1]$$

filtered by the order $\alpha_1 + \cdots + \alpha_g$ of $\partial^\alpha = \partial_1^{\alpha_1} \ldots \partial_g^{\alpha_g}$.

Now N_{ct} can be embedded in $\mathcal{P}$ as a $\mathcal{D}$-submodule, $\varphi \in N_{ct} \mapsto \iota(\varphi) = W_\varphi$, in such a way that $W_{\partial\varphi/\partial x_i} = (\partial W_\varphi/\partial x_i) + W_\varphi \partial_i = \partial_i W_\varphi$ for $1 \leq i \leq g$ and the $\mathcal{D}$-submodule of $\mathcal{P}$, $\mathcal{J}_{ct}(n) = \iota(N_{ct}(n+1))$, satisfies $\mathcal{P}^{(n)} = \mathcal{J}_{ct}(n) \oplus \mathcal{P}(J_{n,ct})$ where $J_{n,ct}$ is a suitable collection of indices from $\mathbb{Z} \times \mathbb{N}^{g-1}$, and

$$\mathcal{P}(J) = \left\{\sum a_\alpha \partial^\alpha \mid a_\alpha = 0 \text{ unless } \alpha \in J\right\}.$$

Lastly, a set of $g!$ suitable $\mathcal{D}$-generators W_α of $\mathcal{J}_{ct}$, $\alpha \notin J_{n,ct}$ for all $n \geq 0$, can be chosen of the form $\partial^\alpha + [$an operator whose terms have multiindices belonging to $J_{|\alpha|,ct}]$ and these satisfy the evolution equations $(\partial W_\alpha/\partial t_\beta) + W_\alpha \partial^\beta \in \mathcal{J}_{ct} = \bigcup_{n=0}^\infty \mathcal{J}_{ct}(n)$, for β in the index set $(m+1,(i))$, with $(m,(i))$ defined above. In [Mironov 2002], it is claimed that the functions in

$$
N_{ct} \cdot \exp\left(-\sum_{i=1}^{g} \sum_{m \geq \delta_{i1}} t_{m,(i)} \frac{(-1)^m}{m!} \left(u_{m,(i)} + d_i(1 - \delta_{i1})u_{m+1,(1)} \right) \right)
$$

are independent of the time variables, but I think this is due to a small oversight, since the first g time variables do enter the argument of ϑ, as (x_1, x'), whereas, as correctly asserted in [Mironov 2002], the higher-time variables are stationary. The commutative ring of PDOs does not undergo a deformation beyond the g-dimensional variety $A^\vee$, which indeed is $\mathrm{Pic}^0 A$.

Nakayashiki does not claim that his equations characterize abelian varieties. Nevertheless, it should be possible to produce them from Barsotti's equations, which characterize theta functions, and it would be very interesting to see how Nakayashiki's formulas are given by constraints on Barsotti's universal polynomials (among these, what Barsotti calls "initial conditions" return the moduli of the each specific abelian variety; see §7 of [Barsotti 1983] for the example of elliptic curves).

3. Sato's Grassmannian, Parshin's flag manifold, Arbarello–De Concini's projective space

Grassmannian for a chosen splitting. Nakayashiki's theory was inspired by Sato's programme [1989], a specific splitting $\mathcal{P} = \mathcal{J} \oplus \mathcal{E}_0$ into $\mathcal{D}$-modules. In one variable, there is a natural splitting and the corresponding $\mathcal{J}$ are exactly the cyclic submodules; the deformations are linear flows on the universal Grassmann manifold modeled on the vector space $\mathcal{P}_{const}: \partial^\alpha \leftrightarrow \partial^\alpha / (\mathcal{P}t_1 + \cdots + \mathcal{P}t_g)$. What is the correct model in several variables? To my knowledge there is no definitive answer known; I provide two different models below, based on Parshin's, respectively, Arbarello–De Concini's constructions, and the project of computing Nakayashiki's flows in both, which should be both doable (in dimension 2) and enlightening.

Parshin's Krichever flag manifold. Parshin [1999] proposed a different construction, based on the theory of higher local fields, in which the commuting partial differential operators are scalar. An n-dimensional local field K (with "last" residue field $\mathbb{C}$) is the field of iterated Laurent series $K = \mathbb{C}((x_1)) \ldots ((x_n))$, with the structure of a complete discrete valuation ring $\mathcal{O} = \mathbb{C}((x_1)) \ldots ((x_{n-1}))[[x_n]]$ having an $(n-1)$-dimensional local field for its residue field. Note that the order of the variables matters, in the sense that $\mathbb{C}((x_1))((x_2))$ does not contain the same elements as $\mathbb{C}((x_2))((x_1))$ — for instance, the former contains elements of unbounded positive degree in x_1 — although they are isomorphic. These are suited to give local coordinates on an n-dimensional manifold, since the inverse of a polynomial in x_1, x_2, say, can be written as the inverse of the highest-order monomial times something entire, so as a Laurent series it is bounded in both variables. Whereas the symbols $\mathbb{C}((x_1, x_2)) = \{\sum_{|i+j|<N} c_{ij} x_1^i x_2^j\}$ cannot be given a ring structure unless we want to define sums of infinitely many complex numbers, because $i + j = N$ involves infinitely many indices unless we bound j (or i) from above. With this definition, Parshin constructs a $2n$-dimensional skew-field $\mathcal{P}$, infinite-dimensional over its center, namely the (formal) pseudodifferential operators

$$\mathcal{P} = \mathbb{C}((x_1)) \ldots ((x_n))((\partial_1^{-1})) \ldots ((\partial_n^{-1})).$$

The order of the variables is also singled out in the definition of the grading: If $L = \sum_{i \leq m} a_i \partial_n^i$ with $a_m \neq 0$, we say that the operator L has order m and write ord $L = m$. If $P_i = \{L \in \mathcal{P} \mid \text{ord } L \leq i\}$, then $\cdots P_{-1} \subset P_0 \subset \cdots$ is a decreasing filtration of $\mathcal{P}$ by subspaces and $\mathcal{P} = P_+ \oplus P_-$, where $P_- = P_{-1}$ and P_+ consists of operators involving only nonnegative powers of ∂_n. The highest term (h.t.) of an operator L is defined by induction on n. If $L = \sum_{i \leq m} a_i \partial_n^i$ and ord $L = m$, then h.t.$(L) = $ h.t.$(a_m) \cdot \partial^m$. If h.t.$(L) = f \partial_1^{m_1} \ldots \partial_n^{m_n}$ with $0 \neq f \in \mathbb{C}((x_1)) \ldots ((x_n))$, then we let $v(L) = (m_1, \ldots, m_n)$. We consider also the subring $E = \mathbb{C}[[x_1, \ldots, x_n]]((\partial_1^{-1})) \ldots ((\partial_n^{-1}))$ of $\mathcal{P}$, and $E_\pm = E \cap \mathcal{P}_\pm$.

In this setting, Parshin's original proposal for a KP hierarchy — which is currently being modified by his former student Dr. A. Zheglov [Zheglov 2005] — makes good on his striking conjugation result, based on [Krichever 1977; Sato 1989] (I omit some technical specifications, for which see [Parshin 1999]):

PROPOSITION. (i) *An operator $L \in E$ is invertible in E if and only if the coefficient f in the highest-order term of L is invertible in the ring $\mathbb{C}[[x_1, \ldots, x_n]]$. If f in $L \in \mathcal{P}$ is an m-th power in $\mathbb{C}((x_1)) \ldots ((x_n))$ (resp., $\mathbb{C}[[x_1, \ldots, x_n]]$ for $L \in E$) then there exists, unique up to multiplication by m-th root of unity, an operator $M \in \mathcal{P}$ (resp. $M \in E$) such that $M^m = L$. Thus, P_0 is a discrete valuation ring in $\mathcal{P}$ with residue field $\mathbb{C}((x_1)) \ldots ((x_n))((\partial_1^{-1})) \ldots ((\partial_{n-1}^{-1}))$.*

(ii) *Let $L_1 \in \partial_1 + E_-, \ldots, L_n \in \partial_n + E_-$. Then $[L_i, L_j] = 0$ for all i, j if and only if there exists an operator $S \in 1 + E_-$ such that $L_i = S^{-1} \partial_i S$, for all i.*

(iii) *For $L = (L_1, \ldots, L_n)$ as in (ii), the flows*

$$\frac{\partial L}{\partial t_M} = ([(L_1^{m_1} \cdots L_n^{m_n})_+, L_1] \ldots [(L_1^{m_1} \cdots L_n^{m_n})_+, L_n]),$$
$$M = (m_1, \ldots, m_n) \in \mathbb{Z}_{\geq 0} \times \ldots \times \mathbb{Z}_{\geq 0}$$

commute, and if $S \in 1 + \mathcal{P}_-$ satisfies

$$\frac{\partial S}{\partial t_M} = -(S \partial_1^{m_1} \cdots \partial_n^{m_n} S^{-1})_- S,$$

then $L = (S \partial_1 S^{-1}, \ldots, S \partial_n S^{-1})$ evolves according to them.

REMARK. Barsotti's field $(k\{u\} := \mathcal{Q}(k\{u\})$ is larger than Sato's ring of pseudodifferential operators. Parshin's E is much larger than Sato's ring

$$\mathbb{C}[[x_1, \ldots, x_n]][[\partial_1^{-1}, \partial_1^{-1} \partial_2, \ldots, \partial_1^{-1} \partial_g]]$$

when $n > 1$. Finally, Barsotti's and Parshin's rings are different, though both adapted to a local description of a (say, if $g = 2$) surface. Parshin's ring is smaller and it is not symmetric in x, y, for instance, while Barsotti's $k\{x, y\}$ is.

Parshin [2001a; 2001b] generalizes the Krichever map, which associates to a local parameter on a curve and other geometric data a point of an infinite-dimensional Grassmannian (via the Baker–Akhiezer function), and to two local parameters, roughly speaking the choice of a curve on a surface and a point on that curve, and geometric data (a sheaf on the surface), associates a point of an infinite-dimensional 2-step flag manifold. This would be an appropriate setting for producing Nakayashiki's (2×2) matrix operators, via a choice of one basis element in a subspace and one in the quotient. This approach has not been taken, but the operators are explicit enough for genus 2 that the plan is concrete. At the same time, the brothers Aloysius and Gerard Helminck [1994a; 1994b; 1995; 2002] put a Fubini–Study metric on the infinite-dimensional projective space of flags, computed the central extension of the restricted linear group that acts on the manifold, and adapted the resulting (Kähler) manifold to flows of completely integrable systems, which include well-known ones. This is a natural setting for linearizing Nakayashiki's and Parshin's generalizations of the KP hierarchy. Sato's result, to the effect that Hirota's bilinear equation is equivalent to the Plücker relations which characterize the image of the Grassmannian in its Plücker embedding, should then be extended to the image of the Parshin flags.

Arbarello–De Concini's Plücker embedding. A different Grassmannian construction for abelian varieties is devised by Arbarello and De Concini [Arbarello and De Concini 1991]. They model a moduli space of abelian varieties on a Grassmannian, making use of one local parameter only, reminiscent of Sato's codirection, though they do not assume that its dual is tangent to the theta divisor, as Sato and Parshin do. They succeed, using classical theta-function theory, in producing enough data to embed the moduli space $\tilde{\mathcal{H}}_g$ (very roughly, a universal family of abelian varieties of dimension g, $\tilde{A}_g$ extended by a Heisenberg action) in $\mathbb{P}B$, where B is the usual Boson space $\mathbb{C}[[t_1, \ldots, t_k, \ldots]]$; they also give a theta-function formula for the τ function. The significant advantage of this construction is that they can compare this embedding with that of Jacobians via the usual Krichever map and they prove that the diagram

$$
\begin{array}{ccccc}
\tilde{\mathcal{A}}_g & \leftarrow & \tilde{\mathcal{H}}_g & \longrightarrow & \mathbb{P}B \\
\uparrow & & \uparrow & \nearrow & \uparrow \\
\tilde{\mathcal{M}}_g & \leftarrow & \tilde{\mathcal{F}}_g & \hookrightarrow \mathrm{Gr}H \hookrightarrow & \mathbb{P}F
\end{array}
$$

is commutative, where $\tilde{\mathcal{M}}_g$ is, again roughly speaking, a moduli space of genus-g curves, and $\tilde{\mathcal{F}}_g$ is fibred over $\tilde{\mathcal{M}}_g$ by the $\mathrm{Pic}^{(g-1)}$'s of the curves, $H = \mathbb{C}((z))$ is the space of formal Laurent series, $F = \Gamma(\mathrm{Gr}H, \det^{-1})^*$.

In their moduli spaces, Arbarello and De Concini use one complex variable z, which suggests that Barsotti's line may provide the embedding equations. This is also the principle of the (formal) work we carried out in [Lee and Previato 2006]. Thanks to Parshin's conjugation result, one "Sato operator" S suffices. One then can write, by the usual boson-fermion correspondence, a Baker function, as done by M. H. Lee and also in [Plaza-Martín 2000]; the function comprises $(z_1, \ldots, z_n)$ but essentially records points of a Grassmannian where the variable z_1 plays a distinguished role (as in Parshin's grading), and one can write a formal inversion of the logarithm and a formal τ function in the following way. (I omit as usual technical provisos; see [Lee and Previato 2006] for those.)

In analogy to the Segal–Wilson construction for the one-variable case, we let $H = L^2(T^n)$ be the Hilbert space consisting of all square-integrable functions on the n-torus

$$
T^n = \{(z_1, \ldots, z_n) \in \mathbb{C}^n \mid |z_1| = \cdots = |z_n| = 1\},
$$

which can be identified with the product of n copies of the unit circle $S^1 \subset \mathbb{C}^n$. Then the Hilbert space H can be written in the form

$$
H = \langle z^\alpha | \alpha \in \mathbb{Z}^n \rangle_{\mathbb{C}}.
$$

The multi-index notation is defined as follows: $z^\alpha = z_1^{\alpha_1} \cdots z_n^{\alpha_n}$, $|\alpha| = \alpha_1 + \cdots + \alpha_n$ if $z = (z_1, \ldots, z_n) \in \mathbb{C}^n$ and $\alpha = (\alpha_1, \ldots, \alpha_n) \in \mathbb{Z}^n$. If $\beta = (\beta_1, \ldots, \beta_n)$ is another element of $\mathbb{Z}^n$, we write $\alpha \leq \beta$ when $\alpha_i \leq \beta_i$ for each $i \in \{1, \ldots, n\}$. We define a splitting $H = H_+ \oplus H_-$ adapted to Parshin's filtration [1999; 2001b; 2001a] and the Krichever map. Then, as in the one-variable case, there is a one-to-one correspondence between certain subspaces of H commensurable to $H_+ := \mathbb{C}[[z_1, \ldots, z_n]]$ and wave functions, given by $\psi \mapsto W$, where a spanning set for W is given by all derivatives $\partial_1^{j_1} \ldots \partial_n^{j_n} \psi$, evaluated at $z = \mathbf{0}$, where $\mathbf{0} = (0, \ldots, 0) \in \mathbb{Z}^n$. We take this to be the Grassmannian $\mathrm{Gr}(H)$. We denote by $p_+ : H \to H_+$ and $p_- : H \to H_-$ the natural projection maps. A subspace W of H is said to be transversal to H_- if the restriction $p_+|_W : W \to H_+$ of

p_+ to W is an isomorphism. For a holomorphic function $g : D^n \to \mathbb{C}$ defined on the closed polydisk

$$D^n = \{(z_1, \ldots, z_n) \in \mathbb{C}^n \mid |z_1| \leq 1, \ldots, |z_n| \leq 1\}$$

with $g(0) = 1$, $g(z) = g(z_1, \ldots, z_n)$ can be written in the form

$$g(z) = \exp\left(\sum_{\alpha \in \mathbb{Z}_+^n} t_\alpha z^\alpha \right)$$

with $t_\alpha \in \mathbb{C}$ for all $\alpha \in \mathbb{Z}_+^n := \{\alpha \in \mathbb{Z}^n \mid \alpha \geq 0, \ \alpha \neq 0\}$. We define the maps $\mu_g, \mu_{g^{-1}} : H \to H$ by

$$(\mu_g f)(z) = g(z) f(z), \quad (\mu_{g^{-1}} f)(z) = g(z)^{-1} f(z)$$

for all $f \in W$ and $z \in \mathbb{C}^n$. Since $\mu_{g^{-1}}(H_+) \subset H_+$, with respect to the decomposition of H, the map $\mu_{g^{-1}}$ can be represented by a block matrix of the form

$$\mu_{g^{-1}} = \begin{bmatrix} a & b \\ 0 & c \end{bmatrix},$$

whose entries are the maps $a : H_+ \to H_+$, $\quad b : H_- \to H_+$, $\quad c : H_- \to H_-$. Given $W \in \mathrm{Gr}(H)$, we set

$$\Gamma_+^W = \{g \in \Gamma_+ \mid \mu_{g^{-1}} W \text{ is transversal to } H_-\}.$$

Thus g belongs to Γ_+^W if and only if the map $p_+|_{\mu_{g^{-1}} W} : \mu_{g^{-1}} W \to H_+$ is an isomorphism.

Let $\mathcal{S}$ be the complex vector space of formal Laurent series in $z_1^{-1}, \ldots, z_n^{-1}$ consisting of series of the form

$$v = \sum_{\alpha \leq \nu} f_\alpha(t) z^\alpha$$

for some $\nu \in \mathbb{Z}^n$ with $t = (t_\alpha)_{\alpha \in \mathbb{Z}_+^n}$. We consider the subspace $\mathcal{S}_-$ of $\mathcal{S}$ consisting of the series which can be written as

$$v = \sum_{k=-\infty}^{k_0} f_k(t; z_1, \ldots, z_{n-1}) z_n^k$$

for some $k_0 \in \mathbb{Z}$ with $k_0 \leq -1$, so that there is a decomposition of the form $\mathcal{S} = \mathcal{S}_+ \oplus \mathcal{S}_-$, where $\mathcal{S}_+$ consists of the series of the form

$$\sum_{k=0}^{\ell_0} f_k(t; z_1, \ldots, z_{n-1}) z_n^k$$

for some nonzero integer ℓ_0. Given an element W of the Grassmannian $\mathrm{Gr}(H)$, the associated Baker function $w_W(g, z)$ is the function defined for $g \in \Gamma_+^W$ and $z \in T^n$ satisfying the conditions

$$w_W(g, z) \in W, \quad \mu_{g^{-1}} w_W(g, z) = 1 + u$$

with $u \in \mathcal{S}_-$. Since each element $g \in \Gamma_+^W$ can be written in the exponential form, the Baker function $w_W(g, z)$ may be regarded as a function for $t = (t_\alpha)_{\alpha \in \mathbb{Z}_+^n}$ and $z \in T^n$. Thus we may write $w_W(g, z) = w_W(t, z)$, $t = (t_\alpha)_{\alpha \in \mathbb{Z}_+^n}$.

Let $W \in \mathrm{Gr}(H)$ be transversal to H_-, so that the map $p_+|_W : W \to H_+$ is an isomorphism, and let g be an element of Γ_+^W. We consider the sequence

$$
\begin{array}{ccccccc}
& (p_+|_W)^{-1} & & \mu_{g^{-1}} & & p_+ & \mu_g \\
H_+ & \to & W & \to & \mu_{g^{-1}} W & \to & H_+ \to H_+
\end{array}
$$

of complex linear maps. Given $g \in \Gamma_+^W$ and an element $W \in \mathrm{Gr}(H)$ transversal to H_-, the associated τ-function $\tau_W(g) = \tau_W(t) = \tau_W((t_\alpha)_{\alpha \in \mathbb{Z}_+^n})$ is the function

$$\tau_W(g) = \det\!\big(\mu_g \circ p_+ \circ \mu_{g^{-1}} \circ (p_+|_W)^{-1}\big)$$

given by the determinant of the composite of the linear maps above. Let $\Lambda : H_+ \to H_-$ be the linear map given by $\Lambda = p_- \circ (p_+|_W)^{-1}$. Then the τ-function can be written in the form

$$\tau_W(g) = \det(1 + a^{-1} b \Lambda),$$

where a and b are as above and 1 denotes the identity map on H_+. We define the rational numbers ε_α for $\alpha \in \mathbb{Z}_+^n$ by requiring

$$\sum_{\alpha \in \mathbb{Z}_+^n} \varepsilon_\alpha x^\alpha = \sum_{k=1}^{\infty} \frac{(-1)^k}{k} \left(\sum_{\beta \in \mathbb{Z}_+^n} x^\beta \right)^k,$$

where $x = (x_1, \dots, x_n)$ is a multivariable.

THEOREM 5 [Lee and Previato 2006]. *Let* $W \in \mathrm{Gr}(H)$ *be transversal, and let* $g : D^n \to \mathbb{C}$ *be an exponential. Then the associated τ-function*

$$\tau_W(g) = \tau_W((t_\alpha)_{\alpha \in \mathbb{Z}_+^n})$$

satisfies

$$\mu_{g^{-1}} w_W(g, z) = \frac{\tau_W\big((t_\alpha + \varepsilon_\alpha z^{-\alpha})_{\alpha \in \mathbb{Z}_+^n}\big)}{\tau_W((t_\alpha)_{\alpha \in \mathbb{Z}_+^n})},$$

where $w_W(g, z)$ *is the Baker function.*

The next step would be to write this formula in terms of theta functions after Arbarello–De Concini.

4. Reducible cases

A project which I believe less trivial than it seems, occurs in the case of reducible abelian varieties. For example, the Schrödinger operator

$$\frac{\partial}{\partial x^2} + \frac{\partial}{\partial y^2} - \wp(x)\wp(y)$$

has commutator which must be isomorphic to the ring $\mathcal{C}(\partial/\partial x^2 - \wp(x)) \otimes \mathcal{C}(\partial/\partial y^2 - \wp(y))$, whose associated "spectral" variety is $E_0 \times E_0$, where E_0 is the spectral variety of $\mathcal{C}(\partial/\partial x^2 - \wp(x))$.

While this case can be regarded as trivial, by analogy, the elliptic (or rational) curve case of the Hitchin system for vector bundles, in which the moduli of vector bundles are simply a product of copies of the curve, is still the only one in which the solutions can be given explicitly ([Nekrasov 1996] is just one earliest reference). In the reducible-potential case, Parshin's flows for L_1, L_2 and Nakayashiki's equations can be written explicitly, but they have not yet been compared; in this case there is a KP hierarchy.

The differential resultant for this case, in which the variety is known, can serve as a toy model for a truly generalized theory. (The model used in [Kasman and Previato 2001], by analogy with the algebraic definition of resultant of polynomial equations in several variables given by Macaulay [1916], falls short because, due to the additional variables at infinity, it is often identically zero.) It can also serve for testing the conjecture made in [Kasman and Previato 2001] that the resultant is independent of the operator variables, up to a factor whose numerical nature (degrees in the variables, e.g., for the case of the Weyl algebra), should be the same in general as in the reducible case. Moreover this reducible case provides a nonexample for Barsotti's theorem.[4] Indeed, if an abelian surface is isogenous to the product $E_1 \times E_2$ of two elliptic curves, as is the case for the Jacobian of a genus-2 curve that covers an elliptic curve, then we can take u to be the direction that projects to one of the tori; the derivatives in the u direction will only produce the elliptic functions in one variable; an explicit calculation is known classically and was reproduced by J. C. Eilbeck (unpublished notes) to input the parametrization of all the genus-2 elliptic covers whose Jacobian is isomorphic (without principal polarization) to the product of two elliptic curves. I briefly provide some motivation and the formula (which does not do sufficient justice to Eilbeck's considerable work in implementing two reduction algorithms, on Siegel matrices and Fourier expansions, given in theory by H. H. Martens and J.-I. Igusa). The motivation was a recent result

---

[4] "Quite often in mathematics, a "nonexample" is as helpful in understanding a concept as an example" — J. A. Gallian, *Contemporary abstract algebra*, Chapter 4.

of C. Earle [2006], who described all the 2×2 matrices in the Siegel upper-half space that correspond to genus-2 curves whose Jacobian is isomorphic to a product of elliptic curves. Note that the expected parameter count should be one and not two (Jacobians that split up to isogeny) since Martens had shown that in the isomorphic case the two elliptic curves must be isomorphic. I asked Eilbeck whether we could do the effectivization of the KdV solutions for all these matrices. The "Earle matrix"[5]

$$Z = \tau \begin{bmatrix} na & nb \\ nb & d \end{bmatrix}$$

with τ in the upper-half plane, and a, b, d, n positive integers such that $ad - nb^2 = 1$, nonsymplectically equivalent to a diagonal:

$$(I, z) = (I, \tau V) \begin{bmatrix} I & 0 \\ 0 & T \end{bmatrix},$$

where $V := \begin{bmatrix} n & 0 \\ 0 & 1 \end{bmatrix}$, $T := \tau \begin{bmatrix} a & b \\ nb & d \end{bmatrix}$ can be symplectically transformed into

$$\begin{bmatrix} -\dfrac{1}{\tau na} & -\dfrac{b}{a} \\ -\dfrac{b}{a} & 1 + \dfrac{\tau}{a} \end{bmatrix}.$$

Eilbeck implemented (by creating Maple routines) a special case, the 2-dimensional abelian variety being (2:1)-isogenous to $E_1 \times E_2$, the two elliptic curves E_i having invariants τ_i, and decomposed the theta function Θ (with characteristics) of A thus: For the matrix

$$\tilde{\tau} = \begin{bmatrix} \dfrac{1}{2}\tau_1 & \dfrac{1}{2} \\ \dfrac{1}{2} & -\dfrac{1}{2(2+\tau_2)} \end{bmatrix},$$

we have

$$\Theta \begin{bmatrix} 0 & 0 \\ \frac{1}{2} & 0 \end{bmatrix} \left(\begin{bmatrix} -\frac{1}{2}v_1 \\ -\frac{v_1 - 2v_2}{2(2+\tau_2)} \end{bmatrix}, \tilde{\tau} \right)$$

$$= \Theta \begin{bmatrix} 0 \\ \frac{1}{2} \end{bmatrix} \left(-\frac{v_1}{2}, \frac{\tau_1}{2} \right) \Theta \begin{bmatrix} 0 \\ 0 \end{bmatrix} \left(\frac{v_1 - 2v_2}{2+\tau_2}, -\frac{2}{2+\tau_2} \right)$$

$$+ \Theta \begin{bmatrix} 0 \\ 0 \end{bmatrix} \left(-\frac{v_1}{2}, \frac{\tau_1}{2} \right) \Theta \begin{bmatrix} \frac{1}{2} \\ 0 \end{bmatrix} \left(\frac{v_1 - 2v_2}{2+\tau_2}, -\frac{2}{2+\tau_2} \right).$$

---

[5] Not all matrices of this form are period matrices of genus-2 curves; Earle gives a criterion. Also, not all the matrices of this form that are period matrices correspond to different curves, as they may come from the same curve via a different choice of homology basis; in two further theorems Earle gives criteria to tell curves apart.

This shows that higher derivatives of Θ in the directions τ_i are elliptic functions in τ_i, thus cannot generate the function field of A.

Appendix

I cannot help mentioning the theory of infinite-genus Riemann surfaces that Henry McKean, originally in collaboration with Eugene Trubowitz, developed for KdV spectral varieties attached to a periodic potential that is not "finite-gap". In a rigorous analytic way, this extends the theory of the Jacobian, and the theta function.[6] There is still scope for a theory of reduction, elliptic solitons, and differential operators with elliptic coefficients; there are many more "Variations on a Theme of Jacobi", in other words, awaiting for Henry's face-altering contributions to the field: one more reason to say, Henry, many, many happy returns!

Acknowledgements

As always, it is a privilege to participate in an MSRI workshop. The invitation extended to me by the organizers, Björn Birnir, Darryl Holm, Charles Newman, Mark Pinsky, Kirill Vaninsky, and Lai-Sang Young, was both humbling and profoundly rewarding. I am also most grateful to J. C. Eilbeck for his permission to reproduce the formula of Section 4.

References

[Arbarello and De Concini 1991] E. Arbarello and C. De Concini, "Abelian varieties, infinite-dimensional Lie algebras, and the heat equation", pp. 1–31 in *Complex geometry and Lie theory* (Sundance, UT, 1989), edited by J. A. Carlson et al., Proc. Sympos. Pure Math. **53**, Amer. Math. Soc., Providence, RI, 1991.

[Baker 1907] H. F. Baker, *An introduction to the theory of multiply-periodic functions*, Cambridge Univ. Press, Cambridge, 1907.

[Barsotti 1970] I. Barsotti, "Considerazioni sulle funzioni theta", pp. 247–277 in *Symposia Mathematica* (Rome, 1968/69), vol. 3, Academic Press, London, 1970.

[Barsotti 1983] I. Barsotti, "Differential equations of theta functions", *Rend. Accad. Naz. Sci. XL Mem. Mat.* (5) **7** (1983), 227–276.

[Barsotti 1985] I. Barsotti, "Differential equations of theta functions. II", *Rend. Accad. Naz. Sci. XL Mem. Mat.* (5) **9**:1 (1985), 215–236.

[Barsotti 1989] I. Barsotti, "A new look for thetas", pp. 649–662 in *Theta functions* (Brunswick, ME, 1987), vol. I, edited by L. Ehrenpreis and R. C. Gunning, Proc. Sympos. Pure Math. **49**, Amer. Math. Soc., Providence, RI, 1989.

---

[6]See [Ercolani and McKean 1990; Feldman et al. 1996; 2003; McKean 1980; 1989; McKean and Trubowitz 1976; 1978; Schmidt 1996.]

[Beauville 1982] A. Beauville, "Diviseurs spéciaux et intersection de cycles dans la jacobienne d'une courbe algébrique", pp. 133–142 in *Enumerative geometry and classical algebraic geometry* (Nice, 1981), edited by P. L. Barz and Y. Hervier, Progr. Math. **24**, Birkhäuser, Mass., 1982.

[Birkenhake and Vanhaecke 2003] C. Birkenhake and P. Vanhaecke, "The vanishing of the theta function in the KP direction: a geometric approach", *Compositio Math.* **135**:3 (2003), 323–330.

[Earle 2006] C. J. Earle, "The genus two Jacobians that are isomorphic to a product of elliptic curves", pp. 27–36 in *The geometry of Riemann surfaces and abelian varieties*, Contemp. Math. **397**, Amer. Math. Soc., Providence, RI, 2006.

[Eilbeck et al. 2001] J. C. Eilbeck, V. Z. Enolskii, and E. Previato, "Varieties of elliptic solitons", *J. Phys. A* **34**:11 (2001), 2215–2227.

[Eilbeck et al. 2007] J. C. Eilbeck, V. Z. Enolski, S. Matsutani, Y. Ônishi, and E. Previato, "Abelian functions for trigonal curves of genus three", *Int. Math. Res. Not.* (2007). Available at arxiv.org/abs/math.AG/0610019.

[Ercolani and McKean 1990] N. Ercolani and H. P. McKean, "Geometry of KdV, IV: Abel sums, Jacobi variety, and theta function in the scattering case", *Invent. Math.* **99**:3 (1990), 483–544.

[Feldman et al. 1996] J. Feldman, H. Knörrer, and E. Trubowitz, "Infinite genus Riemann surfaces", pp. 91–111 in *Canadian Mathematical Society. 1945–1995*, vol. 3, Canadian Math. Soc., Ottawa, ON, 1996.

[Feldman et al. 2003] J. Feldman, H. Knörrer, and E. Trubowitz, *Riemann surfaces of infinite genus*, CRM Monograph Series **20**, American Mathematical Society, Providence, RI, 2003.

[Gunning 1986] R. C. Gunning, "Some identities for abelian integrals", *Amer. J. Math.* **108**:1 (1986), 39–74.

[Helminck and Helminck 1994a] A. G. Helminck and G. F. Helminck, "Holomorphic line bundles over Hilbert flag varieties", pp. 349–375 in *Algebraic groups and their generalizations: quantum and infinite-dimensional methods* (University Park, PA, 1991), vol. 2, edited by W. J. Haboush and B. Parshall, Proc. Sympos. Pure Math. **56**, Amer. Math. Soc., Providence, RI, 1994.

[Helminck and Helminck 1994b] G. F. Helminck and A. G. Helminck, "The structure of Hilbert flag varieties", *Publ. Res. Inst. Math. Sci.* **30**:3 (1994), 401–441.

[Helminck and Helminck 1995] G. F. Helminck and A. G. Helminck, "Infinite-dimensional flag manifolds in integrable systems", *Acta Appl. Math.* **41**:1-3 (1995), 99–121.

[Helminck and Helminck 2002] G. F. Helminck and A. G. Helminck, "Hilbert flag varieties and their Kähler structure", *J. Phys. A* **35**:40 (2002), 8531–8550.

[Jorgenson 1992a] J. Jorgenson, "On directional derivatives of the theta function along its divisor", *Israel J. Math.* **77**:3 (1992), 273–284.

[Jorgenson 1992b] J. Jorgenson, "Some uses of analytic torsion in the study of Weierstrass points", pp. 255–281 in *Curves, jacobians, and abelian varieties* (Amherst, MA, 1990), edited by R. Donagi, Contemp. Math. **136**, Amer. Math. Soc., Providence, RI, 1992.

[Kasman and Previato 2001] A. Kasman and E. Previato, "Commutative partial differential operators", *Phys. D* **152/153** (2001), 66–77.

[Krichever 1977] I. M. Krichever, "The methods of algebraic geometry in the theory of nonlinear equations", *Uspekhi Mat. Nauk* **32**:6 (1977), 183–208.

[Lee and Previato 2006] M. H. Lee and E. Previato, "Grassmannians of higher local fields and multivariable tau functions", pp. 311–319 in *The ubiquitous heat kernel* (Boulder, CO, 2003), edited by J. Jorgenson, Contemp. Math. **398**, Amer. Math. Soc., Providence, RI, 2006.

[Macaulay 1916] F. S. Macaulay, *The algebraic theory of modular systems*, Cambridge Univ. Press, Cambridge, 1916. Revised reprint, 1994.

[Matsutani and Previato 2008] S. Matsutani and E. Previato, "A generalized Kiepert formula for C_{ab} curves", *Israel J. Math.* (2008).

[McKean 1979] H. P. McKean, "Integrable systems and algebraic curves", pp. 83–200 in *Global analysis* (Calgary, AB, 1978), edited by M. Grmela and J. E. Marsden, Lecture Notes in Math. **755**, Springer, Berlin, 1979.

[McKean 1980] H. P. McKean, "Algebraic curves of infinite genus arising in the theory of nonlinear waves", pp. 777–783 in *Proceedings of the International Congress of Mathematicians* (Helsinki, 1978), Acad. Sci. Fennica, Helsinki, 1980.

[McKean 1989] H. P. McKean, "Is there an infinite-dimensional algebraic geometry? Hints from KdV", pp. 27–37 in *Theta functions* (Brunswick, ME, 1987), edited by L. Ehrenpreis and R. C. Gunning, Proc. Sympos. Pure Math. **49**, Amer. Math. Soc., Providence, RI, 1989.

[McKean and Trubowitz 1976] H. P. McKean and E. Trubowitz, "Hill's operator and hyperelliptic function theory in the presence of infinitely many branch points", *Comm. Pure Appl. Math.* **29**:2 (1976), 143–226.

[McKean and Trubowitz 1978] H. P. McKean and E. Trubowitz, "Hill's surfaces and their theta functions", *Bull. Amer. Math. Soc.* **84**:6 (1978), 1042–1085.

[Mironov 2002] A. E. Mironov, "Real commuting differential operators associated with two-dimensional abelian varieties", *Sibirsk. Mat. Zh.* **43**:1 (2002), 126–143. In Russian; translated in *Siberian Math. J.* **43**:1, 97–113.

[Mumford 1984] D. Mumford, *Tata lectures on theta, II*, Progress in Mathematics **43**, Birkhäuser, Boston, 1984.

[Nakayashiki 1991] A. Nakayashiki, "Structure of Baker-Akhiezer modules of principally polarized abelian varieties, commuting partial differential operators and associated integrable systems", *Duke Math. J.* **62**:2 (1991), 315–358.

[Nekrasov 1996] N. Nekrasov, "Holomorphic bundles and many-body systems", *Commun. Math. Phys.* **180**:3 (1996), 587–603.

[Parshin 1999] A. N. Parshin, "On a ring of formal pseudo-differential operators", *Tr. Mat. Inst. Steklova* **224**:Algebra. Topol. Differ. Uravn. i ikh Prilozh. (1999), 291–305. In Russian; translated in *Proc. Steklov Inst. Math.* **224**:1 (1999), 266–280.

[Parshin 2001a] A. N. Parshin, "Integrable systems and local fields", *Comm. Algebra* **29**:9 (2001), 4157–4181.

[Parshin 2001b] A. N. Parshin, "The Krichever correspondence for algebraic surfaces", *Funktsional. Anal. i Prilozhen.* **35**:1 (2001), 88–90. In Russian; translated in *Funct. Anal. Appl.* **35**:1 (2001), 74–76.

[Plaza-Martín 2000] F. Plaza-Martín, "Generalized KP hierarchy for several variables", preprint, 2000. Available at arxiv.org/abs/math.AG/0008004.

[Previato 2008] E. Previato, "Review of *Soliton equations and their algebro-geometric solutions*, vol. I, by Fritz Gesztesy and Helge Holden, Cambridge University Press, 2003", *Bull. Amer. Math. Soc.* (2008). To appear.

[Sato 1989] M. Sato, "The KP hierarchy and infinite-dimensional Grassmann manifolds", pp. 51–66 in *Theta functions* (Brunswick, ME, 1987), edited by L. Ehrenpreis and R. C. Gunning, Proc. Sympos. Pure Math. **49**, Amer. Math. Soc., Providence, RI, 1989.

[Schmidt 1996] M. U. Schmidt, "Integrable systems and Riemann surfaces of infinite genus", *Mem. Amer. Math. Soc.* **122**:581 (1996), viii+111.

[Scorza Dragoni 1988] G. Scorza Dragoni, "Iacopo Barsotti in memoriam", *Rend. Accad. Naz. Sci. XL Mem. Mat.* (5) **12**:1 (1988), 1–2.

[Welters 1986] G. E. Welters, "The surface $C - C$ on Jacobi varieties and 2nd order theta functions", *Acta Math.* **157**:1-2 (1986), 1–22.

[Zheglov 2005] A. Zheglov, "Two dimensional KP systems and their solvability", preprint, 2005. Available at http://www.arxiv.org/abs/math-ph/0503067.

EMMA PREVIATO
DEPARTMENT OF MATHEMATICS AND STATISTICS
BOSTON UNIVERSITY
BOSTON, MA 02215-2411
UNITED STATES
ep@bu.edu

Probability, Geometry and Integrable Systems
MSRI Publications
Volume **55**, 2008

Integrable models of waves in shallow water

HARVEY SEGUR

Dedicated to Henry McKean, on the occasion of his 75th birthday

ABSTRACT. Integrable partial differential equations have been studied because of their remarkable mathematical structure ever since they were discovered in the 1960s. Some of these equations were originally derived to describe approximately the evolution of water waves as they propagate in shallow water. This paper examines how well these integrable models describe actual waves in shallow water.

1. Introduction

Zabusky and Kruskal [1965] introduced the concept of a *soliton* — a spatially localized solution of a nonlinear partial differential equation with the property that this solution always regains its initial shape and velocity after interacting with another localized disturbance. They were led to the concept of a soliton by their careful computational study of solutions of the Korteweg–de Vries (or KdV) equation,

$$\partial_t u + u \partial_x u + \partial_x^3 u = 0. \tag{1}$$

(See [Zabusky 2005] for his summary of this history.) After that breakthrough, they and their colleagues found that the KdV equation has many remarkable properties, including the property discovered by Gardner, Greene, Kruskal and Miura [Gardner et al. 1967]: *the KdV equation can be solved exactly, as an initial-value problem, starting with arbitrary initial data in a suitable space.* This discovery was revolutionary, and it drew the interest of many people. We note especially the work of Zakharov and Faddeev [1971], who showed that the KdV equation is a nontrivial example of an infinite-dimensional Hamiltonian system that is *completely integrable.* This means that under a canonical change of variables, the original problem can be written in terms of action-angle variables, in which the action variables are constants of the motion, while the angle variables evolve according to nearly trivial ordinary differential equations (ODEs). Zakharov and Faddeev showed that GGKM's method to solve the KdV equation amounts to transforming from $u(x, t)$ into action-angle variables at the initial time, integrating the ODEs forward in time, and then transforming back to $u(x, t)$ at any later time. In this way the Korteweg-de Vries equation, (1), became the prototype of a completely integrable partial differential equation, the study of which makes up one facet of this conference.

Zabusky and Kruskal had derived (1) as an approximate model of longitudinal vibrations of a one-dimensional crystal lattice. Later they learned that Korteweg and de Vries [1895] had already derived the same equation as an approximate model of the evolution of long waves of moderate amplitude, propagating in one direction in shallow water of uniform depth. So the KdV equation, (1), is of interest to at least two communities of scientists:

- mathematicians, who are primarily interested in its extraordinary mathematical structure; and
- coastal engineers and oceanographers, who use it to make engineering and environmental decisions related to physical processes in shallow water.

The KdV equation is one of several equations that are known to be completely integrable and that also describe approximately waves in shallow water. Other well known examples include the KP equation, a two-dimensional generalization of the KdV equation due to Kadomtsev and Petviashvili [1970]:

$$\partial_x\left(\partial_t u + u\partial_x u + \alpha \cdot \partial_x^3\right) + \partial_y^2 u = 0; \tag{2}$$

an equation first studied by Boussinesq [1871]:

$$\partial_t^2 u = c^2 \partial_x^2 u + \partial_x^2 (u^2) + \beta \cdot \partial_x^4 u; \tag{3}$$

and the Camassa–Holm equation [1993],

$$\partial_t m + c\partial_x u + u\partial_x m + 2m\partial_x u + \gamma \cdot \partial_x^3 u = 0, \tag{4}$$

where

$$m = u - \delta^2 \partial_x^2 u.$$

Here are four comments about these equations.

- Among all known integrable equations, this subset is particularly relevant for this conference, because of the many important contributions of Henry McKean and his coauthors to the development of the theory for (1), (3) and (4) with periodic boundary conditions. See [Constantin and McKean 1999; McKean

1977; 1978; 1981a; 1981b; McKean and van Moerbeke 1975; McKean and Trubowitz 1976; 1978].

- The coefficients in (1) are not important, because they can be scaled into $\{u, x, t\}$. But in (2), (3), (4), both the physical meaning and the mathematical structure of each equation changes, depending on the signs of $\{\alpha, \beta, \gamma\}$ respectively. Judgmental nicknames like "good Boussinesq" and "bad Boussinesq" indicate the importance of these signs.

- Equation (1) is a special case of (2), after setting $\partial_y u \equiv 0$ and neglecting a constant of integration. Equation (3) is also a special case of (2), after setting $\partial_t u = -c^2 \partial_x u$, rescaling $u \to 2u$, and then interpreting y in (2) as the time-like variable.

- The KP equation is degenerate: for example, it does not have a unique solution. If $u(x, y, 0) \equiv 0$ at $t = 0$, then both $u = 0$ and $u = t$ solve (2) and satisfy this initial condition. In this paper, we remove this degeneracy by requiring that any solution of (2) also satisfy

$$\int_{-\infty}^{\infty} u(x, y, t)\, dx = 0. \tag{5}$$

Every completely integrable equation possesses extraordinary mathematical structure. Each of the equations listed above is completely integrable, and also describes (approximately) waves in shallow water. This paper addresses the question: *Does this extra mathematical structure provide useful information about the behavior of actual, physical waves in shallow water?*

The outline of the rest of this paper is as follows. Section 2 reviews the derivation of (1)–(4) as approximate models for waves in shallow water. Sections 3, 4, 5 all discuss applications of these approximate models to practical problems involving ocean waves. Section 3 discusses an application of (1) on the whole line (or of (2) on the whole plane): the tsunami of December 26, 2004. Section 4 focusses on spatially periodic, travelling wave solutions of (2), and their relation to periodic travelling waves in shallow water. Section 5 relates doubly periodic waves to the phenomenon of rip currents. Finally Section 6 discusses more complicated, quasiperiodic solutions of (2).

2. Derivation of integrable models from the problem of inviscid water waves

The mathematical theory of water waves goes back at least to Stokes [1847], who first wrote down the equations for the motion of an incompressible, inviscid fluid, subject to a constant (vertical) gravitational force, where the fluid is bounded below by an impermeable bottom and above by a free surface. In the

Figure 1. The integrable models of waves in shallow water in (1)–(4) all depend on an ordering of the scales shown here. A fourth scale, the typical length scale out of the page, is not shown.

discussion that follows, we assume that the bottom of the fluid is strictly horizontal (at $z = -h$). In addition to gravity, we also include the effect of surface tension, because including it provides the extra freedom needed to change the signs of $\{\alpha, \beta, \gamma\}$ in (2), (3) and (4).

Without viscosity, we may consider purely irrotational motions. Then the fluid velocity can be written in terms of a velocity potential,

$$\vec{u} = \nabla\phi,$$

and the velocity potential satisfies

$$\nabla^2\phi = 0 \qquad\qquad\qquad\qquad \text{for } -h < z < \zeta(x, y, t),$$

$$\partial_z\phi = 0 \qquad\qquad\qquad\qquad \text{on } z = -h,$$

$$\partial_t\zeta + \partial_x\zeta \cdot \partial_x\phi + \partial_y\zeta \cdot \partial_y\phi = \partial_z\phi \qquad\qquad \text{on } z = \zeta(x, y, t),$$

$$\partial_t\phi + \tfrac{1}{2}|\nabla\phi|^2 + g\zeta = T\frac{\partial_x^2\zeta(1+(\partial_y\zeta)^2)+\partial_y^2\zeta(1+(\partial_x\zeta)^2)-2\partial_x\partial_y\zeta\cdot\partial_x\zeta\partial_y\zeta}{(1+(\partial_x\zeta)^2+(\partial_y\zeta)^2)^{3/2}}$$

$$\text{on } z = \zeta(x, y, t), \qquad (6)$$

where g is the acceleration due to gravity, T represents surface tension, and $z = \zeta(x, y, t)$ is the instantaneous location of the free surface.

These equations, known for more than 150 years, are still too difficult to solve in any general sense. Even well-posedness (for short times) was established only recently; see [Wu 1999; Lannes 2005; Coutand and Shkoller 2005]. Progress has been made by focusing on specific limits, in which the equations simplify. The limit of interest here can be stated in terms of length scales, which must be arranged in a certain order. Three of the four relevant lengths are shown in Figure 1. The derivation of either KdV or KP from (6) is based on four assumptions:

(a) long waves (or shallow water), $h \ll \lambda$;
(b) small amplitude; $a \ll h$;
(c) the waves move primarily in one direction;

(d) All these small effects are comparable in size; for KdV, this means

$$\varepsilon = \frac{a}{h} = O\left(\left(\frac{h}{\lambda}\right)^2\right). \tag{7}$$

If (c) is exactly true, the derivation leads to the KdV equation, (1) If it is approximately true, the derivation can lead to the KP equation, (2).

As we discuss below, assumptions (a)–(d) are also implicit in (3) and (4). One imposes these assumptions on both the velocity potential, $\phi(x, y, z, t)$, and the location of the free surface, $\zeta(x, y, t)$, in (6). (See [Ablowitz and Segur 1981, §4.1] or [Johnson 1997, §3.2] for details.) At leading order ($\varepsilon = 0$ in a formal expansion), the waves in question are infinitely long, infinitesimally small, and the motion at the free surface is exactly one-dimensional. The result is the one-dimensional wave equation:

$$\partial_t^2 \zeta = c^2 \partial_x^2 \zeta, \quad \text{with } c^2 = gh. \tag{8}$$

Inserting D'Alembert's solution of (8) back into the expansion for $\zeta(x, y, t; \varepsilon)$ yields

$$\zeta(x, y, t; \varepsilon) = \varepsilon h\big(F\left(x - ct; y, \varepsilon t\right) + G\left(x + ct; y, \varepsilon t\right)\big) + O(\varepsilon^2), \tag{9}$$

where F and G are determined from the initial data. At leading order, F and G are required only to be bounded, and to be smooth enough for the terms in (6) to make sense.

There are two ways to proceed to the next order. The simpler but more restricted method, used by Johnson [1997], is to ignore one of the two waves in (9). Then we may follow (for example) the F-wave by changing to a coordinate system that moves with that wave, at speed $\sqrt{gh}$. To do so, set

$$\xi = \frac{\sqrt{\varepsilon}}{h}\left(x - t\sqrt{gh}\right). \tag{10a}$$

At leading order, according to (9), F does not change in this coordinate system, so we may proceed to the next order, $O(\varepsilon^2)$. Now small effects that were ignored at leading order — namely, that the wave amplitude is small but not infinitesimal, that the wavelength is long but not infinitely long, and that slow transverse variations are allowed — can be observed. These small effects can build up over a long distance, to produce a significant cumulative change in F. To capture this slow evolution of F, introduce a slow time scale,

$$\tau = \varepsilon t \sqrt{\frac{\varepsilon g}{h}}, \tag{10b}$$

and find that F satisfies approximately the KdV equation,

$$2\partial_\tau F + 3F\partial_\xi F + \left(\frac{1}{3} - \frac{T}{gh^2}\right)\partial_\xi^3 F = 0, \tag{11}$$

if the surface waves are strictly one-dimensional (that is, if $\partial_y F \equiv 0$). Or, if the surface patterns are weakly two-dimensional, we can obtain instead the KP equation,

$$\partial_\xi \left(2\partial_\tau F + 3F\partial_\xi F + \left(\frac{1}{3} - \frac{T}{gh^2}\right)\partial_\xi^3 F\right) + \partial_\eta^2 F = 0. \tag{12}$$

After rescaling the variables in (11) or (12) to absorb constants, (11) becomes (1), and (12) becomes (2). The dimensionless parameter, T/gh^2, which appears in both (11) and (12), is the inverse of the Bond number — its value determines the relative strength of gravity and surface tension. The magnitudes and signs of all coefficients in (11) can be scaled out of the problem, but this is not possible in (12) with any real-valued scaling. The sign of $\frac{1}{3} - T/gh^2$ determines the sign of α in (2), of β in (3) and of γ in (4).

In words, (9) says that one wave (F) propagates to the right, while another wave (G) propagates to the left, both with speed $\sqrt{gh}$. Neither wave changes shape as it propagates, during the short time when (9) is valid without correction. On a longer time-scale, the KdV equation (11) describes how F changes slowly, due to weak nonlinearity ($F\partial_\xi F$) and weak dispersion ($\partial_\xi^3 F$). Or, the KP equation (12) shows how F changes because of these two weak effects and also because of weak two-dimensionality ($\partial_\eta^2 F$).

The KdV and KP equations have been derived in many physical contexts, and they always have the same physical meaning: on a short time-scale, the leading-order equation is the one-dimensional, linear wave equation; on a longer time-scale, each of the two free waves that make up the solution of the 1-D wave equation satisfies its own KdV (or KP) equation, so each of the two waves changes slowly because of the cumulative effect of weak nonlinearity, weak dispersion and (for KP) weak two-dimensionality.

The Boussinesq equation, (3), describes approximately the evolution of water waves under the same assumptions as KdV. It is the basis for several numerical codes to model wave propagation in shallow water; see, for instance, [Wei et al. 1995; Bona et al. 2002; 2004; Madsen et al. 2002]. Equation (3) appears to be more general than (1) because (3) allows waves to propagate in two directions, as (1) does not. But Bona et al. [2002] note that the usual derivation of (3) from (6) includes an assumption that the waves are propagating primarily in one direction, so (1) and (3) are formally equivalent.

Conceptually, the Camassa–Holm equation, (4), is based on the same set of assumptions as (1) and (3). But Johnson [2002] shows that the usual derivations of (4) are logically inconsistent. He then gives a self-consistent derivation of (4) from (6), but the price he pays is that the solution of (4) does not approximate the shape of the free surface — an additional step is needed. The shape of the free surface is an easy quantity to measure experimentally, so this extra cost is not trivial.

The next three sections of this paper compare predictions of (1) and (2) with experimental observations of waves in shallow water.

3. Application: the tsunami of 2004

A dramatic example of a long ocean wave of small amplitude was the tsunami that occurred in the Indian Ocean on December 26, 2004. The tsunami caused terrible destruction in many coastal regions around the Indian Ocean, killing more than 200,000 people. Even so, we show next that until the tsunami neared the shoreline, it was well approximated by the theory leading to (8) and (9). We also show that the nonlinear integrable models, (1)–(4), were *not* relevant for the 2004 tsunami.

The tsunami was generated by a strong, undersea earthquake off the coast of Sumatra. Figure 2 shows a map of the northern Indian Ocean, and the initial shape of the tsunami. (This figure is the first image in an informative animation of the tsunami's propagation, done by Kenji Satake of Japan. To see his entire animation, go to http://staff.aist.go.jp/kenji.satake/animation.html. A comparable simulation by S. Ward can be found at http://www.es.ucsc.edu/~ward.) The fault line of the earthquake, clearly evident in Figure 2, lies on or near the boundary of two tectonic plates, one that carries India and one that holds Burma (Myanmar). Most of the seismic activity in this region occurs because the India plate is slowly sliding beneath the Burma plate.

The original earthquake (near Sumatra) triggered a series of other quakes, which occurred along this fault line, all within about 10 minutes. The north-south distance along this line of quakes was about 900 km. As Figure 2 shows, the effect of the quakes was to *raise* the ocean floor to the west of this (curved) line, and to *lower* it to the east of the line. The lateral extent of this change in the sea floor was about 100 km, on each side of the fault line. The vertical displacement was 1–2 meters or less. The change in level of the sea floor occurred quickly enough that the water above it simply rose (or fell) with the sea floor. These conditions provided the initial conditions for the tsunami. (The estimates quoted here were given by S. Ward. None of the conclusions drawn in this section changes qualitatively if any of these estimates is changed by a factor of 2.)

The ocean depth in the Bay of Bengal (the part of the Indian Ocean west and north of the fault line in Figure 2) is about 3 km. The region east of the fault line, called the Andaman Sea, is shallower: it average depth is about 1 km.

These estimates provide enough information to consider the theory summarized in Section 2. In the Bay of Bengal, the requirements for KdV theory at leading order are:

- *small amplitude*:

$$\frac{a}{h} = \frac{1}{3000} = 3.3 \cdot 10^{-4} \ll 1;$$

- *long waves*:

$$\left(\frac{h}{\lambda}\right)^2 = \left(\frac{3}{100}\right)^2 = 9 \cdot 10^{-4} \ll 1;$$

- *nearly 1-D surface patterns*:

$$\frac{\lambda}{L} = \frac{100}{900} \ll 1;$$

- *comparable scales*:

$$\varepsilon = \frac{a}{h} = O\left(\left(\frac{h}{\lambda}\right)^2\right).$$

These estimates show that the theory for long waves of small amplitude should work well for the tsunami that propagated westward, across the Bay of Bengal. The reader can verify that this conclusion also holds for the eastward-propagating tsunami, in the Andaman Sea. Therefore, the tsunami in the Bay of Bengal propagated with a speed ($\sqrt{gh}$) of about 620 km/hr, while the speed in the Andaman Sea was about 360 km/hr.

(The analysis that follows also assumes constant ocean depth. This is the weakest assumption in the analysis, but it is easily corrected; see [Segur 2007].) We may take the initial shape of the wave to be that shown in Figure 2, with no vertical motion initially. If we neglect variations along the fault line, then according to (8), a wave with this shape and half of its amplitude propagated to the west, and an identical wave propagated to the east. Neither wave changed its shape as it propagated, so the wave propagating towards India and Sri Lanka consisted of a wave of elevation followed by a wave of depression. The wave propagating towards Thailand was the opposite: a wave of depression, followed by a wave of elevation. These conclusions are consistent with reports from survivors in those two regions.

This description applies on a short time-scale. The KdV (or KP) equation applies on the next time-scale, approximately ε^{-1} longer. Equivalently, the distance required for KdV dynamics to affect the wave forms is approximately ε^{-1} longer than a typical length scale in the problem. Using h as a typical length, this suggests that we need about 3000×3 km = 9000 km of propagation distance to see KdV dynamics in the westward propagating wave. But the distance across

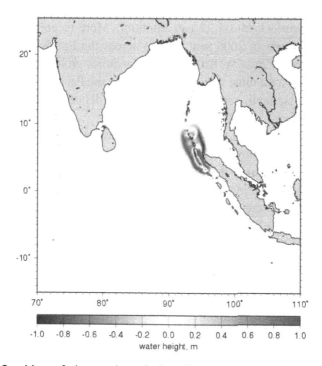

Figure 2. Map of the northern Indian Ocean, showing the shape and intensity of the initial tsunami, 10 minutes after the beginning of the first earthquake. This image is the first frame from a simulation by K. Satake; it shows an elevated water surface west of the island chain and a depressed water surface east of that chain. (For the color coding see this article online at http://www.msri.org/publications/books; Satake's animation can be found at http://staff.aist.go.jp/kenji.satake/animation.html.)

the Bay of Bengal is nowhere larger than about 1500 km, much too short for KdV dynamics to have built up. The numbers for the eastward propagating wave are different, but the conclusion is the same: the distance across the Andaman Sea is too short to see significant KdV dynamics. *Thus, the 2004 tsunami did not propagate far enough for either the KdV or KP equation to apply.*

This conclusion, that propagation distances for tsunamis are too short for soliton dynamics to have an important effect, applies to many tsunamis, but not all. Lakshmanan and Rajasekar [2003] point out that the 1960 Chilean earthquake, the largest earthquake ever recorded (magnitude 9.6 on a Richter scale), produced a tsunami that propagated across the Pacific Ocean. It reached Hawaii after 15 hours, Japan after 22 hours, and it caused massive destruction in both places. This tsunami propagated over a long enough distance that KdV dynamics probably were relevant. For more information about this earthquake and its tsunami, see the reference just cited, [Scott 1999], or http://neic.usgs.gov/ neis/eq_depot/world/1960_05_22_tsunami.html. More recently, a seismic event off the Kuril Islands in 2006 sent a wave across the Pacific that damaged the harbor in Crescent City, CA. KdV (or KP) dynamics probably were relevant for this wave, which took more than 8 hours to cross the Pacific. For more information about this tsunami, see http://www.usc.edu/dept/tsunamis/2005/tsunamis/ Kuril_2006/.

Back to the tsunami of 2004. The discussion presented here might leave the reader wondering how a wave of such small amplitude, satisfying a linear equation, could be responsible for so much damage and so much loss of life. The answer is that the "long waves, small amplitude" model applies only away from shore — near shore the wave changes its nature entirely. To see this change in tsunami dynamics, imagine sitting in a boat in the middle of the Indian Ocean when a tsunami like that in 2004 passes by. The tsunami is 1 m high, 100 km long, and it travels at 620 km/hr. It would take about 10 minutes to pass the boat, so in the course of 10 minutes, the boat would rise 1 meter and then fall 1 meter. Hence in the open ocean, it is difficult for a sensor at the free surface even to detect a passing tsunami.

Near shore, everything changes. The local speed of propagation is still $\sqrt{gh}$, but h decreases near shore, so the wave slows down. More precisely, the front of the wave slows down — the back of the wave, still 100 km out at sea, is not slowing down. The result is that the wave must compress horizontally, as the back of the wave catches up with the front. But water is nearly incompressible, so if the wave compresses horizontally, then it must grow vertically as it approaches shore. The result is that a very long wave that was barely noticeable in the open ocean becomes shorter (horizontally), larger (vertically), and far more destructive near shore.

See articles in *Science*, **308** (2005), 1125–1146 or in [Kundu 2007] for more discussion of the 2004 tsunami.

4. Application: periodic ocean waves

Among ocean waves, tsunamis are anomalous. The vast majority of ocean waves are approximately periodic, and they are generated by winds and storms [Munk et al. 1962]. The water surface is two-dimensional, so the KP equation, (2), is a natural place to seek solutions that might describe approximately periodic waves in shallow water. Gravity dominates surface tension except for very short waves, so we may set $\alpha = 1$ in (2).

The simplest periodic solution of (2) is a (one-dimensional) plane wave, of the form

$$u = 12k^2 m^2 \operatorname{cn}^2 \{kx + ly + wt + \phi_0; m\} + u_0, \tag{13}$$

where $\text{cn}\{\theta; m\}$ is a Jacobian elliptic function with elliptic modulus m. If we impose (5), then the solution in (13) has four free parameters; for example $\{k, l, \phi_0, m\}$. If $l = 0$, then (13) solves the KdV equation; this solution was first discovered by Korteweg and de Vries [1895], who named it a *cnoidal wave*. Wiegel [1960] brought cnoidal waves to the attention of coastal engineers, who now use them regularly for engineering calculations. (See Chapter 2 of the *Shore Protection Manual* [SPM 1984] for this viewpoint.)

Figure 3 shows ocean waves photographed by Anna Segur near the beach in Lima, Peru. These plane, periodic waves have broad, flat troughs and narrow, sharp crests — typical of cnoidal waves with elliptic modulus near 1, and also typical of plane, periodic waves in shallow water of nearly uniform depth. It is unusual to see such a clean example of a cnoidal wave train, but that might be because few beaches are as flat as the beach in Lima.

Cnoidal waves are appealing because of their simplicity, but they are degenerate in the sense that the water surface is two-dimensional, while cnoidal waves vary only in the direction of propagation. One might wish for a wave pattern that is nontrivially periodic in two spatial directions, and that travels as a wave of permanent form in water of uniform depth.

Figure 4 shows a photograph of a wave pattern photographed by Terry Toedtemeier off the coast of Oregon. This photo can be interpreted in two different ways, each with some validity. The first interpretation is that Figure 4 shows two plane solitary waves, interacting obliquely in shallow water of nearly uniform depth. A basic rule of soliton theory is that the interaction of two solitons results in a phase shift. A phase shift is evident in Figure 4: each wave crest is shifted beyond the interaction region from where it would have been without the interaction. The KP equation admits a 2-soliton solution that looks very much like the wave pattern in Figure 4. Equivalently, one can identify this wave pattern with a 2-soliton solution of the Boussinesq equation, (3).

Figure 3. Periodic plane waves in shallow water, off the coast of Lima, Peru. (Photographs courtesy of A. Segur)

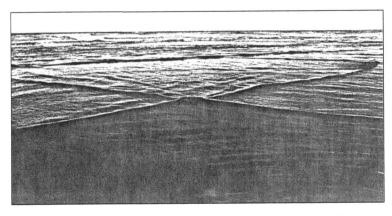

Figure 4. Oblique interaction of two shallow water waves, off the coast of Oregon. (Photograph courtesy of T. Toedtemeier)

The other interpretation is that Figure 4 shows the oblique, nonlinear interaction of two (plane) cnoidal wave trains. Each wave train exhibits the flat-trough, sharp-crest pattern seen in Figure 3, but in Figure 4 successive wave crests in the same train are so far apart that each crest acts nearly like a solitary wave. Even so, if one looks carefully at Figure 4, one can see the next crest after the prominent crest in each wave train. With this interpretation, Figure 4 shows a wave pattern that propagates with permanent form, and is nontrivially periodic in two horizontal directions. Each wave crest in one wavetrain undergoes a phase shift every time it interacts with a wave crest from the other wavetrain. The result is a two-dimensional, periodic wave pattern, in which the basic "tile" of the pattern is a hexagon: two parallel sides of the hexagon are crests from one wavetrain, two sides are crests from the other wavetrain, and the last two sides are two, short interaction regions. (Only one interaction region is evident in Figure 4.)

The KP equation admits an 8-parameter family of real-valued solutions like this — each KP solution is a travelling wave of permanent form, nontrivially periodic in two spatial directions; see [Segur and Finkel 1985] for details. In terms of Riemann surface theory, every cnoidal wave solution of KdV or KP corresponds to a Riemann surface of genus 1; each of the KP solutions considered here corresponds to a Riemann surface of genus 2. These two-dimensional, doubly periodic wave patterns are the simplest periodic or quasiperiodic solutions of the KP equation beyond cnoidal waves.

In a series of experiments, Joe Hammack and Norm Scheffner created waves in shallow water with spatial periodicity in two directions, in order to test the KP model of such waves [Hammack et al. 1989; 1995]. Figure 5 shows overhead photographs of three of their propagating wave patterns. Each wave pattern in these photos is generated by a complicated set of paddles at one end of a long wave tank. The photos are oriented so that the waves propagate downward in each pair of photos. The pattern in the top photos is symmetrical, and it propagates directly away from the paddles (so straight down in Figure 5). The other two patterns are asymmetric; these patterns propagate with nearly permanent form but not directly away from the paddles — there is also a uniform drift to the left or right for each wave pattern. The corresponding KP solution predicts this direction of propagation, along with the detailed shape of the two-dimensional, doubly periodic wave pattern. Hammack et al. [1995] showed experimentally that for each wave pattern they generated, the appropriate KP solution of genus 2 predicts the detailed shape of that pattern with remarkable accuracy. See their paper for these comparisons.

In addition to these photographs, they also made videos of the experiments. There was no convenient way to present those videos in 1989, but now they are archived at the MSRI website. Go to http://phoebe.msri.org:8080/vicksburg/vicksburg.mov to see the first video. The experiments were conducted in a large (30 m x 56 m) wave tank at the US Army Corps of Engineers Waterways Experiment Station, in Vicksburg, MS. A segmented wavemaker, consisting of 60 piston-type paddles that spanned the tank width, is shown in the first scene. In this scene the paddles all move together, and they generate a train of plane, periodic (*i.e.*, cnoidal) waves that propagate to the other end of the tank, where they are absorbed.

In the second scene, the camera looks down on the tank from above; the paddles are visible along the end of the tank at the upper right. The paddles were programmed to create approximately a KP solution of genus 2, with a specific set of choices of the free parameters. Hence the wave pattern coming off the paddles is periodic in two spatial directions. The experiment shows that the entire two-dimensional pattern propagates as a wave of nearly permanent form. As in Figures 4 and 5, the basic tile of the periodic pattern is a hexagon, but the long, straight, dominant crests seen in the video are the interaction regions, which are quite short in Figure 4. The relatively narrow, zigzag region that connects adjacent cells contains the other four edges of the hexagonal tile. Wave amplitudes in the zigzag region are smaller than those of the long, dominant crests, and the KP solution shows that horizontal velocities are smaller in this region as well.

The relative length of the long, straight, dominant crests within a hexagonal tile is a parameter that one can chose by choosing properly the free parameters of the KP solution. This freedom of choice is demonstrated in the third scene in the video. In this experiment, all of the parameters of the KP solution are the

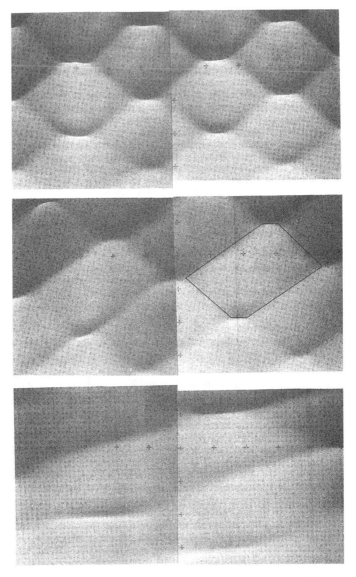

Figure 5. Mosaics of two overhead photos, showing three surface patterns of travelling waves of nearly permanent form, with periodicity in two spatial directions, in shallow water of uniform depth. The basic tile of each pattern is a hexagon; one hexagon is drawn in the middle photos. Each wave pattern is generated by a set of paddles, above the top of each pair of photos. The wave pattern propagates away from the paddles, so downward in these photos. The waves are illuminated by a light that shines towards the paddles, so a bright region identifies the front of a wave crest, a dark region lies behind the crest, and a sharp transition from bright to dark represents a steep wave crest. (Figure taken from [Hammack et al. 1995].)

same as those in the previous experiment, except that the length of the dominant crest is shorter.

Almost all real-valued genus 2 solutions of the KP equation define wave patterns like this: travelling waves of permanent form that are periodic in two spatial directions. The existence of these KP solutions does not guarantee that such solutions are stable within the KP equation, or that the corresponding water waves are stable. The video shows that these water waves look stable. See Section 6 for more discussion of stability.

Wiegel [1960] pointed out the practical, engineering value of KdV (or KP) solutions of genus 1, and the experiments shown here demonstrate the practical value of KP solutions of genus 2 — both sets of KP solutions describe accurately waves of nearly permanent form in shallow water of uniform depth. See Section 6 for a discussion of KP solutions of higher genus.

We end this discussion of spatially periodic waves of permanent form by noting the work of Craig and Nicholls [2000; 2002]. Motivated in part by the experimental results shown here, these authors proved directly that the equations of inviscid water waves, (6), admit travelling wave solutions that are spatially periodic in two horizontal directions, like those shown here. The parameter range of their family of solutions is not identical with the parameter range of KP solutions of genus 2, but the two overlap. (The KP equation approximates water waves only in shallow water, but Craig and Nicholls find solutions in water of any depth. In the other direction, they have not yet found asymmetric solutions, like those shown in the middle and bottom photos of Figure 5.) One value of an approximate model, like KP, is that it provides hypotheses about what might be true in the unapproximated problem. The success of Craig and Nicholls demonstrates how effective that strategy has been in this particular problem.

5. Application: rip currents

The material in this section can be considered an application of an application. The KP equation predicts the existence of spatially periodic wave patterns of permanent form, approximately like those shown in Figure 5, in shallow water of uniform depth. As these waves propagate into a region near shore where the water depth decreases (to zero at the shoreline), the KP equation no longer applies. But the waves themselves persist, and their behavior near shore can have important practical consequences, as we discuss next.

A rip current is a narrow jet that forms in shallow water near shore under certain circumstances. It carries water away from shore, through the "surf zone" (the region of breaking waves), out to deeper water. A typical rip current flows directly away from the shoreline, and it remains a strong, narrow jet through

Figure 6. Rip currents on two beaches on the Pacific coast. Left: Rosarita Beach in Baja California, Mexico. Starting at the lower left, the dark region shows vegetation, the white strip above it is sandy beach, the shoreline runs from upper left to lower right, the white water beyond that is the surf zone, with deeper water at the upper right. Three separated rip currents are shown. Right: Sand City, California. The land is to the lower left, the Pacific Ocean is to the upper right, with a white surf zone in between. Here the entire coastline is filled with an approximately periodic array of rip currents.

the surf zone. Beyond the surf zone, the current "blossoms" into a wider flow and loses its strength. Figure 6, left, shows three rip currents, while the right-hand part shows an approximately periodic array of rip currents all along the coastline. Rip currents can be dangerous because even a good swimmer cannot overcome the high flow rate in a strong rip current. As a result, every year rip currents carry swimmers out to deep water, where they drown.

What causes rip currents? They form in the presence of breaking waves, but breaking waves alone do not guarantee their formation. Some rip currents are stationary, while others migrate slowly along the beach. Some persist for a few hours, while others last longer.

There is a standard explanation for how rip currents form, which can be found at http://www.ripcurrents.noaa.gov/science.shtml and elsewhere. According to this explanation, rip currents require a long sandbar, parallel to the beach and just beyond the surf zone. Incoming waves break in the surf zone, and then a return flow carries that water back out to sea. Where can the return flow go? The easiest place for the return flow to carry water past the sandbar is where the height of the sandbar has a local minimum. So the return flow goes through this (initially small) pass. In doing so, its flow scours out more sand at the local minimum, which makes the height even lower there. Then more water can go through, and a feedback loop carves a larger hole in the sandbar. As it carves this hole, the return flow strengthens until it forms into a narrow jet — a rip current.

The rip current is located at the hole in the sandbar, and the width of the current is the width of the hole.

This is a sensible explanation, and it is probably correct sometimes. But the left half of Figure 6 shows three rip currents, with some spacing between them. What determines their spacing? The right half shows a long array of rip currents, with an approximately constant spacing between adjacent rips. What determines their spacing? Not all rip currents appear in these approximately periodic arrays, but they often do. And the standard explanation, summarized above, provides no insight into why rips appear in these regular arrays, and no way to predict the spacing between adjacent rips. Separately, this explanation seems to imply that a rip current cannot migrate slowly along the beach, even though some do.

An alternative mechanism to create rip currents was proposed by Hammack et al. [1991]. It requires no sand bars. Recall the doubly periodic wave patterns from their movie, which travel with nearly permanent form (see http://phoebe.msri.org:8080/vicksburg/vicksburg.mov). What would happen to such a spatially periodic wave pattern, as it traveled up a sloping beach? It is easy to imagine that the large, dominant wave crests would break first, while the smaller crests in the narrow zigzag region would break later, or not at all. After the waves break, a return flow must carry that water back out to deep water. Where can the return flow go? The return flow is likely to go where the incoming flow is the weakest. Where is that?

[Once waves break, the KP equation no longer applies. The next two paragraphs, therefore, define a conjecture, with little mathematical justification at this time.] The derivation of either (1) of (2) from (6) shows that the horizontal velocity of the water is proportional to the wave height, so the strongest horizontal flow occurs where the waves are highest. For the wave patterns shown in the video, therefore, the long, dominant wave crests are also regions of large forward velocity, where a return flow would be resisted by a strong incoming flow. In the narrow zigzag regions, wave heights are smaller, horizontal flows are also smaller, and here the return flow would meet less resistance.

If the return flow travelled through the narrow zigzag region of the incoming flow field, then the return flow would acquire a spatial structure determined by the structure of the incoming waves. Specifically, suppose the incoming wave pattern were one of the hexagonal patterns seen in the previous video. Then:

• The return flow would appear in narrow jets (*i.e.*, rip currents) because incoming flow contains narrow zigzag regions, where the incoming velocities are smaller.

• These narrow jets would be periodically arranged along the beach because the incoming wave patterns are periodic in the direction along the beach.

- The spacing between adjacent rips would be determined by the spacing between adjacent zigzag regions in the incoming waves.
- The width of the jets would be related to the width of the zigzag region in the incoming wave pattern.

Hammack et al. [1991] tested this conjecture with another set of experiments, after some modifications of their wave tank. For the experiments on rip currents, the set of paddles was moved from the end of the tank to the side, so the waves now propagate across the tank instead of along it. In addition, they installed a sloping beach in the tank. The result was that the waves propagated across a region of uniform depth, and then up a uniformly sloping beach. The beach was made of concrete, so there were no sandbars.

The experiments can be viewed in another video, which is available at http://phoebe.msri.org:8080/ripcurrent84/ripcurrent84.mov. As the video opens, one sees dry beach along the bottom of the screen, with quiescent water higher up the screen. A strip of dye (food coloring) has been poured along the water's edge, to mark where the return flow goes. As the camera scans the entire tank, one sees the array of paddles on the far side of the tank. These paddles then create a set of hexagonally shaped, periodic waves of nearly permanent form, which propagate away from the paddles and towards the beach. As the hexagonal wave patterns climb the beach, their crests begin to break. Then a few seconds later, the dye begins to move from the shoreline, away from the beach, in a narrow jet. The jet is clearly in the zigzag region of the incoming flow, because we see the dye zigzagging out to deep water in the video. After the experiment has run for a while, the dye shows the spatial pattern of the rip currents: a periodic array of jets that remain narrow through the surf zone, and then spread out and stop beyond the surf zone. Both the spacing between jets and the width of an individual jet are determined by the spatial structure of the incoming wave pattern. Sand bars are irrelevant for these rip currents.

(In addition to the array of narrow rip currents marked by dye at the end of the experiment, one also sees a weaker, roughly circular glob of dye between each pair of rip currents. This secondary flow of dye might be due to undertow, which exists because the bottom boundary layer of the incoming waves is another place where the incoming flow is weaker. But this is a separate conjecture, with little experimental support at this time.)

6. Quasiperiodic solutions of the KP equation

Krichever [1977a] showed that the KP equation, (2), admits a large family of quasiperiodic solutions of the form

$$u(x, y, t) = 12\partial_x^2 \ln \theta, \tag{14}$$

where θ is a Riemann theta-function, associated with a (compact, connected) Riemann surface of some finite genus. Such a formulation was already known for KdV: (13) with $u_0 = 0$ can be written in this form, with genus 1. The Riemann surface is necessarily hyperelliptic (so only square-root branch points are allowed) for KdV, but Krichever [1976; 1977b] proved that any Riemann surface would do for KP.

As described in detail in [Dubrovin 1981; Belokolos et al. 1994], a Riemann theta function with g phases is defined by a g-fold Fourier series; the coefficients in this series are defined by a $g \times g$ Riemann matrix. If one starts with a Riemann surface of genus g, then by a standard procedure one generates a $g \times g$ Riemann matrix associated with that surface. Krichever showed that (14), with the theta function obtained in this way, solves the KP equation. Then S. P. Novikov conjectured that the connection between the KP equation and Riemann surfaces is even stronger, and that (14) solves KP *only* if its theta function is associated with some Riemann surface. In other words, out of all possible theta functions, of any finite genus, the KP equation can identify those associated with some compact Riemann surface. The conjecture seems remarkable, but Shiota [1986] proved it, after earlier work by Mulase [1984] and by Arbarello and de Concini [1984]. Thus the KP equation, (2), can be studied from at least three perspectives:

- as a completely integrable partial differential equation;
- as an approximate model of waves in shallow water; and
- because of its deep connection with the theory of Riemann surfaces.

An objective of this paper is to relate the extra mathematical structure of the KP equation to the behavior of physical waves in shallow water, if possible. The theory for the KP equation is much less developed than the corresponding theory for the KdV equation, especially for the quasiperiodic KP solutions given by (14). This final section summarizes our current knowledge of four aspects of these solutions: (a) the qualitative nature of quasiperiodic KP solutions; (b) effective methods to construct KP solutions of a given genus; (c) solving the initial-value problem; and (d) stability of quasiperiodic KP solutions.

(a) the qualitative nature of quasiperiodic KP solutions. Here is a summary of what is known about bounded, real-valued, KP and KdV solutions of various genera. (See [Dubrovin 1981] for details.)

$g = 1$: Any KP solution of genus 1 is a cnoidal wave — a plane wave that travels with permanent form, given by (13). The cnoidal wave solutions of KdV are special cases of this, with $l = 0$ in (13).

$g = 2$: A KP solution of genus 2 has two phase variables: $\{k_j x + l_j y + w_j t + \phi_j\}$ for $j = 1, 2$. There are two possibilities.

- If $k_1 l_2 = k_2 l_1$, then all lines of constant phase are parallel, so the spatial pattern of the wave is always one-dimensional. Any such solution of KP can be transformed ("rotated") into a solution of KdV, also of genus 2. These KdV solutions are necessarily time-dependent, in any Galilean coordinate system.

- Otherwise $k_1 l_2 \neq k_2 l_1$, and the solution is spatially periodic — the basic tile of the pattern is a hexagon, as discussed in Section 4. The wave patterns shown in Figure 5 approximate KP solutions of genus 2. Each such solution travels as a wave of permanent form in an appropriately translating coordinate system. Almost all KP solutions of genus 2 have $k_1 l_2 \neq k_2 l_1$, so they are travelling waves of permanent form.

$g \geq 3$: Almost all KP solutions of genus 3 or higher are time-dependent, in every Galilean coordinate system. Hence these KP solutions can describe physical processes that are time-dependent, including energy transfer among modes. Because of this nontrivial time-dependence, snapshots at a particular time, like those in Figure 5, are inadequate to view these solutions.

$g \geq 3$: For $g \geq 3$, the only KP solutions that are waves of permanent form are those that also solve the Boussinesq equation, (3). For $g = 3$, the KP equation admits a 12-parameter family of bounded, real-valued, quasiperiodic solutions, each with three independent phases. The Boussinesq solutions with $g = 3$ comprise an 11-parameter subfamily, so almost all KP solutions with $g = 3$ are intrinsically time-dependent. Even so, this 11-parameter sub-family is much larger than the 8-parameter family of KP solutions of genus 2. Which of these solutions are stable, and in what sense, are open questions.

$g \to \infty$: The development of "finite-gap" solutions of the KdV equation by Novikov [1974], Lax [1975], McKean and van Moerbeke [1975] and others can be considered a nonlinear generalization of a finite Fourier series (which contains only a finite number of terms). Each such finite-gap solution of KdV is based on a hyperelliptic Riemann surface of finite genus. The genus determines the number of open gaps in the spectrum of Hill's equation; it corresponds to the number of terms in a finite Fourier series. McKean and Trubowitz [1976] made this correspondence legitimate, by developing a theory of hyperelliptic curves with infinitely many branch points. In this context, one can discuss convergence of a sequence of finite-gap solutions of KdV, as the number of gaps (and the genus of the Riemann surface) increases without bound.

$g \to \infty$: When we switch from the KdV to the KP equation, we also switch from hyperelliptic to general Riemann surfaces, and things become more complicated. The recent book by Feldman et al. [2003] (see *Bull. Amer. Math. Soc.* **42** (2004), 79–87 for McKean's review) explores Riemann surfaces of infinite genus, generalizing [McKean and Trubowitz 1976]. When KP solutions of finite genus are understood well enough to consider questions of convergence, one can hope that this recent work will provide a suitable framework in which to address such questions.

$g \geq 3$: Back to finite genus. Time-dependent or not, KP solutions with $g \geq 3$ are typically not periodic in space, but only quasiperiodic.

The fact that KP solutions of higher genus are typically only quasiperiodic in space or time is worth discussing. Physically, it is common to see water waves that are approximately periodic, but truly periodic water waves seem to be rare. In this sense, a mathematical model that naturally produces quasiperiodic solutions is an advantage. In terms of scientific computations, people often use periodic boundary conditions, not because the physical problem is periodic but for computational simplicity. How to build numerical codes that compute efficiently in a space of almost periodic functions seems to be an open problem. In terms of mathematical theory, Dubrovin et al. [1976] developed a theory to construct solutions of the KdV equation that need be only almost periodic. Their task was simpler than for the corresponding problem for KP, because the KdV equation only allows hyperelliptic Riemann surfaces, which are better understood than general Riemann surfaces.

(b) effective methods to construct KP solutions of a given (finite) genus. The method of inverse scattering, to solve the initial-value problem for a completely integrable evolution equation, typically has three parts: map the initial data into scattering data, evolve the scattering data forward in time, and then map back. For solutions of the form (14), one can identify the "scattering data" with the Riemann surface plus a divisor on that surface. Hence, one part of the method of inverse scattering is to produce an explicit Riemann theta function from these scattering data. This procedure is carried out for KP and several other integrable problems in [Belokolos et al. 1994].

Earlier, Bobenko and Bordag [1989] started with different "scattering data", and demonstrated that their method is effective by producing KP solutions. In principle their method can generate solutions of any genus, and they exhibit a solution of genus 4, among others.

Any method that uses the underlying Riemann surface as scattering data faces inherent difficulties related to our inadequate knowledge of Riemann surface theory. A Riemann surface can be defined by an algebraic curve: a polynomial relation of finite degree between two complex variables, $P(w, z) = 0$. But this relation might have singularities, where $\partial P(w, z)/\partial w = 0$ and $\partial P(w, z)/\partial z = 0$ simultaneously. One does not obtain a Riemann surface until all such singularities are resolved. Separately, a given Riemann surface can have more than one such representation, and it can be difficult to tell whether two such polynomial relations represent the same surface.

Consequently it has been necessary to build computational machinery, to make the abstract theory of Riemann surfaces concrete and effective. See [Deconinck and van Hoeij 2001; Deconinck et al. 2004] for some of this machinery. At this time, computing the ingredients in a Riemann theta function from a representation of its algebraic curve is still not straightforward.

Dubrovin [1981] proposed another approach, which is effective for genus 1, 2 or 3, and only for them. He observed that for these low genera, *any* Riemann matrix that is "irreducible" can be associated with some Riemann surface. Hence one can skip the Riemann surface altogether, and work directly with the Riemann matrix. The papers [Segur and Finkel 1985; Dubrovin et al. 1997] were both based on this approach. These authors demonstrated the effectiveness of their method not only with example solutions, but also with publicly available computer codes that allow an interested reader to compute and to visualize real-valued KP solutions of genus 1, 2 or 3. The limitation of their method is that it fails for any genus larger than 3.

(c) solving the initial-value problem. Let us focus on two published methods to solve the KP equation as an initial-value problem, starting with either periodic or quasiperiodic initial data: by [Krichever 1989; Deconinck and Segur 1998]. Both methods rely on (14) to describe KP solutions of some finite genus.

Krichever requires that the initial data be periodic in x and in y, with fixed periods in each direction (i.e., in a fixed rectangle). Any KP solution that evolves from these initial data then retains that periodicity. He establishes the formal existence of a sequence of KP solutions, each of finite genus and in the form (14), which provide better and better approximations to the given initial data at $t = 0$. An important accomplishment in this work is his approximation theorem, which shows that these finite-genus solutions are dense in a suitable space of KP solutions with the given periods in x and in y.

Our approach in [Deconinck and Segur 1998] differed from that of Krichever in several respects. We considered initial data that are quasiperiodic in space, rather than requiring strict periodicity in x and in y. Even at genus 2, requiring that waves be periodic in x and in y is overly restrictive. Mathematically, the family of real-valued KP solutions of genus 2 that are periodic in x and in y has 5 free parameters, while the full family of real-valued KP solutions of genus 2 has 8. Physically, all three patterns of water waves photographed in Figure 5 are spatially periodic, but only the top pattern is periodic in x and in y.

We paid for the extra flexibility of allowing initial data that are quasiperiodic in space, by requiring that their initial data have the form (14), with some finite number (g) of phases. Then we gave a constructive procedure to determine

g (the number of phases and the genus of the Riemann surface), the Riemann surface itself and the divisor on that surface. Unfortunately, we have no approximation theorem, so we cannot prove that the KP solutions obtained in this way are dense in any suitable space of KP solutions.

The opinion of this writer is that more work is needed to produce a constructive method to solve the initial-value problem for the KP equation with quasiperiodic initial data.

(d) stability of quasiperiodic KP solutions. If one views the KP equation as a mathematical model of a physical system, like waves in shallow water, then the stability of its solutions is an essential piece of information about the model. The video at http://phoebe.msri.org:8080/vicksburg/vicksburg.mov shows water waves that are well approximated by KP solutions of genus 2, and that appear to be stable as they propagate. But one cannot prove stability experimentally — the video only shows that if there is an instability, then its growth rate must be slow enough that it does not appear within the test section of the tank.

At this time, almost nothing is known about the stability of quasiperiodic solutions of the KP equation. The problem is even more difficult than usual because standard numerical methods to test for stability/instability are based on codes with periodic boundary conditions, and these are not suitable for KP solutions that are only quasiperiodic. The problem seems to be completely open at this time.

Acknowledgements

The author congratulates Henry McKean for his lifetime of accomplishments, including those in mathematics. He is grateful to Joe Hammack and Norm Scheffner, who carried out the careful and insightful experiments shown in the videos cited and in Figure 5. Joe Hammack died unexpectedly in 2004. The author thanks Bernard Deconinck for many helpful comments about Section 6, and Kenji Satake, Anna Segur and Terry Toedtemeier for permission to show their work in Figures 2, 3 and 4 respectively. Finally, he is grateful to MSRI for hosting the conference that led to this volume, and for posting the two videos cited herein on its website.

References

[Ablowitz and Segur 1981] M. J. Ablowitz and H. Segur, *Solitons and the inverse scattering transform*, SIAM Studies in Applied Mathematics **4**, Soc. Ind. App. Math., Philadelphia, PA, 1981.

[Arbarello and De Concini 1984] E. Arbarello and C. De Concini, "On a set of equations characterizing Riemann matrices", *Ann. of Math.* (2) **120**:1 (1984), 119–140.

[Belokolos et al. 1994] E. D. Belokolos, A. I. Bobenko, V. Z. Enol'skii, A. R. Its, and V. B. Matveev, *Algebro-geometric approach to nonlinear integrable equations*, Springer, Berlin, 1994.

[Bobenko and Bordag 1989] A. I. Bobenko and L. A. Bordag, "Periodic multiphase solutions of the Kadomsev–Petviashvili equation", *J. Phys. A* **22**:9 (1989), 1259–1274.

[Bona et al. 2002] J. L. Bona, M. Chen, and J.-C. Saut, "Boussinesq equations and other systems for small-amplitude long waves in nonlinear dispersive media. I. Derivation and linear theory", *J. Nonlinear Sci.* **12**:4 (2002), 283–318.

[Bona et al. 2004] J. L. Bona, M. Chen, and J.-C. Saut, "Boussinesq equations and other systems for small-amplitude long waves in nonlinear dispersive media. II. The nonlinear theory", *Nonlinearity* **17**:3 (2004), 925–952.

[Boussinesq 1871] J. Boussinesq, "Théorie de l'intumescence liquide appellée onde solitaire ou de translation, se propageant dans un canal rectangulaire", *Comptes Rendus Acad. Sci. Paris* **72** (1871), 755–759.

[Camassa and Holm 1993] R. Camassa and D. D. Holm, "An integrable shallow water equation with peaked solitons", *Phys. Rev. Lett.* **71**:11 (1993), 1661–1664.

[Constantin and McKean 1999] A. Constantin and H. P. McKean, "A shallow water equation on the circle", *Comm. Pure Appl. Math.* **52**:8 (1999), 949–982.

[Coutand and Shkoller 2005] D. Coutand and S. Shkoller, "Wellposedness of the free surface incompressible Euler equations with or without a free surface", preprint, 2005.

[Craig and Nicholls 2000] W. Craig and D. P. Nicholls, "Travelling two and three dimensional capillary gravity water waves", *SIAM J. Math. Anal.* **32**:2 (2000), 323–359.

[Craig and Nicholls 2002] W. Craig and D. P. Nicholls, "Traveling gravity water waves in two and three dimensions", *Eur. J. Mech. B Fluids* **21**:6 (2002), 615–641.

[Deconinck and Segur 1998] B. Deconinck and H. Segur, "The KP equation with quasiperiodic initial data", *Phys. D* **123**:1-4 (1998), 123–152.

[Deconinck and van Hoeij 2001] B. Deconinck and M. van Hoeij, "Computing Riemann matrices of algebraic curves", *Phys. D* **152/153** (2001), 28–46. Advances in nonlinear mathematics and science.

[Deconinck et al. 2004] B. Deconinck, M. Heil, A. Bobenko, M. van Hoeij, and M. Schmies, "Computing Riemann theta functions", *Math. Comp.* **73**:247 (2004), 1417–1442.

[Dubrovin 1981] B. A. Dubrovin, "Theta functions and nonlinear equations", *Uspekhi Mat. Nauk* **36**:2 (1981), 11–80. In Russian; translated in *Russ. Math. Surveys*, **36** (1981), 11-92.

[Dubrovin et al. 1976] B. A. Dubrovin, V. B. Matveev, and S. P. Novikov, "Nonlinear equations of Korteweg-de Vries type, finite-band linear operators and Abelian varieties", *Uspehi Mat. Nauk* **31**:1 (1976), 55–136. In Russian; translated in *Russ. Math. Surveys* **31** (1976), 59-146.

[Dubrovin et al. 1997] B. A. Dubrovin, R. Flickinger, and H. Segur, "Three-phase solutions of the Kadomtsev–Petviashvili equation", *Stud. Appl. Math.* **99**:2 (1997), 137–203.

[Feldman et al. 2003] J. Feldman, H. Knörrer, and E. Trubowitz, *Riemann surfaces of infinite genus*, CRM Monograph Series **20**, American Mathematical Society, Providence, RI, 2003.

[Gardner et al. 1967] C. S. Gardner, J. M. Greene, M. D. Kruskal, and R. M. Muira, "Method for solving the Korteweg-de Vries equation", *Phys. Rev. Lett.* **19** (1967), 1095–1097.

[Hammack et al. 1989] J. Hammack, N. Scheffner, and H. Segur, "Two-dimensional periodic waves in shallow water", *J. Fluid Mech.* **209** (1989), 567–589.

[Hammack et al. 1991] J. L. Hammack, N. W. Scheffner, and H. Segur, "A note on the generation and narrowness of periodic rip currents", *J. Geophys. Res.* **96** (1991), 4909–4914.

[Hammack et al. 1995] J. Hammack, D. McCallister, N. Scheffner, and H. Segur, "Two-dimensional periodic waves in shallow water, II: Asymmetric waves", *J. Fluid Mech.* **285** (1995), 95–122.

[Johnson 1997] R. S. Johnson, *A modern introduction to the mathematical theory of water waves*, Cambridge University Press, Cambridge, 1997.

[Johnson 2002] R. S. Johnson, "Camassa–Holm, Korteweg–de Vries and related models for water waves", *J. Fluid Mech.* **455** (2002), 63–82.

[Kadomtsev and Petviashvili 1970] B. B. Kadomtsev and V. I. Petviashvili, "On the stability of solitary waves in weakly dispersive media", *Sov. Phys. Doklady* **15** (1970), 539–541.

[Korteweg and Vries 1895] D. J. Korteweg and G. D. Vries, "On the change of form of long waves advancing in a rectangular canal, and on a new type of long stationary waves", *Phil. Mag.* (5) **39** (1895), 422–443.

[Krichever 1976] I. M. Krichever, "An algebraic-geometric construction of the Zaharov–Shabat equations and their periodic solutions", *Dokl. Akad. Nauk SSSR* **227**:2 (1976), 291–294.

[Krichever 1977a] I. M. Krichever, "Integration of nonlinear equations by the methods of algebraic geometry", *Funkcional. Anal. i Prilozen.* **11**:1 (1977), 15–31, 96. In Russian; translated in *Funct. Anal. Appl.* **11** (1977), 12–26.

[Krichever 1977b] I. M. Krichever, "Methods of algebraic geometry in the theory of nonlinear equations", *Uspehi Mat. Nauk* **32**:6 (1977), 183–208, 287. In Russian; translated in *Russ. Math. Surveys.* **32** (1977), 185–213.

[Krichever 1989] I. M. Krichever, "Spectral theory of two-dimensional periodic operators and its applications", *Uspekhi Mat. Nauk* **44**:2 (1989), 121–184. In Russian; translated in *Russ. Math. Surveys.* **44** (1989), 145-225.

[Kundu 2007] A. Kundu (editor), *Tsunami and nonlinear waves*, Springer, Berlin, 2007.

[Lakshmanan and Rajasekar 2003] M. Lakshmanan and S. Rajasekar, *Nonlinear dynamics: Integrability, chaos and patterns*, Springer, Berlin, 2003.

[Lannes 2005] D. Lannes, "Well-posedness of the water-waves equations", *J. Amer. Math. Soc.* **18**:3 (2005), 605–654.

[Lax 1975] P. D. Lax, "Periodic solutions of the KdV equation", *Comm. Pure Appl. Math.* **28** (1975), 141–188.

[Madsen et al. 2002] P. A. Madsen, H. B. Bingham, and H. Liu, "A new Boussinesq method for fully nonlinear waves from shallow to deep water", *J. Fluid Mech.* **462** (2002), 1–30.

[McKean 1977] H. P. McKean, "Stability for the Korteweg-de Vries equation", *Comm. Pure Appl. Math.* **30**:3 (1977), 347–353.

[McKean 1978] H. P. McKean, "Boussinesq's equation as a Hamiltonian system", pp. 217–226 in *Topics in functional analysis (essays dedicated to M. G. Kreĭn on the occasion of his 70th birthday)*, Adv. in Math. Suppl. Stud. **3**, Academic Press, New York, 1978.

[McKean 1981a] H. P. McKean, "Boussinesq's equation on the circle", *Physica* **3D** (1981), 294–305.

[McKean 1981b] H. P. McKean, "Boussinesq's equation on the circle", *Comm. Pure Appl. Math.* **34**:5 (1981), 599–691.

[McKean and Trubowitz 1976] H. P. McKean and E. Trubowitz, "Hill's operator and hyperelliptic function theory in the presence of infinitely many branch points", *Comm. Pure Appl. Math.* **29**:2 (1976), 143–226.

[McKean and Trubowitz 1978] H. P. McKean and E. Trubowitz, "Hill's surfaces and their theta functions", *Bull. Amer. Math. Soc.* **84**:6 (1978), 1042–1085.

[McKean and van Moerbeke 1975] H. P. McKean and P. van Moerbeke, "The spectrum of Hill's equation", *Invent. Math.* **30**:3 (1975), 217–274.

[Mulase 1984] M. Mulase, "Cohomological structure in soliton equations and Jacobian varieties", *J. Differential Geom.* **19**:2 (1984), 403–430.

[Munk et al. 1962] W. H. Munk, G. R. Miller, F. E. Snodgrass, and N. F. Barber, "Directional recording of swell from distant storms", *Phil. Trans. A* **255** (1962), 505–583.

[Novikov 1974] S. P. Novikov, "A periodic problem for the Korteweg-de Vries equation. I", *Funkcional. Anal. i Priložen.* **8**:3 (1974), 54–66.

[Scott 1999] A. Scott, *Nonlinear science*, Oxford Texts in Applied and Engineering Mathematics **1**, Oxford University Press, Oxford, 1999. Emergence and dynamics of coherent structures, With contributions by Mads Peter Sørensen and Peter Leth Christiansen.

[Segur 2007] H. Segur, "Waves in shallow water, with emphasis on the tsunami of 2004", pp. 3–30 in *Tsunami and nonlinear waves*, edited by A. Kundu, Springer, Berlin, 2007.

[Segur and Finkel 1985] H. Segur and A. Finkel, "An analytical model of periodic waves in shallow water", *Stud. Appl. Math.* **73**:3 (1985), 183–220.

[Shiota 1986] T. Shiota, "Characterization of Jacobian varieties in terms of soliton equations", *Invent. Math.* **83**:2 (1986), 333–382.

[SPM 1984] *Shore Protection Manual*, U. S. Army Corps of Engineers, Waterways Experimental Station, Vicksburg, MS, 1984.

[Stokes 1847] G. G. Stokes, "On the theory of oscillatory waves", *Trans. Camb. Phil. Soc.* **8** (1847), 441–455.

[Wei et al. 1995] G. Wei, J. T. Kirby, S. T. Grilli, and R. Subramanya, "A fully nonlinear Boussinesq model for surface waves. I. Highly nonlinear unsteady waves", *J. Fluid Mech.* **294** (1995), 71–92.

[Wiegel 1960] R. L. Wiegel, "A presentation of cnoidal wave theory for practical application", *J. Fluid Mech.* **7** (1960), 273–286.

[Wu 1999] S. Wu, "Well-posedness in Sobolev spaces of the full water wave problem in 3-D", *J. Amer. Math. Soc.* **12**:2 (1999), 445–495.

[Zabusky 2005] N. J. Zabusky, "Fermi–Pasta–Ulam, solitons and the fabric of nonlinear and computational science: history, synergetics, and visiometrics", *Chaos* **15**:1 (2005), 015102, 16.

[Zabusky and Kruskal 1965] N. J. Zabusky and M. D. Kruskal, "Interactions of solitons in a collisionless plasma and the recurrence of initial states", *Phys. Rev. Lett.* **15** (1965), 240–243.

[Zakharov and Faddeev 1971] V. E. Zakharov and L. D. Faddeev, "Korteweg-de Vries equation, a completely integrable Hamiltonian system", *Funct. Anal. Appl.* **5** (1971), 280–287.

HARVEY SEGUR
DEPARTMENT OF APPLIED MATHEMATICS
UNIVERSITY OF COLORADO
BOULDER, CO 80309-0526
UNITED STATES
segur@colorado.edu

Probability, Geometry and Integrable Systems
MSRI Publications
Volume **55**, 2008

Nonintersecting Brownian motions, integrable systems and orthogonal polynomials

PIERRE VAN MOERBEKE

To Henry, teacher and friend, with admiration and gratitude

ABSTRACT. Consider n nonintersecting Brownian motions on $\mathbb{R}$, which leave from p definite points and are forced to end up at q points at time $t = 1$. When $n \to \infty$, the equilibrium measure for these Brownian particles has its support on p intervals, for $t \sim 0$, and on q intervals, for $t \sim 1$. Hence it is clear that, when t evolves, intervals must merge, must disappear and be created, leading to various *phase transitions* between times $t = 0$ and 1.

Near these moments of phase transitions, there appears an infinite-dimensional diffusion, *a Markov cloud*, in the limit $n \nearrow \infty$, which one expects to depend only on the nature of the singularity associated with this phase change. The transition probabilities for these Markov clouds satisfy nonlinear PDE's, which are obtained from taking limits of the Brownian motion model with finite particles; the finite model is closely related to Hermitian matrix integrals, which themselves satisfy nonlinear PDE's. The latter are obtained from investigating the connection between the Karlin-McGregor formula, moment matrices, the theory of orthogonal polynomials and the associated integrable systems. Various special cases are provided to illustrate these general ideas. This is based on work by Adler and van Moerbeke.

Mathematics Subject Classification: Primary: 60G60, 60G65, 35Q53; secondary: 60G10, 35Q58.

Keywords: Dyson's Brownian motion, Pearcey process, random matrices coupled in a chain, random matrices with external potential, infinite-dimensional diffusions.

This work was done while PvM was a Miller visiting Professor at the University of California, Berkeley. The support of National Science Foundation grant #DMS-04-06287, a European Science Foundation grant (MISGAM), a Marie Curie Grant (ENIGMA), FNRS, Francqui Foundation and IUAP grants is gratefully acknowledged.

1. Introduction

This lecture in honor of Henry McKean forms a step in the direction of understanding the behavior of nonintersecting Brownian motions on $\mathbb{R}$ (Dyson's Brownian motions), when the number of particles tends to ∞. It explains a novel interface between diffusion theory, integrable systems and the theory of orthogonal polynomials. These subjects have been at the center of Henry McKean's oeuvre. I am delighted to dedicate this paper to Henry, teacher and friend, with admiration for his pioneering work in these fields.

Consider n Brownian particles leaving from points $a_1 < \cdots < a_p$ and forced to end up at $b_1 < \cdots < b_q$ at time $t = 1$. It is clear that, when $n \to \infty$, the equilibrium measure for $t \sim 0$ has its support on p intervals and for $t \sim 1$ on q intervals. It is also clear that, when t evolves, intervals must merge, must disappear and be created, leading to various *phase transitions*, depending on the respective fraction of particles leaving from the points a_i and arriving at the points b_j. Therefore the region $\mathcal{R}$ in the space-time strip (x, t) formed by the support ($\subset \mathbb{R}$) of the equilibrium measure as a function of time $0 \le t \le 1$ will typically present singularities of different types.

Near the moments, where a phase transition takes place, one would expect to find in the limit $n \nearrow \infty$ an infinite-dimensional diffusion, *a Markov cloud*, having some universality properties. Universality here means that the infinite-dimensional diffusion is to depend on the type of singularity only. These Markov clouds are infinite-dimensional diffusions, which 'in principle' could be described by an infinite-dimensional Laplacian with a drift term. We conjecture that each of the Markov clouds obtained in this fashion is related to some *integrable system*, which enables one to derive a nonlinear (finite-dimensional) PDE, satisfied by the joint probabilities. The purpose of this lecture is to show the intimate relationship between these subjects: nonintersecting Brownian motions and integrable systems, via the theory of orthogonal polynomials. Special cases have also shown an intimate connection between the integrable system and the Riemann-Hilbert problem associated with the singularity. These ideas will then be applied to a simple model, where we show that the transition probabilities for the infinite-dimensional Brownian motions near a cusp satisfy a nonlinear PDE. The interrelations between all such equations, "initial" and "final" ($t \to \pm\infty$) conditions, are interesting and challenging open problems. Universality in this context is a largely open field. For references, see later.

2. Biorthogonal polynomials and the 2-component KP hierarchy

Consider the inner product for the weight $\rho(x, y)$ on $\mathbb{R}^2$,

$$\langle f \mid g \rangle := \iint_{\mathbb{R}^2} f(x)g(y)\rho(x, y)\,dx\,dy.$$

and an inner product for this weight, augmented with an extra-exponential factor, depending on "time" parameters $t := (t_1, t_2, \dots)$ and $s := (s_1, s_2, \dots)$,

$$\langle f \mid g \rangle_{t,s} := \iint_{\mathbb{R}^2} f(x)g(y)\rho(x, y)e^{\sum_1^\infty (t_i y^i - s_i x^i)}\,dx\,dy.$$

Construct monic biorthogonal polynomials $p_m^{(1)}(y)$ and $p_n^{(2)}(x)$ (also depending on the parameters t and s) with regard to this deformed weight,

$$\left\langle p_n^{(2)}(x)e^{-\sum_1^\infty s_i x^i} \;\middle|\; p_m^{(1)}(y)e^{\sum_1^\infty t_i y^i} \right\rangle$$

$$= \iint_{\mathbb{R}^2} p_n^{(2)}(x)p_m^{(1)}(y)\rho(x, y)e^{\sum_1^\infty (t_i y^i - s_i x^i)}\,dx\,dy$$

$$= \delta_{nm}h_n,$$

and let τ_n be the determinant of the moment matrix

$$\tau_n(t, s) := \det\left(\left\langle x^k e^{-\sum_1^\infty s_i x^i} \;\middle|\; y^\ell e^{\sum_1^\infty t_i y^i} \right\rangle\right)_{0 \le k, \ell \le n-1}.$$

The following theorem and its corollary, due to Adler and van Moerbeke [1997; 1999b] and inspired by Sato's theory, establishes a link between the functions τ_n and the biorthogonal polynomials:

THEOREM 2.1. *Given these data, the determinant $\tau_n(t, s)$ and the biorthogonal polynomials are related by the following relations, where we have set $[\alpha] :=$ $(\alpha, \frac{1}{2}\alpha^2, \frac{1}{3}\alpha^3, \dots)$ for $\alpha \in \mathbb{C}$:*

$$z^n \frac{\tau_n(t - [z^{-1}], s)}{\tau_n(t, s)} = p_n^{(1)}(z),$$

$$z^n \frac{\tau_n(t, s + [z^{-1}])}{\tau_n(t, s)} = p_n^{(2)}(z),$$

$$z^{-n-1} \frac{\tau_{n+1}(t + [z^{-1}], s)}{\tau_n(t, s)} = \iint_{\mathbb{R}^2} \frac{p_n^{(2)}(x)}{z - y} e^{\sum_1^\infty (t_i y^i - s_i x^i)}\rho(x, y)\,dx\,dy,$$

$$z^{-n-1} \frac{\tau_{n+1}(t, s - [z^{-1}])}{\tau_n(t, s)} = \iint_{\mathbb{R}^2} \frac{p_n^{(1)}(y)}{z - x} e^{\sum_1^\infty (t_i y^i - s_i x^i)}\rho(x, y)\,dx\,dy, \quad (2\text{-}1)$$

with the $\tau_n(t, s)$ satisfying bilinear equations, for all integers $n, m \geq 0$ and all $t, t', s, s' \in \mathbb{C}^\infty$:

$$\oint_{z=\infty} \tau_{n-1}(t - [z^{-1}], s)\tau_{m+1}(t' + [z^{-1}], s')e^{\sum_1^\infty (t_i - t'_i)z^i} z^{n-m-2} dz$$

$$= \oint_{z=\infty} \tau_n(t, s - [z^{-1}])\tau_m(t', s' + [z^{-1}])e^{\sum_1^\infty (s_i - s'_i)z^i} z^{m-n} dz.$$

Two-component KP hierarchy. Define the *Hirota symbol* between functions $f = f(t_1, t_2, \dots)$ and $g = g(t_1, t_2, \dots)$ by

$$p\left(\frac{\partial}{\partial t_1}, \frac{\partial}{\partial t_2}, \dots\right) f \circ g := p\left(\frac{\partial}{\partial y_1}, \frac{\partial}{\partial y_2}, \dots\right) f(t + y)g(t - y)\Big|_{y=0}.$$

The elementary Schur polynomials S_ℓ are defined by $e^{\sum_1^\infty t_i z^i} := \sum_{i \geq 0} S_i(t)z^i$ for $\ell \geq 0$ and $S_\ell(t) = 0$ for $\ell < 0$; moreover, set for later use

$$S_\ell(\tilde{\partial}_t) := S_\ell\left(\frac{\partial}{\partial t_1}, \frac{1}{2}\frac{\partial}{\partial t_2}, \frac{1}{3}\frac{\partial}{\partial t_3}, \dots\right).$$

Finally, recall that the Wronskian $\{f, g\}_x$ of f and g is given by

$$\frac{\partial f}{\partial x}g(x) - \frac{\partial g}{\partial x}f(x).$$

COROLLARY. *From Theorem 2.1, one deduces the equations*

$$S_j\left(\frac{\partial}{\partial t_1}, \frac{1}{2}\frac{\partial}{\partial t_2}, \dots\right)\tau_{n+1} \circ \tau_{n-1} = -\tau_n^2 \frac{\partial^2}{\partial s_1 \partial t_{j+1}} \log \tau_n,$$

$$S_j\left(\frac{\partial}{\partial s_1}, \frac{1}{2}\frac{\partial}{\partial s_2}, \dots\right)\tau_{n-1} \circ \tau_{n+1} = -\tau_n^2 \frac{\partial^2}{\partial t_1 \partial s_{j+1}} \log \tau_n,$$

(2-2)

and finally a single partial differential equation for τ_n in terms of Wronskians,

$$\left\{\frac{\partial^2 \log \tau_n}{\partial t_1 \partial s_2}, \frac{\partial^2 \log \tau_n}{\partial t_1 \partial s_1}\right\}_{t_1} + \left\{\frac{\partial^2 \log \tau_n}{\partial s_1 \partial t_2}, \frac{\partial^2 \log \tau_n}{\partial t_1 \partial s_1}\right\}_{s_1} = 0.$$

(2-3)

SKETCH OF PROOF OF THEOREM 2.1 AND ITS COROLLARY. The following double integral can be expanded in two different ways with regard to the parameters $a := (a_1, a_2, \dots)$:

$$\tau_n(t,s)\tau_{n+1}(t',s')$$

$$\iint_{\mathbb{R}^2} dx\, dy\, p^{(2)}_{n+1}(t',s';x)p^{(1)}_n(t,s;y)e^{\sum_1^\infty (t_i y^i - s_i' x^i)}\rho(x,y)\Bigg|_{\substack{t\mapsto t-a \\ t'\mapsto t+a \\ s'=s}}$$

$$= \left(\sum_{j=0}^\infty -2a_{j+1}S_j\left(\frac{\partial}{\partial t_1},\frac{1}{2}\frac{\partial}{\partial t_2},\frac{1}{3}\frac{\partial}{\partial t_3},\dots\right)\tau_{n+2}\circ\tau_n + O(a^2)\right)$$

$$= \left(\sum_{j=0}^\infty 2a_{j+1}\tau^2_{n+1}\frac{\partial^2}{\partial s_1\partial t_{j+1}}\log\tau_{n+1} + O(a^2)\right), \tag{2-4}$$

using the fact that the space $\mathcal{H} := \operatorname{span}\{z^i,\ i\in\mathbb{Z}\}$ can be equipped with two (formal) inner products:

(i) $\langle f,g\rangle = \displaystyle\int_{\mathbb{R}} f(z)g(z)\,dz,$

(ii) a residue pairing about $z=\infty$ between $f = \sum_{i\geq 0} a_i z^i \in \mathcal{H}^+$ and $h = \sum_{j\in\mathbb{Z}} b_j z^{-j-1} \in \mathcal{H}$:

$$\langle f,h\rangle_\infty = \oint_{z=\infty} f(z)h(z)\frac{dz}{2\pi i} = \sum_{i\geq 0} a_i b_i.$$

The two inner products are related by

$$\langle f,g\rangle = \int_{\mathbb{R}} f(z)g(z)\,dz = \left\langle f,\int_{\mathbb{R}}\frac{g(u)}{z-u}du\right\rangle_\infty.$$

Then the two expansions (2-4) are obtained, using the τ-function representation (2-1) of the biorthogonal polynomials, transforming the double integral (2-4) into a contour integral about ∞ and finally computing the residues. Upon equating the two series in (2-4) for arbitrary a_j, one finds the first identity (2-2). Application of a similar shift $s \mapsto s - a$, $s' \mapsto s + a$, $t' = t$ yields the second identity (2-2). Then combining the identities (2-2) for $j = 0$ and 1 leads to the PDE (2-3). $\qquad\square$

3. Orthogonal polynomials with regard to several weights and the n-component KP hierarchy

Now considering two sets of weights,

$$\psi_1,\dots,\psi_q \quad\text{and}\quad \varphi_1,\dots,\varphi_p,$$

and deform each weight with its own set of times:

$$\psi_k^{-s}(x) := \psi_k(x)e^{-\sum_1^\infty s_{ki}x^i} \quad\text{and}\quad \varphi_k^t(y) := \varphi_k(y)e^{\sum_1^\infty t_{ki}y^i},$$

the time parameters being

$$s_k = (s_{k1}, s_{k2}, \dots) \text{ for } 1 \le k \le q \quad \text{and} \quad t_k = (t_{k1}, t_{k2}, \dots) \text{ for } 1 \le k \le p.$$

Take a moment matrix consisting of $p \times q$ blocks of sizes $m_i \times n_j$, formed of moments with regard to all the different combinations of ψ_i and φ_j's; of course for the full matrix to be a square matrix, the integers $m_1, m_2, \dots \ge 0$ and $n_1, n_2, \dots \ge 0$ must satisfy $\sum_1^q m_i = \sum_1^p n_i$. Define the determinant τ_{mn} of these moment matrices (the inner product is the same as in Section 2):

$$\tau_{m_1,\dots,m_q;n_1,\dots,n_p}(s_1,\dots,s_q;t_1,\dots,t_p) :=$$

$$\det \begin{pmatrix} \left(\langle x^k \psi_1^{-s_1}(x) \mid y^\ell \varphi_1^{t_1}(y) \rangle \right)_{\substack{0 \le k < m_1 \\ 0 \le \ell < n_1}} & \cdots & \left(\langle x^k \psi_1^{-s_1}(x) \mid y^\ell \varphi_p^{t_p}(y) \rangle \right)_{\substack{0 \le k < m_1 \\ 0 \le \ell < n_p}} \\ \vdots & & \vdots \\ \left(\langle x^k \psi_q^{-s_q}(x) \mid y^\ell \varphi_1^{t_1}(y) \rangle \right)_{\substack{0 \le k < m_q \\ 0 \le \ell < n_1}} & \cdots & \left(\langle x^k \psi_q^{-s_q}(x) \mid y^\ell \varphi_p^{t_p}(y) \rangle \right)_{\substack{0 \le k < m_q \\ 0 \le \ell < n_p}} \end{pmatrix}.$$

$$(3\text{-}1)$$

Notice that Section 1 is a special case of this situation, where $p = q = 1$. In this general setup, the analogue of Theorem 2.1 is the following statement, due to [Adler et al. 2006]. (The precise signs $\pm$, which we omit here, can be found in that reference. The symbol e_α stands for $0, \dots, 0, 1, 0, \dots$), with 1 at the α-th place. The meaning of $\tau_{mn}(t_\ell - [z^{-1}])$ is that only the t_ℓ variable gets shifted and no other, i.e., reference to the unshifted variables is omitted.)

I. The expressions

$$z^{n_\ell} \frac{\tau_{mn}(t_\ell - [z^{-1}])}{\tau_{mn}} := Q_{mn}^{(\ell\ell)}(z) = z^{n_\ell} + \cdots,$$

$$z^{n_\alpha - 1} \frac{\tau_{m,n+e_\ell-e_\alpha}(t_\alpha - [z^{-1}])}{\tau_{mn}} = Q_{mn}^{(\ell\alpha)}(z) = c_\alpha z^{n_\alpha - 1} + \cdots \quad \text{for } \alpha \ne \ell$$

are polynomials (involving $\sum_1^p n_\alpha$ coefficients), satisfying $\sum_1^q m_\alpha$ orthogonality conditions

$$\left\langle x^j \psi_\alpha^{-s}(x) \,\middle|\, \sum_{i=1}^p Q_{mn}^{(\ell i)}(y) \varphi_i^t(y) \right\rangle = 0 \quad \text{for} \quad \begin{cases} 1 \le \alpha \le q, \\ 0 \le j \le m_\alpha - 1. \end{cases}$$

II. Similarly, the expressions

$$\pm z^{m_\alpha - 1} \frac{\tau_{m-e_\alpha, n-e_\ell}(s_\alpha + [z^{-1}])}{\tau_{mn}} = P_{nm}^{(\ell\alpha)}(z) \quad \text{of degree} < m_\alpha$$

are polynomials (involving $\sum_1^q m_\alpha$ coefficients), satisfying $\sum_1^p n_\alpha$ orthogonality conditions:

$$\left\langle \sum_{i=1}^q P_{nm}^{(\ell i)}(x)\psi_i^{-s}(x) \,\middle|\, y^j \varphi_\alpha^t(y) \right\rangle = 0 \quad \text{for} \begin{cases} 1 \le \alpha \le p, \ 0 \le j \le n_\alpha - 1 \\ \text{except } \alpha = \ell, \ j = n_\ell - 1, \end{cases}$$

$$\left\langle \sum_{i=1}^q P_{nm}^{(\ell i)}(x)\psi_i^{-s}(x) \,\middle|\, y^{n_\ell-1} \varphi_\ell^t(y) \right\rangle = 1.$$

III. The Cauchy transforms of the polynomials in II are

$$z^{-n_\ell} \frac{\tau_{mn}(t_\ell + [z^{-1}])}{\tau_{mn}} := \left\langle \sum_{i=1}^q P_{nm}^{(\ell i)}(x)\psi_i^{-s}(x) \,\middle|\, \frac{\varphi_\ell^t(y)}{z-y} \right\rangle,$$

$$\pm z^{-n_\ell-1} \frac{\tau_{m,n+e_\ell-e_\alpha}(t_\ell + [z^{-1}])}{\tau_{mn}} = \left\langle \sum_{i=1}^q P_{nm}^{(\alpha i)}(x)\psi_i^{-s}(x) \,\middle|\, \frac{\varphi_\ell^t(y)}{z-y} \right\rangle.$$

IV. The Cauchy transforms of the polynomials in I are

$$\pm z^{-m_\alpha-1} \frac{\tau_{m+e_\alpha,n+e_\ell}(s_\alpha - [z^{-1}])}{\tau_{mn}} = \left\langle \frac{\psi_\alpha^{-s}(x)}{z-x} \,\middle|\, \sum_{i=1}^p Q_{mn}^{(\ell i)}(y)\varphi_i^t(y) \right\rangle.$$

The orthogonality conditions for these polynomials lead to the following statement:

PROPOSITION 3.1. *The determinants τ_{mn} defined in (3-1) satisfy the $(p+q)$-KP hierarchy; that is,*

$$\sum_{\beta=1}^p \oint_\infty \tau_{m,n-e_\beta}(t_\beta-[z^{-1}])\tau_{m',n'+e_\beta}(t'_\beta+[z^{-1}])e^{\sum_1^\infty (t_{\beta i}-t'_{\beta i})z^i} z^{n_\beta-n'_\beta-2}\,dz =$$

$$\pm \sum_{\alpha=1}^q \oint_\infty \tau_{m+e_\alpha,n}(s_\alpha-[z^{-1}])\tau_{m'-e_\alpha,n'}(s'_\alpha+[z^{-1}])e^{\sum_1^\infty (s_{\alpha i}-s'_{\alpha i})z^i} z^{m'_\alpha-m_\alpha-2}\,dz,$$

where $\sum m'_\alpha = \sum n'_\alpha + 1$ and $\sum m_\alpha = \sum n_\alpha + 1$.

These polynomials happen to be the so-called multiple orthogonal polynomials of mixed type, introduced in [Daems and Kuijlaars 2007] in the context of nonintersecting Brownian motions; they generalize multiple orthogonal polynomials, introduced in [Aptekarev 1998; Aptekarev et al. 2003; Adler and van Moerbeke 1999a]. This will now be used in the next section.

4. Nonintersecting Brownian motions

If the transition density for standard Brownian motion $x(t)$ in $\mathbb{R}$, leaving from x and arriving at y, is given by

$$p(t, x, y) = \frac{1}{\sqrt{\pi t}} e^{-(x-y)^2/t},$$

then the probability that N nonintersecting Brownian motions $x_1(t), \ldots, x_N(t)$ in $\mathbb{R}$, leaving at $\alpha := (\alpha_1, \ldots, \alpha_N)$ and arriving at $\beta := (\beta_1, \ldots, \beta_N)$, belong to E at time t, is given by the Karlin–McGregor formula [1959]:

$$\int_{E^N} \det \left[p(t, \alpha_i, x_j) \right]_{1 \leq i, j \leq N} \det \left[p(1-t, x_i, \beta_j) \right]_{1 \leq i, j \leq N} \prod_{i=1}^{N} dx_i.$$

Considering the particular case where several points coincide, i.e., where

$$\alpha := a = (\overbrace{a_1, a_1, \ldots, a_1}^{m_1}, \overbrace{a_2, a_2, \ldots, a_2}^{m_2}, \ldots, \overbrace{a_q, a_q, \ldots, a_q}^{m_q}) \in \mathbb{R}^N$$

$$\beta := b = (\overbrace{b_1, b_1, \ldots, b_1}^{n_1}, \overbrace{b_2, b_2, \ldots, b_2}^{n_2}, \ldots, \overbrace{b_p, b_p, \ldots, b_p}^{n_p}) \in \mathbb{R}^N,$$

(4-1)

one verifies that the probability below can be expressed as a determinant of a moment matrix of the form (3-1) with $p \times q$ blocks,

$$\mathbb{P} \left(\text{all } x_i(t) \in E \, \middle| \, \begin{array}{l} (x_1(0), \ldots, x_N(0)) = \alpha \\ (x_1(1), \ldots, x_N(1)) = \beta \end{array} \right) \quad (0 < t < 1)$$

$$= \lim_{\substack{(\alpha_1, \ldots, \alpha_N) \to a \\ (\beta_1, \ldots, \beta_N) \to b}}$$

$$\frac{1}{Z_N} \int_{E^N} \det[p(t, \alpha_i, x_j)]_{1 \leq i, j \leq N} \det[p(1-t, x_i, \beta_j)]_{1 \leq i, j \leq N} \prod_{i=1}^{N} dx_i$$

$$= \frac{N!}{Z'_N} \det \left(\left(\int_{\tilde{E}} dy \, e^{-\frac{y^2}{2}} y^{i+j} e^{(\tilde{a}_\alpha + \tilde{b}_\beta) y} \right)_{\substack{0 \leq i < m_\alpha \\ 0 \leq j < n_\beta}} \right)_{\substack{1 \leq \alpha \leq q \\ 1 \leq \beta \leq p}}, \quad (4\text{-}2)$$

where

$$\tilde{E} = E \sqrt{\frac{2}{t(1-t)}}, \quad \tilde{a}_i = \sqrt{\frac{2(1-t)}{t}} a_i, \quad \tilde{b}_i = \sqrt{\frac{2t}{1-t}} b_i.$$

PROOF. It is based on the matrix identity

$$\det (A_{ik})_{1 \leq i, k \leq n} \det (B_{ik})_{1 \leq i, k \leq n} = \sum_{\sigma \in S_n} \det \left(A_{i,\sigma(j)} \, B_{j,\sigma(j)} \right)_{1 \leq i, j \leq n}. \qquad \square$$

Upon adding extra-time parameters

$$t_\beta = (t_{\beta,1}, t_{\beta,2}, \dots) \quad \text{and} \quad s_\alpha = (s_{\alpha,1}, s_{\alpha,2}, \dots)$$

to

$$\det\left(\left(\int_{\tilde{E}} dy \, e^{-\frac{y^2}{2}} y^{i+j} e^{(\tilde{a}_\alpha + \tilde{b}_\beta)y}\right)_{\substack{0 \le i < m_\alpha \\ 0 \le j < n_\beta}}\right)_{\substack{1 \le \alpha \le q \\ 1 \le \beta \le p}},$$

it follows automatically from Section 3 that the expression

$$\tau_{m_1,\dots,m_q;n_1,\dots,n_p}(t_1,\dots,t_p;s_1,\dots,s_q)$$

$$= \det\left(\left(\int_{\tilde{E}} dy \, e^{-\frac{y^2}{2}} y^{i+j} e^{(\tilde{a}_\alpha + \tilde{b}_\beta)y + \sum_1^\infty (t_{\beta,k} - s_{\alpha,k})y^k}\right)_{\substack{0 \le i < m_\alpha \\ 0 \le j < n_\beta}}\right)_{\substack{1 \le \alpha \le q \\ 1 \le \beta \le p}}$$

satisfies the $p + q$-KP hierarchy, where p denotes the number of starting points and q the number of end points of the Brownian motions; see (4-1). Nonintersecting Brownian motions have been studied in [Karlin and McGregor 1959; Dyson 1962; Grabiner 1999; Johansson 2001; Bleher and Kuijlaars 2004b; 2004a; Daems and Kuijlaars 2007; Tracy and Widom 2004; 2006; Adler and van Moerbeke 2005; 2006].

In the next section, I work out the example where the Brownian motions all depart from 0 and end up at the points $-a$ and a.

5. Nonintersecting Brownian motions leaving from the origin and forced to end up at two points

Consider $n = n_1 + n_2$ nonintersecting Brownian motions on $\mathbb{R}$, all leaving from the origin, with n_1 paths forced to go to $-a$ and n_2 paths forced to go to a, at time $t = 1$. The probability that all the particles belong to the set E at time $0 < t < 1$ can be expressed as a Gaussian Hermitian random matrix "with external potential", specified by the diagonal matrix

$$A := \begin{pmatrix} \alpha & & & & \\ & \ddots & & \mathbf{O} & \\ & & \alpha & & \\ & & & -\alpha & \\ & \mathbf{O} & & & \ddots \\ & & & & & -\alpha \end{pmatrix} \begin{matrix} \updownarrow n_1 \\ \\ \updownarrow n_2 \end{matrix} \quad \text{with} \quad \alpha = a\sqrt{\frac{2t}{1-t}},$$

but also as a determinant of a moment matrix, a consequence of Section 4. This gives (with $n = n_1 + n_2$),

$$\mathbb{P}_0^{\pm a}\left(\text{all } x_i(t) \in E \left|\begin{array}{l} \text{all } x_j(0) = 0, \\ n_1 \text{ left paths end up at } -a \text{ at time } t = 1, \\ n_2 \text{ right paths end up at } +a \text{ at time } t = 1 \end{array}\right.\right)$$

$$= \mathbb{P}_n\left(a\sqrt{\frac{2t}{1-t}}; E\sqrt{\frac{2}{t(1-t)}}\right), \quad (5\text{-}1)$$

with $\mathbb{P}_n$ being an integral over the space $\mathcal{H}_n(E')$ of Hermitian matrices with spectrum belonging to the set $E' \subseteq \mathbb{R}$:

$$\mathbb{P}_n(\alpha; E') := \frac{1}{Z_n}\int_{\mathcal{H}_n(E')} dM \, e^{-\text{Tr}(\frac{1}{2}M^2 - AM)}$$

$$= \frac{1}{Z_n}\det\left(\begin{array}{l}\left(\displaystyle\int_{E'} z^{i+j-1}e^{-z^2/2+\alpha z}dz\right)_{\substack{1\le i\le n_1, \\ 1\le j\le n_1+n_2}} \\[3mm] \left(\displaystyle\int_{E'} z^{i+j-1}e^{-z^2/2-\alpha z}dz\right)_{\substack{1\le i\le n_2, \\ 1\le j\le n_1+n_2}}\end{array}\right) \quad (5\text{-}2)$$

THEOREM 5.1 [Adler and van Moerbeke 2007]. *The log of the probability* $\mathbb{P}_n(\alpha; E)$ *satisfies a fourth-order PDE in* α *and in the boundary points* $b_1, \ldots,$ b_{2r} *of the set* E, *with quartic nonlinearity*:

$$\det\begin{pmatrix} F^+ & F^- & 0 \\ \mathcal{B}_{-1}F^+ & \mathcal{B}_{-1}F^- & F^-G^+ + F^+G^- \\ \mathcal{B}_{-1}^2F^+ & \mathcal{B}_{-1}^2F^- & F^-\mathcal{B}_{-1}G^+ + F^+\mathcal{B}_{-1}G^- \end{pmatrix} = 0, \quad (5\text{-}3)$$

where $\mathcal{B}_k := \sum_{i=1}^{2r} b_i^{k+1}\partial/\partial b_i$ *and*

$$F^+ := 2\mathcal{B}_{-1}\left(\frac{\partial}{\partial\alpha} - \mathcal{B}_{-1}\right)\log\mathbb{P}_n - 4n_1, \qquad F^- = F^+\big|_{\substack{\alpha\to-\alpha \\ n_1\leftrightarrow n_2}}$$

$$2G^+ := \{H_1^+, F^+\}_{\mathcal{B}_{-1}} - \{H_2^+, F^+\}_{\partial/\partial\alpha}, \qquad G^- = G^+\big|_{\substack{\alpha\to-\alpha, \\ n_1\leftrightarrow n_2}}$$

with

$$H_1^+ := \frac{\partial}{\partial\alpha}\left(\mathcal{B}_0 - \alpha\frac{\partial}{\partial\alpha} - \alpha\mathcal{B}_{-1}\right)\log\mathbb{P}_n + \left(\mathcal{B}_0\mathcal{B}_{-1} + 4\frac{\partial}{\partial\alpha}\right)\log\mathbb{P}_n + 4n_1\left(\alpha + \frac{n_2}{\alpha}\right),$$

$$H_2^+ := \frac{\partial}{\partial\alpha}\left(\mathcal{B}_0 - \alpha\frac{\partial}{\partial\alpha} - \alpha\mathcal{B}_{-1}\right)\log\mathbb{P}_n - (\mathcal{B}_0 - 2\alpha\mathcal{B}_{-1} - 2)\,\mathcal{B}_{-1}\log\mathbb{P}_n.$$

SKETCH OF PROOF. In view of the results in Section 3, we add extra parameters $t_1, t_2, \ldots, s_1, s_2, \ldots$ and β to the integrals in the moment matrix above (5-2). In terms of the Vandermonde determinants $\Delta_k(x) = \prod_{1\le i<j\le k}(x_i - x_j)$ for the

variables $x_1, \ldots, x_{n_1}$ and $\Delta_n(x, y)$ for all variables $x_1, \ldots, x_{n_1}, y_1, \ldots, y_{n_2}$, we obtain from the results in Section 4 that (again with $n = n_1 + n_2$) the function

$$\tau_{n_1 n_2}(t, s, u; \alpha, \beta; E) := \det \begin{pmatrix} (\mu_{ij}^+(t, s, \alpha, \beta, E))_{1 \le i \le n_1,\, 1 \le j \le n_1 + n_2} \\ (\mu_{ij}^-(t, u, \alpha, \beta, E))_{1 \le i \le n_2,\, 1 \le j \le n_1 + n_2} \end{pmatrix}$$

$$= \frac{1}{n_1! \, n_2!} \int_{E^n} \Delta_n(x, y) \prod_{j=1}^{n_1} e^{\sum_1^\infty t_i x_j^i} \prod_{j=1}^{n_2} e^{\sum_1^\infty t_i y_j^i}$$

$$\times \left(\Delta_{n_1}(x) \prod_{j=1}^{n_1} e^{-x_j^2/2 + \alpha x_j + \beta x_j^2} e^{-\sum_1^\infty s_i x_j^i} dx_j \right)$$

$$\times \left(\Delta_{n_2}(y) \prod_{j=1}^{n_2} e^{-y_j^2/2 - \alpha y_j - \beta y_j^2} e^{-\sum_1^\infty u_i y_j^i} dy_j \right) \quad (5\text{-}4)$$

satisfies the 3-component KP equation, since $p + q = 2 + 1 = 3$, since this matrix corresponds to $p = 2$, $q = 1$. The function $\tau_{n_1 n_2}(t, s, u; \alpha, \beta; E)$ also satisfies Virasoro constraints, to be explained below.

(i)   The three-component KP bilinear equations of Proposition 3.1 imply, using a standard residue computation on the bilinear equation (equations of the type (2-2) for $j = 0$ and $j = 1$, except that the three-component KP bilinear equations give rise to τ-functions depending on two integer indices)

$$\frac{\partial^2 \log \tau_{n_1, n_2}}{\partial t_1 \partial s_1} = -\frac{\tau_{n_1+1, n_2} \tau_{n_1-1, n_2}}{\tau_{n_1, n_2}^2} \quad (5\text{-}5)$$

and

$$\frac{\partial}{\partial t_1} \log \frac{\tau_{n_1+1, n_2}}{\tau_{n_1-1, n_2}} = \frac{(\partial^2/\partial t_2 \partial s_1) \log \tau_{n_1, n_2}}{(\partial^2/\partial t_1 \partial s_1) \log \tau_{n_1, n_2}} \quad (5\text{-}6)$$

$$-\frac{\partial}{\partial s_1} \log \frac{\tau_{n_1+1, n_2}}{\tau_{n_1-1, n_2}} = \frac{(\partial^2/\partial t_1 \partial s_2) \log \tau_{n_1, n_2}}{(\partial^2/\partial t_1 \partial s_1) \log \tau_{n_1, n_2}}. \quad (5\text{-}7)$$

(ii)   The Virasoro equations are as follows: The integral $\tau_{n_1 n_2}(t, s, u; \alpha, \beta; E)$, as defined in (5-4), satisfies

$$\mathcal{B}_m \tau_{n_1, n_2} = \mathbb{V}_m^{n_1, n_2} \tau_{n_1, n_2} \quad \text{for} \quad m \ge -1, \quad (5\text{-}8)$$

where $\mathcal{B}_m$ and $\mathbb{V}_m$ are differential operators:

$$\mathcal{B}_m = \sum_1^{2r} b_i^{m+1} \frac{\partial}{\partial b_i}, \quad \text{for} \quad E = \bigcup_1^{2r} [b_{2i-1}, b_{2i}] \subset \mathbb{R}$$

and (with the convention that t_i is omitted whenever it appears for $i = 0, -1, \ldots$)

$$
\mathbb{V}_m^{n_1 n_2} := \frac{1}{2} \sum_{i+j=m} \left(\frac{\partial^2}{\partial t_i \partial t_j} + \frac{\partial^2}{\partial s_i \partial s_j} + \frac{\partial^2}{\partial u_i \partial u_j} \right)
$$

$$
+ \sum_{i \geq 1} \left(i\, t_i \frac{\partial}{\partial t_{i+m}} + i\, s_i \frac{\partial}{\partial s_{i+m}} + i\, u_i \frac{\partial}{\partial u_{i+m}} \right)
$$

$$
+ (n_1 + n_2) \left(\frac{\partial}{\partial t_m} + (-m) t_{-m} \right) - n_1 \left(\frac{\partial}{\partial s_m} + (-m) s_{-m} \right)
$$

$$
- n_2 \left(\frac{\partial}{\partial u_m} + (-m) u_{-m} \right) + (n_1^2 + n_1 n_2 + n_2^2) \delta_{m0}
$$

$$
+ \alpha(n_1 - n_2) \delta_{m+1,0} + \frac{m(m+1)}{2} (t_{-m} + s_{-m} + u_{-m})
$$

$$
- \frac{\partial}{\partial t_{m+2}} + \alpha \left(- \frac{\partial}{\partial s_{m+1}} + \frac{\partial}{\partial u_{m+1}} + (m+1)(s_{-m+1} - u_{-m+1}) \right)
$$

$$
+ 2\beta \left(\frac{\partial}{\partial u_{m+2}} - \frac{\partial}{\partial s_{m+2}} \right).
$$

These Virasoro equations are obtained by setting

$$
x_i \mapsto x_i + \varepsilon x_i^{m+1},
$$
$$
y_i \mapsto y_i + \varepsilon y_i^{m+1}
$$

in the integral (5-4) and observing that this substitution does not change the value of the integral, provided the boundary is changed infinitesimally as well.

The Virasoro constraints (5-8) above for $m = -1$ and $m = 0$ lead to the following equations for $f = \log \tau_{n_1 n_2}(t, s, u; \alpha, \beta; E)$ along the locus $\mathcal{L}$ of points where $t = s = u = 0$, $\beta = 0$:

$$
\frac{\partial f}{\partial t_1} = -\mathcal{B}_{-1} f + \alpha(n_1 - n_2),
$$

$$
\frac{\partial f}{\partial s_1} = \frac{1}{2} \left(\mathcal{B}_{-1} - \frac{\partial}{\partial \alpha} \right) f + \frac{\alpha}{2} (n_2 - n_1),
$$

$$
2 \frac{\partial^2 f}{\partial t_1 \partial s_1} = \mathcal{B}_{-1} \left(\frac{\partial}{\partial \alpha} - \mathcal{B}_{-1} \right) f - 2 n_1,
$$

$$
2 \frac{\partial^2 f}{\partial t_1 \partial s_2} = \left(\alpha \frac{\partial}{\partial \alpha} + \frac{\partial}{\partial \beta} - \mathcal{B}_0 + 1 \right) \mathcal{B}_{-1} f - 2 \frac{\partial f}{\partial \alpha} - 2\alpha(n_1 - n_2),
$$

$$
2 \frac{\partial^2 f}{\partial t_2 \partial s_1} = \frac{\partial}{\partial \alpha} (\mathcal{B}_0 - \alpha \frac{\partial}{\partial \alpha} + \alpha \mathcal{B}_{-1}) f - \mathcal{B}_{-1}(\mathcal{B}_0 - 1) f - 2\alpha(n_1 - n_2). \quad (5\text{-}9)
$$

From the differential equations (5-6)–(5-7) and from the two first two Virasoro equations (5-9) it follows that, along the locus $\mathcal{L}$, and for the indices $n_1 \pm 1, n_2$,

$$
\frac{\dfrac{\partial^2}{\partial t_2 \partial s_1} \log \tau_{n_1 n_2}}{\dfrac{\partial^2}{\partial t_1 \partial s_1} \log \tau_{n_1 n_2}} = \frac{\partial}{\partial t_1} \log \frac{\tau_{n_1+1,n_2}}{\tau_{n_1-1,n_2}} = -\mathcal{B}_{-1} \log \frac{\tau_{n_1+1,n_2}}{\tau_{n_1-1,n_2}} + 2\alpha,
$$

$$
-\frac{\dfrac{\partial^2}{\partial t_1 \partial s_2} \log \tau_{n_1 n_2}}{\dfrac{\partial^2}{\partial t_1 \partial s_1} \log \tau_{n_1 n_2}} = \frac{\partial}{\partial s_1} \log \frac{\tau_{n_1+1,n_2}}{\tau_{n_1-1,n_2}} = \frac{1}{2}\left(\mathcal{B}_{-1} - \frac{\partial}{\partial \alpha}\right) \log \frac{\tau_{n_1+1,n_2}}{\tau_{n_1-1,n_2}} - \alpha.
$$

From these two equations, the logarithmic expression on the right can be eliminated, by acting on the first equation with the operator $\frac{1}{2}\left(\mathcal{B}_{-1} - (\partial/\partial\alpha)\right)$ and on the second with $-\mathcal{B}_{-1}$ and subtracting, thus yielding

$$
\frac{1}{2}\left(\mathcal{B}_{-1} - \frac{\partial}{\partial \alpha}\right)\left(\frac{\dfrac{\partial^2}{\partial t_2 \partial s_1} \log \tau_{n_1 n_2}}{\dfrac{\partial^2}{\partial t_1 \partial s_1} \log \tau_{n_1 n_2}} - 2\alpha\right) = \mathcal{B}_{-1}\left(\frac{\dfrac{\partial^2}{\partial t_1 \partial s_2} \log \tau_{n_1 n_2}}{\dfrac{\partial^2}{\partial t_1 \partial s_1} \log \tau_{n_1 n_2}} - \alpha\right)
$$

or, equivalently,

$$
\mathcal{B}_{-1}\left(\frac{\left(\dfrac{\partial^2}{\partial t_2 \partial s_1} - 2\dfrac{\partial^2}{\partial t_1 \partial s_2}\right) \log \tau_{n_1 n_2}}{\dfrac{\partial^2}{\partial t_1 \partial s_1} \log \tau_{n_1 n_2}}\right)
$$

$$
- \frac{\partial}{\partial \alpha}\left(\frac{\left(\dfrac{\partial^2}{\partial t_2 \partial s_1} - 2\alpha\dfrac{\partial^2}{\partial t_1 \partial s_1}\right) \log \tau_{n_1 n_2}}{\dfrac{\partial^2}{\partial t_1 \partial s_1} \log \tau_{n_1 n_2}}\right) = 0. \quad (5\text{-}10)
$$

Using the remaining Virasoro relations (5-9), one obtains along $\mathcal{L}$ the equalities

$$
4\frac{\partial^2}{\partial t_1 \partial s_1} \log \tau_{n_1 n_2} = F^+, \quad 2\left(\frac{\partial^2}{\partial t_2 \partial s_1} - 2\alpha\frac{\partial^2}{\partial t_1 \partial s_1}\right) \log \tau_{n_1 n_2} = H_2^+,
$$

$$
2\left(\frac{\partial^2}{\partial t_2 \partial s_1} - 2\frac{\partial^2}{\partial t_1 \partial s_2}\right) \log \tau_{n_1 n_2} = H_1^+ - 2\mathcal{B}_{-1}\frac{\partial}{\partial \beta} \log \tau_{n_1 n_2}
$$

where we have set[1]

$$
F^+ := 2\mathcal{B}_{-1}\left(\frac{\partial}{\partial \alpha} - \mathcal{B}_{-1}\right) \log \tau_{n_1 n_2} - 4n_1 = 2\mathcal{B}_{-1}\left(\frac{\partial}{\partial \alpha} - \mathcal{B}_{-1}\right) \log \mathbb{P}_n - 4n_1,
$$

---

[1] One checks that $\tau_{n_1 n_2}(t, s, u; \alpha, \beta, \mathbb{R})|_{\mathcal{L}} = (-2)^{n_1 n_2} (2\pi)^{-\frac{n_1+n_2}{2}} \prod_0^{n_1-1} j! \prod_0^{n_2-1} j! \, \alpha^{n_1 n_2} e^{-\frac{n_1+n_2}{2}\alpha^2}$.

$$H_1^+ := \frac{\partial}{\partial \alpha}\left(\mathcal{B}_0 - \alpha\frac{\partial}{\partial \alpha} - \alpha\mathcal{B}_{-1}\right)\log \tau_{n_1 n_2} + \left(\mathcal{B}_0\mathcal{B}_{-1} + 4\frac{\partial}{\partial \alpha}\right)\log \tau_{n_1 n_2} + 2\alpha(n_1 - n_2)$$

$$= \frac{\partial}{\partial \alpha}\left(\mathcal{B}_0 - \alpha\frac{\partial}{\partial \alpha} - \alpha\mathcal{B}_{-1}\right)\log \mathbb{P}_n + \left(\mathcal{B}_0\mathcal{B}_{-1} + 4\frac{\partial}{\partial \alpha}\right)\log \mathbb{P}_n + 4\alpha n_1 + \frac{4n_1 n_2}{\alpha}$$

$$H_2^+ := \frac{\partial}{\partial \alpha}\left(\mathcal{B}_0 - \alpha\frac{\partial}{\partial \alpha} - \alpha\mathcal{B}_{-1}\right)\log \tau_{n_1 n_2} + (2\alpha\mathcal{B}_{-1} - \mathcal{B}_0 + 2)\mathcal{B}_{-1}\log \tau_{n_1 n_2} + 2\alpha(n_1 + n_2)$$

$$= \frac{\partial}{\partial \alpha}\left(\mathcal{B}_0 - \alpha\frac{\partial}{\partial \alpha} - \alpha\mathcal{B}_{-1}\right)\log \mathbb{P}_n + (2\alpha\mathcal{B}_{-1} - \mathcal{B}_0 + 2)\mathcal{B}_{-1}\log \mathbb{P}_n.$$

Further define

$$F^- = F^+\big|_{\substack{\alpha \to -\alpha \\ n_1 \leftrightarrow n_2}}, \qquad H_i^- = H_i^+\big|_{\substack{\alpha \to -\alpha \\ n_1 \leftrightarrow n_2}}.$$

With this notation, equation 5-10 becomes

$$\left\{\mathcal{B}_{-1}\frac{\partial}{\partial \beta}\log \tau_{n_1 n_2}\Big|_{\mathcal{L}}, F^+\right\}_{\mathcal{B}_{-1}} = \left\{H_1^+, \tfrac{1}{2}F^+\right\}_{\mathcal{B}_{-1}} - \left\{H_2^+, \tfrac{1}{2}F^+\right\}_{\partial/\partial\alpha} =: G^+,$$

yielding automatically a second equation, using the involution $\alpha \mapsto -\alpha$, $\beta \mapsto -\beta$, $n_1 \leftrightarrow n_2$ (which leaves (5-4) unchanged):

$$-\left\{\mathcal{B}_{-1}\frac{\partial}{\partial \beta}\log \tau_{n_1 n_2}\Big|_{\mathcal{L}}, F^-\right\}_{\mathcal{B}_{-1}} = \left\{H_1^-, \tfrac{1}{2}F^-\right\}_{\mathcal{B}_{-1}} - \left\{H_2^-, \tfrac{1}{2}F^-\right\}_{-\partial/\partial\alpha} =: G^-.$$

The last two displays yield a linear system of equations in

$$\mathcal{B}_{-1}\frac{\partial \log \tau_{n_1 n_2}}{\partial \beta}\Big|_{\mathcal{L}} \quad \text{and} \quad \mathcal{B}_{-1}^2\frac{\partial \log \tau_{n_1 n_2}}{\partial \beta}\Big|_{\mathcal{L}}$$

from which

$$\mathcal{B}_{-1}\frac{\partial \log \tau_{n_1 n_2}}{\partial \beta}\Big|_{\mathcal{L}} = \frac{G^- F^+ + G^+ F^-}{-F^-(\mathcal{B}_{-1}F^+) + F^+(\mathcal{B}_{-1}F^-)},$$

$$\mathcal{B}_{-1}^2\frac{\partial \log \tau_{n_1 n_2}}{\partial \beta}\Big|_{\mathcal{L}} = \frac{G^-(\mathcal{B}_{-1}F^+) + G^+(\mathcal{B}_{-1}F^-)}{-F^-(\mathcal{B}_{-1}F^+) + F^+(\mathcal{B}_{-1}F^-)}.$$

Subtracting the second equation from $\mathcal{B}_{-1}$ of the first equation yields the differential equation

$$\left(F^+\mathcal{B}_{-1}G^- + F^-\mathcal{B}_{-1}G^+\right)\left(F^+\mathcal{B}_{-1}F^- - F^-\mathcal{B}_{-1}F^+\right)$$
$$- \left(F^+G^- + F^-G^+\right)\left(F^+\mathcal{B}_{-1}^2 F^- - F^-\mathcal{B}_{-1}^2 F^+\right) = 0, \quad (5\text{-}11)$$

which can be rewritten as

$$F^+ F^- \det\begin{pmatrix} F^+ & F^- & 0 \\[4pt] \mathcal{B}_{-1}F^+ & \mathcal{B}_{-1}F^- & \dfrac{G^+}{F^+} + \dfrac{G^-}{F^-} \\[8pt] \mathcal{B}_{-1}^2 F^+ & \mathcal{B}_{-1}^2 F^- & \dfrac{\mathcal{B}_{-1}G^+}{F^+} + \dfrac{\mathcal{B}_{-1}G^-}{F^-} \end{pmatrix} = 0, \quad (5\text{-}12)$$

establishing (5-3) for $\log \mathbb{P}_n$.   □

6. The Pearcey process

As in section 5, consider $n = 2k$ nonintersecting Brownian motions on $\mathbb{R}$ (Dyson's Brownian motions), all starting at the origin, such that the k left paths end up at $-a$ and the k right paths end up at $+a$ at time $t = 1$.

Also as observed in section 5, the transition probability can be expressed in terms of the Gaussian Hermitian random matrix probability $\mathbb{P}_n(\alpha; E)$ with external source, for which the PDE (5-3) was deduced.

Let now the number $n = 2k$ of particles go to infinity, and let the points a and $-a$, properly rescaled, go to $\pm\infty$. This forces the left k particles to $-\infty$ at $t = 1$ and the right k particles to $+\infty$ at $t = 1$. Since the particles all leave from the origin at $t = 0$, it is natural to believe that for small times the equilibrium measure (mean density of particles) is supported by one interval, and for times close to 1, the equilibrium measure is supported by two intervals. With a precise scaling, $t = 1/2$ is critical in the sense that for $t < 1/2$, the equilibrium measure for the particles is supported by one, and for $t > 1/2$, it is supported by two intervals. The Pearcey process $\mathcal{P}(t)$ is now defined [Tracy and Widom 2006] as the motion of an infinite number of nonintersecting Brownian paths, just around time $t = 1/2$ near $x = 0$, with the precise scaling (upon introducing the scaling parameter z):

$$n = 2k = \frac{2}{z^4}, \quad \pm a = \pm\frac{1}{z^2}, \quad x_i \mapsto x_i z, \quad t \mapsto \frac{1}{2} + tz^2, \quad \text{for } z \to 0. \quad (6\text{-}1)$$

The Pearcey process has also arisen in the context of various growth models [Okounkov and Reshitikhin 2005]. Even though the pathwise interpretation of $\mathcal{P}(t)$ is unclear and deserves investigation, it is natural to define the following probability for $t \in \mathbb{R}$, in terms of the probability (5-1),

$$\mathbb{P}(\mathcal{P}(t) \cap E = \varnothing) := \lim_{z \to 0} \mathbb{P}_0^{\pm 1/z^2} \left(\text{all } x_j \left(\tfrac{1}{2} + tz^2\right) \notin zE; \ 1 \leq j \leq n \right)\Big|_{n=2/z^4}.$$

The results of Brézin and Hikami [1996; 1997; 1998b; 1998a] for the Pearcey kernel and Tracy and Widom [2006] for the extended kernels show that this limit exists and equals a Fredholm determinant:

$$\mathbb{P}(\mathcal{P}(t) \cap E = \varnothing) = \det\left(I - K_t \chi_E\right),$$

where $K_t(x, y)$ is the Pearcey kernel, defined as follows:

$$K_t(x, y) := \frac{p(x)q''(y) - p'(x)q'(y) + p''(x)q(y) - tp(x)q(y)}{x - y}$$

$$= \int_0^\infty p(x + z)q(y + z)\, dz, \quad (6\text{-}2)$$

where (note that $\omega = e^{i\pi/4}$)

$$p(x) := \frac{1}{2\pi} \int_{-\infty}^{\infty} e^{-u^4/4 - tu^2/2 - iux} du,$$

$$q(y) := \frac{1}{2\pi i} \int_X e^{u^4/4 - tu^2/2 + uy} du$$

$$= \mathrm{Im}\left(\frac{\omega}{\pi} \int_0^{\infty} du\, e^{-u^4/4 - (it/2)u^2} \left(e^{\omega uy} - e^{-\omega uy} \right) \right)$$

satisfy the differential equations

$$p''' - tp' - xp = 0 \quad \text{and} \quad q''' - tq' + yq = 0.$$

The contour X is given by the ingoing rays from $\pm\infty e^{i\pi/4}$ to 0 and the outgoing rays from 0 to $\pm\infty e^{-i\pi/4}$, i.e., X stands for the contour

For compact $E = \bigcup_{i=1}^r [x_{2i-1}, x_{2i}] \subset \mathbb{R}$, define the gradient and the Euler operator with regard to the boundary points of E,

$$\mathcal{B}_{-1} = \sum_1^{2r} \frac{\partial}{\partial x_i}, \quad \mathcal{B}_0 = \sum_1^{2r} x_i \frac{\partial}{\partial x_i}. \tag{6-3}$$

THEOREM 6.1 [Adler and van Moerbeke 2007].

$$Q(t; x_1, \ldots, x_{2r}) := \log \mathbb{P}\left(\mathcal{P}(t) \cap E = \varnothing \right) = \log \det \left(I - K_t \chi_E \right) \tag{6-4}$$

satisfies a fourth-order, third-degree PDE, which can be written as a single Wronskian:

$$\left\{ \frac{1}{2} \frac{\partial^3 Q}{\partial t^3} + (\mathcal{B}_0 - 2)\mathcal{B}_{-1}^2 Q + \frac{1}{16} \left\{ \mathcal{B}_{-1} \frac{\partial Q}{\partial t}, \mathcal{B}_{-1}^2 Q \right\}_{\mathcal{B}_{-1}}, \ \mathcal{B}_{-1}^2 \frac{\partial Q}{\partial t} \right\}_{\mathcal{B}_{-1}} = 0. \tag{6-5}$$

REMARK. A similar PDE can be written for the transition probability involving several times; see [Adler and van Moerbeke 2006]. Such equations can be used to compute the asymptotic behavior of the Pearcey process for $t \to -\infty$.

SKETCH OF PROOF. Consider the function $Q_z(s; x_1, \ldots, x_{2r})$, defined in terms of the probabilities $\mathbb{P}_0^{\pm a}$, defined in (5-1) and $\mathbb{P}_n$, defined in (5-2), as follows:

$$Q_z(s; x_1, \ldots, x_{2r}) := \log \mathbb{P}_0^{\pm a}(t; b_1, \ldots, b_{2r}) \Big|_{\substack{n=2/z^4,\ a=1/z^2, \\ b_i = x_i z,\ t = \frac{1}{2} + sz^2}}$$

$$= \log \mathbb{P}_n\left(a\sqrt{\frac{2t}{1-t}}\,;\, b_1\sqrt{\frac{2}{t(1-t)}},\,\ldots,\,b_{2r}\sqrt{\frac{2}{t(1-t)}}\,\right)\Bigg|_{\substack{n=2/z^4,\ a=1/z^2,\\ b_i=x_i z,\ t=\frac{1}{2}+sz^2}}$$

$$= \log \mathbb{P}_{2/z^4}\left(\frac{\sqrt{2}}{z^2}\sqrt{\frac{\frac{1}{2}+sz^2}{\frac{1}{2}-sz^2}}\,;\, \frac{x_1 z\sqrt{2}}{\sqrt{\frac{1}{4}-s^2z^4}},\,\ldots,\, \frac{x_{2r}z\sqrt{2}}{\sqrt{\frac{1}{4}-s^2z^4}}\right),$$

from which it follows, by inversion, that

$$Q_z\left(\frac{u^2z^4-2}{2z^2\left(u^2z^4+2\right)}\,;\, \frac{v_1 uz}{u^2z^4+2},\,\ldots,\, \frac{v_{2r}uz}{u^2z^4+2}\right)$$

$$= \log \mathbb{P}_{2/z^4}(u; v_1,\ldots,v_{2r}). \quad (6\text{-}6)$$

This expression satisfies the PDE (5-3), with α and $b_1,\ldots,b_{2r}$ replaced by u and $v_1,\ldots,v_{2r}$. Therefore all the partials of $\log \mathbb{P}$ with regard to these variables u and $v_1,\ldots,v_r$, as appears in the PDE (5-3), can be expressed, by virtue of (6-6), by partials of Q_z with regard to s and $x_1,\ldots,x_{2r}$.

For this, we need to compute the expressions $F^{\pm}, \tilde{\mathcal{B}}_{-1}F^{\pm}, \tilde{\mathcal{B}}^2_{-1}F^{\pm}, G^{\pm}$ and $\tilde{\mathcal{B}}_{-1}G^{\pm}$ appearing in (5-3) (where we use tildes in contrast to the operators defined in (6-3)), in terms of

$$Q_z(s; x_1,\ldots,x_{2r})$$

$$= \log \mathbb{P}_{2/z^4}\left(\frac{\sqrt{2}}{z^2}\sqrt{\frac{\frac{1}{2}+sz^2}{\frac{1}{2}-sz^2}}\,;\, x_1\frac{z\sqrt{2}}{\sqrt{\frac{1}{4}-s^2z^4}},\,\ldots,\, x_{2r}\frac{z\sqrt{2}}{\sqrt{\frac{1}{4}-s^2z^4}}\right)$$

$$= Q(s; x_1,\ldots,x_{2r}) + O(z), \quad (6\text{-}7)$$

with

$$Q(s; x_1,\ldots,x_{2r}) = \log \det\left(I - K_s \chi_{E^c}\right). \quad (6\text{-}8)$$

Without taking the limit $z \to 0$ on $Q_z(s; x_1,\ldots,x_{2r})$ yet, one computes, upon setting $\varepsilon := \pm$,

$$F^{\varepsilon} = -\frac{4}{z^4} - \frac{1}{4z^2}\mathcal{B}^2_{-1}Q_z + \frac{\varepsilon}{4z}\mathcal{B}_{-1}\frac{\partial Q_z}{\partial s} + O(z),$$

$$\frac{1}{\sqrt{2}}\tilde{\mathcal{B}}_{-1}F^{\varepsilon} = -\frac{1}{16z^3}\mathcal{B}^3_{-1}Q_z + \frac{\varepsilon}{16z^2}\mathcal{B}^2_{-1}\frac{\partial Q_z}{\partial s} - \frac{\varepsilon s}{8}\mathcal{B}^2_{-1}\frac{\partial Q_z}{\partial s} + O(z),$$

$$\tilde{\mathcal{B}}^2_{-1}F^{\varepsilon} = -\frac{1}{32z^4}\mathcal{B}^4_{-1}Q_z + \frac{\varepsilon}{32z^3}\mathcal{B}^3_{-1}\frac{\partial Q_z}{\partial s} - \frac{\varepsilon s}{16z}\mathcal{B}^3_{-1}\frac{\partial Q_z}{\partial s} + O(1),$$

$$G^{\varepsilon} = \frac{3\varepsilon}{8z^9}\mathcal{B}^3_{-1}Q_z + \frac{\varepsilon s}{4z^7}\mathcal{B}^3_{-1}Q_z$$

$$- \frac{1}{128z^6}\left(\left(\mathcal{B}_{-1}\frac{\partial Q_z}{\partial s}\right)(\mathcal{B}^3_{-1}Q_z) + 32\mathcal{B}_0\mathcal{B}^2_{-1}Q_z\right.$$

$$\left. - (\mathcal{B}^2_{-1}Q_z + 64s)\mathcal{B}^2_{-1}\frac{\partial Q_z}{\partial s} - 64\mathcal{B}^2_{-1}Q_z + 16\frac{\partial^3 Q_z}{\partial s^3}\right) + O\left(\frac{1}{z^5}\right),$$

$$\frac{1}{\sqrt{2}}\tilde{\mathcal{B}}_{-1}G^{\varepsilon} = \frac{3\varepsilon}{32z^{10}}\mathcal{B}^4_{-1}Q_z + \frac{\varepsilon s}{16z^8}\mathcal{B}^4_{-1}Q_z$$

$$+ \frac{1}{512z^7}\left(-\left(\mathcal{B}_{-1}\frac{\partial Q_z}{\partial s}\right)(\mathcal{B}^4_{-1}Q_z) - 32\mathcal{B}_0\mathcal{B}^3_{-1}Q_z\right.$$

$$+ (\mathcal{B}^2_{-1}Q + 64s)\mathcal{B}^3_{-1}\frac{\partial Q_z}{\partial s}$$

$$\left. + 32\mathcal{B}^3_{-1}Q_z - 16\mathcal{B}_{-1}\frac{\partial^3 Q_z}{\partial s^3}\right) + O\left(\frac{1}{z^6}\right).$$

Using these expressions, one easily deduces for small z,

$$0 = \left(F^+\tilde{\mathcal{B}}_{-1}G^- + F^-\tilde{\mathcal{B}}_{-1}G^+\right)\left(F^+\tilde{\mathcal{B}}_{-1}F^- - F^-\tilde{\mathcal{B}}_{-1}F^+\right)$$

$$- \left(F^+G^- + F^-G^+\right)\left(F^+\tilde{\mathcal{B}}^2_{-1}F^- - F^-\tilde{\mathcal{B}}^2_{-1}F^+\right)$$

$$= -\frac{\varepsilon}{2z^{17}}\left(\left\{\mathcal{B}^2_{-1}\frac{\partial Q_z}{\partial s}, \frac{1}{2}\frac{\partial^3 Q_z}{\partial s^3} + (\mathcal{B}_0 - 2)\mathcal{B}^2_{-1}Q_z\right\}_{\mathcal{B}_{-1}}\right.$$

$$\left. + \frac{1}{16}\mathcal{B}_{-1}\frac{\partial Q_z}{\partial s}\left\{\mathcal{B}^3_{-1}Q_z, \mathcal{B}^2_{-1}\frac{\partial Q_z}{\partial s}\right\}_{\mathcal{B}_{-1}}\right) + O\left(\frac{1}{z^{15}}\right)$$

$$= -\frac{\varepsilon}{2z^{17}}(\text{the same expression for } Q(s; x_1, \ldots, x_{2r})) + O\left(\frac{1}{z^{16}}\right),$$

using (6-8) in the last equality. Taking the limit when $z \to 0$ yields equation 6-5 of Theorem 6.1. □

7. The Airy process

Consider n nonintersecting Brownian motions on $\mathbb{R}$, all leaving from the origin and forced to return to the origin. According to formula (4-2), this probability,

$$\Pi := \mathbb{P}^0_0\left(\text{all } x_i(t) \in E \mid \text{all } x_j(0) = x_j(1) = 0\right),$$

can be expressed in terms of the determinant of a moment matrix and further as an integral over Hermitian matrices, both with rescaled space, for $0 \le t \le 1$. To do this we let $\mathcal{H}_n(E)$ denote the space of $n \times n$ Hermitian matrices with spectrum in the set $E \subset \mathbb{R}$, and one checks that

$$\Pi = \frac{1}{Z_n} \det \left(\int_{E(\sqrt{2}/\sqrt{t(1-t)})} dy \, y^{i+j} e^{-y^2/2} \right)_{0 \le i, j \le n-1}$$

$$= \frac{1}{Z_n'} \int_{\mathcal{H}_n(E(1/\sqrt{t(1-t)}))} e^{-\operatorname{Tr} M^2} dM.$$

The Airy process $A(\tau)$ describes the nonintersecting Brownian motions above for large n, but viewed from the (right-hand) edge $\sqrt{2nt(1-t)}$ of the set of particles, with time and space properly rescaled, so that the new time scale τ equals 0 when $t = 1/2$. Random matrix theory suggests the following time and space rescaling (edge rescaling):

$$t = \frac{1}{1 + e^{-2\tau/n^{1/3}}}, \qquad E = \frac{\sqrt{2n} + \dfrac{(-\infty, x)}{\sqrt{2}n^{1/6}}}{2 \cosh \dfrac{\tau}{n^{1/3}}}.$$

Taking the limit when $n \to \infty$, one finds that the rescaled motion becomes time-independent (stationary),

$$P(A(\tau) \le x)$$

$$:= \lim_{n \to \infty} \mathbb{P}_0^0 \left(\text{all } x_i \left(\frac{1}{1 + e^{-2\tau/n^{1/3}}} \right) \in \frac{\sqrt{2n} + \dfrac{(-\infty, x)}{\sqrt{2}n^{1/6}}}{2 \cosh(\tau/n^{1/3})} \, \middle| \, \text{all } x_j(0) = x_j(1) = 0 \right)$$

$$= \lim_{n \to \infty} \frac{1}{Z_n} \int_{\mathcal{H}_n\left(\sqrt{2n} + ((-\infty, x)/\sqrt{2}n^{1/6})\right)} e^{-\operatorname{Tr} M^2} dM$$

$$= \lim_{n \to \infty} \operatorname{Prob} \left((\text{all eigenvalues of } M) \le \sqrt{2n} + \frac{x}{\sqrt{2}n^{1/6}} \right)$$

$$= \exp\left(-\int_x^\infty (\alpha - x) g^2(\alpha) d\alpha \right)$$

$$=: F_2(x) = \text{Tracy–Widom distribution},$$

with $g(\alpha)$ the unique solution of

$$\begin{cases} g'' = \alpha g + 2g^3 \\ g(\alpha) \cong -\dfrac{e^{-(2/3)\alpha^{3/2}}}{2\sqrt{\pi}\alpha^{1/4}} \text{ for } \alpha \nearrow \infty. \end{cases} \qquad \textbf{(Painlevé II)}. \qquad (7\text{-}1)$$

This is to say the outmost particle in the nonintersecting Brownian motions fluctuates according to the Tracy–Widom distribution [1994] for $n \to \infty$.

Since the Airy process is stationary, the joint distribution for two times $t_1 < t_2$ in $[0, 1]$ is of interest; here one checks that

$\mathbb{P}_0^0 \left(\text{all } x_i(t_1) \in E_1, \text{all } x_i(t_2) \in E_2 \mid \text{all } x_j(0) = x_j(1) = 0 \right)$

$$= \mathbb{P}_n \left(\sqrt{\frac{t_1(1-t_2)}{t_2(1-t_1)}}; E_1 \sqrt{\frac{2t_2}{(t_2-t_1)t_1}}, E_2 \sqrt{\frac{2(1-t_1)}{(1-t_2)(t_2-t_1)}} \right), \quad (7\text{-}2)$$

where

$$\mathbb{P}_n(c; E_1', E_2') := \frac{1}{Z_n} \iint_{\mathcal{H}(E_1') \times \mathcal{H}(E_2')} dM_1\, dM_2\, e^{-\frac{1}{2}\mathrm{Tr}(M_1^2 + M_2^2 - 2cM_1M_2)}$$

$$= c_N' \iint_{E^N} \Delta_N(x)\Delta_N(y) \prod_{k=1}^N e^{-\frac{1}{2}(x_k^2 + y_k^2 - 2cx_k y_k)}\, dx_k\, dy_k.$$

According to [Adler and van Moerbeke 1999b], given

$$E = E_1 \times E_2 := \bigcup_{i=1}^r [a_{2i-1}, a_{2i}] \times \bigcup_{i=1}^s [b_{2i-1}, b_{2i}] \subset \mathbb{R}^2, \quad (7\text{-}3)$$

$\log \mathbb{P}_n(c; E_1, E_2)$ satisfies a nonlinear third-order partial differential equation (in terms of the Wronskian $\{f, g\}_X = g(Xf) - f(Xg)$, with regard to the first order differential operator X):

$$\left\{ \mathcal{B}_2 \mathcal{A}_1 \log \mathbb{P}_n,\ \mathcal{B}_1 \mathcal{A}_1 \log \mathbb{P}_n + \frac{nc}{c^2 - 1} \right\}_{\mathcal{A}_1}$$

$$- \left\{ \mathcal{A}_2 \mathcal{B}_1 \log \mathbb{P}_n,\ \mathcal{A}_1 \mathcal{B}_1 \log \mathbb{P}_n + \frac{nc}{c^2 - 1} \right\}_{\mathcal{B}_1} = 0. \quad (7\text{-}4)$$

in terms of the differential operators, depending on the coupling term c and the boundary of E,

$$\mathcal{A}_1 = \frac{1}{c^2 - 1} \left(\sum_1^r \frac{\partial}{\partial a_j} + c \sum_1^s \frac{\partial}{\partial b_j} \right), \quad \mathcal{B}_1 = \frac{1}{1 - c^2} \left(c \sum_1^r \frac{\partial}{\partial a_j} + \sum_1^s \frac{\partial}{\partial b_j} \right),$$

$$\mathcal{A}_2 = \sum_{j=1}^r a_j \frac{\partial}{\partial a_j} - c \frac{\partial}{\partial c}, \qquad \mathcal{B}_2 = \sum_{j=1}^s b_j \frac{\partial}{\partial b_j} - c \frac{\partial}{\partial c}. \quad (7\text{-}5)$$

Using the same rescaled space and time variables, as before, introduce new times $\tau_1 < \tau_2$ and points $x, y \in \mathbb{R}$, defined as

$$t_i = \frac{1}{1 + e^{-2\tau_i/n^{1/3}}}, \quad E_1 = \frac{\sqrt{2n} + \dfrac{(-\infty, x)}{\sqrt{2}n^{1/6}}}{2\cosh \dfrac{\tau_1}{n^{1/3}}}, \quad E_2 = \frac{\sqrt{2n} + \dfrac{(-\infty, y)}{\sqrt{2}n^{1/6}}}{2\cosh \dfrac{\tau_2}{n^{1/3}}}.$$

One verifies, in view of (7-2), that

$$E_1 \sqrt{\frac{2t_2}{(t_2-t_1)t_1}} = \frac{\sqrt{2}\left(\sqrt{2n} + \frac{(-\infty, x)}{\sqrt{2}n^{1/6}}\right)}{\sqrt{1 - e^{-2(\tau_2-\tau_1)/n^{1/3}}}},$$

$$E_2 \sqrt{\frac{2(1-t_1)}{(1-t_2)(t_2-t_1)}} = \frac{\sqrt{2}\left(\sqrt{2n} + \frac{(-\infty, y)}{\sqrt{2}n^{1/6}}\right)}{\sqrt{1 - e^{-2(\tau_2-\tau_1)/n^{1/3}}}},$$

$$c = \sqrt{\frac{t_1(1-t_2)}{t_2(1-t_1)}} = e^{-(\tau_2-\tau_1)/n^{1/3}}.$$

Defining

$$\mathbb{Q}(\tau_2 - \tau_1; x, y) :=$$

$$\log \mathbb{P}_n\left(e^{-(\tau_2-\tau_1)/n^{1/3}}; \frac{\left(2\sqrt{n} + \frac{x}{n^{1/6}}\right)}{\sqrt{1 - e^{-2(\tau_2-\tau_1)/n^{1/3}}}}, \frac{\left(2\sqrt{n} + \frac{y}{n^{1/6}}\right)}{\sqrt{1 - e^{-2(\tau_2-\tau_1)/n^{1/3}}}}\right),$$

one checks, setting $z = n^{-1/6}$ and using the inverse map, that

$$\log \mathbb{P}_n(c; a, b) = \mathbb{Q}\left(-z^{-2} \log c; az^{-1}\sqrt{1-c^2} - 2z^{-4}, bz^{-1}\sqrt{1-c^2} - 2z^{-4}\right).$$

But $\log \mathbb{P}_n(c; E_1, E_2)$ satisfies the PDE (7-4), which induces a PDE for $\mathbb{Q}$; then letting $z \to \infty$, the leading term in this series must be $= 0$. One finds thus the following PDE for the Airy joint probability, namely

$$H(t; x, y) := \log P\left(A(\tau_1) \le y + x, A(\tau_2) \le y - x\right),$$

takes on the following simple form in x, y and t^2, with $t = \tau_2 - \tau_1$, also involving a Wronskian (see [Adler and van Moerbeke 2005])

$$2t \frac{\partial^3 H}{\partial t \partial x \partial y} = \left(\frac{t^2}{2}\frac{\partial}{\partial x} - x\frac{\partial}{\partial y}\right)\left(\frac{\partial^2 H}{\partial x^2} - \frac{\partial^2 H}{\partial y^2}\right) + \left\{\frac{\partial^2 H}{\partial x \partial y}, \frac{\partial^2 H}{\partial y^2}\right\}_y, \quad (7\text{-}6)$$

with initial condition

$$\lim_{t \searrow 0} H(t; x, y) = \log F_2\left(\min(y + x, y - x)\right).$$

The edge $\sup A(t)$ of the cloud is non-Markovian, as is the largest particle in the finite nonintersecting Brownian problem. As $t = \tau_2 - \tau_1 \to \infty$, the edges $\sup A(\tau_1)$ and $\sup A(\tau_2)$ become independent. This poses the question: *How much does the process remember from the remote past?* The following asymptotics for the covariance of the edge of the cloud, for large $t = \tau_2 - \tau_1$, is deduced from the PDE:

$$E(\sup A(\tau_2) \sup A(\tau_1)) - E(\sup A(\tau_2)) E(\sup A(\tau_1))$$
$$= \frac{1}{t^2} + \frac{2}{t^4} \iint_{\mathbb{R}^2} \Phi(u, v) \, du \, dv + \cdots,$$

where

$$\Phi(u, v) := F_2(u) F_2(v) \left(\frac{1}{4} \left(\int_u^\infty g^2 \, d\alpha \right)^2 \left(\int_v^\infty g^2 \, d\alpha \right)^2 \right.$$
$$+ g^2(u) \left(\frac{1}{4} g^2(v) - \frac{1}{2} \left(\int_v^\infty g^2 \, d\alpha \right)^2 \right)$$
$$\left. + \int_v^\infty d\alpha \left(2(v - \alpha) g^2 + g'^2 - g^4 \right) \int_u^\infty g^2 \, d\alpha \right).$$

(Here $g = g(\alpha)$ is the function (7-1) and $F_2(u)$ is the Tracy–Widom distribution.)

The Airy process was introduced by Spohn and Prähofer [2002] in the context of polynuclear growth models. It has been further investigated by Johansson [2001; 2003; 2005], by Tracy and Widom [2004] and by Adler and van Moerbeke [2005]; see also [Widom 2004].

References

[Adler and van Moerbeke 1997] M. Adler and P. van Moerbeke, "String-orthogonal polynomials, string equations, and 2-Toda symmetries", *Comm. Pure Appl. Math.* **50**:3 (1997), 241–290. Available at arxiv.org/abs/hep-th/9706182.

[Adler and van Moerbeke 1999a] M. Adler and P. van Moerbeke, "Generalized orthogonal polynomials, discrete KP and Riemann-Hilbert problems", *Comm. Math. Phys.* **207**:3 (1999), 589–620. Available at arxiv.org/abs/nlin.SI/0009002.

[Adler and van Moerbeke 1999b] M. Adler and P. van Moerbeke, "The spectrum of coupled random matrices", *Ann. of Math.* (2) **149**:3 (1999), 921–976.

[Adler and van Moerbeke 2005] M. Adler and P. van Moerbeke, "PDEs for the joint distributions of the Dyson, Airy and sine processes", *Ann. Probab.* **33**:4 (2005), 1326–1361. Available at arxiv.org/abs/math.PR/0302329.

[Adler and van Moerbeke 2006] M. Adler and P. van Moerbeke, "Joint probability for the Pearcey process", preprint, 2006. Available at arxiv.org/abs/math.PR/0612393.

[Adler and van Moerbeke 2007] M. Adler and P. van Moerbeke, "PDEs for the Gaussian ensemble with external source and the Pearcey distribution", *Comm. Pure Appl. Math.* **60**:9 (2007), 1261–1292. Changed page range following MathSciNet; please check.

[Adler et al. 2006] M. Adler, P. van Moerbeke, and P. Vanhaecke, "Moment matrices and multi-component KP, with applications to random matrix theory", preprint, 2006. Available at arxiv.org/abs/math-ph/0612064.

[Aptekarev 1998] A. I. Aptekarev, "Multiple orthogonal polynomials", *J. Comput. Appl. Math.* **99**:1-2 (1998), 423–447.

[Aptekarev et al. 2003] A. I. Aptekarev, A. Branquinho, and W. Van Assche, "Multiple orthogonal polynomials for classical weights", *Trans. Amer. Math. Soc.* **355**:10 (2003), 3887–3914.

[Bleher and Kuijlaars 2004a] P. Bleher and A. B. J. Kuijlaars, "Large *n* limit of Gaussian random matrices with external source. I", *Comm. Math. Phys.* **252**:1-3 (2004), 43–76.

[Bleher and Kuijlaars 2004b] P. M. Bleher and A. B. J. Kuijlaars, "Random matrices with external source and multiple orthogonal polynomials", *Int. Math. Res. Not.* no. 3 (2004), 109–129. Available at arxiv.org/abs/math-ph/0307055.

[Brézin and Hikami 1996] E. Brézin and S. Hikami, "Correlations of nearby levels induced by a random potential", *Nuclear Phys.* B **479**:3 (1996), 697–706.

[Brézin and Hikami 1997] E. Brézin and S. Hikami, "Extension of level spacing universality", *Phys. Rev. E* **56** (1997), 264–269.

[Brézin and Hikami 1998a] E. Brézin and S. Hikami, "Level spacing of random matrices in an external source", *Phys. Rev. E* (3) **58**:6, part A (1998), 7176–7185.

[Brézin and Hikami 1998b] E. Brézin and S. Hikami, "Universal singularity at the closure of a gap in a random matrix theory", *Phys. Rev. E* (3) **57**:4 (1998), 4140–4149.

[Daems and Kuijlaars 2007] E. Daems and A. B. J. Kuijlaars, "Multiple orthogonal polynomials of mixed type and non-intersecting Brownian motions", *J. Approx. Theory* **146**:1 (2007), 91–114. Available at arxiv.org/abs/math.CA/0511470.

[Dyson 1962] F. J. Dyson, "A Brownian-motion model for the eigenvalues of a random matrix", *J. Mathematical Phys.* **3** (1962), 1191–1198.

[Grabiner 1999] D. J. Grabiner, "Brownian motion in a Weyl chamber, non-colliding particles, and random matrices", *Ann. Inst. H. Poincaré Probab. Statist.* **35**:2 (1999), 177–204.

[Johansson 2001] K. Johansson, "Universality of the local spacing distribution in certain ensembles of Hermitian Wigner matrices", *Comm. Math. Phys.* **215**:3 (2001), 683–705.

[Johansson 2003] K. Johansson, "Discrete polynuclear growth and determinantal processes", *Comm. Math. Phys.* **242**:1-2 (2003), 277–329.

[Johansson 2005] K. Johansson, "The arctic circle boundary and the Airy process", *Ann. Probab.* **33**:1 (2005), 1–30. Available at arxiv.org/abs/math.PR/0306216.

[Karlin and McGregor 1959] S. Karlin and J. McGregor, "Coincidence probabilities", *Pacific J. Math.* **9** (1959), 1141–1164.

[Okounkov and Reshitikhin 2005] A. Okounkov and N. Reshitikhin, "Random skew plane partitions and the Pearcey process", preprint, 2005. Available at arxiv.org/abs/math.CO/0503508.

[Prähofer and Spohn 2002] M. Prähofer and H. Spohn, "Scale invariance of the PNG droplet and the Airy process", *J. Statist. Phys.* **108**:5-6 (2002), 1071–1106. Dedicated to David Ruelle and Yasha Sinai on the occasion of their 65th birthdays.

[Tracy and Widom 1994] C. A. Tracy and H. Widom, "Level-spacing distributions and the Airy kernel", *Comm. Math. Phys.* **159**:1 (1994), 151–174.

[Tracy and Widom 2004] C. A. Tracy and H. Widom, "Differential equations for Dyson processes", *Comm. Math. Phys.* **252**:1-3 (2004), 7–41. Available at arxiv.org/abs/math.PR/0309082.

[Tracy and Widom 2006] C. A. Tracy and H. Widom, "The Pearcey process", *Comm. Math. Phys.* **263**:2 (2006), 381–400. Available at arxiv.org/abs/math.PR/0412005.

[Widom 2004] H. Widom, "On asymptotics for the Airy process", *J. Statist. Phys.* **115**:3-4 (2004), 1129–1134. Available at arxiv.org/abs/math.PR/0308157.

PIERRE VAN MOERBEKE
DEPARTMENT OF MATHEMATICS
UNIVERSITÉ CATHOLIQUE DE LOUVAIN
1348 LOUVAIN-LA-NEUVE
BELGIUM
and
BRANDEIS UNIVERSITY
WALTHAM, MA 02454
UNITED STATES
pierre.vanmoerbeke@uclouvain.be

Probability, Geometry and Integrable Systems
MSRI Publications
Volume **55**, 2008

Homogenization of random Hamilton–Jacobi–Bellman Equations

S. R. SRINIVASA VARADHAN

ABSTRACT. We consider nonlinear parabolic equations of Hamilton–Jacobi–Bellman type. The Lagrangian is assumed to be convex, but with a spatial dependence which is stationary and random. Rescaling in space and time produces a similar equation with a rapidly varying spatial dependence and a small viscosity term. Motivated by corresponding results for the linear elliptic equation with small viscosity, we seek to find the limiting behavior of the solution of the Cauchy (final value) problem in terms of a homogenized problem, described by a convex function of the gradient of the solution. The main idea is to use the principle of dynamic programming to write a variational formula for the solution in terms of solutions of linear problems. We then show that asymptotically it is enough to restrict the optimization to a subclass, one for which the asymptotic behavior can be fully analyzed. The paper outlines these steps and refers to the recently published work of Kosygina, Rezakhanlou and the author for full details.

Homogenization is a theory about approximating solutions of a differential equation with rapidly varying coefficients by a solution of a constant coefficient differential equation of a similar nature. The simplest example of its kind is the solution u^ε of the equation

$$u_t^\varepsilon = \tfrac{1}{2}a\Big(\frac{x}{\varepsilon}\Big)u_{xx}^\varepsilon; \quad u^\varepsilon(0, x) = f(x)$$

on $[0, \infty]\times\mathbb{R}$. The function $a(\cdot)$ is assumed to be uniformly positive, continuous and periodic of period 1. The limit u of u^ε exists and solves the equation

$$u_t = \frac{\bar{a}}{2}u_{xx}; \quad u(0, x) = f(x)$$

where $\bar{a}$ is the harmonic mean

$$\bar{a} = \Big(\int_0^1 \frac{dx}{a(x)}\Big)^{-1}.$$

Although this is a result about solutions of PDE's it can be viewed as a limit theorem in probability. If we consider the Markov process $x(t)$ with generator

$$\tfrac{1}{2} a(x) D_x^2$$

starting from 0 at time 0, as $t \to \infty$ the limiting distribution of $y(t) = \frac{x(t)}{\sqrt{t}}$ is Gaussian with mean 0 and variance $\bar{a}$. The actual variance of $y(t)$ is

$$E\left[\frac{1}{t} \int_0^t a(x(s)) \, ds \right].$$

The result on the convergence of u^ε to u is seen to follow from an ergodic theorem of the type

$$\lim_{t \to \infty} \frac{1}{t} \int_0^t a(x(s)) \, ds = \bar{a}.$$

From the theory of Markov processes one can see an ergodic theorem of this type with

$$\bar{a} = \int a(x) \phi(x) \, dx,$$

where $\phi(x)$ is the normalized invariant measure on $[0, 1]$ with end points identified. This is seen to be

$$\phi(x) = \left(\int_0^1 \frac{dx}{a(x)} \right)^{-1} \frac{1}{a(x)},$$

so that

$$\bar{a} = \int_0^1 a(x) \phi(x) \, dx = \left(\int_0^1 \frac{dx}{a(x)} \right)^{-1}.$$

We can consider the situation where $a(x) = a(x, \omega)$ is a random process, stationary with respect to translations in x. We can formally consider a probability space (Ω, Σ, P), and an ergodic action τ_x of $\mathbb{R}$ on Ω. We also have a function $a(\omega)$ satisfying $0 < c \le a(\omega) \le C < \infty$. The stationary process $a(x, \omega)$ is given by $a(x, \omega) = a(\tau_x \omega)$. Now the solution u^ε of

$$u_t^\varepsilon(t, x, \omega) = \tfrac{1}{2} a(x, \omega) u_{xx}^\varepsilon(t, x, \omega); \quad u^\varepsilon(0, x, \omega) = f(x)$$

can be shown to converge again, in probability, to the nonrandom solution u of

$$u_t(t, x) = \frac{\bar{a}}{2} u_{xx}(t, x); \quad u^\varepsilon(0, x) = f(x)$$

with

$$\bar{a} = \left(\int \frac{1}{a(\omega)} dP \right)^{-1}.$$

This is also an ergodic theorem for

$$\frac{1}{t} \int_0^t a(\omega(s)) \, ds,$$

but the actual Markov process $\omega(t)$ for which the ergodic theorem is proved is one that takes values in Ω with generator

$$L = \tfrac{1}{2} a(\omega) D^2,$$

where D is the generator of the translation group τ_x on Ω. The invariant measure is seen to be

$$dQ = \frac{\bar{a}}{a(\omega)} dP,$$

where

$$\bar{a} = \left(\int \frac{1}{a(x)} dP \right)^{-1}.$$

We will try to adapt this type of approach to some nonlinear problems of Hamilton-Jacobi–Bellman type. One part of the work that we outline here was done jointly with Elena Kosygina and Fraydoun Rezakhanlou and has appeared in print [Kosygina et al. 2006], while another part, carried out with Kosygina, has been submitted for publication.

The problems we wish to consider are of the form

$$u_t^\varepsilon + \frac{\varepsilon}{2} \Delta u^\varepsilon + H\left(\frac{x}{\varepsilon}, \nabla u^\varepsilon, \omega\right) = 0; \quad u(T, x) = f(x)$$

for $[0, T] \times \mathbb{R}^d$. Here f is a continuous function with at most linear growth. (Ω, Σ, P) is a probability space on which $\mathbb{R}^d$ acts ergodically as measure preserving transformations τ_x. $H(0, p, \omega)$ is a function on $\mathbb{R}^d \times \Omega$ which is a convex function of p for every ω and $H(x, p, \omega) = H(0, p, \tau_x \omega)$. It satisfies some bounds and some additional regularity. The problem is to prove that $u^\varepsilon \to u$ as $\varepsilon \to 0$, where u is a solution of

$$u_t + \overline{H}(\nabla u) = 0; \quad u(T, x) = f(x)$$

for some convex function $\overline{H}(p)$ of p and determine it.

The analysis consists of several steps. We might as well assume $T = 1$ and concentrate on $u^\varepsilon(0, 0, \omega)$. First we note that, by rescaling, the problem can be reduced to the behavior of

$$\lim_{t \to \infty} \frac{1}{t} u^t(0, 0, \omega),$$

where u is the solution in $[0, t] \times \mathbb{R}^d$, of

$$u_s + \tfrac{1}{2} \Delta u + H(x, \nabla u, \omega); \quad u(t, x) = t f\left(\frac{x}{t}\right).$$

The second step is to use the principle of dynamic programming to write a variational formula for $u^t(s, x, \omega)$. Denote by $L(\tau_x \omega, q)$ the convex dual

$$L(x, q, \omega) = \sup_p (\langle p, q \rangle - H(x, p, \omega))$$

Let $b(s, x)$ be a function $b : [0, t] \times \mathbb{R}^d \to \mathbb{R}^d$. Let $\mathcal{B}$ denote the space all such bounded functions. For each $b \in \mathcal{B}$, we consider the linear equation

$$v_s^b + \tfrac{1}{2} \Delta v^b + \langle b(s, x), \nabla v^b \rangle - L(\tau_x \omega, b(s, x)) = 0, v(t, x) = tf\left(\frac{x}{t}\right);$$

then the solution $u(s, x)$ is $\sup_b v^b(s, x)$. If we denote by Q^b the Markov process with generator

$$\mathcal{L}_s^b = \tfrac{1}{2} \Delta + \langle b(s, x), \nabla \rangle$$

starting from $(0, 0)$, then

$$v^b(0, 0, \omega) = E^{Q^b}\left[tf\left(\frac{x(t)}{t}\right) - \int_0^t L(x(s), b(s, x(s)), \omega) \, ds \right]$$

and

$$u = \sup_{b \in \mathcal{B}} v^b$$

The third step is to consider a subclass of $\mathcal{B}$ of the form $b(t, x) = c(\tau_x \omega)$ with $c : \Omega \to \mathbb{R}^d$ chosen from a reasonable class $\mathcal{C}$. The solution v^b with this choice of $b(t, x) = b(x) = c(\tau_x \omega)$ will be denoted by v^c. We will show that for our choice of $\mathcal{C}$, the limit

$$\lim_{t \to \infty} \frac{1}{t} v^c(0, 0, \omega) = g(c)$$

will exist for every $c \in \mathcal{C}$. It then follows that

$$\liminf_{t \to \infty} \frac{1}{t} u^t(0, 0) \geq \sup_{c \in \mathcal{C}} g(c).$$

Given c there is a Markov process $Q^{c, \omega}$ on Ω starting from ω with generator

$$\mathcal{A}_c = \tfrac{1}{2} \Delta + \langle c(\omega), \nabla \rangle.$$

Here ∇ is the infinitesimal generator of the $\mathbb{R}^d$ action $\{\tau_x\}$ and $\Delta = \nabla \cdot \nabla$. This process can be constructed by solving

$$dx(t) = c(\tau_{x(t)} \omega) \, dt + \beta(t); \quad x(0) = 0$$

Then one lifts it to Ω by defining $\omega(t) = \tau_{x(t)} \omega$. Such a process with generator $\mathcal{A}_c$ could have an invariant density P_c and it could (although it is unlikely) be mutually absolutely continuous with respect to P, having density Φ_c. Φ_c will be a weak solution of

$$\tfrac{1}{2} \Delta \Phi_c = \nabla \cdot c(\cdot) \, \Phi_c.$$

We can then expect

$$g(c) = f\left(\int c(\omega)\, dP_c\right) - \int L(\omega, c(\omega))\, dP_c.$$

In general the existence of such a Φ for a given c is nearly impossible to prove. On the other hand for a given Φ finding a c is easy. For instance,

$$c = \frac{\nabla\Phi}{2\Phi}$$

will do. More generally one can have

$$c = \frac{\nabla\Phi}{2\Phi} + c',$$

so long as $\nabla \cdot c'\Phi = 0$. So pairs (c, Φ) such that

$$\tfrac{1}{2}\Delta\Phi_c = \nabla \cdot c(\cdot)\, \Phi_c$$

exist. Our class $\mathcal{C}$ will be those for which Φ exists. It is not hard to show, using the ergodicity of $\{\tau_x\}$ action, that Φ is unique for a given c when it exists and the Markov process with generator $\mathcal{A}_c$ is ergodic with $dP_c = \Phi_c dP$ as invariant measure. We will denote by $\mathcal{C}$ the class of pairs (c, Φ) satisfying the above relation. So we have a lower bound

$$\liminf_{t\to\infty} \frac{1}{t} u^c(0,0) \ge \sup_{m\in\mathbb{R}^d} [f(m) - I(m)]$$

where

$$I(m) = \inf_{\substack{c,\Phi:(c,\Phi)\in\mathcal{C}_0 \\ \int c\Phi\, dP = m}} \int L(c(\omega),\omega)\Phi\, dP$$

Now we turn to proving upper bounds. Fix $\theta \in \mathbb{R}^d$. If we had a "nice" test function $W(x, \omega)$ such that for almost all ω

$$|W(x,\omega) - \langle\theta, x\rangle| \le o(|x|)$$

and

$$\tfrac{1}{2}\Delta W + H(x, \nabla W, \omega) \le \lambda$$

Then, by convex duality with $\tilde{W} = W(x, \omega) - \lambda(s - t)$, we have

$$\tilde{W}_s + \tfrac{1}{2}\Delta\tilde{W} + \langle b(s, x), \nabla\tilde{W}\rangle - L(b(s,x), \omega) \le 0.$$

If $\overline{H}(\theta)$ is defined as

$$\overline{H}(\theta) = \inf\{\lambda : W \text{ exists}\}$$

then under some control on the growth of L, it is not hard to deduce that with $f(x) = \langle \theta, x \rangle$,

$$\limsup_{t \to \infty} \frac{1}{t} u^t(0, 0, \omega) \leq \overline{H}(\theta)$$

If we can prove that

$$\overline{H}(\theta) = \sup_m [\langle \theta, m \rangle - I(m)],$$

we are done. We would match the upper and lower bounds. We reduce this to a minmax equals maxmin theorem.

$$
\begin{aligned}
\sup_m [\langle \theta, m \rangle - I(m)] &= \sup_{(c, \Phi) \in \mathcal{C}} \int \left(\langle c(\omega), \theta \rangle - L(c(\omega), \omega) \right) \Phi \, dP \\
&= \sup_{(c, \Phi)} \inf_W \int \left(\langle c(\omega), \theta \rangle + \mathcal{A}_c W - L(c(\omega), \omega) \right) \Phi \, dP \\
&= \inf_W \sup_{(c, \Phi)} \int \left(\langle c(\omega), \theta \rangle + \mathcal{A}_c W - L(c(\omega), \omega) \right) \Phi \, dP \\
&= \inf_W \sup_\Phi \int \left(\tfrac{1}{2} \Delta W + H(\theta + \nabla W, \omega) \right) \Phi \, dP \\
&= \inf_W \sup_\omega \int \left(\tfrac{1}{2} \Delta W + H(\theta + \nabla W, \omega) \right) \Phi \, dP \\
&= \overline{H}(\theta).
\end{aligned}
$$

While W may not exist, ∇W will exist. We can integrate on $\mathbb{R}^d$, then ergodic theorem will yield an estimate of the form $W(x) = o(|x|)$ and

$$\langle \theta, x \rangle + W(x)$$

will work as a test function. There are some technical details on the issues of growth and regularity. The details have appeared in [Kosygina et al. 2006] along with additional references. Similar results on the homogenization of random Hamilton–Jacobi–Bellman equations have been obtained by Lions and Sougani-dis [2005], using different methods.

Now we examine the time dependent case. If we replace $\mathbb{R}^d$ action by $\mathbb{R}^{d+1}$ action with (t, x) denoting time and space, then the stationary processes H and L are space time processes. The lower bound works more or less in the same manner. In addition to ∇ we now have D_t the derivative in the time direction. The $\omega(t)$ process is the space-time process. Its construction for a given c is slightly different. We start with $b(t, x) = c(\tau_{t,x} \omega)$ and construct a diffusion on $\mathbb{R}^d$ corresponding to the time dependent generator

$$\mathcal{A}_s^c = \tfrac{1}{2} \Delta + \langle b(s, x), \nabla \rangle$$

and then lift it by $\omega(s) = \tau_{s,x(s)}\omega$. The invariant densities are solutions of

$$-D_t\Phi + \tfrac{1}{2}\Delta\Phi = \nabla \cdot c\Phi.$$

The lower bound works the same way. But for obtaining the upper bound, a test function W has to be constructed that satisfies

$$W_t + \tfrac{1}{2}\Delta W + H(t, x, \nabla W, \omega) \le \overline{H}(\theta)$$

In the time independent case there was a lower bound on the growth of the convex function H that provided estimates on ∇W. Here one has to work much harder in order to control in some manner W_t. The details will appear in [Kosygina and Varadhan 2008].

References

[Kosygina and Varadhan 2008] E. Kosygina and S. R. S. Varadhan, "Homogenization of Hamilton–Jacobi–Bellman equations with respect to time-space shifts in a stationary ergodic medium", *Comm. Pure Appl. Math.* **61**:6 (2008).

[Kosygina et al. 2006] E. Kosygina, F. Rezakhanlou, and S. R. S. Varadhan, "Stochastic homogenization of Hamilton–Jacobi–Bellman equations", *Comm. Pure Appl. Math.* **59**:10 (2006), 1489–1521.

[Lions and Souganidis 2005] P.-L. Lions and P. E. Souganidis, "Homogenization of "viscous" Hamilton–Jacobi equations in stationary ergodic media", *Comm. Partial Differential Equations* **30**:1-3 (2005), 335–375.

S. R. Srinivasa Varadhan
Courant Institute
New York University
251 Mercer Street
New York, NY 10012
United States
varadhan@cims.nyu.edu